Physical Geology

Physical Geology

Carla W. Montgomery

Northern Illinois University

wcb

Wm. C. Brown Publishers
Dubuque, Iowa

Book Team

Edward G. Jaffe
Executive Editor

Lynne M. Meyers
Senior Developmental Editor

Mark Elliot Christianson
Designer

Carol A. Kozlik
Production Editor

Carol M. Smith
Photo Research Editor

Carla D. Arnold
Permissions Editor

Matt Shaughnessy
Product Manager

Wm. C. Brown
Chairman of the Board

Mark C. Falb
President and Chief Executive Officer

Cover photo © David Muench.

Photographs for figures: 1.13, 2.8–2.16, 2.22, 3.1, 3.9, 3.10, 3.12–3.22, Box 3.2.1, 3.2.2, 3.28B, 4.2, 4.3, 4.5B, 4.7–4.9, 4.21–4.24B, Box 5.1.1, 5.2, 5.5, 5.7, 5.8, 5.10–5.12, 11.12A, 20.4, 20.5B: © William C. Brown Company Publishers/Photography by Bob Coyle.

All photographs unless otherwise credited © Carla Montgomery.

wcb group

Wm. C. Brown
Chairman of the Board

Mark C. Falb
President and Chief Executive Officer

wcb

Wm. C. Brown Publishers, College Division

G. Franklin Lewis
Executive Vice-President, General Manager

E. F. Jogerst
Vice-President, Cost Analyst

George Wm. Bergquist
Editor in Chief

Beverly Kolz
Director of Production

Chris C. Guzzardo
Vice-President, Director of Marketing

Bob McLaughlin
National Sales Manager

Craig S. Marty
Manager, Marketing Research

Marilyn A. Phelps
Manager of Design

Colleen A. Yonda
Production Editorial Manager

Faye M. Schilling
Photo Research Manager

Library of Congress Catalog Card Number: 86–70955

ISBN 0–697–00851–7

Printed in the United States of America
10 9 8 7 6 5 4 3 2 1

Contents

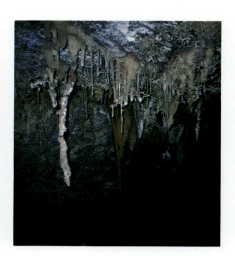

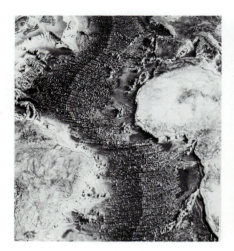

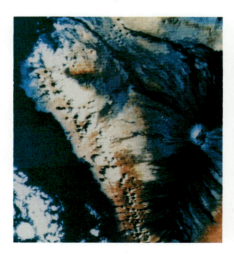

11

The Continental Crust 211

12

The Ocean Basins 235

15

Coastal Zones and Processes   303

16

Glaciers   321

Part Four

Contemporary Topics   399

20

Mineral Resources   401

21

Energy Resources   423

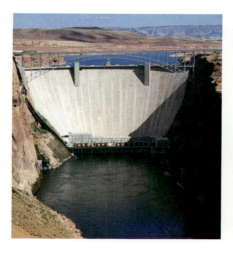

22

Planets and Other Strangers: The Solar System   453

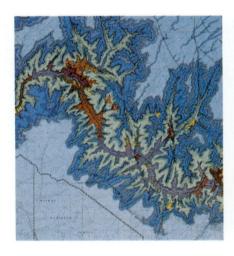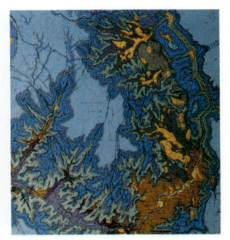

A biologist colleague once remarked that everyone should take one course each in geology, botany, and astronomy, for everywhere one goes there are rocks, plants, and stars. A strong case could be made for physical geology as that one geology course. The broad subject of physical geology comprises the study of the various geologic features and materials of the earth as well as the nature of the processes by which they are formed and modified. It therefore gives us a greater understanding of the planet upon which we live. As humans have come increasingly to depend on geologic resources, knowledge of geology has been important in realizing the origins and the limitations of those resources. We have also come to recognize not only that we can anticipate—and therefore avoid—some of the disasters that are a natural consequence of certain geologic processes; in a few cases, we may even be able to modify those processes for our benefit.

In practice, most students who study physical geology fall into one of two groups. Some are prospective geology majors, for whom the wide perspective of physical geology is commonly the foundation on which more advanced coursework builds. A larger number are nonmajors, and indeed not prospective scientists of any kind, who are prompted to take the course by some mixture of interest in the subject and the need to satisfy a science distribution requirement. Although the needs of these two groups are not identical, every effort has been made to devise a text with sufficient versatility and appropriate learning aids so that it can serve both kinds of students.

About the Book

The book is intended for an introductory-level college course in physical geology. It assumes no prior exposure to college-level mathematics or science. For the most part, metric units are used throughout, except where other units are conventional within a discipline. For the convenience of students not yet comfortable with the metric system, a unit conversion table is given in Appendix C, and English-unit equivalents are occasionally given within the text also.

The subject matter is divided into four parts. In the first part, Fundamentals, the student is introduced to the nature of geology as a science; to the broad outline of the history of the earth by way of historical perspective; to minerals and rocks, the building blocks of geologic features; and to the concept of time in geology.

Once this groundwork has been laid, the second part, Internal Processes and Structures, explores the major features and processes of the earth's interior. It begins with plate tectonics, which provides the conceptual framework for understanding much about seismicity and volcanism. Chapters on earthquakes and volcanoes follow. The data of seismology and volcanology are the major source of information about the earth's interior, the subject of the next chapter. Part Two concludes with chapters on the continental crust, including crustal structures and mountain-building, and the ocean basins and crust.

Superimposed on these large-scale features are the effects of the Surface Processes, the subject of Part Three. These include the various ways in which water, ice, wind, and gravity act to modify the earth's surface features and forms. It is at the surface that we see the interaction between the internal heat that drives the internal processes discussed in Part Two and the solar energy that drives the winds and the water cycle.

The subjects of Part Four, Contemporary Topics, go somewhat beyond the traditional concerns of classical physical geology, yet they are closely linked to it. There is growing preoccupation with availability of resources. Mineral and many energy sources are formed or concentrated by geologic processes, and these are explored in chapters 20 and 21. The final chapter briefly surveys our fellow travelers in the solar system—other planets, moons, asteroids, meteorites, and comets. Advances in knowledge and understanding resulting from various space missions of the past two decades make it possible to compare many of the features of planetary objects with possible terrestrial counterparts. Appendix A presents a short introduction to the nature and interpretation of topographic and geologic maps and to satellite imagery. Appendix B provides an identification chart for common minerals and some guidelines for recognizing different rock types.

To the Instructor

The organization described above differs somewhat from most physical geology texts, primarily in the placement of internal processes ahead of surface processes, and the appearance of plate tectonics about a third of the way into the text. This organization puts discussion of the large-scale processes ahead of the more localized surface processes that act on the resulting features. The far-reaching concepts of plate tectonics are introduced in some detail as soon as the students have a background in rocks and time on which to draw for understanding. This allows plate-tectonic concepts to be reinforced by references in subsequent chapters, as relevant to the topics under discussion. There has also seemed to be a certain logic in building features by tectonic processes before tearing them down by surface processes. However, Parts Two and Three are relatively independent overall, so an instructor who prefers to do so can cover surface processes first with minimal difficulty. An effort has also been made to keep individual chapters within each part as self-contained as possible (without

undue repetition), so that, again, the order can be adjusted by the instructor if desired. The glossary should help to bridge any gaps arising from re-ordering of blocks of the text.

"Environmental geology" has been added to some texts as a separate chapter. However, there are environmental aspects to many of the topics in the text, from volcanoes to streams to resources. Therefore, in this book environmental and human-impact considerations are woven into various chapters as appropriate. This may help students to see the present relevance of the subject matter while mastering the corresponding facts, theories, and vocabulary.

A variety of pedagogical aids and features are included. Each chapter begins with an outline of the subject headings to follow, by way of overview. Terms are printed in boldface and defined at first encounter; these boldfaced terms are collected as Terms to Remember at the end of each chapter. These same terms are defined in the glossary for quick reference. At the end of each chapter are Questions for Review to help the study efforts of students, and in nearly every case there are a small number of questions or problems For Further Thought that go beyond basic review of text material. There are also Suggestions for Further Reading that include several kinds of material: up-to-date (but often relatively sophisticated) references in the subject area of the chapter; materials that may be more readable for the nonspecialist (including some older but fundamentally accurate works); and, occasionally, "classic" works by prominent geologists.

Most chapters contain one or more boxed inserts. These are of several types. Some describe tools of the geologic trade (e.g., thin sections, mass spectrometry). Some present case studies related to chapter material (flood recurrence-interval projection for a particular stream, groundwater depletion in the Ogallala aquifer system, shoreline stabilization efforts along the Texas shore). A few present tangential material of current or historical interest (subduction zones as candidates for waste-disposal sites, the nature of El Niño events), or somewhat more advanced concepts that might be appreciated most by better prepared students (paired metamorphic belts and plate tectonics, an introduction to simple binary phase diagrams). In all cases, the desire is to include the material for enrichment or information without disrupting the flow of the main body of text and presentation of fundamental concepts. Individual boxes may be included or omitted at the discretion of the instructor. Occasional smaller boxes also appear within the text. These could be viewed somewhat as long parenthetical remarks, minor digressions not lengthy enough to justify a major boxed insert, and usually lacking associated figures.

Appendices A and B will probably be of most use to those students whose physical geology course does not include a required laboratory. They may also be helpful in cases in which lecture and laboratory sections proceed independently, so that the lecture may get ahead of the corresponding subject in the lab.

Acknowledgments

A great many individuals have contributed to this project. The text owes its very existence in large part to the persistence and persuasiveness of Ed Jaffe, and in part to the more than 1000 beginning geology students I have taught. Those students continually remind me of the fascination of discovering geology for the first time, as well as of the need not to leap ahead too fast in teaching it to someone else.

I would like to extend my sincere thanks to the reviewers of the manuscript: Richard A. Marston, University of Texas—El Paso; Erwin J. Mantei, Southwest Missouri State University; Lawrence L. Malinconico, Jr., Southern Illinois University; Robert A. Gastaldo, Auburn University; Collette Dick Burke, Wichita State University; John M. Burket, Tyler Junior College; Robert H. Filson, Green River Community College; and Scott Burns, Louisiana Tech University. Their genuine interest in the project came through clearly in their reviews. The added perspective provided by the diversity of their viewpoints and experience, and the many thoughtful suggestions they have made along the way, have led to significant improvements in the final text. Naturally, any remaining shortcomings must be my responsibility.

Thanks also go to Marge Dalechek and Carole Edwards at the U.S. Geological Survey Photographic Library, for their enthusiastic help with the photo research (in spite of the exigencies imposed by the Gramm-Rudman-Hollings amendment), to Donald M. Davidson, Jr., for approving the loan of several departmental rock and mineral specimens for preparation of illustrations, and to fellow author and deanly colleague Jerrold H. Zar for his interest and encouragement. A very large share of the credit for the quality of the finished book goes to the book team at WCB, a delightful group of dedicated, talented, and hardworking professionals with whom it has truly been a pleasure to work. And last, but certainly not least, I would like to thank my husband, Warren, for his supportiveness throughout this project—for his encouragement, for his patience with the undeniably large intrusion that this book has made into our life, for his willingness (and even eagerness) to have ''vacations'' diverted into book-photo-taking trips, and just for being there.

Physical Geology

Part opener photo © Jeff Gnass

As the title implies, these first six chapters are intended to provide a foundation in basic concepts upon which later chapters will draw. The first chapter serves a twofold purpose. Because many students have only a vague idea of what geology is, the chapter gives an introduction to the nature of geology as a science, and it explains how geology differs in certain respects from other physical sciences. The same chapter introduces our subject, the earth, with a brief summary of key aspects of its origin, development, and present structure.

The next four chapters deal with earth materials. Rocks are built of minerals; chapter 2 describes the various types of minerals in some detail. The chapter includes some general discussion of physical properties applicable to all rocks; for example, the ways in which rocks respond to stress, which is relevant to rock structure and to earthquakes. All rocks, too, are related by the rock cycle, a summary of which concludes chapter 2. Chapters 3, 4, and 5 concern, respectively, igneous, sedimentary, and metamorphic rocks, the three classes of rocks. Within each chapter, the characteristics, classification, and mode of formation of the corresponding rock type are discussed. It will become apparent that each class of rocks represents rock formation under a distinctly different set of conditions, which leads to distinctive characteristics in the resulting rocks.

The last chapter of this part, chapter 6, deals with the concept of time in geology. Among the physical sciences, time is of special relevance to geology. Long spans of time are a key element in the formation of many geologic features and materials, and a grasp of the magnitude of those time periods is essential to understanding the earth and its development. The chapter includes both older methods of estimating the length of earth history and modern methods for accurate dating of rock materials.

Chapter 1

An Invitation to Geology

Introduction

Geology is the study of the earth and the processes that shape it. **Physical geology,** in particular, is concerned with the materials and physical features of the earth, changes in those features, and the processes that bring them about. Intellectual curiosity about the way the earth works is one reason for the study of geology. Piecing together the history of a mountain range (figure 1.1) or even a single rock can be exciting.

There are also practical aspects. Certain geologic processes and events can be hazardous (figures 1.2, 1.3), and a better understanding of such phenomena may help us to minimize the risks. We have also come to depend heavily on certain earth materials for energy (figure 1.4) or as raw materials for manufacturing (figure 1.5), and knowing how and where those resources are to be found can be very useful to modern society. Before entering into the detailed study of physical geology, it may be useful to survey briefly geology as a discipline, and the history of the earth that is its subject.

Figure 1.1

By carefully applying geologic principles the history of an entire mountain range can be deciphered. The Beartooth Mountains near Yellowstone National Park.

Figure 1.2

Aftermath of the 1964 earthquake, Anchorage, Alaska: remains of Government Hill School.

Photo courtesy U.S. Geological Survey.

Sandbags turn streets into canals in Salt Lake City, May 1983.

Photo by P. Blanchard, courtesy U.S. Geological Survey.

An open-pit iron mine: Empire Mine, Palmer, Michigan. Each of the steplike ledges is approximately 15 meters high.

A   B

In pursuit of petroleum. *(A)* Liquid ''black gold'' was so highly prized at the turn of the century that drilling rigs proliferated even in residential areas. Long Beach, California, 1901. *(B)* Sunset over a Californian oil field.

Photos: *(A)* C. W. Hayes, courtesy U.S. Geological Survey. *(B)* Belinda Rain, courtesy EPA/National Archives.

Geology as a Discipline

In some sense, geology is a particularly broad-based discipline, for it draws on many other sciences. A knowledge of physics helps us understand rock structures and deformation and supplies tools with which we can investigate the earth's deep interior indirectly. The chemistry of geologic materials provides clues to their origins and history. Modern biological principles are important in studying ancient life forms. Mathematics provides a quantitative framework within which geologic processes can be described and analyzed. Physical geographers study earth's surface features much as some geologists do. What makes geology a distinctive discipline is in part that it focuses all these approaches, and others, on the study of the earth. Moreover, having the earth as a subject introduces some special complexities not common to most other sciences.

The Issue of Time

The modern earth has been billions of years in the making. As will be seen in later chapters, many of the processes shaping it are extremely slow on a human time scale, some barely detectable even with sensitive instruments. It is therefore difficult to observe or to demonstrate directly, in detail, how certain materials or features have formed.

Furthermore, materials may respond differently to the same forces depending on how those forces are applied. This can be seen, for example, in the phenomenon of fatigue of machine parts, in which a material quite strong enough to withstand a certain level of sustained stress fails during the application of a much smaller stress that has been repeated many times. As a practical matter, then, it may be impossible to duplicate some geologic processes in the laboratory because the human lifetime is simply too short.

The Matter of Scale

Likewise, some natural systems are just too large to duplicate in the laboratory. One can study a single crystal or small piece of volcanic rock in great detail. But it is hardly possible to build a volcano (figure 1.6), or a whole continent, in a laboratory in order to carry out experiments on them. By way of compromise, scale-model experiments can sometimes be made, in which the materials and the forces applied to them are scaled down proportionately. This is done, for example, in studies of designs for earthquake-resistant buildings. Model structures are shaken by small vibrations with the hope that the results will mimic the response of large buildings to great earthquakes. As another example, one can agitate a large tank of water artificially to study wave motion or the effects of currents. Scale modeling is an inexact science, however, and not all natural systems lend themselves to such studies.

Complexity in Natural Systems

When a laboratory scientist sets out to conduct an experiment, one common objective is to minimize the number of variables so as to obtain as clear a picture as possible of the effect of any one change—whether of temperature, pressure, or the quantity of some particular substance present. Typically, the experimental materials are kept simple: a single rock type or mineral, chemically quite pure. Natural geologic systems, however, are rarely so simple. Natural rocks and minerals invariably contain chemical impurities and physical imperfections; many different rocks and minerals may be mingled together; and temperature and pressure may change simultaneously, while gases, water, or other chemicals flow in and out. Extrapolating from carefully controlled laboratory experiments to the real world becomes correspondingly difficult.

Also, the laboratory scientist can perform an experiment in stages, examining the results after each step. The geologist may be confronted by rocks or other materials or structures that have been altered or reworked several times, perhaps dozens of times, each time under different conditions, over millions or even billions of years. That history can be difficult to decipher from the present end product. This is particularly true because the same end product can often be formed from several different possible starting materials, via different combinations of geologic processes—just as one can arrive at a given spot by traveling in various ways from various starting points.

Geology and the Scientific Method

The scientific method is a means of discovering basic scientific principles. One begins with a set of observations and/or a body of data, based on measurements of natural phenomena or on experiments. One or more hypotheses are formulated to explain the observations or data. It is also possible that no systematic relationship exists among the observations (*null hypothesis*). A hypothesis can take many forms, ranging from a general conceptual framework or model describing the functioning of a natural system, to a very precise mathematical formula relating several kinds of numerical data. What all hypotheses have in common is that they must be susceptible to testing.

Figure 1.6

Mount St. Helens in eruption, May 1980.
Photo by P. W. Lipman, courtesy U.S. Geological Survey.

In the classical conception of the scientific method, one uses a hypothesis to make a set of predictions. Then one devises and conducts experiments to test each hypothesis, seeing whether experimental results agree with predictions based on the hypothesis. If so, the hypothesis gains credibility. If not, if unexpected results are found, the hypothesis must be modified to account for the new data. Several cycles of modifying and retesting hypotheses may be required before a satisfactory hypothesis is arrived at that is consistent with all the observations and experiments that one can conceive. A hypothesis that is repeatedly supported by new experiments advances in time to the status of a **theory,** a generally accepted explanation for a set of data or observations.

It should be evident that this approach is not strictly applicable to many geologic phenomena because of the difficulty of experimenting with natural systems. In such cases, hypotheses are often tested entirely through further observations and modified as necessary until they accommodate all the relevant observations. This broader conception of the scientific method is well illustrated by the development of the theory of plate tectonics, discussed in chapter 7. It should be noted, too, that even a theory may ultimately be proved incorrect. In the case of geology, the most common cause of complete rejection of an older theory is the development of new analytical or observational techniques, which make available wholly new kinds of data unknown at the time the original theory was formulated.

In addition to hypotheses and theories, there is a smaller body of scientific **laws,** fundamental, typically simple principles or formulas that are invariably found to be true. In this category would be Newton's law of gravity, or the principle of physical chemistry that states that heat always flows from a warmer body to a colder one, never the reverse.

Key Concepts in the History of Geology

Humans have wondered about the earth in some way for thousands of years. The ancient Greeks measured it and recognized fossils preserved in its rocks as remains of ancient life forms. Theologians, philosophers, and scientists have speculated on its age for centuries. The systematic study of the earth that constitutes the science of geology, however, has existed as an organized discipline for only about 250 years. In its early years, it was predominantly a descriptive subject. Two principal opposing schools of thought emerged in the eighteenth and nineteenth centuries to explain geologic observations.

One, popularized by James Hutton and later named by Charles Lyell, was the concept of **uniformitarianism.** Sometimes condensed to the phrase, "The present is the key to the past," it comprises the ideas that the surface of the earth has been continuously and gradually changed and modified over the immense span of geologic time, and that by studying the geologic processes now active in shaping the earth we can understand how it has evolved through time. It is not assumed that the *rates* of all processes have been the same throughout time, but that the nature of the processes is similar, the same physical principles operating on the earth in the past as in the present.

continuous gradual

James Hutton was a remarkably versatile individual—physician, farmer, and only part-time geologist. In the early days of the science, many advances in understanding were made by nonspecialists capable of careful observation and logical deduction. As various disciplines become more advanced and more sophisticated, the amateur is far less able to play a major role. Too much accumulated knowledge must be assimilated to arrive at the forefront of research.

The second, contrasting theory was **catastrophism.** The catastrophists, led by Georges Cuvier, believed that a series of immense, worldwide upheavals were the agents of change and that between catastrophes the earth was static. Violent volcanic eruptions

Episodic

followed by torrential rains and floods were said to form mountains and valleys and to bury animal populations that later became fossilized. In between those episodic global devastations, the earth's surface did not change. Entire plant and animal populations were created anew after each such event, to be wholly destroyed by the next.

Is the Present the Key to the Past?

More detailed observations and calculations, greater use of the allied sciences, and development of increasingly sophisticated instruments have collectively provided overwhelming evidence in support of the uniformitarian view. The great length of the earth's history makes it entirely plausible that processes that seem gradual and even insignificant on a human time scale could, over those long spans of time, create the modern earth in all its geologic complexity.

This is not to say that earth history has not been punctuated occasionally by sudden, violent events that have had a substantial impact on a regional or global scale. We see evidence of past collisions of large meteorites with the earth (figure 1.7) and find volcanic debris from ancient eruptions that would dwarf that of Mount St. Helens. But these events are unusual and, for the most part, of only temporary significance in the context of the earth's long history. By and large, the modern earth is indeed the product of uncounted small and gradual changes, repeated or continued over very long spans of time.

The same physical and chemical laws can be presumed to have operated throughout earth's history, and thus by observing modern geologic processes we can learn much about how those same processes might have shaped the earth in the past. It is necessary to keep in mind, however, that we cannot assume that the relative importance of each process has always been just the same as it is now, or even that all processes operated in detail just as they do now. Certain irreversible changes in the earth have no doubt changed the nature and intensity of corresponding geologic processes. For example, the earth has been slowly losing heat ever since it formed. Present internal temperatures must be substantially lower than they were several billion years ago, especially near the surface. The earth's internal heat plays a key role in melting rocks and thus in volcanic activity. It is reasonable to infer that melting in the interior was more extensive in the past than it now is, that the products of that more extensive melting might be somewhat different from modern volcanic rocks, and that volcanic activity was more extensive in the past also. Earth's atmosphere, too, has undergone profound changes, as will be explored further shortly. Briefly, it has gone

Figure 1.7

Meteor Crater, Arizona.

from oxygen-poor to oxygen-rich, and this change would in turn necessarily affect the chemical details of such processes as weathering of rocks by interaction with air and water. However, it is perfectly possible to determine, experimentally and theoretically, how rocks would react with the kind of atmosphere deduced for the early earth and thus to characterize ancient weathering processes even though we cannot observe their exact equivalents in nature today. The earth may change; the physical laws do not. This concept is fundamental to modern uniformitarianism.

The Earth in Its Universe

The progress of astronomy in recent decades has made it possible for scientists to construct an ever clearer picture of the origins of the solar system and of the universe itself. From observations that the stars are all moving apart from one another came the recognition of an expanding universe. Clearly that expansion cannot have been going on forever: if one looks backward in time and extrapolates the stars' movements backwards too, a point is reached at which all matter was apparently together in one place. Most astronomers now accept a cataclysmic explosion, or "big bang," as the origin of the modern universe. At that time, enormous quantities of matter were synthesized and flung violently apart, hurtling out across an ever larger volume of space. The time of the so-called big bang can be estimated in several ways. Perhaps the most direct is to calculate the universe's expansion backward to its apparent beginning, by extrapolating the present motions of the stars backward in time until they converge. Other methods depend on astrophysical models of creation of the elements, or the rate of evolution of different types of stars. The age estimates overlap in the range of 15–20 billion years. This, then, may be taken as the approximate age of the modern universe.

Whether 15–20 billion years is the *ultimate* age of the universe depends on one's choice of astrophysical model. There are two principal schools of thought. One holds that the universe will expand indefinitely, the other that it will continue expanding only up to a point, then reverse and begin to collapse to another big bang, continuing to oscillate in this fashion. In the latter case, the date of 15 to 20 billion years would be the time of the last big bang and subsequent expansion.

The Birth and Death of Stars

The matter in the expanding universe was not uniformly distributed. Locally, higher concentrations of mass were collected together by gravity, and some became large enough and dense enough that energy-releasing atomic reactions were set off deep within them. These were the earliest stars. Stars are not permanent objects. Their radiance is evidence that they are constantly losing energy, and thus mass, as they burn their nuclear fuel. The principal product of the big bang was hydrogen, the lightest element. The atomic reactions that occur in stars are responsible for the production of most of the other heavier chemical elements, starting from that abundant hydrogen.

The mass of material that initially formed the star determines how rapidly it burns, and its ultimate fate. In time, each star will either collapse and cool to a small black dwarf or, if it is more massive, explode as a supernova. Some stars burned out or exploded billions of years ago. Others are still forming from the original matter of the universe mixed with the debris of older stars. The chemical complexity of our sun and the earth indicates that the solar system incorporates matter that has cycled through several earlier generations of stars since the big bang.

The Early Solar System

Our sun is a middle-aged star. That is, given its mass and rate of energy output, this star should be able to shine for about ten billion years. The solar system has existed for close to half that length of time. The sun and its system of circling planets, including the earth, are believed to have formed from a single rotating cloud of gas and dust, starting nearly five billion years ago. It may be that a shock wave from a nearby exploding supernova pushed enough of the material in the cloud together that it began to collapse under its own gravitational pull. Whatever the cause, most of the mass coalesced to form what would eventually become the sun. Like the rest of the universe, the early sun consisted mostly of hydrogen. The inner parts of this ball of gas were so compressed by its enormous mass that they became hot and dense enough for nuclear reactions to start. The ball of gas became a star, radiating light and other forms of energy. The sun should continue to shine for about five billion years more, before it has used up so much of its fuel that it collapses to a cold dwarf, and the earth's solar energy is turned off.

While the proto-sun developed, the remaining matter settled into a rotating disk around it. Dust began to condense from the gas, and planets to form from the dust, continuing to circle the sun as they formed

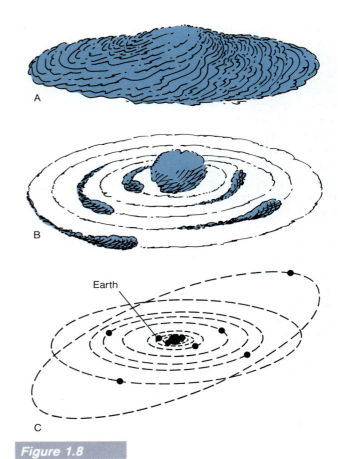

Figure 1.8

So-called nebular model of solar system formation. *(A)* Beginnings as a rotating gas cloud. *(B)* Most of the mass becomes concentrated at the center to form the sun; remaining material condenses and accumulates to form planets. *(C)* Present solar system. Earth is the third planet from the sun. It is one of the smaller planets, but it has a unique combination of composition (including surface water) and climate that makes possible life as we know it.

(figure 1.8). Modern methods of dating rock material (see chapter 6) have shown the oldest fragments of meteorites and moon rocks to be close to 4.6 billion years old. Formation of the solar system is thus believed to have been substantially complete over 4 1/2 billion years ago.

The Planets

The composition of each planet depended strongly on how near it was to the hot young sun. Very close to the sun, temperatures were so high that nothing solid could exist at all, at first. As cooling progressed, the solids that condensed nearest the sun contained mainly high-temperature materials: metallic iron and a few other minerals with very high melting temperatures. The nearest planets to the sun, then, consist mostly of these materials. Somewhat farther out, where temperatures were lower, the developing planets incorporated much larger amounts of lower-temperature materials, including some that contain water locked within their crystal structures. (This eventually made it possible for the earth to have liquid water at its surface.) Still farther from the sun, temperatures were so low that nearly all of the materials in the original gas cloud condensed—even materials like methane and ammonia, which are gases at normal earth surface temperatures and pressures. Each planet, then, is believed to have formed as an accumulation of bits of these condensed materials drawn together by gravity. Uncondensed gases were swept out of the interplanetary spaces by streams of matter and energy radiating from the young sun.

The compositions of the planets are explored individually in chapter 22. For the present, it is sufficient to note that the solar-system-forming processes outlined on the previous page would lead to a series of planets with a variety of compositions, mostly quite different from that of earth. This is something to keep in mind when it is proposed that some day we might mine other planets for needed minerals. Both the basic chemistry of these other bodies, and the kinds of ore-forming or other resource-forming processes that might occur on them, probably differ considerably from those on earth and may not lead to products we would find useful. (This is leaving aside the economics or technical practicality of such mining activities!) Also, our principal present energy sources required living organisms in order to form, and we have so far found no life on other planets or moons.

The Early Earth

The earth has changed considerably since its formation. It underwent some particularly profound changes in its early history.

Heating and Differentiation

Like most of the planets, the earth is believed to have begun as a sort of dust ball of small bits of condensed material collected together by gravity—no free water, no atmosphere, and a very different surface from the present. The dust ball was heated up in several ways. The impact of the colliding particles as they came together to form the earth provided some heat. Much of this heat was radiated out into space, but some was trapped in the interior of the accumulating earth. As the dust ball grew, compression of the interior by gravity heated it also. (The fact that materials heat up when compressed can be demonstrated by pumping up a bicycle tire, then feeling the barrel of the pump.) Furthermore, the earth contains small amounts of several naturally radioactive elements that decay, releasing energy (see chapter 6). These three heat

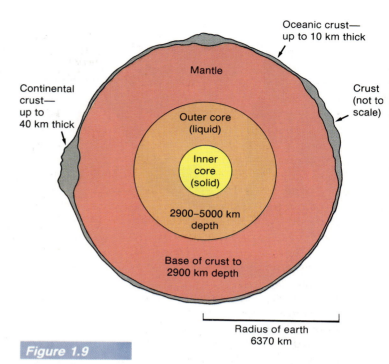

Figure 1.9

A chemically differentiated earth (crust not drawn to scale). Oceanic crust, which forms the sea floor, has a composition somewhat like that of the mantle but richer in silicon. Continental crust is thicker and contains more of the low-density minerals rich in calcium, sodium, potassium, and aluminum. It rises above both the sea floor and the ocean surface.

sources combined to raise the earth's internal temperature enough that parts of it, eventually most of it, melted.

When accretion of material was complete, slow cooling set in. Metallic iron, being very dense, sank into the middle of the earth. As cooling progressed, and the remaining melt began to crystallize, lighter, low-density minerals floated out toward the surface. The eventual result was an earth differentiated into several major compositional zones: a large, iron-rich core; a thick surrounding mineral mantle; and a thin, low-density crust at the surface (see figure 1.9). Chapter 10 examines how we know the structure and composition of the earth's interior. It can be shown that the core is made up mostly of iron, with some nickel and a few minor elements, and that the mantle consists mainly of the elements iron, magnesium, silicon, and oxygen combined in varying proportions in several different minerals. The crust is much more varied in composition and very different chemically from the average composition of the earth, as will also be seen in chapter 10. The differentiation process was complete at least 4 billion years ago.

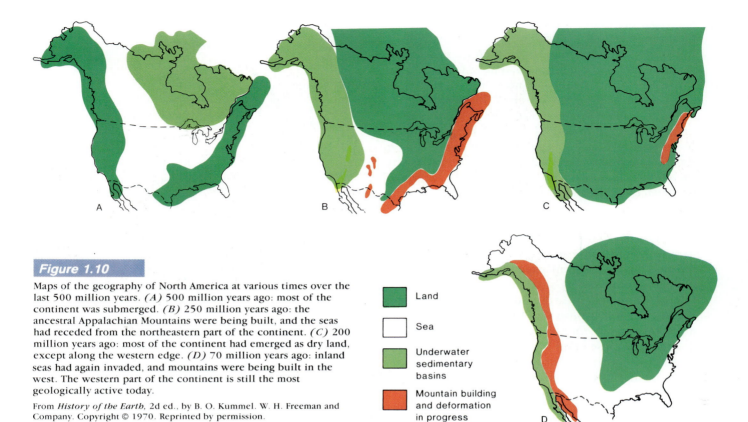

Figure 1.10

Maps of the geography of North America at various times over the last 500 million years. *(A)* 500 million years ago: most of the continent was submerged. *(B)* 250 million years ago: the ancestral Appalachian Mountains were being built, and the seas had receded from the northeastern part of the continent. *(C)* 200 million years ago: most of the continent had emerged as dry land, except along the western edge. *(D)* 70 million years ago: inland seas had again invaded, and mountains were being built in the west. The western part of the continent is still the most geologically active today.

From *History of the Earth*, 2d ed., by B. O. Kummel. W. H. Freeman and Company. Copyright © 1970. Reprinted by permission.

Legend:

- Land
- Sea
- Underwater sedimentary basins
- Mountain building and deformation in progress

The Early Atmosphere and Oceans

The heating and subsequent differentiation of the early earth led to another important result: formation of the atmosphere and oceans. Many minerals that had contained water or gases locked in their crystals released them during the heating and melting. It is not clear whether the earth was ever wholly molten at the surface; at the very least, it was much hotter than at present and subject to more extensive volcanic activity, with water among the gases thus released. As the earth's surface cooled, the water could condense to form the oceans. Without this abundant surface water, which in the solar system is unique to earth, most life as we know it could not exist.

The early atmosphere was quite different from the modern one, even disregarding the effects of modern pollution. The first atmosphere had little or no free oxygen in it. Humans could not have survived in it. Oxygen-breathing life could not exist before the first simple plants, the single-celled, blue-green algae, appeared in large numbers to modify the atmosphere. Their remains are known in rocks as old as several billion years. They manufacture food by photosynthesis, using sunlight for energy and releasing oxygen as a byproduct. In time, enough oxygen accumulated that the atmosphere could support oxygen-breathing organisms.

Subsequent History

Differentiation produced the most profound internal changes in the earth and also established a combination of land, air, and water at the surface. Significant changes have continued at and near the surface through time.

The Changing Face of Earth

After the early differentiation, the crust with its continents and ocean basins did not look the way we know it today. For one thing, the continents have moved (see chapter 7). The continents have also not always been the same size and shape. Rocks that were deposited in ocean basins can now be found high and dry on land, revealing the former presence of inland seas followed by great uplift. New pieces have been added to the edges of the continents. Volcanoes once erupted where none now exist, leaving behind evidence of their earlier activity in ancient volcanic rocks. Tall mountains have been built up and then eroded away, sometimes several times in the same place, over billions of years.

We can to some extent reconstruct the distribution of land, water, and surface features as they were at times in the past, and identify geologically active

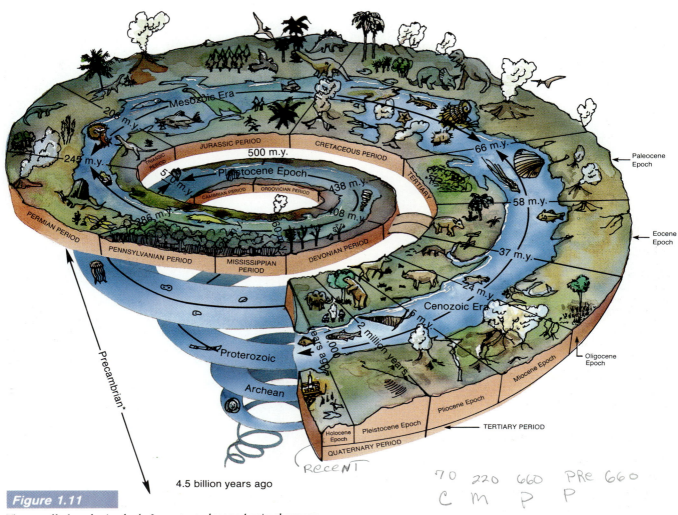

70 220 660 PRe 660
C   M   P   P

ReCeNT

Figure 1.11

The so-called geologic clock. Important plant and animal groups are shown where they first appear in significant numbers. All complex organisms have developed relatively recently in the geologic sense, humans especially so.

After U.S. Geological Survey publication, "Geologic Time."

areas such as developing mountain ranges. This is done on the basis of the kinds of rocks or fossils of each age that we find and what we know of how such rocks formed or in what setting the fossilized creatures lived. Such reconstruction becomes more difficult the farther back we try to go in time, for the oldest rocks have often been covered by younger ones or disrupted by more recent geologic events.

Figure 1.10 is a series of simplified maps showing the distribution of stable dry land, seas, and mountains for North America at several times over the last 500 million years. Even over this geologically short period, dramatic changes in the face of the continent have occurred. Aside from its academic interest, reconstruction of ancient geography and geology can be of practical value. If we know that certain kinds of

needed mineral or energy resources have formed in particular geologic settings, we can look for those resources not only in the appropriate modern geologic environments but in rocks that formed in similar environments in the past.

Life on Earth

We can look back in the rock record to see when different plant and animal groups appeared. Some are represented schematically in figure 1.11. The earliest creatures left very limited remains, because they had no hard skeletons, teeth, shells, or other hard parts that could be preserved in rocks. The photosynthetic microorganisms are known principally from remains

Figure 1.12

These layered structures can be identified by a trained eye as *stromatolites, fossilized algal mats,* here found in rocks more than 2 1/2 billion years old.

Figure 1.13

By several hundred million years ago, a variety of organisms with hard, preservable bones, shells, and other parts had become widespread.

of their communities—algal mats, for instance (figure 1.12). The first multicelled oxygen-breathing creatures are believed to have developed about one billion years ago, after sufficient oxygen had accumulated in the atmosphere. However, they are poorly preserved, probably because they were entirely soft bodied. By about 600 million years ago, marine animals with shells had become widespread.

The development of organisms with preservable hard parts greatly increased the number of preserved animal remains in the rock record, so we understand biological developments since that time far better than we do for the earth's first few billion years (see figure 1.13). Dry land was still barren of large plants or animals half a billion years ago. In rocks about 400 million years old we find the first evidence of animals with backbones—the fish—and also some land plants. Insects appeared approximately 300 million years ago. Later, reptiles and amphibians moved onto the continents. The dinosaurs appeared about 200 million years ago and the first mammals at nearly the same time. With the development of birds, warm-blooded animals took to the air about 150 million years ago, and by 100 million years ago both birds and mammals were well established. Again, such information has current applications. Certain of our energy sources

have formed from plant or animal remains (see chapter 21). Knowing the times at which particular groups of organisms appeared and flourished is helpful in assessing the probable amounts of these energy sources available and in concentrating the search for these fuels on rocks of appropriate ages.

On a time scale of billions of years, human beings have just arrived. The most primitive human-type remains are no more than 3–4 million years old, and modern rational man (*Homo sapiens*) developed only about half a million years ago. Half a million years may sound like a long time, and indeed it is compared to a single human lifetime. In a geologic sense, though, it is a very short time. Nevertheless, humans have had an enormous impact on the earth, at least at its surface, an impact far out of proportion to the length of time we have occupied the planet.

The Modern Dynamic Earth

Although it has been cooling for billions of years, the earth still retains enough internal heat to drive large-scale mountain-building processes, to cause volcanic eruptions, to make continents mobile, and indirectly to trigger earthquakes. At the same time, the continual supply of solar energy to the surface drives many of the surface processes: water is evaporated from the oceans to descend again as rain and snow to feed the rivers and glaciers that sculpture the surface; differential heating of the surface leads to formation of warmer and colder air masses above it, which in turn produces atmospheric instability and wind.

Dynamic Equilibrium in Geologic Processes

Natural systems tend toward a balance or equilibrium among opposing forces. As new dissolved minerals are washed into the sea by rivers, sediments are deposited in the ocean basins, removing dissolved chemicals from solution. Internal forces push up a mountain; gravity, wind, water, and ice collectively act to tear it down again. When one factor changes, other compensating changes occur in response. If the disruption of a system is relatively small and temporary, the system may in time return to its original condition, and evidence of the disturbance will be erased. A coastal storm may wash away beach vegetation and destroy colonies of marine organisms living in a tidal flat, but when the storm has passed, new organisms will start to move back into the area, new grasses to establish roots in the dunes. The violent eruption of a volcano like Krakatoa or Mount St. Helens may spew ash high into the atmosphere, partially blocking sunlight and causing the earth to cool, but within a few years the ash will have settled back to the ground and normal temperatures will be restored.

This is not to say that permanent changes never occur in natural systems. The size of a river's channel depends in part on the maximum amount of water it normally carries. If long-term climatic or other conditions change so that the volume of water regularly reaching the stream increases, the larger quantity of water will in time carve out a correspondingly larger channel to accommodate it. The soil carried downhill by a landslide certainly does not begin moving back upslope after the landslide is over; the face of the land is irreversibly changed. Even so, a hillside forest uprooted and destroyed by the slide may within decades be replaced by fresh growth in the soil newly deposited at the bottom of the hill.

The Impact of Human Activities; Earth as a Closed System

For all practical purposes, the earth is a **closed system,** meaning that the amount of matter in and on the earth is fixed. No new elements are being added. There is, therefore, an ultimate limit to how much of any metal we can exploit. There is also only so much land to live on. Conversely, any harmful elements we create remain in our geologic environment, unless we take some extraordinary step, such as expending the funds, materials, and energy required to cast something out into space via spacecraft.

Human activities can cause or accelerate permanent changes in natural systems. The impact of humans on the global environment is broadly proportional to the size of the population, as well as to the level of technological development. This can be illustrated readily within the context of pollution. The smoke from one campfire pollutes only the air in the immediate vicinity. By the time that smoke is dispersed through the atmosphere, its global impact is negligible. The collective smoke from a century of industrialized society, on the other hand, has caused measurable increases in several atmospheric pollutants worldwide, and those pollutants continue to pour into the air from many sources.

Likewise, in the context of resources, the recent impact of human activities has been dramatic. The world's population has soared from 2 billion to more than 5 billion in little more than 50 years. That fact, combined with the desire for ever higher standards of living around the world, has created a voracious demand for mineral and energy resources. Yet although the population is growing, the earth is not. Therefore our resources are in finite supply, and some could be exhausted within our lifetimes. This problem will be considered in chapters 20 and 21. The very presence of humans, and their building and farming activities, alters the landscape, and does so increasingly as the population grows (figure 1.14). Superimposed on the natural geologic processes of change on the earth, then, are further changes, deliberate or unconscious, caused by human activities. Many of the latter are occurring more rapidly than the compensating natural processes.

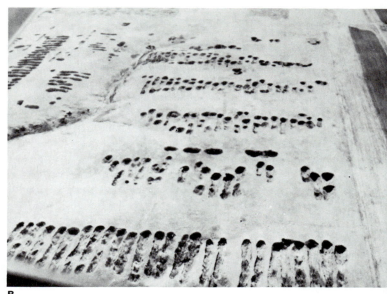

A

B

Figure 1.14

Human impacts on the earth's surface. *(A)* Seasonal clearing of land for farming allows accelerated soil erosion. *(B)* Years after the completion of underground coal mining here near Mercer County, North Dakota, the surface is collapsing into the mine workings.

Photos: *(A)* Courtesy U.S.D.A. Soil Conservation Service. *(B)* C. R. Dunrud, courtesy U.S. Geological Survey.

The principal focus of this text is natural geologic processes. However, from time to time it will highlight particularly significant human impacts on geologic systems, as well as the reverse, geologic impacts on human activities (such as the hazards associated with floods, earthquakes, and other geologic processes; see figure 1.15).

Figure 1.15

Houses buried by volcanic ash eruption of Heimaey: Vestmannejar, Iceland, 1973.

Photo courtesy U.S. Geological Survey.

Summary

Geology is the study of the earth. It is not an isolated discipline, but draws on the principles of many other sciences. The earth is a challenging subject, for it is old, complex in composition and structure, and large in scale. Physical geology focuses particularly on the physical features of the earth and how they have formed. Observations suggest that, for the most part, those features are the result of many individually small, gradual changes continuing over long periods of time, punctuated by occasional unusual, cataclysmic events.

The solar system formed from a cloud of gas and dust close to 5 billion years ago. The compositions of the resulting planets reflect their proximity to the evolving sun. Shortly after its formation, the earth underwent melting and compositional differentiation into core, mantle, and crust. The early atmosphere and oceans formed at the same time. Heat from within the earth and from the sun together drive many of the internal and surface processes that have shaped and modified the earth throughout its history, and continue to do so. The earliest life forms date back several billion years. Organisms with hard parts became widespread about 600 million years ago. Humans only appeared 3 to 4 million years ago, but their large and growing numbers and technological advances have had significant impacts on natural systems, some of which may not readily be erased by slower-paced geologic processes.

Terms to Remember

Remember – look up & put on cards.

catastrophism	crust	mantle	theory
closed system	hypothesis	physical geology	uniformitarianism
core	law	scientific method	

Questions for Review

1. What is *physical geology*?

2. Describe how the time factor complicates our attempts to understand geologic processes.

3. What is the scientific method? To what extent is it applicable to geology?

4. Compare and contrast the concepts of uniformitarianism and catastrophism.

5. How is the time of the "big bang" determined?

6. Briefly summarize the process by which the solar system formed. Does this process lead to planets that are similar or quite different in composition? Explain.

7. How were the earth's atmosphere and oceans first formed? How and why has the atmosphere changed through time?

8. What are the principal compositional zones of the earth?

9. The earth is a closed system. Explain this concept in the context of resource use or of pollution.

For Further Thought

1. If the whole 4 1/2-billion-year span of earth's history were represented by one 24-hour day, how much time would correspond to (a) the 600 million years of existence of complex organisms with hard parts; (b) the half-million years that *Homo sapiens* has existed?

2. Many geologic processes proceed at rates of the order of one centimeter per year (1 inch = 2.54 centimeters). At that rate, how long would it take you to travel from home to school, or from your room to class?

Suggestions for Further Reading

Cameron, A. G. W. "The Origin and Evolution of the Solar System." *Scientific American* 233 (September 1975): 32–41.

Head, J. W.; Wood, C. A.; and Mutch, T. "Geological Evolution of the Terrestrial Planets." *American Scientist* 65 (1976): 21–29.

Kummel, B. *History of the Earth.* 2d ed. San Francisco: W. H. Freeman and Co., 1970.

Lewis, J. "The Chemistry of the Solar System." *Scientific American* 230 (March 1974): 60–65.

Pilbeam, D. "The Descent of Hominoids and Hominids." *Scientific American* 250 (March 1984): 84–96.

Siever, R. "The Dynamic Earth." *Scientific American* 249 (September 1983): 46–65.

Simpson, G. G. *This View of Life.* New York: Harcourt, Brace & World, 1974.

Wood, J. A. *The Solar System.* Englewood Cliffs, N.J.: Prentice-Hall, 1979.

York, D. *Planet Earth.* New York: McGraw-Hill Book Co., 1975.

Minerals and Rocks

Introduction

It is difficult to talk at length about geology without talking about rocks or the minerals of which they are composed. Some minerals are economically valuable. All provide clues to the processes by which they formed, and thus to the geologic setting in which they can be found.

Just eight elements account for over 98% of the earth's crust, yet there are dozens of common minerals and more than a thousand recognized minerals in all, with a corresponding diversity of physical properties. Fortunately for the beginning student, it is necessary to become familiar with only a small number of the more common minerals and mineral groups in order to discuss a wide range of rock types and geologic processes. This chapter introduces those minerals and some underlying chemical and mineralogical concepts, then finishes with a brief look at rocks considered as assemblages of minerals.

Atoms, Elements, and Isotopes

All natural and most synthetic substances on earth are made from the 90 naturally occurring chemical elements. An **element** is the simplest kind of chemical: it cannot be broken down further by ordinary chemical or physical processes such as heating or reaction with other chemicals. An **atom** is the smallest particle into which an element can be subdivided while still retaining its distinctive chemical characteristics.

Basic Atomic Structure

A simplified sketch of the structure of an atom is shown in figure 2.1. Atoms contain several types of subatomic particles: **protons**, which have a positive electrical charge; **electrons**, which have a negative electrical charge; and **neutrons**, which, as their name implies, are electrically neutral (have no charge). The **nucleus**, at the center of the atom, contains the protons and neutrons; the electrons circle around the nucleus.

The number of protons in the nucleus determines what chemical element that atom is. Every atom of hydrogen contains one proton in its nucleus; every carbon atom, 6 protons; every oxygen atom, 8 protons; every uranium atom, 92 protons. The characteristic number of protons is the **atomic number** of the element.

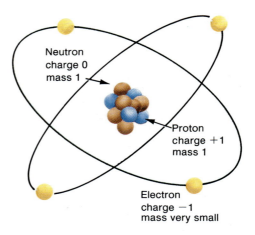

Neutron
charge 0
mass 1

Proton
charge +1
mass 1

Electron
charge −1
mass very small

Basic atomic structure: protons and any neutrons in the nucleus, electrons circling outside it.

Isotopes

Most nuclei contain neutrons as well as protons, and the number of neutrons is similar to or somewhat greater than the number of protons. The number of neutrons in different atoms of one element is not always the same. The sum of the number of protons and the number of neutrons in a nucleus is that atom's **atomic mass number.** (Protons and neutrons have similar masses; electrons are much lighter and contribute little to the mass of the whole atom.) Atoms of a given element with different atomic mass numbers—in other words, atoms with the same number of protons but different numbers of neutrons—are distinct **isotopes** of that element (see figure 2.2). Some elements have only a single isotope, while others may have ten or more. (The reasons for this are complex, involving principles of nuclear physics, and we will not pursue them here.)

A particular isotope is designated by the element name (which by definition specifies the atomic number, or number of protons) and the atomic mass number (protons plus neutrons). Carbon, for example, has three natural isotopes. By far the most abundant is carbon–12, the isotope with 6 neutrons in the nucleus in addition to the 6 protons common to all carbon atoms. The two rarer isotopes are carbon—13 (6 protons + 7 neutrons) and carbon–14 (6 protons + 8 neutrons). For most applications, we need be concerned only with the elements involved, not with specific isotopes. Chemically, all isotopes of one element behave alike. The human body cannot, for instance, distinguish between sugar containing carbon—12 and sugar containing carbon–13.

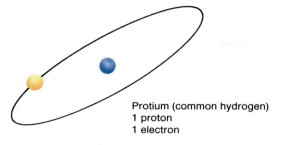

Protium (common hydrogen)
1 proton
1 electron

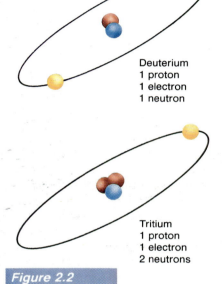

Deuterium
1 proton
1 electron
1 neutron

Tritium
1 proton
1 electron
2 neutrons

Figure 2.2

Isotopes of hydrogen: each has one proton in the nucleus, but there may be zero, one, or two neutrons. The isotope with two neutrons, tritium, is unstable and naturally radioactive.

The principal practical difference between isotopes is in the stability of the different nuclei. Certain combinations of protons and neutrons are inherently unstable, and in time those unstable nuclei will decay, or break down spontaneously. This is the nature of **radioactivity.** The existence of naturally occurring radioactive isotopes makes accurate dating of many materials possible, as will be seen in chapter 6. Differences in the nuclear properties of two uranium isotopes are relevant to our nuclear-power options (chapter 21).

Ions and Bonding

In an electrically neutral atom, the number of protons and the number of electrons is the same. The negative charge of one electron just equals the positive charge of one proton. However, it is possible for most atoms to gain or lose electrons. When this happens, the atom has a positive or negative electrical charge and is termed an **ion.** If it loses electrons it becomes a positively-charged **cation,** as the number of protons exceeds the number of electrons. If it has gained electrons, the resulting ion has a negative electrical charge and is termed an **anion.**

Causes of Ion Formation

The tendency to gain or lose electrons is related to the number of electrons present in the neutral atom. Electrons are organized into shells, or orbitals, each of which can accommodate up to a specific maximum number of electrons (figure 2.3). Electrons fill these shells in order, beginning with the one closest to the nucleus, which can hold up to 2 electrons. For most atoms, the last shell will be only partially filled. Added chemical stability is associated with having all shells exactly filled. Therefore atoms whose outermost electron shells are only partially filled will tend to gain, lose, or share electrons with other atoms to achieve evenly filled shells. Atoms that have only a few "left-over" electrons in their outermost shell will tend to lose them. A neutral sodium atom, for example, has one electron in its outermost shell, so the simplest way to achieve a configuration consisting of all filled shells is to lose that one electron, forming a sodium cation with a $+1$ charge (Na^+). A neutral chlorine atom, by contrast, has an outermost shell containing 7 electrons, though that shell can hold 8. Chlorine then tends to gain one electron, to form the chloride anion (Cl^-). Carbon has 4 electrons in its outermost shell, which can hold 8 electrons. It also has 6 protons in the nucleus. Gain or loss of 4 electrons would create, proportionately, a large imbalance between protons and electrons. However, carbon can also share electrons with neighboring atoms to achieve filled shells for all. Indeed this is what occurs in diamond, a mineral consisting of pure carbon in which each carbon atom shares electrons with adjacent atoms (figure 2.4).

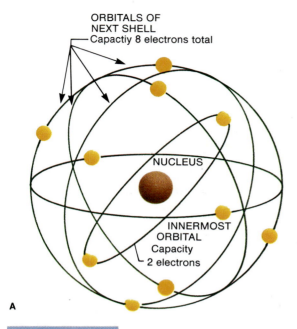

ORBITALS OF
NEXT SHELL
Capactiy 8 electrons total

NUCLEUS

INNERMOST
ORBITAL
Capacity
2 electrons

A

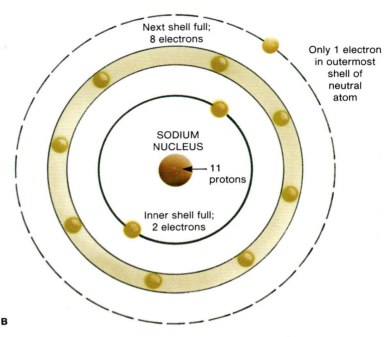

Next shell full;
8 electrons

Only 1 electron
in outermost
shell of
neutral
atom

SODIUM
NUCLEUS

11
protons

Inner shell full;
2 electrons

B

Figure 2.3

Electronic structure and ion formation. *(A)* The electron shell
structure of atoms. *(B)* Sodium can best achieve a filled
outermost shell by losing one electron, making the Na^+ ion.
(C) A neutral chlorine atom lacks one electron in its outermost
shell, so it forms an ion by gaining one electron.

Bonding and Compounds

Opposite electrical charges attract. Cations and anions may become chemically bonded together by this
attraction, forming an **ionic bond.** Sodium and chloride ions in table salt (sodium chloride) display ionic
bonding. The sharing of electrons between atoms creates another kind of bond, a **covalent bond.** The
carbon atoms in diamond display covalent bonding.
(The chemist Kekulé once envisioned covalent
bonding among carbon atoms as a kind of hand-
holding between atoms.) Ionic and covalent bonding
are the two most common kinds of bonding in minerals. Actually, few if any chemical bonds in nature
are either wholly ionic or wholly covalent, but many
are predominantly one or the other kind, and it is convenient to think of them that way.

There are a few elements in which neutral atoms have
all electron shells exactly filled already. They have,
therefore, no need to gain, lose, or share electrons in
order to achieve chemical stability. These elements
are described as chemically **inert,** for they do not generally bond with other elements. They exist on earth
as gases, with individual atoms floating freely and independently. Helium and neon are among the inert
gases.

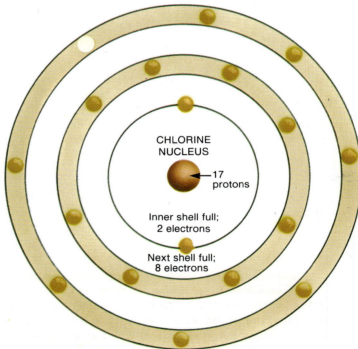

CHLORINE
NUCLEUS

17
protons

Inner shell full;
2 electrons

Next shell full;
8 electrons

Outer shell (capacity 8)
lacks 1 electron in
neutral atom

C

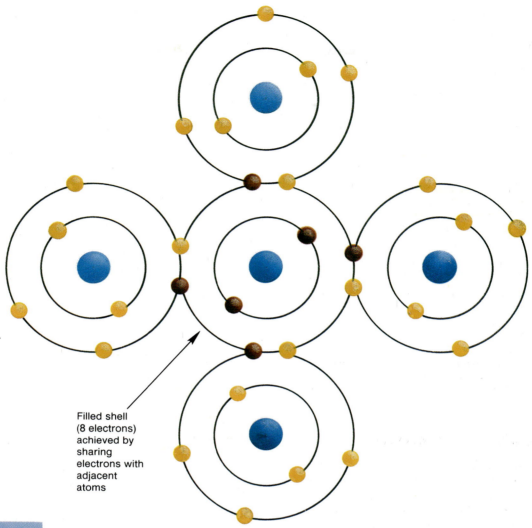

Filled shell
(8 electrons)
achieved by
sharing
electrons with
adjacent
atoms

Figure 2.4

In diamond, carbon forms covalent bonds with other carbon
atoms to achieve filled outermost shells for all.

When atoms or ions of different elements bond
together they form a **compound.** A compound is a
chemical combination of two or more elements, in
particular proportions, and it has a distinct set of
physical properties, usually very different from those
of the individual elements in it. Sodium is a silvery-
colored metal; chlorine is a greenish gas that is poisonous in large doses. When equal numbers of sodium and chlorine atoms combine to form table salt,
sodium chloride (NaCl), the resulting compound
consists of colorless crystals with properties that do
not resemble those of either of the constituent elements.

Whatever the nature of the bonding in a solid, it
is overall electrically neutral, the positive charges of
the cations just balanced by the negative charges of
the anions. Any ions present are firmly bonded in
place, not free to drift through the solid. Liquids, too,
are overall electrically neutral, but within a liquid
compounds may break up into individual free ions that
can move independently. When table salt goes into solution, it does so by breaking up into sodium and
chloride ions.

Box 2.1

The Periodic Table

Some idea of the probable chemical behavior of elements can be gained from a knowledge of the **periodic table** (figure 1). The Russian scientist Dmitri Mendeleyev first observed that certain groups of elements showed similar chemical properties, which seemed to be related in a regular way to their atomic numbers. At the time (1869) that Mendeleyev published the first periodic table, in which elements were arranged so as to reflect these similarities of behavior, not all the elements had even been discovered, so there were some gaps. The addition of elements identified later confirmed the basic concept, and in fact some of the missing elements were found more easily because their properties could to some extent be anticipated from their expected position in the periodic table.

We now can relate the periodicity of chemical behavior to the electronic structures of the elements. For example, those elements in the first column, known as the alkali metals, have one electron in the outermost shell of the neutral atom. Thus they all tend to form cations of +1 charge by losing that odd electron. Outermost electron shells become increasingly full from left to right across a row. The next-to-last column, the halogens, are those elements lacking only one electron in the outermost shell, and they thus tend to gain one electron to form anions of charge −1. In the right-hand column are the inert gases, whose neutral atoms contain all fully filled electron shells.

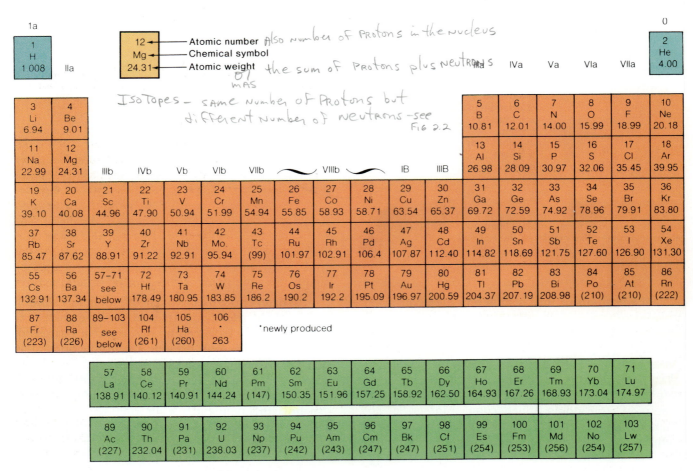

Values in parentheses are approximate.

Figure 1
The periodic table.

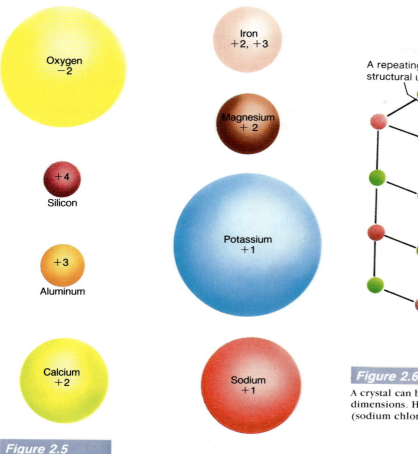

Figure 2.5

Charges and relative sizes of the most common ions in the crust.

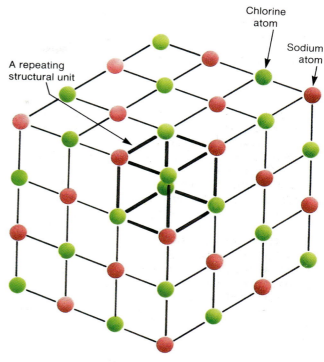

Figure 2.6

A crystal can be thought of as a structural unit repeated in three dimensions. Here, the basic structural unit of a crystal of halite (sodium chloride, NaCl).

Minerals—General

A **mineral,** by definition, is a naturally occurring, inorganic solid element or compound, with a definite composition and a regular internal crystal structure. This definition can be clarified by considering its parts separately, with examples. *Naturally occurring*, as distinguished from synthetic, means that the minerals do not include the thousands of chemical substances invented by humans. *Inorganic* means not produced solely by living organisms, by biological processes. The fact that minerals must be solid means that the ice of a glacier is a mineral but liquid water is not.

Chemically, minerals consist either of one element—like a diamond, which is pure carbon—or they are compounds of two or more elements. Some mineral compositions are very complex indeed, consisting of ten elements or more. It must be possible to specify the chemical composition of a mineral, or a compositional range within which it falls. The presence of certain elements in certain proportions is one of the defining characteristics of each mineral (see box

2.2). The reason for the qualifying phrase "or a compositional range" is the phenomenon of solid solution observed in many minerals, in which two or more ions of similar size and charge may substitute more or less interchangeably for each other in the same mineral structure. Magnesium (Mg^{2+}) and iron (Fe^{2+}) ions are nearly the same size (see figure 2.5), so any mineral that contains one tends to contain some of the other, with the proportions varying from sample to sample. Although they are less similar in size, sodium (Na^+) and potassium (K^+) can show similar behavior, as can many other sets of elements. The effect of this on the mineral's chemical formula can be seen in box 2.2.

Finally, minerals are crystalline, at least on the microscopic scale. **Crystalline** materials are solids in which the atoms are arranged in regular, repeating patterns (figure 2.6). These patterns may not be apparent to the naked eye, but most solid elements and compounds in fact are crystalline, and their crystal structures can be recognized and studied using X rays and other techniques.

Box 2.2

Chemical Symbols and Mineral Names

Each of the chemical elements has a one- or two-letter symbol by which it can be represented. Many of these symbols make sense in terms of the English name for the element—O for oxygen, He for helium, Si for silicon, and so on. Others reflect the fact that in earlier centuries scientists were generally versed in Latin or Greek: the symbols Fe for iron and Pb for lead are derived from *ferrum* and *plumbum* respectively, the Latin names of these elements.

The chemical symbols of the elements can be used to express the compositions of substances very precisely. Subscripts after a symbol indicate the number of atoms/ions of that element that are present in proportion to the other elements in the formula. For example, the formula $Fe_3Al_2Si_3O_{12}$ represents a compound in which, for every twelve oxygen atoms, there are three iron atoms, two of aluminum, and three of silicon. (This happens to be a variety of the mineral garnet.) The chemical formula is much briefer than describing the composition in words. It is also more exact than the mineral name "garnet," for there are several compositions of garnets with the same basic kind of formula and crystal structure. Other examples include a calcium-aluminum garnet with formula $Ca_3Al_2Si_3O_{12}$, a calcium-chromium garnet, $Ca_3Cr_2Si_3O_{12}$, and a magnesium-aluminum garnet, $Mg_3Al_2Si_3O_{12}$. Moreover, chemical formulas are understood by all scientists, while mineral names are known primarily to geologists.

Sample formulas of some common minerals are shown below.

Formulas can become very complex, especially when different elements can substitute for each other in the same site in the crystal structure, the phenomenon of *solid solution*. Biotite, a common dark-colored mica, may be iron-rich, with formula $KFe_3AlSi_3O_{10}(OH)_2$, or magnesium-rich, with formula $KMg_3AlSi_3O_{10}(OH)_2$, or, more commonly, may contain some iron and some magnesium, where both together total three atoms per formula. We would then write the generalized formula as $K(Fe,Mg)_3AlSi_3O_{10}(OH)_2$. Or consider the mineral tourmaline, with a generalized formula of $Na(Mg,Fe,Mn,Li,Al)_3Al_6Si_6O_{18}(BO_3)_3(OH,F)_4$! Other examples of solid solution are indicated in the text.

Mineral	Chemical composition	Chemical formula
quartz	silicon dioxide	SiO_2
microcline (a potassium feldspar)	potassium aluminum silicate	$KAlSi_3O_8$
calcite	calcium carbonate	$CaCO_3$
hematite	ferric iron oxide	Fe_2O_3
pyrite	iron disulfide	FeS_2

Crystals

Crystals form because they are the most stable arrangement of atoms in a solid. Consider, for example, a solid containing various cations and anions. Opposite electrical charges attract, but like charges repel, and the repulsion is minimized when positive and negative charges are symmetrically distributed through the solid. A crystal structure can be thought of as a single structural and compositional unit, repeated over and over in three dimensions. The basic unit of the sodium chloride crystal is indicated on figure 2.6. The more complex the mineral, the larger and more complex the repeating unit (see, for example, figure 2.7).

In order for crystals of a given mineral to form, the requisite elements must be present in appropriate proportions, and there must be time for the atoms to arrange themselves into the regular pattern. The importance of time can be seen by a simple analogy. Imagine a roomful of people milling about, who are suddenly told to arrange themselves in rows of six. If they are given ten seconds in which to do this, some will succeed in forming small organized groups, but many will still be out of position when time is up. Given half a minute, the results will be more orderly. Given several minutes, the entire group can probably arrange itself properly. Likewise with crystals growing in a melt: if the liquid cools and solidifies very quickly, there may not be time for regular crystals to grow. The

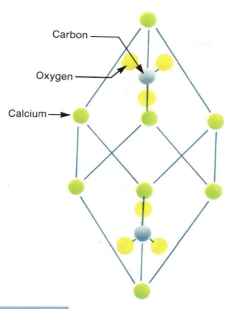

Figure 2.7

The more complex structural unit of calcite (calcium carbonate, CaCO₃).

Figure 2.8

All of these solids are internally crystalline, though not all show well-developed crystal forms.

resulting solid, in which atoms are randomly arranged, is a noncrystalline glass. With slower cooling, there is more time for the orderly crystals to grow.

Some solids satisfy all the conditions of the definition of a mineral, except that they lack a regular crystal structure. Such solids are termed mineraloids. Opal is an example of a mineraloid. Also, some true minerals can be produced by organic as well as inorganic means. For example, seashells are obviously produced by organisms, but the materials that make up seashells are nevertheless minerals, for they can also be found in nature formed inorganically.

The True Identifying Characteristics of Minerals

The two fundamental characteristics of a mineral that together distinguish it from all other minerals are its chemical composition and its crystal structure. No two minerals will be identical in both respects, though they may be similar in one. For example, diamond and graphite (the "lead" in a lead pencil) are chemically the same, both made up of pure carbon. Their physical properties, however, are vastly different, because their internal crystalline structures differ. In a diamond, each carbon atom is firmly bonded to every adjacent carbon atom in three dimensions. In graphite, the carbon atoms are bonded strongly in two dimensions, into sheets, but the sheets are only weakly held together in the third dimension. Diamond is clear,

transparent, colorless, and very hard; it can be cut by a jeweler into brilliant precious gemstones. Graphite is black, opaque, and soft, its sheets of carbon atoms tending to slide apart as the weak bonds between them are broken.

Diamond and graphite are examples of the phenomenon of *polymorphism* in minerals. **Polymorphs** are minerals having the same composition but distinctly different crystal structures. There may be more than two polymorphs of a single composition. Silica (SiO_2), which most often occurs as the mineral quartz, has at least half a dozen known polymorphs, though most are geologically rare, formed only under unusual conditions. Under a given set of physical and chemical conditions, one polymorph of a set will be the stable one; under different conditions other atomic arrangements may be more stable. The very compact crystal structure of diamond is stable under high-pressure conditions; at low pressure, graphite forms instead.

Composition and crystal structure may be the truly diagnostic characteristics of a mineral, but both can usually be determined only by using sophisticated laboratory equipment. When a mineral happens to have formed large crystals with well-developed shapes, a trained mineralogist may be able to infer some characteristics of its internal atomic arrangement, but most of the bits of minerals one encounters do not show large symmetric crystal forms by which they can be recognized with the naked eye (figure 2.8). Certainly no one can look at a lump of some

Figure 2.9

Color is often a poor guide in identification: all of these are samples of quartz.

mineral and know its chemical composition without first recognizing what mineral it is. Thus when the scientific instruments are not at hand, mineral identification must be based on a variety of other physical properties that in some way reflect the mineral's composition and structure. These other properties are often what make the mineral commercially valuable. However, they are rarely unique to one mineral and often are deceptive. A few examples of such diagnostic properties are outlined in the next section.

Other Physical Properties of Minerals

The most immediately obvious characteristic of most mineral samples is *color.* One might suppose that color would be a good way to identify minerals. It is true that some minerals always appear the same color, but many vary from specimen to specimen. Variation in color is usually due to the presence of small amounts of chemical impurities in the mineral that have nothing to do with its basic, characteristic composition. Such color variation is especially common when the pure mineral is light-colored or colorless. The mineral quartz, for instance, is colorless in its pure form. However, quartz that looks milky white is commonly found, and there are many occurrences of brightly colored quartz—rose pink, golden yellow, smoky brown, purple, and others (figure 2.9). Clearly, then, quartz cannot always be recognized by its color (or lack of it).

Unusual coloring may make a particular mineral specimen especially valuable. Many colored quartz samples are used as semi-precious gems. The purple variety is amethyst. Other valued colors include citrine, a golden quartz named from the Latin *citrus,* and smoky quartz, sometimes deceptively marketed as "smoky topaz," although topaz is an entirely different mineral.

Another example of this phenomenon of color variation is the mineral corundum, a simple compound of aluminum and oxygen. In pure form it too is colorless, and also quite hard, which makes it a good abrasive. It is often used for the grit on sandpaper. Yet a little color from trace impurities can transform this common, utilitarian material into highly prized gems: blue-tinted corundum is also known as sapphire, and red corundum is ruby. Even when the color shown by a mineral sample *is* the true color of the pure mineral, it is probably not unique. With more than a thousand minerals known to occur on earth, there are usually a great many of any one particular color.

Perhaps surprisingly, streak, the color of the powdered mineral, is more consistent from sample to sample than is the color of a bulk mineral. Streak is conventionally tested by scraping the sample across a piece of unglazed tile or porcelain, then examining the color of the mark. There are two principal limitations to the usefulness of streak as an identification tool. First, like color, it is not unique. Many different minerals will leave a brown streak, or white, or grey. Second, those minerals that are harder than ceramic will simply gouge the test surface, not be powdered onto it.

Hardness, the ability to resist scratching, is another physical property that can be of help in mineral identification. Classically, hardness is measured using the Mohs hardness scale (see table 2.1), on which ten minerals are arranged in order of hardness, from talc (the softest, assigned a hardness of 1) to diamond (10). Unknown minerals are used to scratch, or are shown to be scratched by, pieces of the reference minerals of the scale. A mineral that can both scratch and be scratched by quartz would be assigned a hardness of 7. One that scratches gypsum but is scratched by calcite would have a hardness of 2 1/2 (similar to an average fingernail). Diamond is the hardest natural substance known, and corundum the second-hardest mineral, so these might be readily identified by their hardnesses. However, among the thousands of softer, or more easily scratched, minerals, there are again many of any particular hardness.

Table 2.1

The Mohs hardness scale

Mineral	Assigned hardness
Talc	1
Gypsum	2
Calcite	3
Fluorite	4
Apatite	5
Orthoclase	6
Quartz	7
Topaz	8
Corundum	9
Diamond	10

handwritten: 2 —Fingernail 2.5; 3 —Penny 3.5; 5 — Glass

For comparison, here are the approximate hardnesses of some common objects, measured on the same scale: fingernail, 2 1/2; copper penny, 3 1/2; glass, 5 to 6; pocketknife blade, 5 to 6.

The exceptional hardness of diamond makes it not only a particularly durable gemstone but also a valuable industrial commodity. Large, clear, gem-quality diamonds are relatively rare. The smaller and less attractive industrial-grade diamonds are nevertheless useful in abrasives for grinding and polishing and in saw blades and drill bits where the material being shaped is itself hard and the tools used must be correspondingly durable.

Crystal form, the shape of well-developed crystals of a mineral, is a very useful clue to mineral identification, because it is related to the (invisible) internal geometric arrangement of atoms in the crystal structure. In some cases the relationship is readily apparent (figure 2.10): the ions in sodium chloride are arranged in a cubic structure, and table salt tends to crystallize in cubes; so-called dogtooth crystals of the mineral calcite are similar in shape to the rhombohedral atomic units from which they are built. However, variations in the conditions under which crystals grow cause different planes of atoms to be more or less prominent in the overall crystal form. A given mineral with a single internal structure can grow in several quite distinctive crystal shapes (figure 2.11). Each one necessarily reflects in some way the internal symmetry of the crystal structure, but all crystals of a specific mineral do not necessarily look alike in form.

handwritten: photos in hall

A

B

Figure 2.10

Here, crystal form mimics the geometry of the internal crystal structure. *(A)* Halite (compare with figure 2.6). *(B)* Calcite (compare with figure 2.7).

Conversely, different minerals may show the same crystal form (figure 2.12), so crystal form is often not uniquely diagnostic of minerals. Moreover, as noted earlier, well-developed crystal forms are relatively uncommon.

Another property that is controlled by internal crystal structure is **cleavage,** the tendency of minerals to break preferentially in certain directions, corresponding to zones of weakness in the crystal structure. Cleavage can be investigated simply by striking a mineral sample with a hammer. It can be tested even on irregular chunks of mineral that show

Figure 2.13

The cleavage directions of halite parallel the cube faces of its crystals.

Figure 2.11

All of these silvery crystals are crystals of the mineral galena, and internally their atomic arrangement (crystal structure) is identical. However, the crystal forms can vary considerably from sample to sample.

Figure 2.14

Fluorite forms cubic crystals also but cleaves into octahedra.

Figure 2.12

Many minerals can share the same crystal form. Here, pyrite (fool's gold, FeS_2), galena (PbS), halite (NaCl), and fluorite (CaF_2) all form cubes.

Figure 2.15

Calcite cleaves into rhombohedral fragments.

no well-developed crystals. In some cases, the prominent cleavage directions are the same as the prominent faces of well-formed crystals. Sodium chloride cleaves well in three mutually perpendicular directions that correspond to the faces of its cubic crystals (figure 2.13). The cleavage fragments of other minerals do not necessarily resemble their well-grown crystals (figures 2.14, 2.15). Invariably, however, cleavage directions are directly related to the arrangement of atoms/ions in the crystal structure.

For many minerals, there are no well-defined planes along which the crystals break cleanly. These minerals exhibit more irregular **fracture** rather than cleavage (figure 2.16). A distinctive type of fracture is the *conchoidal,* or shell-like, fracture shown by volcanic glass, quartz, and a few other minerals.

The surface sheen, or **luster,** is another diagnostic property. A very few terms are used to describe this quality, and they are for the most part self-explanatory. Examples include *metallic* (bright and shiny like metal), pearly (like pearl), and *vitreous* (glassy), and *earthy* (figure 2.16).

A mineral's **specific gravity** is related to its density. Specific gravity is the ratio of the mass of a given volume of the mineral to the mass of an equal volume of water. A mineral having the same density as liquid water, then, has a specific gravity of 1 by definition. The higher the specific gravity, the denser the mineral. For comparison, sodium chloride has a specific gravity of about 2.16; garnet, 3.1 to 4.2 depending on its exact composition; metallic copper, 8.9; gold, 19.3. The precise determination of specific gravity requires specialized equipment, and many of the most common minerals fall in a narrow range of specific gravity, from about 2.5 to 4. For the nonspecialist, specific gravity is a qualitative tool, most useful for identifying those minerals of unusually high specific gravity (density), such as gold.

There are many other properties useful mainly in identifying those very few minerals that possess them. The iron mineral magnetite, as its name suggests, is strongly magnetic, which is very rare among minerals. Sodium chloride, known mineralogically as halite, naturally tastes like table salt. The similar salt potassium chloride (sometimes used in salt substitutes for those on low-sodium diets) also tastes salty but is more bitter. A few minerals effervesce (fizz) when acid is dripped onto them. These are carbonate minerals (see next section), which react with the acid and release carbon dioxide gas in the process.

Lacking exact knowledge of a mineral sample's composition and crystal structure, one must rely on other less distinctive properties such as color, hardness, and cleavage. These individual physical properties are only rarely unique to one mineral. But considering a set of properties collectively may narrow the possibilities down to one or a very few minerals. That is, while there may be many green minerals, there are few that are green *and* show conchoidal fracture *and* glassy luster *and* have a hardness of 6.5 to 7, and so on. The more properties that are considered at once, the fewer minerals will fit them all.

Figure 2.16

Examples of conchoidal fracture (left) and earthy luster.

Types of Minerals

As was indicated earlier, minerals can be grouped or subdivided on the basis of their two fundamental characteristics, composition and crystal structure. In this section some of the basic mineral groups will be introduced briefly, together with a few examples that may be familiar. Most of these mineral groups, or representatives of them, will be mentioned again elsewhere in the text. A comprehensive survey of minerals is well beyond the scope of this book, and the interested reader is referred to standard mineralogy texts for more information.

Silicates

The two most common elements in the earth's crust, by far, are silicon and oxygen, which together make up over 70% of the crust. It should therefore come as no surprise that by far the largest group of minerals is the silicate group, all of which are compounds containing silicon and oxygen, and most of which contain other elements as well. Because this group of minerals is so large, it is subdivided on the basis of crystal structure, by the ways in which the silicon and oxygen atoms are linked together.

The basic structural unit of all the silicates is the *silica tetrahedron* (figure 2.17), a compact building-block formed by a **silicon cation** (Si^{4+}) closely surrounded by four oxygen anions (O^{2-}). The different silicate mineral groups are distinguished by the way in which these tetrahedra are assembled. A number of specific minerals of various compositions fall into each structural group. We will note only some of the more common representatives of each structural type.

The simplest arrangement of tetrahedra is as isolated units (figure 2.18). It can be seen arithmetically that an individual SiO_4 tetrahedron will have a net negative charge of -4: $[1(+4) + 4(-2)] = -4$. Thus in such a silicate, additional cations are necessary to balance the electrical charge in order for the solid to be electrically neutral overall, and to bond the negatively charged tetrahedra together. For each tetrahedron, cations totaling 4 positive charge units are required. In the most common single-tetrahedron silicate, **olivine**, the charges are balanced by iron (Fe^{2+}) and/or magnesium (Mg^{2+}), which requires a total of two $+2$ ions per formula unit: $(Fe,Mg)_2SiO_4$. Another fairly common single-tetrahedron silicate is *garnet,* for which a variety of chemical compositions are possible (see box 2.2).

Other silicates are known as **chain silicates,** as their silica tetrahedra are arranged in chains, formed by the sharing of oxygen atoms between adjacent tetrahedra in one dimension. The most common arrangements are single chains (figure 2.19A) and double chains (figure 2.19B). When tetrahedra share oxygen atoms, the net effect is less total negative charge from the silica tetrahedra to be balanced by additional cations.

Consider one tetrahedron in a single-chain silicate (figure 2.20). Two of its oxygen atoms are associated only with that tetrahedron, while two are shared with other tetrahedra in the chain, so that effectively only half of each of the shared atoms is associated with

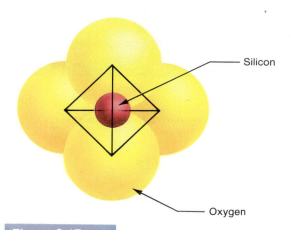

Figure 2.17

The basic silica tetrahedron, building block of all silicate minerals.

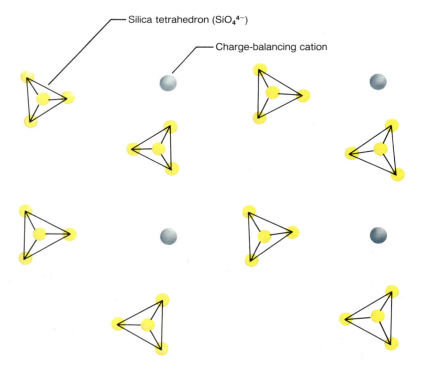

Silica tetrahedron ($SiO_4{}^{4-}$)

Charge-balancing cation

Figure 2.18

Silicate formed of individual, isolated silica tetrahedra, with charge-balancing cations symmetrically distributed among the tetrahedra.

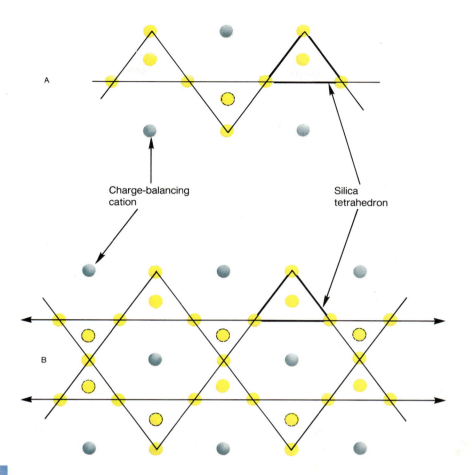

Charge-balancing
cation

Silica
tetrahedron

A

B

Figure 2.19

In chain silicates, tetrahedra are linked in one dimension by
shared oxygen atoms. *(A)* Single chains, characteristic of
pyroxenes. *(B)* Double chains, characteristic of amphiboles.

that tetrahedron. The net ratio of Si to O is then 1 :
$[2(1) + 2(1/2)]$, or 1:3 rather than 1:4. The net negative charge on an SiO_3 unit is $[1(+4) + 3(-2)] =$
-2, so that cations totaling only 2 positive charges
will suffice to balance each tetrahedron. One large
group of single-chain silicates is collectively known
as **pyroxenes.** This group comprises many specific
minerals, differing in the cations involved in the
charge-balancing (and in details of crystal structure).
A compositionally simple example in which charges
are again balanced by iron and magnesium would have
this formula: $(Fe,Mg)SiO_3$. The common pyroxene
augite is compositionally more complex, containing
calcium and aluminum in addition to iron and magnesium.

The chemical formulas for the double-chain **amphiboles** are somewhat more complex still. This is
true in part because the amphiboles contain some
water in their crystal structures, in the form of hydroxyl ions (OH^-) substituting for oxygen in the
tetrahedra. The structural geometry and nature of

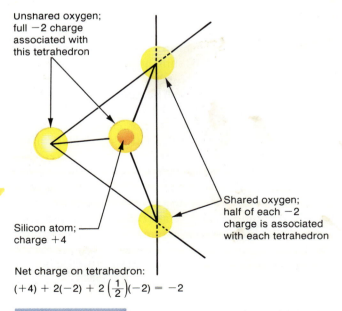

Unshared oxygen;
full -2 charge
associated with
this tetrahedron

Shared oxygen;
half of each -2
charge is associated
with each tetrahedron

Silicon atom;
charge $+4$

Net charge on tetrahedron:

$$(+4) + 2(-2) + 2\left(\frac{1}{2}\right)(-2) = -2$$

Figure 2.20

The net negative charge on a single tetrahedron in a single chain
is -2.

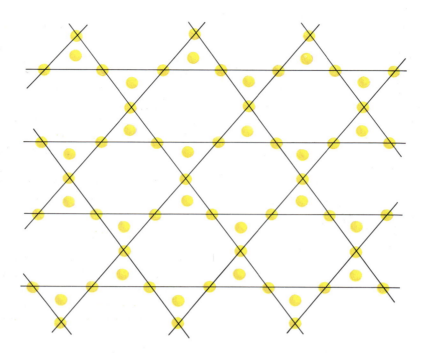

The two-dimensional linking of tetrahedra in a sheet silicate.

the charge-balancing are correspondingly more complex. A very common amphibole, *hornblende*, illustrates this well, with its general formula of $(Ca,Na)_3(Mg,Fe,Al,Ti)_5(Si,Al)_8O_{22}(OH,F)_2$.

The chain silicates represent sharing of oxygen between adjacent tetrahedra in one dimension. The **sheet silicates** are those in which tetrahedra are linked by shared oxygen atoms in two dimensions (figure 2.21). The net charge on each tetrahedral unit is $-1\frac{1}{2}$, so still fewer charge-balancing cations are required. In fact, most sheet silicates contain some water (as hydroxyl ions) also, substituting for oxygen as in the amphiboles, which further reduces the need for charge-balancing cations. The cations commonly fit between stacked sheets, bonding them together. However, typically there are not enough cations to hold the sheets very tightly. This is reflected in the macroscopic properties of many sheet silicates. The **micas** are a compositionally diverse group of sheet silicates that have in common excellent cleavage parallel to the weakly bonded sheets of tetrahedra (figure 2.22). Common examples are the light-colored mica *muscovite*, often found in granite, and the dark mica *biotite*, rich in iron and magnesium. **Clay minerals are also sheet silicates**, and their often slippery feeling can likewise be attributed to the sliding apart of such sheets of atoms.

Figure 2.22

Excellent cleavage in micas results from splitting of the crystals between weakly bonded sheets.

When tetrahedra are firmly linked in all three dimensions by shared oxygen atoms, the result is a **framework silicate** (figure 2.23). If the framework consists entirely of silicon-oxygen tetrahedra, the net charge on each tetrahedron is $[1(+4) + 4(\frac{1}{2})(-2)] = 0$ and there is no need for further charge-balancing cations in the structure. This is the case with the mineral *quartz*, compositionally the simplest silicate (SiO_2).

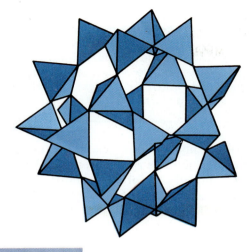

Figure 2.23

When silica tetrahedra are linked in three dimensions, the result is a framework silicate.

There are other framework silicates besides quartz and its polymorphs, however. This is so because it is possible—in silicates of virtually all structural types—for some aluminum atoms (Al^{3+}) to substitute in tetrahedra for silicon (Si^{4+}). This results in a charge imbalance that must be compensated by additional cations. The most abundant minerals in the crust, the **feldspars,** are such framework silicates in which the charge-balancing cations are sodium (Na^+), potassium (K^+), or calcium (Ca^{2+}).

The feldspars illustrate some of the limitations of solid solution in minerals. Sodium and calcium are very close in size, though slightly different in charge, so they can substitute effectively for each other in a series of sodium-calcium feldspars known as the *plagioclase* feldspars (generalized formula $(Na,Ca)(Al,Si)_2Si_2O_8$). Potassium is so much larger that it does not substitute in plagioclase to any great extent. Conversely, in potassium feldspar ($KAlSi_3O_8$), some sodium may substitute because the charges of sodium and potassium ions are the same (both $+1$) although they differ somewhat in size. But calcium differs significantly in both size and charge from potassium, and thus these two elements are not readily interchangeable in the potassium feldspar structure.

Other structural variants occur in the silicates, but less commonly—for example, structures in which tetrahedra are linked into rings, or those in which pairs of tetrahedra are joined. It is also true that in some contexts it may be useful to refer collectively to a set of silicate minerals of different structures that share a compositional characteristic. For example, one might refer to the **ferromagnesian** silicates. This group, as the name suggests, comprises those silicates that contain significant amounts of iron (Fe) and/or magnesium (Mg). (They may or may not contain other cations also.) The group would include olivines, pyroxenes, amphiboles, some micas, and other silicates as well. The ferromagnesians are commonly characterized by dark colors, most often black, brown, or green. Another possible compositional group would be that of the **hydrous** (water-bearing) silicates, which would include micas, clays, amphiboles, and others. A silicate mineral may thus fall within one or more particular compositional groupings in addition to its structural category.

Nonsilicates

Just as the silicates, by definition, all contain silicon plus oxygen as part of their chemical compositions, each nonsilicate mineral group is defined by some chemical constituent or characteristic that all members of the group have in common. Most often the common component is the same negatively charged ion, or group of atoms (somewhat analogous to a silica tetrahedron—a structural unit). Below is a partial compilation of some of the nonsilicate mineral groups, with examples of more common or familiar members of each. See the partial listing in table 2.2.

The chemical formulas of the **carbonates** all contain carbon and oxygen combined in the proportions of 1 atom of carbon to 3 of oxygen (CO_3). The carbonate minerals all dissolve relatively easily, particularly in acids. The oceans also contain a great deal of dissolved carbonate. Geologically, the most important and most abundant carbonate mineral is *calcite,* which is calcium carbonate ($CaCO_3$). Another common carbonate mineral is *dolomite,* which contains both calcium and magnesium in approximately equal proportions ($CaMg(CO_3)_2$). Carbonates may contain many other elements—iron, manganese, or lead, for example.

The **sulfates** all contain sulfur and oxygen in a ratio of 1 to 4 (SO_4). A calcium sulfate, *gypsum,* is the most important, for it is both relatively abundant and commercially useful. Sulfates of many other elements, including barium, lead, and strontium, are also found.

When sulfur is present without oxygen, the resultant minerals are called **sulfides.** A common and well-known sulfide mineral is the iron sulfide *pyrite* (FeS_2). Pyrite has also been called "fool's gold," because its metallic golden color often deceived early gold miners and prospectors into thinking that they had struck it rich.

Table 2.2

Some nonsilicate mineral groups

Compositional class	Compositional characteristic	Examples
carbonates	metal(s) plus carbonate (1 carbon + 3 oxygen atoms, CO_3)	calcite (calcium carbonate, $CaCO_3$); dolomite (calcium-magnesium carbonate, $CaMg(CO_3)_2$)
sulfates	metal(s) plus sulfate (1 sulfur + 4 oxygen atoms, SO_4)	gypsum (calcium sulfate, with water, $CaSO_4 \cdot 2H_2O$); barite (barium sulfate, $BaSO_4$)
sulfides	metal(s) plus sulfur, without oxygen	pyrite (iron sulfide, FeS_2); galena (lead sulfide, PbS); cinnabar (mercury sulfide, HgS)
oxides	metal(s) plus oxygen	magnetite (iron oxide, Fe_3O_4); hematite (iron oxide, Fe_2O_3); corundum (aluminum oxide, Al_2O_3); spinel (magnesium-aluminum oxide, $MgAl_2O_4$)
hydroxides	metal(s) plus hydroxyl (1 oxygen + 1 hydrogen, OH)	gibbsite (aluminum hydroxide, $Al(OH)_3$, found in aluminum ore); brucite (magnesium hydroxide, $Mg(OH)_2$, an ore of magnesium)
halides	metal(s) plus halogen element (fluorine, chlorine, bromine, or iodine)	halite (sodium chloride, NaCl); fluorite (calcium fluoride, CaF_2)
phosphates	metal(s) plus phosphate group (1 phosphorous + 4 oxygen, PO_4)	apatite ($Ca_5(PO_4)_3F$)
native elements	mineral consists of a single element	gold (Au), silver (Ag), copper (Cu), sulfur (S), graphite (C)

Other groups exist, and some complex minerals contain components of several groups (carbonate and hydroxyl groups, for example).

Had the miners who struck pyrite tested its specific gravity, of course, they would have known better: the specific gravity of pyrite is about 5, while that of gold is 19.3! Pyrite and gold also have different crystal structures and streak differently, gold having a golden streak, while that of pyrite is dark greenish-black.

Pyrite is not a commercially useful source of iron, because there are richer ores of this metal, but many economically important metallic ore minerals are sulfides. An example that may be familiar is the lead sulfide mineral *galena* (PbS), which often forms in silver-colored cubes. The rich lead ore deposits near Galena, Illinois, gave the town its name. Sulfides of copper, zinc, and numerous other metals may also form valuable ore deposits.

The **halides** are the minerals composed of metal(s) plus one or more of the halogen elements (the gaseous elements in the next-to-last column of the periodic table: fluorine, chlorine, iodine, and bromine). The most common halide is *halite* (NaCl), the most abundant salt dissolved in the oceans.

Minerals that contain just one or more metals combined with oxygen, and that lack the other elements necessary for them to be classified as silicates, sulfates, carbonates, and so on, are the **oxides.** We earlier referred to the mineral *corundum;* it is an aluminum oxide mineral (Al_2O_3). Iron forms more than one oxide mineral, containing iron and oxygen in different proportions. *Magnetite* (Fe_3O_4) is one of these. Another iron oxide, *hematite* (Fe_2O_3), is sometimes silvery black, but often shows a red color and gives a reddish tint to many rocks and soils. (All colors of hematite have the same characteristic reddish-brown streak, however, illustrating again the usefulness of streak as a diagnostic property in mineral identification.) Many other oxide minerals are also known.

The **native elements** are even simpler chemically. Native elements are minerals that are each made up of a single chemical element. The minerals' names are usually the same as the names of corresponding elements. Not all elements can be found, even rarely, as native elements. However, the group does include some of our most highly prized materials. Gold, silver, and platinum often occur as native elements. Diamond and graphite are both examples of native carbon; here two mineral names are needed to distinguish these two very different crystalline forms of the same element. Sulfur may occur as a native element, either with or without associated sulfide minerals. Some of the richest copper ores contain native copper. Other metals that can occur as native elements include tin, iron, and antimony.

Rocks as Mineral Aggregates

A **rock** is a solid, cohesive aggregate of one or more minerals (or mineral materials, including volcanic glass). This means that a rock consists of many individual mineral grains, not necessarily all of the same mineral, or of mineral grains plus glass, all firmly held together in a solid mass. A beach sand consists of many mineral grains, but they fall apart when handled, so sand is not a rock, although in time sand grains may be cemented together to make a rock. Rocks' physical properties can be important in determining their suitability for particular applications, such as use as construction materials or as the base for building foundations. Each rock also contains within it a record of at least a part of its history, in the nature of the minerals in it and in the way in which the mineral grains fit together. Before proceeding to the detailed consideration of different rock types in the next three chapters, we can consider rocks simply as masses of mineral grains and survey briefly a few rock properties that will be important in later discussions.

Porosity and Permeability

Porosity and permeability are related to the ability of rocks, soils, or other geologic materials to contain fluids and to allow fluids to pass through them. **Porosity** is the proportion of void space in the material—holes or cracks, unfilled by solid material, whether within or between mineral grains. Porosity can be expressed either as a percentage (e.g., 1.5%) or as an equivalent decimal fraction (0.015). The pore spaces may be occupied by gas, liquid, or a combination of the two. **Permeability** is a measure of how readily fluids pass through the material. It is related to the extent to which pores or cracks are interconnected. Both the porosity and permeability of geological materials are influenced by the shapes of mineral grains or rock fragments in the material, the range of grain sizes present, and the ways in which the grains fit together (figure 2.24). Rocks that consist of tightly interlocking crystals usually have both little porosity and low permeability, unless they have been broken up by fracturing or weathering. In contrast, materials consisting of well-rounded, equidimensional grains of similar size may have quite high porosity and permeability. (This can be illustrated by pouring water through a pile of marbles.) In materials containing a wide range of grain sizes, finer materials can fill the gaps between coarser grains, reducing porosity, though permeability may remain high. Conversely, a rock in which mineral grains are platy or slab shaped

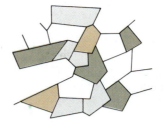

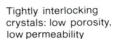

Tightly interlocking crystals: low porosity, low permeability

Round grains, similar in size: high porosity, high permeability

Round grains, varying sizes: lower porosity, still high permeability.

Stacked platy grains: high porosity, low permeability

Figure 2.24

Porosity and permeability vary with grain shapes and the ways in which grains fit together.

may have those grains arranged in such a way that porosity is high, but the pores are poorly connected, so permeability is low. (Imagine a stack of closed, empty boxes.)

Some representative values of porosity and permeability are given in table 2.3. These properties will be relevant to later discussion of such matters as stream flooding, groundwater availability and use, and petroleum resources.

Stress, Strain, and the Strength of Geological Materials

An object is under **stress** when force is being applied to it. The stress may be **compressive,** tending to squeeze the object, or compress it, or **tensile,** tending to pull the object apart. A **shearing stress** is one that tends to cause different parts of the object to move in different directions across a plane, or to slide past one another, as when a deck of cards is spread out on a tabletop by a sideways sweep of the hand.

Strain is deformation resulting from stress. It may be either temporary or permanent, depending on the amount and type of stress, and the strength of the material to resist it. If the deformation is **elastic,** the amount of deformation is proportional to the stress applied, and the material will return to its original size and shape when the stress is removed. A gently stretched rubber band will show elastic behavior.

Table 2.3

Representative porosities and permeabilities of geological materials

Material	Porosity (%)	Permeability (m/day)
Unconsolidated		
clay	45–55	less than 0.01
fine sand	30–52	0.01–10
gravel	25–40	1,000–10,000
glacial till	25–45	0.001–10
Consolidated (rock)		
sandstone and conglomerate	5–30	0.3–3
limestone (crystalline, unfractured)	1–10	0.00003–0.1
granite (unweathered)	less than 1–5	0.0003–0.003
lava	1–30; mostly less than 10	0.0003–3, depending on presence or absence of fractures or interconnected gas bubbles

Source: Data from *Water in Environmental Planning*, by Dunne, T., and L. B. Leopold. W. H. Freeman and Company. Copyright © 1978. Reprinted by permission.

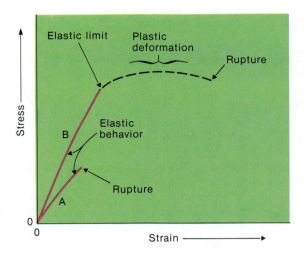

Figure 2.25

Stress-strain relationships for elastic *(A)* and plastic *(B)* materials. Failure or breaking occurs when stress exceeds rupture strength *(C)*.

Rocks too may behave elastically, although much greater stress will be needed to produce detectable strain. Once the **elastic limit** of a material is reached, the material may go through a phase of **plastic deformation** with increasing stress. During this stage, relatively small added stresses will yield large corresponding strains, and the changes of shape are permanent—the material will not return to its original dimensions after removal of the stress. A glassblower, an artist shaping clay, a carpenter fitting caulk into cracks around a window, and a blacksmith shaping a bar of hot iron into a horseshoe are all making use of plastic behavior of materials.

If stress is increased further, solids will eventually break, or **rupture.** In **brittle** materials, rupture may occur before any plastic deformation occurs. Brittle behavior is characteristic of most rocks at near-surface conditions, and is illustrated by line A in figure

2.25. At greater depths, where temperatures are higher and rocks are confined (in a sense, supported by surrounding rocks at high pressure), rocks may indeed behave plastically (dashed section of curve B). The effect of temperature can be seen in the behavior of warm and cold glass. A rod of cold glass is brittle and snaps under stress before appreciable plastic deformation occurs, while a sufficiently warmed glass rod may be bent and twisted without breaking. Plastic deformation of rocks is often reflected in the production of folds, such as those illustrated in figure 2.26 or described in chapter 11.

As indicated above, the physical behavior of a rock is affected by external factors such as temperature or confining pressure as well as by the intrinsic characteristics of the rock itself. Also, rocks respond differently to different types of stress. Most are far stronger under compression than tension: a given rock may rupture under a tensile stress only one-tenth as large as the compressive stress required to break it at the same pressure and temperature. Consequently, the term *strength* has no single simple meaning when applied to rocks unless all these variables are specified. Even when they are, rocks in their natural settings do not always exhibit the same behavior as carefully selected samples tested in the laboratory. The natural samples may, for example, be weakened by fracturing or weathering. Several rock types of quite different properties may be present at the same site. Considerations such as these complicate the work of the geological engineer.

Figure 2.26

Intricate folds in rocks illustrate plastic deformation.

The Rock Cycle

In chapter 1, we noted that the earth is a constantly changing body. Mountains come and go, seas advance and retreat over the faces of continents, surface processes and processes occurring deep in the crust or mantle are constantly altering the planet. One aspect of this continual change is the idea that rocks, too, are always subject to change. We do not have a single sample of rock that has remained unchanged since the earth formed, and many rocks have been changed many times.

Rocks are divided into three classes on the basis of the way in which they form. The igneous rocks (chapter 3) are those that crystallize from hot silicate melts. The sedimentary rocks (chapter 4) form at the very low temperatures near the earth's surface, as a result of the weathering of pre-existing rocks. The metamorphic rocks (chapter 5) are those that have been changed (deformed, recrystallized) by the application of moderate heat and/or pressure in the

crust. But a rock of one type may be transformed into a rock of another type (or a different rock of the same type) by various geologic processes. A sedimentary rock, for example, may be squeezed, heated, and changed into a metamorphic rock. It may even be heated so much that it melts; when the melt cools and crystallizes, the result is a new igneous rock.

The idea that rocks are continually subject to change through time is the essence of the rock cycle (figure 2.27). The following chapters will treat individual rock types in some detail. It is important to keep in mind, however, that when we label a rock as "igneous," "sedimentary," or "metamorphic," we are necessarily describing it only as it has emerged from its last cycle of formation or change. Many earlier stages through which that material has passed will have been wholly or partially obliterated by later changes. Hence some of the difficulty in piecing together the 4 1/2-billion-year history of the earth from rocks we collect today.

IGNEOUS Rocks
Sedimentary Rocks
Metamorphic Rocks

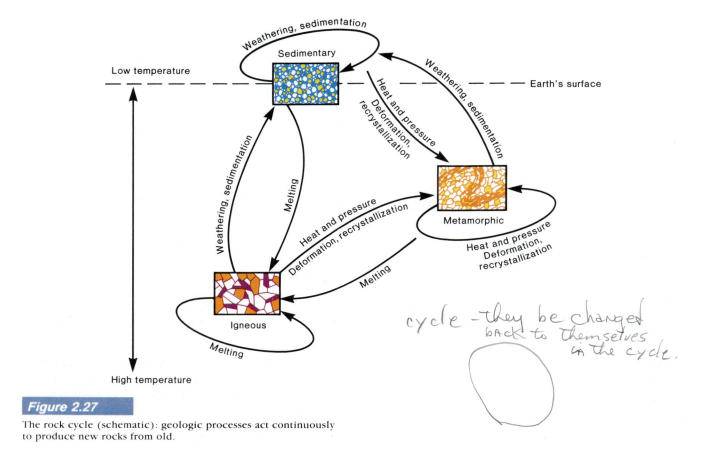

cycle — they be changed
back to themselves
in the cycle.

Figure 2.27

The rock cycle (schematic): geologic processes act continuously to produce new rocks from old.

Summary

Chemical elements consist of atoms, which are in turn composed of protons, neutrons, and electrons. Isotopes are atoms of one element (having, therefore, the same number of protons) with different numbers of neutrons; chemically, isotopes of one element are indistinguishable. Ions are atoms that have gained or lost electrons and thus acquired a positive or negative charge. Atoms of the same or different elements may bond together. The most common kinds of bonding in minerals are ionic (resulting from the attraction between oppositely charged ions) and covalent (involving sharing of electrons between atoms). When atoms of two or more different elements bond together, they form a compound.

A *mineral* is a naturally occurring, inorganic solid element or compound, with a definite composition (or range in composition) and a regular internal crystal structure. When appropriate instruments for determining composition and crystal structure are unavailable, minerals can be identified from a set of physical properties, including color, crystal form, cleavage or fracture, hardness, luster, specific gravity, and others. Minerals are broadly divided into silicates and nonsilicates. The silicates are subdivided into structural types (e.g., chain silicates, sheet silicates, framework silicates) on the basis of how the silica tetrahedra are linked in each mineral. Silicates can alternatively be grouped by compositional characteristics. The nonsilicates are subdivided into groups, each of which has some compositional characteristic in common. Examples include the carbonates (each containing the CO_3 group), sulfates (SO_4), and sulfides (S).

Rocks are cohesive mineral aggregates. Certain of their physical properties—for example, porosity and permeability—are a consequence of the ways in which their constituent mineral grains are assembled. Rocks subjected to stress may undergo temporary strain (elastic deformation), permanent strain (plastic deformation), or rupture, depending on rock type and the temperature and pressure conditions. All rocks are part of the rock cycle, through which old rocks are continually being transformed into new ones. A consequence of this is that no rocks have been preserved throughout earth's history, and many early stages in the development of one rock may have been erased by subsequent events.

Terms to Remember

amphibole	elastic	luster	pyroxene
anion	elastic limit	mica	radioactivity
atom	electron	mineral	rock
atomic mass number	element	mineraloid	rock cycle
atomic number	feldspars	native element	rupture
brittle	ferromagnesian	neutron	shearing stress
carbonates	fracture	nonsilicates	sheet silicates
cation	framework silicates	nucleus	silicates
chain silicates	glass	olivine	solid solution
clay minerals	halides	oxide	specific gravity
cleavage	hardness	periodic table	strain
compound	hydrous	permeability	streak
compressive stress	inert	plastic	stress
covalent bond	ion	polymorph	sulfate
crystal form	ionic bond	porosity	sulfide
crystalline	isotope	proton	tensile stress

Questions for Review

1. What are *isotopes*?

2. Compare and contrast ionic and covalent bonding.

3. How is a *mineral* defined? What are the two key identifying characteristics of a mineral?

4. What is the phenomenon of solid solution, and how does it affect the definition of a mineral?

5. Explain the limitations of using color as a tool in mineral identification. Cite and explain any three other physical characteristics that might aid in mineral identification.

6. What is the basic structural unit of all the silicate minerals? Describe the basic structural arrangements of the following: chain silicates; sheet silicates; framework silicates.

7. Give the compositional characteristic common to each of these nonsilicate mineral groups: carbonates, sulfides, oxides.

8. What is a *rock*?

9. Define and compare the properties of *porosity* and *permeability*.

10. What is the *elastic limit* of a solid under stress? What happens when a material is stressed beyond its elastic limit?

11. The behavior of rocks under stress varies with depth. Explain briefly.

12. Describe the basic concept of the rock cycle.

For Further Thought

1. It is generally true that more ancient rocks are less widely found and more difficult to interpret than are younger rocks. In the context of the rock cycle, consider why this is so.

2. Choose one of the following minerals or mineral groups and investigate the range of physical properties it shows and its applications: quartz; calcite; garnet; clay.

Suggestions for Further Reading

Berry, L. G.; Mason, B.; and Dietrich, R. V. *Mineralogy*. 2d ed. San Francisco: W.H. Freeman and Co., 1983.

Dana, J. D. *Manual of Mineralogy*. 19th ed.; revised by C.S. Hurlbut, Jr., and C. Klein. New York: John Wiley and Sons, 1977.

Deer, W. A.; Howie, R. A.; and Zussman, J. *Rock-forming Minerals*. 2d ed. New York: Halstead Press, 1978.

Dietrich, R. V., and Skinner, B. J. *Rocks and Rock Minerals*. New York: John Wiley and Sons, 1979.

Ernst, W. G. *Earth Materials*. Englewood Cliffs, N.J.: Prentice-Hall, 1969.

Hurlbut, C. S., Jr. *Minerals and Man*. New York: Random House, 1968.

Kirklady, J. F. *Minerals and Rocks in Color*. New York: Hippocrene Books, 1972.

O'Donoghue, M. *VNR Color Dictionary of Minerals and Gemstones*. New York: Van Nostrand Reinhold Co., 1976.

Philips, W. J., and Philips, N. *An Introduction to Mineralogy for Geologists*. New York: John Wiley and Sons, 1980.

Pough, F. *A Field Guide to Rocks and Minerals*. 3d ed. Cambridge, Mass.: The Riverside Press, 1960.

Sinkankas, J. *Mineralogy for Amateurs*. Princeton, N.J.: D. Van Nostrand and Co., 1964.

Watson, J. *Rocks and Minerals*. 2d ed. Boston, Mass.: George Allen and Unwin, 1979.

Zoltai, T., and Stout, J. H. *Mineralogy: Concepts and Principles*. Minneapolis, Minn.: Burgess, 1984.

Igneous Rocks and Processes

Introduction

The term **igneous** comes from the Latin *ignis,* meaning "fire." The name is given to rocks formed at very high temperatures, crystallized from a molten silicate material known as **magma.** This chapter will introduce magmas and aspects of their formation and crystallization; basic characteristics and classification of igneous rocks; and some of the structures magmas form in the crust. A later chapter (9) will focus on the surface manifestations of igneous activity—volcanoes, volcanic rocks, and related phenomena.

Origin of Magmas

It is evident that high temperatures are required to melt rock. Temperatures increase with depth in the earth. Magmas originate at depths where temperatures are high enough to melt rock. The majority of magmas originate in the upper mantle, at depths between about 50 and 250 kilometers. However, the required temperatures (and depths) vary considerably, depending on several other factors.

Effects of Pressure

As the temperature of a solid is increased, the individual atoms vibrate ever more vigorously, until their energy is sufficient to break the bonds holding them in place in the solid structure. They then flow freely in a disordered liquid.

For most substances, the crystalline solid is more dense than the liquid. Therefore an increase of pressure favors the more compact, solid arrangement of atoms. So at high pressure, a correspondingly higher temperature is required to impart enough energy to the atoms to cause melting. This explains why most of the earth's interior is not molten. Temperatures are indeed high enough through most of the earth to melt the rocks if they were at atmospheric (surface) pressures. But the weight of overlying rock puts sufficient pressure on these rocks that most remain solid even at temperatures of thousands of degrees. If that pressure is lessened—for example, by the opening of fractures in rocks above—the solid may begin to melt.

The effects of pressure can be illustrated by two commonplace examples involving water. The first is the difference in boiling temperature with altitude. Vaporization of water involves a transition from a liquid to a less dense, disorderly gas. At sea level, water boils at 100° C (212° F). On a high mountain, where the air is thinner and pressure reduced, water will boil

This magma contained gas bubbles when it solidified.

at temperatures several degrees lower; less heat is required for the molecules to break away from the liquid at lower pressure. Ice skating provides a second example of the effects of pressure. Ice is unusual in that, at least close to its melting point, it is *less* dense than water (note that ice cubes and icebergs float). In this case, an increase of pressure will again favor the denser form, in this case water. This is what makes ice skating possible: concentrating the weight of the body on a narrow skate blade puts tremendous pressure on the ice below, and a little of it melts, providing a watery layer on which to glide smoothly.

Effects of Volatiles

Natural magmas contain dissolved water and various gases (among them oxygen, carbon dioxide, hydrogen sulfide, and others). Sometimes this is obvious from the resultant rock (figure 3.1). The general effect of dissolved volatiles is to lower the melting temperatures of silicate minerals. This is illustrated in figure 3.2. Increasing the water pressure drives more water into the melt and lowers the melting temperature of sodium plagioclase, most dramatically for small additions of water; the effect decreases for high water pressure. Exactly how much melting temperatures are lowered in the presence of volatiles varies with the nature of the volatiles and the minerals being melted, but the general principle holds in any case: more volatiles, lower melting temperatures.

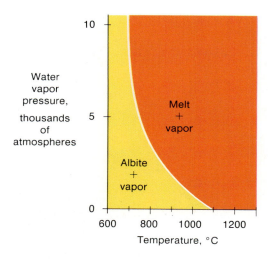

Figure 3.2

The effect of increasing water pressure on the melting temperature of sodic plagioclase feldspar (albite).

Effects of Other Solids Present

When two or more different minerals are in contact, the presence of each lowers the melting temperature of the other, to a point. This is why salt can be used to melt ice on a sidewalk in winter. The presence of the salt lowers the melting temperature of the ice, so it melts below the freezing point of pure water (0° C, or 32° F). (Presumably the ice lowers the melting temperature of the salt, too, but salt has such an extremely high melting point that it remains solid unless, of course, it simply dissolves in the melted ice.) Another common example is provided by lead-tin solders of various compositions (figure 3.3). Nearly pure lead and tin melt at close to 600° F and 500° F respectively. But adding some tin to the lead lowers the latter's melting point, and vice versa. As molten solder cools, the temperature at which it begins to solidify varies with composition. For a mix of 63% tin, 37% lead, crystallization does not occur at temperatures above 361° F.

Similar effects are observed with minerals. Figure 3.4 is a similar diagram for mixtures of quartz and sodic plagioclase (albite). While the quartz in this case appears to have little effect on the crystallization of albite, the effect of adding albite to the quartz is dramatic, reducing its melting temperature from 1713° C (3115° F) to about 1100° C (2010° F) for a mix of 40% quartz, 60% albite. Since the vast majority of rocks contain many minerals, it is safe to say that one can expect rocks to melt at temperatures somewhat below those at which the pure minerals would melt, but the relationships are likely to be complex and it may not

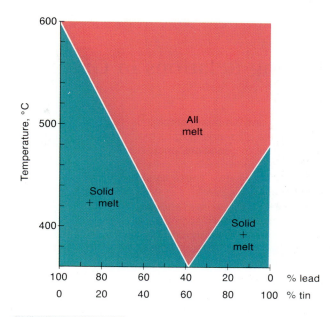

Figure 3.3

Mixing of tin and lead in solder lowers melting temperatures.

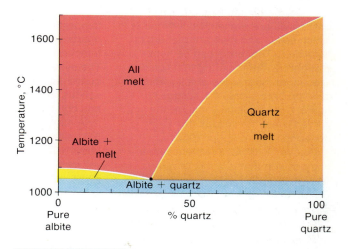

Figure 3.4

A mix of quartz and albite melts at a lower temperature than either mineral alone.

be possible to predict the exact melting temperature of each rock. Note also that melting in natural rocks typically occurs over a range of temperatures, with different minerals melting at different temperatures. The transition from a wholly solid to completely molten material may span several hundred degrees, and in fact complete melting may not be necessary in order for the magma to mobilize and flow. Many magmas are a sort of mush of crystals suspended in silicate liquid.

Box 3.1

Melting Relations in Graphical Form

Figures 3.3 and 3.4 are examples of simple *phase diagrams,* graphs that show a range of conditions under which particular minerals, or minerals plus fluids, are stable. These figures plot temperature against composition. One can also choose other sets of variables, such as pressure against temperature. Let us consider the interpretation of figure 3.4 in greater detail (figure 1).

There are two kinds of lines on this graph. A line separating conditions under which the system is entirely molten (liquid) from conditions under which some solid and some liquid are present is called a *liquidus*. Analogously, a line separating conditions under which only solids are present from those in which solids and melt are present is a *solidus*. (The terms can be kept straight by remembering that liquid is present on both sides of a liquidus, solid on both sides of a solidus.) In figure 1, the liquidus lines are the curved lines, the solidus the horizontal line at about 1060° C.

Above 1713° C, a system with a composition that is a mix of quartz and albite will be entirely molten. Below 1060° C, it will be entirely solid, consisting of crystals of albite and crystals of quartz. At intermediate temperatures, whether solid will be present, and which one, depends on the bulk composition of the system. If, for example, the composition is as represented by point A (about 70% quartz, 30% albite), the system will be molten above the liquidus temperature at that composition (about 1470° C). A melt of that composition will begin to crystallize quartz at that liquidus

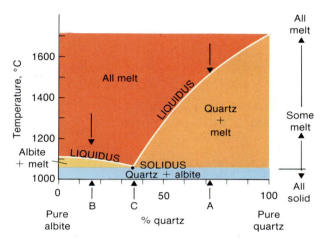

Figure 1
The albite-quartz phase diagram.

temperature. With more cooling more quartz will crystallize, until the solidus temperature (1060° C) is reached. The system will then crystallize quartz and albite, staying at the solidus temperature until all the melt has crystallized.

A melt of composition B (about 83% albite, 17% quartz) will cool to about 1075° C before any crystals will form. These first crystals will be albite rather than quartz, and albite will continue to crystallize as the melt cools to the solidus temperature. Then quartz and albite will crystallize as before until all the melt has solidified.

Melting relationships are the reverse of the above. That is, a mix of quartz and albite crystals will be entirely solid until the solidus temperature is reached. With continued heating, quartz and albite will both melt, in the proportions indicated by point C (about 35% quartz, 65% albite) until one or the

other mineral is used up. If the starting composition corresponds to A (or any point on the quartz-rich side of C), the albite will be used up first during melting. With still more heating, the remaining quartz will gradually melt until, at the liquidus temperature for that composition (1470° C for composition A), all the solid has melted. With an albite-rich composition such as B, melting of quartz plus albite at the solidus will proceed until the quartz is used up. Then with more heating, the remaining albite will melt until, at about 1075° C (for composition B), the whole sample will have melted.

This is actually quite a simple example compared to a natural system. Such diagrams, and the corresponding melting and crystallization relationships, become more complex when more than two minerals are involved, or if one or more of them is a solid solution.

Effects of Solid Solution

Many solid solutions can be regarded as a mix of two or more endmember, or limiting, compositions. For example, olivine ($(Fe,Mg)_2SiO_4$) can be thought of as intermediate in composition between the compounds Fe_2SiO_4 and Mg_2SiO_4. These endmembers do not melt at the same temperatures: the pure iron olivine melts at 1205° C, while the pure magnesium olivine melts at 1890° C. An intermediate composition will melt in between these extremes, at temperatures determined by its exact composition (proportion of iron to magnesium), along with pressure and other factors already noted.

Other Sources of Heat

The most basic source of elevated temperatures is depth, as mentioned earlier. Locally, other factors may also be important in producing heat. We have noted the existence of naturally radioactive elements. Radioactive decay produces heat, and in rocks containing unusually high concentrations of radioactive elements, this can be a significant supplementary source of heat. Friction produces heat too. Rubbing two rocks together will not cause sufficient heating to initiate melting, but rubbing two continents together (figuratively speaking) may add enough extra heat to warm rocks at depth to start them melting. Even the movement of existing magma can be a factor: if a hot mass of magma rises up in the crust, it will heat surrounding rocks and under some conditions may begin to melt them.

Crystallization of Magmas

Once an appropriate combination of factors has produced a quantity of melt, the melt tends to flow away from where it was produced. Usually it flows upward, into rocks at lower temperatures and pressures. As it moves upward, it cools. The cooling in turn leads to crystallization, as the atoms slow down and eventually settle into the orderly arrays of crystals. The details of the rock thus produced will vary with the composition of the melt and the conditions under which it crystallized. However, some generalizations about the resulting mineralogy and texture are possible.

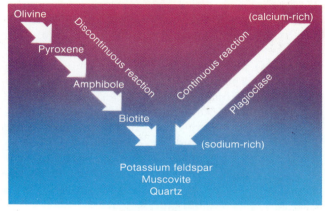

HIGH TEMPERATURE

LOW TEMPERATURE

Figure 3.5

Bowen's reaction series.

Sequence of Crystallization

We have already observed that a mix of minerals will melt over a range of temperatures. Likewise a magma will crystallize over a range of temperatures, or, in other words, over some period of time during cooling as different minerals begin to crystallize at different temperatures/times. Moreover, because most magmas originate in the upper mantle they are in a very general sense similar in composition, consisting predominantly of silica (SiO_2), with varying lesser proportions of aluminum, iron, magnesium, calcium, sodium, potassium, and additional minor elements. Magmas therefore tend to follow a predictable crystallization sequence in terms of the principal minerals forming, although the proportions of these minerals in the ultimate rock will vary. This sequence was recognized more than half a century ago by geologist Norman L. Bowen, who combined careful laboratory studies of compositionally simple silicate systems with wide-ranging field observations of the more complex natural rocks.

The result, known as **Bowen's reaction series,** is illustrated in figure 3.5. Those minerals that tend to crystallize at higher temperatures are shown nearer the top of the series; the later, low-temperature minerals are near the bottom. In general, the earliest minerals to crystallize are relatively low in silica, so that the residual magma remaining after their crystallization is more enriched in silica relative to its starting composition. The high-temperature portion of the sequence is also subdivided into a ferromagnesian branch and a branch involving plagioclase feldspar.

The plagioclase branch is termed a **continuous reaction series.** This refers to interaction between crystals already formed and the remaining melt. Recall from chapter 2 that plagioclase is a solid solution between a calcium-rich endmember (anorthite, $CaAl_2Si_2O_8$) and a sodium-rich endmember (albite, $NaAlSi_3O_8$). The more calcic compositions are the higher-temperature end of the series: pure anorthite melts at about 1550° C, pure albite at 1100° C. Sodium and calcium are freely interchangeable in the plagioclase crystal structure. The first plagioclase to crystallize from a magma, at high temperatures, will be a rather calcic one; but as the magma cools the crystals will react continuously with the melt, with more and more sodium entering into the plagioclase but no changes in basic crystal structure. (Note that the sodic plagioclase also contains a higher proportion of Si, so the later plagioclases are more silica-rich too.) If cooling is too rapid for complete reaction between crystals and melt during cooling, the resultant crystals will show concentric compositional zones, with Ca-rich cores grading outwards through increasingly sodium-rich compositions.

The ferromagnesian side of the crystallization sequence is a **discontinuous reaction series.** Olivine is the first of the ferromagnesians to crystallize. After a period of crystallization, the olivine is so out of balance chemically with the increasingly silica-rich residual melt that olivine and melt react to form pyroxene. (Recall that the ratio of iron plus magnesium to silicon in olivine is 2:1, while in pyroxene it is about 1:1.) Assuming that there is sufficient silica available, all of the olivine will be converted to pyroxene at that point. After an interval of pyroxene crystallization the pyroxene, again out of chemical balance with the remaining melt, will react with it, and pyroxenes will be converted to amphiboles, and so on. The last of the common ferromagnesians to crystallize is biotite mica. This discontinuous reaction series, then, is marked by several changes in mineralogy/crystal structure during cooling and crystallization.

At the tail end of the crystallization sequence, at lowest temperatures, any last, silica-rich residual melt crystallizes potassium feldspar, muscovite mica, and quartz. Note that the hydrous silicates—amphiboles and micas—are relatively late in the sequence. At very high temperatures hydrous minerals are unstable, and any water stays in the melt. Also, not every magma will progress through the whole sequence. A more **mafic** magma (rich in magnesium and iron, poorer in silicon) will be completely crystallized before the lattermost stages of the sequence are reached and will not have started with sufficient silica to make

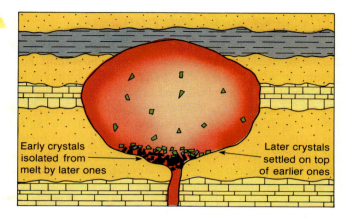

Figure 3.6

Crystal settling in a magma chamber as a mechanism of fractional crystallization. Early crystals are isolated from the residual melt by later ones.

quartz at the end. A very **silicic** (silica-rich, iron- and magnesium-poor) magma will indeed reach the final stages, by which time reactions with the melt will have eliminated early-formed olivine, pyroxene, and calcic plagioclase. Mafic magmas, then, produce rocks rich in the minerals near the top of the diagram in figure 3.5; silicic magmas produce rocks dominated by the minerals near the bottom and poor in ferromagnesians. The latter rocks are typically rich in feldspar and quartz (silica) and are therefore also termed **felsic.**

Modifying Melt Composition

The above discussion tacitly assumed that each crystallizing magma behaves as a *closed system,* neither gaining nor losing matter. This is often not the case in natural systems. Magma compositions may be modified after the melt is formed. The result is a product somewhat different from what would be expected on the basis of the original melt composition.

Fractional crystallization is one way in which melt composition can be changed. In this process, early-formed crystals are physically removed from the remaining magma and so are prevented from reacting with it. One way in which this can happen is if the early crystals settle out of the crystallizing magma (figure 3.6) and are isolated from the melt by later crystals settling above them. The remaining melt may even move away through zones of weakness in surrounding rocks, leaving the early crystals behind altogether. The result is that the average composition of the melt, minus its early, low-silica minerals, is shifted toward a more silica-rich, iron- and magnesium-poor composition. That melt can then progress further down the crystallization sequence than would have been possible if all crystals had remained in the melt as it cooled.

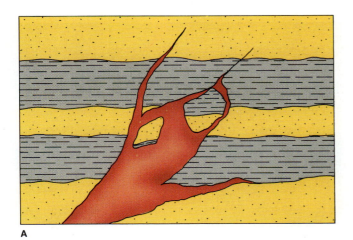

A

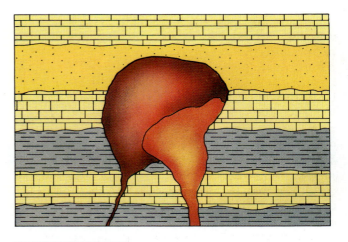

Figure 3.8

Magma mixing: two rising magmas meet and mix to produce an intermediate composition.

B

Figure 3.7

Assimilation by magma. *(A)* Schematic: the piece of wallrock is partially melted by and mixed into the magma. *(B)* Assimilation was in progress here when the melt solidified.

Figure 3.9

A coarse-grained igneous rock, slowly cooled.

A magma can **assimilate** the rock around it, incorporating pieces of it, melting and mixing it in so that the overall magma composition is again changed (figure 3.7). Given the relative temperatures at which mafic and silicic rocks crystallize, assimilation occurs most readily when the initial melt is a (hotter) mafic magma and the assimilated material is more silicic. The result is a modified magma of somewhat more silicic composition, which would move correspondingly further down the crystallization sequence.

Magma mixing (figure 3.8) is another possibility, in which two melts combine to produce a hybrid melt intermediate in composition between them. This is less common, for unusual geologic conditions are needed to produce two distinctly different magmas in nearly the same place at the same time.

Textures of Igneous Rocks

The most noticeable textural feature of most igneous rocks is the *grain size,* the size of the individual mineral crystals. An important control on grain size, as indicated in chapter 2, is cooling rate. If a magma cools slowly, there is more time for atoms to move through the melt and attach themselves at suitable points on growing crystals. The resulting rock will be coarser grained (figure 3.9). In a rapidly cooled magma there is less time for crystal growth, so the rock will be finer grained (figure 3.10). In extreme cases, the result will be a *glassy* rock, with no obvious crystals (figure

Figure 3.10

A fine-grained igneous rock, cooled quickly.

Figure 3.12

The phenocrysts in this porphyry formed early, during a period of slow cooling. Later, rapid cooling caused the remaining melt to crystallize in a mass of smaller crystals.

Figure 3.11

A glassy igneous rock. This melt cooled so rapidly that there was no time for crystals to form.

3.11). Some quickly cooled magmas also trap bubbles or pockets of gas, which are termed **vesicles;** the resulting texture is described as *vesicular* (recall figure 3.1).

Some igneous rocks have a two-stage cooling history, with an initial stage of slow cooling allowing some large crystals to form, followed by rapid cooling leaving the rest of the rock (**groundmass**) finer grained. The resulting rock is called a **porphyry;** the texture is described as *porphyritic* (figure 3.12). The coarse crystals embedded in the finer groundmass are termed **phenocrysts,** the prefix *pheno-* coming from the Greek for "to show." In other words, these are very obvious crystals.

Other factors besides cooling rate can affect grain size. Melt composition is one. Silicic melts are more viscous, or thicker, than mafic ones. In all silicate melts, silica tetrahedra exist in the melt even before actual crystallization. In a mafic melt, most of these tetrahedra float independently. In more silica-rich melts, the tetrahedra are more extensively linked, and as a result atoms move less freely through the melt. More time is required for atoms to move into position on growing crystals. Most volcanic glasses are silicic in composition for this reason (even when dark colored). The melts were so stiff and viscous that rapid cooling produced, not small crystals, but virtually no crystals at all (figures 3.11, 3.13).

On the other hand, some late-stage, residual melts have accumulated high concentrations of dissolved volatiles, so that they are quite fluid, and atoms move easily through them. Even with moderate cooling rates, very large crystals can grow in a volatile-rich, fluid magma. The resultant extremely coarse-grained rock is termed a **pegmatite** (figure 3.14).

Classification of Igneous Rocks

The most fundamental division of igneous rocks is made on the basis of depth of crystallization. The **plutonic** igneous rocks are those crystallized at some depth below the surface. They take their name from Pluto, the Greek god of the lower world. Rocks are poor heat conductors, so magmas at depth are well

Figure 3.13

Obsidian Cliff, Yellowstone National Park. This silicic melt was so viscous that no visible crystals formed in this great mass of magma before it solidified. (White bands mark trails of trapped gas bubbles.)

Figure 3.14

Pegmatite, a very coarse-grained igneous rock.

Photo by A. L. Kimball, courtesy U.S. Geological Survey.

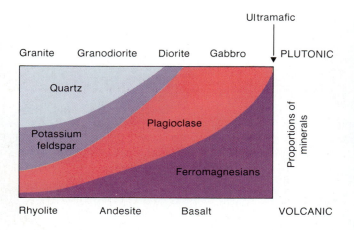

A simplified classification of igneous rocks, based on mineralogy (related to chemical composition) and origin (plutonic vs. volcanic).

An ultramafic rock (*dunite*), made up of olivine (light green) and some pyroxene (black).

Plutonic Rocks

The coarse grain sizes of plutonic rocks make their preliminary identification relatively simple even without special equipment. They are classified on the basis of relative proportions of certain light and dark (ferromagnesian) minerals. A rock consisting entirely of ferromagnesians and dark, calcic plagioclase feldspar is a **gabbro** (figure 3.16). In the extreme case where feldspar is virtually absent and the rock consists almost wholly of olivine and pyroxene, it is termed **ultramafic** (figure 3.17). A rock that is somewhat more silica-rich than gabbro will contain some lighter-colored sodic plagioclase and potassium feldspar, the mix of light and dark minerals giving it a salt-and-pepper appearance. This is **diorite** (figure 3.18). If the rock is sufficiently silica-rich that appreciable quartz is present, and the proportion of ferromagnesians correspondingly less, the rock is a **granite** (figure 3.19).

Although igneous rocks are put in distinct categories, they in fact span a continuum of chemical and mineralogical compositions (see figure 3.15). The boundaries between categories are somewhat arbitrary. One can indicate that a rock belongs in an intermediate compositional group by using a hybrid name for it: for example, a dioritic rock containing just a little quartz could be termed a *granodiorite,* to indicate that its composition lies between the quartz-rich granite and quartz-free diorite classes.

Gabbro, a mafic plutonic rock.

insulated and cool slowly. Plutonic rocks, then, are generally recognized on the basis of their coarser grain sizes, with individual crystals readily visible to the naked eye. The **volcanic** rocks are those formed from magmas cooled at or near the surface. They are typically fine grained or glassy, with individual crystals not easily seen with the naked eye. Porphyritic rocks with very fine-grained groundmass would also be considered volcanic. Further subdivisions within the textural classes are made primarily on the basis of composition. A summary of principal igneous rock types and their corresponding compositions is shown in figure 3.15 and explained in the text that follows.

Figure 3.18

Diorite, an intermediate plutonic rock with salt-and-pepper coloring.

Figure 3.20

Basalt, the volcanic equivalent of gabbro and the most common volcanic rock type.

Figure 3.19

Granite, a silica-rich plutonic rock.

Many rock types were named before sophisticated chemical analytical methods were developed, or before there was general agreement on nomenclature. This has left some confusing contradictions among the rock names. The name *gabbro* comes from the locality Gabbro, Italy. But by modern classification, the plutonic rocks there would be called diorite, not gabbro!

Volcanic Rocks

Most plutonic rock types have fine-grained volcanic compositional equivalents, as can be seen in figure 3.15. They are more difficult to identify definitively in handsample, for the individual mineral grains are by definition tiny. Color is the most frequently used means of identification in the absence of magnifying lenses or other equipment, but as with minerals, it is not an entirely reliable guide.

The most common volcanic rock is **basalt** (figure 3.20), the fine-grained equivalent of gabbro. This dark-colored (usually black) rock makes up the sea floor and is erupted by many volcanoes on the continents as well. **Andesite** is the name given to volcanic rocks of intermediate composition. They are typically lighter in color than basalt, often green or grey (figure 3.21). The volcanic equivalent of granite is **rhyolite** (figure 3.22). Some rhyolites are pinkish in color, but many are not. It can be difficult to distinguish the light-colored andesites and rhyolites, unless the rhyolite is porphyritic and quartz is visible among the phenocrysts.

A few volcanic rock names lack compositional implications. An example is **obsidian,** or volcanic glass. (See figures 3.11 and 3.13.) The term can be applied to glassy rock of any composition, although in fact most obsidians are rhyolitic in composition. As

Figure 3.21

Andesite, volcanic rock of intermediate composition.

Figure 3.22

Rhyolite, a silicic volcanic rock chemically equivalent to granite. (Sample on left is porphyritic.)

with plutonic rocks, one can use mixed names to denote rocks that are close to the boundary between two compositional classes (e.g., basaltic andesite). One can also combine a textural and a compositional term to provide a more complete description of the rock (e.g., porphyritic andesite, vesicular basalt).

Some Further Considerations

Specialists in igneous rocks naturally use a far more elaborate nomenclature, with many more specialized terms. There are many unusual rock types that could

not properly be put in any of the simple categories named. The foregoing scheme, however, covers the vast majority of igneous rocks and provides enough of a working vocabulary for discussion of the geologic processes to be dealt with later in the text.

> The reader should be warned that nongeologists who work with geologic materials have appropriated some of these terms but given them different meanings. It is common engineering practice, for instance, to call all coarse-grained igneous rocks "granite" and all fine-grained ones "basalt," regardless of composition—because for many engineering problems the texture is the critical factor in determining the rock properties. Landscape architects also use the non-term "lava rock" to denote volcanic rocks of various kinds. In this text, however, we will adhere to the more precise geologic definitions as outlined in the text.

Intrusive Rock Structures

Magma moving through the crust is intruding the surrounding rock, which is commonly termed the **country rock,** or **wallrock.** If the magma erupts at the earth's surface, it has become **extrusive.** Extrusive rocks and structures (e.g., volcanoes) will be discussed in some detail in chapter 9. Here we will briefly review the more common forms and features of **intrusions,** igneous rock masses formed by magma crystallizing below the earth's surface.

Factors Controlling the Geometry of Intrusions

Both the properties of the magma and the properties of the surrounding rocks play a role in determining the shape of the intrusion formed. Magmas vary in density, with the iron-rich mafic magmas the more dense. Denser magmas will be less buoyant with respect to the country rock, and their extra mass may even cause the country rocks to sag around the magma body. The viscosity of the magma will influence how readily it flows through narrow cracks or other openings in the country rock. This can be demonstrated in the kitchen using a sieve: water will pour rapidly through the sieve; honey or molasses, which are more viscous, will seep through more slowly; and honey that has partially crystallized (analogous to the crystal-liquid mush of a partly solidified magma) may not pass through the sieve at all, instead remaining as a coherent mass within it. Similarly, very fluid magmas can move through quite narrow cracks in rocks, while thick, viscous magmas are more likely to remain in a compact mass.

Box 3.2

Looking at Rocks Another Way: Thin Sections

Volcanic rocks are so fine-grained that it is commonly impossible to distinguish individual grains with the naked eye. With the majority of rocks of all types, even if individual mineral grains can be seen, it is hard to examine in detail the grain shapes, internal characteristics of individual crystals, and so forth. Geologists therefore rely on microscopic examination of **thin sections,** paper-thin slices of rock mounted on glass slides, to learn more than they can from looking at a chunk of rock unaided.

A thin section (figure 2) is ground down to a thickness of about 0.03 millimeters (0.0012 inches), at which thickness most minerals are transparent or translucent. The special microscopes used to examine thin sections use polarized light (light oriented so that all rays vibrate in parallel). The light rays are deflected or rotated in passing through crystals, different minerals producing different deflections. If the thin section is sandwiched between two polarizers at right angles to each other, *interference colors* diagnostic of the different minerals are seen. (Compare figures 3 and 4.) A suitably trained geologist would be able to recognize plagioclase (grey-and-white striped) and pyroxene (colored) phenocrysts in this volcanic rock, as well as smaller crystals of both and some glass and magnetite (the latter opaque even in plain light) in the groundmass.

Figure 1
Porphyritic andesite (handsample).

Figure 2
Thin section of the same rock. Note how transparent it is.

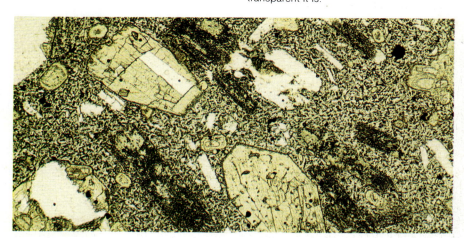

Figure 3
Microscopic view of the thin section, in polarized light.

Figure 4
The same thin section viewed between crossed polarizers.

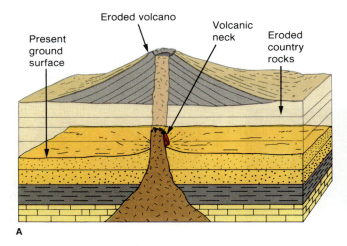

Present ground surface

Eroded volcano

Volcanic neck

Eroded country rocks

A

Figure 3.23

A volcanic neck or pipe—all that remains of a long-eroded volcano. *(A)* Schematic diagram: note cylindrical shape and discordant character. *(B)* An example exposed at the surface by erosion: Devil's Tower, Wyoming.

B

The strength of the wallrock, and whether or not it is fractured, will also play a part. Fractures in rock are zones of weakness through which magmas can pass more easily than through solid, unbroken rock. Zones of weakness may also exist at a contact between dissimilar rock types in the wallrock. The shape of an intrusive body may be controlled by the geometry of zones of weakness in the wallrock through which the magma flows preferentially. Sometimes, too, the magma is under unusual pressure from trapped gases within it, which will allow it forcibly to intrude wallrocks that it might ordinarily be unable to penetrate.

Forms of Intrusive Bodies

Pluton is a general term for any body of plutonic rock, that is, any igneous rock mass crystallized below the earth's surface. The term has no particular geometric significance. Plutons are classified according to their shapes and their relationship to structures in the country rock. A pluton is said to be **concordant** if its contacts are approximately parallel to any structure (such as compositional layering or folds) in the country rock; it is **discordant** if its contacts cut across the structure of the country rock. A few examples should clarify the distinction.

A cylindrical pluton, elongated in one dimension, is a **pipe** or **neck** (figure 3.23). These are believed to be the feeders or conduits leading magma up to a volcano, which once existed above them but has since been eroded away. Pipes are typically discordant.

Tabular, relatively two-dimensional plutons commonly result from magma emplacement along planar cracks or zones of weakness. They are termed **dikes** if they are discordant, **sills** if they are concordant (figure 3.24).

Concordant plutons that are more nearly equidimensional are less common. Those that have flat floors, or bottoms, and have caused doming or arching of the rocks above them are termed **laccoliths.** Those that have floors that are concave upward (their tops may not be visible) are **lopoliths** (figure 3.25). It is observed in the field that laccoliths are commonly formed by silicic rocks (magmas), lopoliths by mafic ones. This suggests that the density of the magma has played a significant role in shaping the pluton, the lopolith perhaps showing sagging of the country rock under the weight of dense mafic rocks.

Discordant equidimensional plutons are somewhat arbitrarily divided on the basis of the area of rock exposed at the surface. A **stock** is exposed over less than 100 square kilometers (about 35 square miles), while a **batholith** is anything larger. Actually many stocks may just be small bulges of magma fingering upward from a batholith below, not yet exposed at the surface by erosion (figure 3.26), so the distinction is not particularly important. Very large batholiths, which can cover thousands of square kilometers of

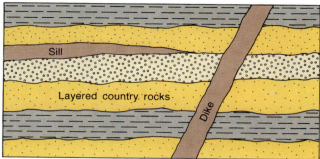

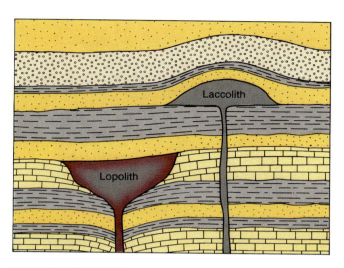

Figure 3.25

Concordant three-dimensional plutons: a laccolith and a lopolith.

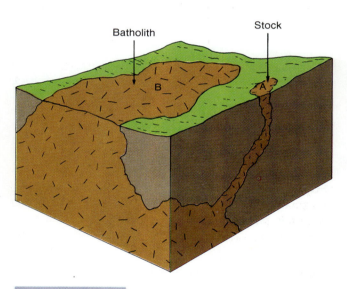

Figure 3.26

The distinction between a stock *(A)* and a batholith *(B)* is purely one of scale of exposure at the surface. Both are discordant more equidimensional plutons.

Other Features of Intrusions

Various textural features may be observed within a pluton, regardless of its overall geometry.

When a hot magma intrudes much cooler country rock, the melt near the contacts is quenched, or rapidly cooled. The resultant **chilled margins** are recognizable because they are finer grained than the interior of the pluton (figure 3.27). Chilled margins are more commonly found in shallower plutons (and volcanic rocks) because the temperature contrast between magma and country rock is greater at shallower depths in the crust.

Figure 3.24

Tabular plutons. *(A)* Sketches of a concordant sill, discordant dike. *(B)* Example of a dike.

outcrop area and extend to depths of 5 kilometers (3 miles) or more, are multiple intrusions. Many batches of magma were emplaced to form them, and often many smaller plutons can be distinguished within the large batholith.

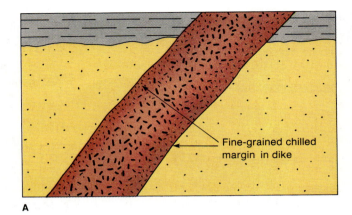

A

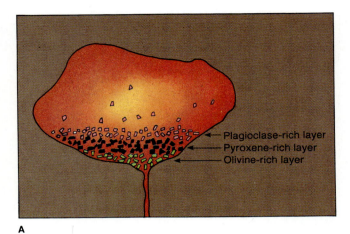

Plagioclase-rich layer
Pyroxene-rich layer
Olivine-rich layer

A

B

Figure 3.27

Chilled margins. *(A)* Rapid cooling at the edges of a pluton results in a fine-grained chilled margin. *(B)* Example of a natural chilled margin.

Fine-grained chilled margin in dike

B

Figure 3.28

Formation of compositional layering in a pluton by gravitational crystal settling. *(A)* Schematic: layers of denser minerals accumulate in succession on the floor of the magma chamber. *(B)* Example of a layered plutonic rock.

Compositional layering is sometimes observed (figure 3.28). It usually develops as a result of gravitational settling-out of denser crystals to the floor of the magma chamber as crystallization proceeds. Layering is more frequently observed in mafic plutons, probably for two reasons. First, in the early stages of crystallization, mafic magmas crystallize abundant dense ferromagnesian minerals, which are especially prone to settling. Second, settling will occur more readily in a less viscous liquid, and mafic magmas are generally less viscous. (To illustrate the effect of viscosity, drop a pebble or dried pea into a glass of water and a glass of molasses and compare the results.)

If some crystals have formed before the magma stops flowing actively, evidence of that flow may be found in the parallel alignment of platy or elongated crystals (figure 3.29). Near the edges of a pluton at least, the flow direction indicated is generally parallel to the contacts. Bubbles, like crystals, may also be aligned during flow (see figure 3.13).

Batholiths and the Origin of Granite

Most batholiths are granitic to granodioritic in composition. This prompts the question: what is the source of such large volumes of granitic magma?

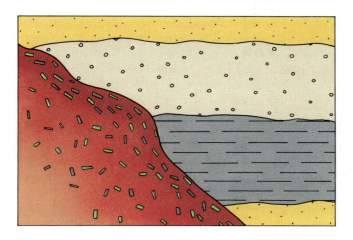

Figure 3.29

Flow texture may be shown by parallel alignment of elongated crystals in an igneous rock. See also figure 3.13.

Partial Melting

We have noted that most magma originates in the upper mantle, where temperatures are high enough and pressures simultaneously low enough to permit melting. What occurs is actually **partial melting,** in which those minerals with the lowest melting points do melt, while higher-temperature minerals do not. Inspection of Bowen's reaction series (figure 3.5) indicates that the first minerals to melt with increasing temperature would indeed be the major constituents of typical granite. However, the mantle is ultramafic in overall composition. Therefore, while the very earliest melt would be granitic, by the time a significant fraction of the mantle had melted the melt would be basaltic in composition through the addition of melted ferromagnesians. If we assume that only a very small percentage of melt forms, there are two difficulties in maintaining a more silicic composition: one, a huge volume of mantle would have to be involved to account for the immense volume of granite in a major

batholith; and two, it is difficult, mechanically, to squeeze a percent or two of melt out of almost completely solid mantle in order to emplace it into the overlying crust.

Fractional Crystallization

An alternative way to make granitic magma would be to start with the basaltic magma expected with significant melting of an ultramafic mantle, then subject it to fractional crystallization at depth. This could proceed, the more mafic constituents being removed until the residual melt was granitic, at which point the melt could intrude the crust. Again there is a volume problem. To end up with such large volumes of granitic melt as batholiths represent, it would be necessary to start with much more enormous volumes of basaltic melt. There is little evidence of such extensive melting in the upper mantle.

A Crustal Contribution

A third possibility is to involve some crust in the melt. For instance, the continental crust is, on average, granodioritic in composition. Partial or complete melting of continental crust could produce magma of the required composition. So would assimilation of considerable continental crust by a rising basaltic, mantle-derived magma. In either case the problem is one of heat budget, for the continental crust is not normally hot enough to melt, even at depth. It is unclear whether rising hot basaltic magma could cause enough heating and melting of continental crust to create a batholith.

Geochemical and other evidence suggests that in fact no single simple model accounts for all granitic batholiths. The mechanism(s) involved vary from batholith to batholith, and sometimes from pluton to pluton within one batholith. We will return again to the question of the origin of batholiths in connection with the subject of the growth of continents (chapter 11).

Summary

Igneous rocks are those crystallized from magma, a silicate melt. Most magmas are produced in the upper mantle, where temperatures are high enough, and pressures low enough, to allow melting. The amount of melting is increased, and the melting temperatures of minerals are reduced, by the presence of volatiles and by the presence of several different minerals in one rock. Beyond the normal increase in temperature

with depth in the earth, additional sources of heat for melting include radioactive decay, friction between large rock masses, and emplacement of hot magma from depth into shallower crustal levels. Once formed, a cooling magma normally crystallizes principal silicate minerals in a sequence predicted by Bowen's reaction series, beginning with olivine and calcic plagioclase and ending with quartz and potassium feldspar. The composition of a magma may be changed by various processes, including fractional crystallization, assimilation, and magma mixing. The grain size of an igneous rock is determined fundamentally by cooling rate: all other factors being equal, slower cooling means larger crystals. Very rapid cooling produces fine-grained or even glassy rocks. Melt composition and the presence or absence of dissolved volatiles also influence grain size. Typically, the plutonic rocks, which crystallize at depth, are coarser grained than volcanic rocks.

Igneous rocks are classified on the basis of grain size and chemical composition. For most compositions there are plutonic and volcanic equivalents, with different names reflecting their depths of crystallization. The mafic volcanic rock basalt is the principal rock type of the sea floor; continental crust is mainly granitic or granodioritic. Intrusive igneous rocks form plutons that may be pipelike, sheetlike, or fairly equidimensional. They are classified on the basis of their shape and whether they are concordant or discordant with respect to the country rock. The shape of a pluton is controlled by the physical properties both of the magma and the country rock. Plutons may exhibit such additional features as chilled margins, compositional layering, and flow textures. The origin of the large plutonic complexes termed batholiths is problematical, for no single mechanism (partial melting of mantle, fractional crystallization of basalt, melting of continental crust) seems able to account readily for the production of the necessary volume of granitic or granodioritic magma.

Terms to Remember

andesite	discontinuous reaction	laccolith	pluton
assimilate	series	lopolith	plutonic
basalt	discordant	mafic	porphyry
batholith	extrusive	magma	rhyolite
Bowen's reaction series	felsic	magma mixing	silicic
chilled margin	fractional crystallization	neck	sill
concordant	gabbro	obsidian	stock
continuous reaction series	granite	partial melting	thin section
country rock	groundmass	pegmatite	ultramafic
dike	igneous	phenocrysts	vesicles
diorite	intrusion	pipe	volcanic
			wallrock

Questions for Review

1. What is an igneous rock?

2. Explain briefly the effect of each of the following on the melting temperature of rock: (a) changes in pressure; (b) presence of water vapor.

3. Cite three possible heat sources that might contribute to melting.

4. What is Bowen's reaction series? How do the continuous and discontinuous branches differ?

5. Describe two ways in which the composition of a magma can be modified.

6. How is the grain size of an igneous rock related to its cooling rate? What does a porphyritic texture indicate?

7. Plutonic rocks may be more readily identifiable in handsample than volcanic rocks. Why?

8. Most volcanic glasses are rhyolitic in composition. Compositional layering is more often observed in mafic than in felsic plutons. What property of a magma may have a bearing on both these observations? Explain.

9. What is a discordant pluton? Give two examples.

10. What is a batholith? Suggest two ways in which the necessary granitic magma might be formed and discuss any problem with each idea.

11. Assimilation of felsic rocks by mafic magmas is more common than the reverse. Why?

For Further Thought

1. Considering magma viscosity and density, would you expect volcanic rocks more often to be mafic or silicic in composition? Why?

2. The minerals quartz, albite, and orthoclase (a potassium feldspar) are sometimes called, collectively, the "residua system" in the study of igneous rocks. How do you suppose this name arose?

Suggestions for Further Reading

Barker, D. S. *Igneous Rocks*. Englewood Cliffs, N.J.: Prentice-Hall, 1983.

Bowen, N. L. *The Evolution of the Igneous Rocks*. New York: Dover Publications, 1956.

Carmichael, I. S.; Turner, F. J.; and Verhoogen, J. *Igneous Petrology*. New York: McGraw-Hill Book Co., 1974.

Cox, K. G.; Bell, J. D.; and Pankhurst, R. J. *The Interpretation of Igneous Rocks*. London: George Allen and Unwin, 1979.

Ehlers, E. G., and Blatt, H. *Petrology: Igneous, Sedimentary, and Metamorphic*. San Francisco: W. H. Freeman and Co., 1982.

Ernst, W. G. *Earth Materials*. Englewood Cliffs, N.J.: Prentice-Hall, 1969.

Hughes, C. J. *Igneous Petrology*. New York: Elsevier, 1982.

Maaloe, S. *Principles of Igneous Petrology*. New York: Springer-Verlag, 1985.

MacKenzie, W. S.; Donaldson, C. H.; and Guilford, C. *Atlas of Igneous Rocks and Their Textures*. New York: Halstead Press, 1982.

Sediment and Sedimentary Rocks

Introduction

Igneous rocks, discussed in the previous chapter, represent one end of the temperature spectrum. Sedimentary rocks represent the other end. They form at and near the earth's surface from sediments—unconsolidated accumulations of rock and mineral grains and organic matter that have been transported and deposited by wind, water, or ice. When the sediments become consolidated into a cohesive mass, they become sedimentary rock. The composition, texture, and other features of a sedimentary rock can provide clues about its origin, source materials, and the setting in which the sediment was deposited.

Classification of Sediments and Sedimentary Rocks

The ultimate source of sediment is the weathering of rock, which will be described in greater detail in chapter 18. Rocks exposed at the surface can be physically broken up into fragments; they can also undergo chemical reactions with air and water, reactions that produce new minerals and dissolve others. Chemicals thus dissolved in streams, lakes, and oceans can later be precipitated out of solution, with or without the aid of organisms. Sediments are classified, first, on the basis of the way in which they are formed, and secondarily on the basis of composition or particle size.

Clastic Sediments

Clastic sediments take their name from the Greek word *klastos,* meaning "broken." These are the sediments composed of broken-up fragments of preexisting rocks and minerals. Individual fragments in a clastic sediment may be single mineral grains or bits of rock comprising several minerals.

Clastic sediments, and the rocks formed from them, are named on the basis of the sizes of the fragments in them (see table 4.1). Gravels and boulders are the most coarse-grained clastic sediments. When they are transformed into rock, the corresponding rock is called either conglomerate or breccia, depending on whether the fragments are well rounded or angular (figure 4.1). The next finer sediments are sands, which when consolidated make sandstone (figure 4.2). Particles smaller than sand size are so fine that in sediment they usually don't feel gritty to the touch; individual grains are also indistinct or invisible to the naked eye. These very fine sediments are the silts and clays. Properly, the corresponding rocks

Table 4.1

Simplified classification scheme for clastic sediments and rocks

Particle size range	Sediment	Rock
over 256 mm (10 in)	boulder	conglomerate (if fragments are rounded) or breccia (angular fragments)
2–256 mm (0.08–10 in)	gravel	
1/16–2 mm (0.025–0.08 in)	sand	sandstone
1/256–1/16 mm (0.00015–0.025 in)	silt	siltstone*
less than 1/256 mm (less than 0.00015 in)	clay	claystone*

*Both siltstone and claystone are also known as mudstone, commonly called *shale* if the rock shows a tendency to split on parallel planes.

A

B

Figure 4.1

Coarse-grained clastic sedimentary rocks. *(A)* Conglomerate—rounded fragments. *(B)* Breccia—angular fragments.

Figure 4.2

Examples of sandstone, a medium-grained clastic rock.

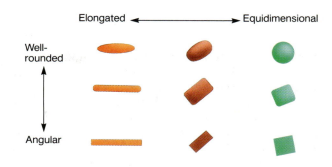

Figure 4.4

Roundedness versus grain shape: a grain can be elongated but well rounded, or blocky (equidimensional) but angular.

Figure 4.3

Shale, a variety of mudstone. Note that individual grains are too fine to be seen with the naked eye.

are *siltstones* and *claystones*. However, since the grains cannot be seen with the naked eye, siltstones and claystones cannot readily be distinguished in handsample. They are often lumped together under the general heading of **mudstone,** a rock made from fine, muddy sediment. It is common for mudstones to show a tendency to break roughly along planes corresponding to depositional layers, in which case they are given the name **shale** (figure 4.3).

Notice that none of these terms carries any implications about the composition of the fragments, or even about whether the fragments are mineral grains or chunks of rock. Even the terms *clay* and *claystone* don't necessarily mean that the sediments consist of clay minerals. It is true that clay minerals usually form only very fine crystals, and many mudstones and shales do indeed consist predominantly of clay minerals. However, it is perfectly possible to have clay-sized fragments of quartz, feldspar, obsidian, or other materials.

In the case of the coarsest sediments, a distinction in rock name is made on the basis of roundness or angularity of fragments. These properties are illustrated in figure 4.4. Angular fragments have sharp corners; the more well rounded they become, the smoother the corners. Fragments need not be spheroidal, like little balls, in order to be considered well rounded; they must simply have smooth, rounded corners and edges. With finer-grained sediments, distinctions are not made on the basis of roundedness. In the very finest sediments, roundedness cannot, after all, be determined with the naked eye because individual grains cannot be seen. Also, by the time rock and mineral fragments have been broken down into very fine pieces, they are usually somewhat rounded by abrasion anyway.

Another textural characteristic of clastic sediments and rocks is the degree of **sorting** they show. Sorting is a measure of how great a range in grain sizes is present. A well-sorted sediment is one in which the grains are all similar in size (figure 4.5); a poorly sorted sediment shows a wide variation in grain size (figure 4.6). The finer material filling in the pores between coarser grains in a poorly sorted sediment is termed the **matrix.** Logically, more potential for poor sorting exists among the coarser sediments: there is a limit to how much grain size variation there can be among clay-sized fragments, for example, when all must by definition be smaller than 1/256 millimeter in diameter! Sorting in turn influences other properties of sediment, such as porosity, as noted in chapter

A

A

B

Figure 4.5

Well-sorted sediments. (A) Schematic: grains are similar in size. (B) Example of a well-sorted sandstone.

B

Figure 4.6

Poorly sorted sediments. (A) Schematic: sediment comprises a broad range of grain sizes. (B) Example of a sedimentary rock with poorly sorted fragments.

2. Sorting also provides clues to the mode of sediment transport and environment of deposition. For example, sediments deposited by melting ice are typically poorly sorted; those deposited by wind or flowing water are more often well sorted.

Chemical Sediments

The **chemical sediments** are those precipitated from solution. The rocks formed from them are named, for the most part, on the basis of composition, not grain size. Several major types are summarized in table 4.2. Volumetrically, the most important is **limestone**, a rock composed of carbonate minerals (figure 4.7). (Any rock consisting mostly of calcium and magnesium carbonate can properly be called limestone in the broad sense. However, the term is sometimes restricted to rocks consisting of calcite (calcium carbonate), while a rock made of the calcium-magnesium

Figure 4.7

Limestones, examples of chemical sedimentary rocks.

Table 4.2

Some chemical sedimentary rocks and constituent minerals

Rock type	Composed of
carbonate	
limestone	calcite, aragonite (both polymorphs of $CaCO_3$)
dolomite	dolomite ($CaMg(CO_3)_2$)
evaporite	variable—possibilities include halite (NaCl), gypsum (hydrated calcium sulfate, $CaSO_4 \cdot 2H_2O$), anhydrite ($CaSO_4$)
chert	silica (SiO_2)
banded iron formation	hematite, magnetite (iron oxides, Fe_2O_3, Fe_3O_4); usually with quartz or calcite

Figure 4.8

Rock salt, an evaporite.

Figure 4.9

Banded iron formation, an iron-rich sedimentary rock.

carbonate mineral **dolomite** may itself be called dolomite.) Carbonates may be precipitated from either fresh or salt waters. The oceans are an enormous reservoir of dissolved carbonate, and marine limestones hundreds of meters thick are known. Judging from observations of modern sedimentation, it appears that the majority of carbonate sediments are precipitated as calcium carbonate. Most dolomite is believed to form from later chemical changes brought about through the addition of magnesium dissolved in pore waters circulating through the sediment.

When salt water is trapped in a shallow sea and dries up, dissolved minerals precipitate out as an **evaporite.** Evaporites can contain a variety of minerals. Halite (sodium chloride) is an important one, since the oceans contain considerable dissolved salt (figure 4.8). Other important minerals in evaporites are the calcium sulfates gypsum and anhydrite.

Under other conditions, dissolved silica (SiO_2) may precipitate out of solution in a noncrystalline solid or even a watery gel. When solidified, such a deposit forms the rock *chert*. Some ancient sedimentary rocks are found in which layers of iron oxides alternate with layers of chert or carbonate. These *banded ironstones* (figure 4.9) are important iron ores today (see chapter 20). Clays may sometimes precipitate from solution; some mudstones, therefore, are chemical sediments also.

Biological Sediments

The **biological sediments** do not really constitute a distinct class of sediments except insofar as organisms are involved in their formation. Most are really chemical sediments. A particular mineral will be precipitated by an organism living in water, generally in order to build a shell or a skeleton. When the organism dies, and the soft body parts decay away, the mineral material remains. Coral reefs form carbonate rocks. In other cases, rocks are formed from sediments that consist of the remains of millions of microorganisms that have accumulated on the sea floor.

A

B

Figure 4.10

Microorganisms that make up many marine sediments.
(A) Radiolaria, which have silica skeletons. *(B)* Foraminifera, with calcium carbonate skeletons.

(A) D. L. Jones, courtesy U.S. Geological Survey. *(B)* K. O. Emery, courtesy U.S. Geological Survey.

Figure 4.11

A limestone rich in shell fragments.

Photo by E. B. Hardin, courtesy U.S. Geological Survey.

Cherts can be formed from the silica skeletons of diatoms and radiolarians (figure 4.10A). Skeletons of calcareous (calcite-secreting) microorganisms can accumulate into a sediment from which limestone will form (figure 4.10B). In the modern oceans, such biogenic carbonate sediments may be more common than inorganically precipitated ones. Less commonly, coarser biological debris such as carbonate shells will be transported and accumulated by wave action or currents in a deposit of carbonate sediment that could produce a clastic limestone (figure 4.11). Other biological sedimentary materials, which will be discussed in more detail in chapter 21, are **coal** and the related substances *peat* and *lignite*. All of these are formed from the remains of land plants, which have been converted through heat and pressure and accompanying chemical reactions to a carbon-rich solid that can be burned as fuel.

The organic matter in coal is not mineral material, for it is produced only through organic means. Thus by strict definition, coal might not be considered a rock. However, it is a cohesive solid that occurs in beds or layers within sedimentary rock sequences, shaped by many of the same sedimentary processes. Most coals also contain significant amounts of clastic sediment mixed in with the plant remains. Therefore, by convention, coal *is* regarded as a rock.

From Sediment into Rock: Diagenesis and Lithification

Consolidation is the transformation of loose sediments into a coherent sedimentary mass. In order to become rock, a sediment must be **lithified** (from the Greek *lithos,* "stone"). The whole set of processes by which this is accomplished is collectively termed **diagenesis.** Examples of diagenetic processes are given below. It should be kept in mind that diagenesis occurs at rather low temperatures and pressures. Higher temperatures and pressures would lead to metamorphism, as described in the next chapter, and the resultant rock would be a metamorphic rock. The distinction between diagenesis and weak metamorphism is not sharp, but by convention diagenetic processes are regarded as those occurring in the upper few kilometers of the crust, at temperatures below about 200° C.

Compaction

As sediments pile atop other sediments, the earliest ones become ever more deeply buried. The weight of sediments above puts pressure on those below. Sediment under pressure will tend to undergo **compaction** (figure 4.12). It will be squeezed more tightly together, porosity will decrease, pore water will be squeezed out, and the grains may begin to stick together. Permeability is commonly decreased also. Platy grains may align themselves perpendicular to the direction of principal stress (horizontally, if the pressure is mainly from the weight of rocks above). Such alignment is common in shales and accounts for their tendency to split preferentially in one direction, parallel to and between the flat mineral grains. Compaction alone is generally insufficient to make the sediment cohesive enough to qualify as rock, though it is more effective on fine-grained sediments than coarser ones. Cementation is almost always involved as well.

Cementation

Cementation, as the name suggests, is the sticking together of mineral grains by additional material between the grains. Most sediments are deposited in water and have fluid in the pore spaces between grains. That fluid in turn contains dissolved minerals. As pore fluids flow through buried sediment, silica, calcite, or other dissolved mineral materials may begin to precipitate on grain surfaces (figure 4.13). In time, enough of this intergranular material may precipitate

Figure 4.12

Compaction (schematic): grains are packed more tightly together, often during sediment burial.

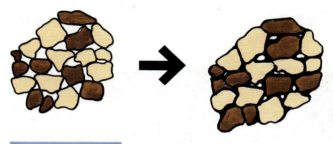

Figure 4.13

Cementation (schematic): precipitation of calcite between grains helps to stick them together, promoting lithification.

that it effectively glues the sediment together into a rigid, cohesive mass: a sedimentary rock. Like compaction, cementation also tends to decrease porosity, as cracks and pores are filled in by precipitating minerals. Compaction and cementation often occur simultaneously. However, any cementing material must be introduced into the rock before compaction has proceeded so far that the corresponding decrease in permeability severely restricts the flow of the cement-bearing fluid through the rock.

Other Diagenetic Changes

Burial subjects sediment not only to added pressure but to increased temperature, due to the increase in temperature with depth in the earth. The combination of increased pressure and temperature may bring about recrystallization of minerals that are stable only at very low temperatures. Pore waters flowing through sediment can dissolve soluble materials in addition to adding minerals as cement. Solution tends to increase the rock porosity, creating more space in the rock to be occupied by water, oil, or other fluids. This process thus has some practical as well as theoretical significance. Chemicals dissolved in circulating pore waters can also react with minerals in the sediment to form new minerals: *dolomitization,* the conversion of calcium carbonate to calcium-magnesium carbonate (mentioned earlier), would be an example.

Diagenesis thus comprises a variety of processes. Many of them change the texture or composition of the material and thus make it more difficult to draw inferences about the original sediment from the rock eventually produced.

Structures and Other Features of Sedimentary Rocks

Sedimentary rocks can contain various kinds of structures. Some are useful in identifying the setting in which the sediments were deposited; some indicate the direction of flow of the wind or water transporting the sediment; some help in determining which side was originally up in cases where the sedimentary rocks have been deformed or displaced after lithification. We will survey several important examples.

Structures Related to Bedding

Virtually all sediments, clastic or chemical, are deposited in layers. For many this layering, or **bedding,** is their most prominent characteristic. The bedding may be very fine (in which case it is often termed **lamination;** see figure 4.14), or a single bed may be many meters thick (figure 4.15). When sediments are deposited in water the bedding is almost always horizontal or nearly so.

Figure 4.14

Lamination, very fine bedding.

Figure 4.15

Bedding on a grand scale.

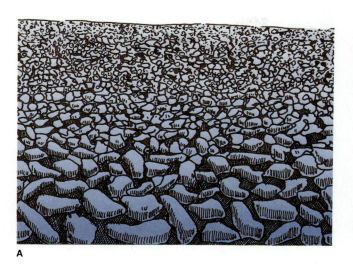

A

B

Figure 4.16

Graded bedding. *(A)* Schematic: graded bedding becoming finer near the top. *(B)* Example of a graded bed.

Bedding does not always take the form of a stack of distinct layers, each of which is internally homogeneous. Particularly in clastic sediments, **graded bedding** may be observed. A graded bed is one in which grain size grades from coarse to fine, or vice versa, vertically within the bed (figure 4.16). Most commonly, the grading goes from coarse at the bottom of the bed to fine at the top. Graded bedding may develop if a mass of poorly sorted sediment is suspended in water suddenly, as for example by a submarine landslide churning up gravel, sand, and silt. The coarser particles will tend to settle fastest, while very fine material may remain suspended for some time and settle out only slowly. (The process can be demonstrated by stirring a handful of poorly sorted sediment or soil into a large jar of water, then letting it settle.) Sediment transport by wind and water is also related to velocity of flow, in that the faster these agents flow, the larger (coarser) the sediments they can move. A sediment-laden current of water that gradually slowed down would drop the coarsest sediments first, then successively finer ones.

Nor is bedding always horizontal. Wind or water flowing across sloping surfaces during sediment deposition will drape layers of sediment over the existing topography. This inclined bedding caused by the currents is known as **cross-bedding** (figure 4.17).

Should the direction of flow change, so will the orientation of subsequent cross-beds. In lithified desert sands, in which many windblown dunes were superimposed one atop another, or in stream sediments subjected to many changes of flow pattern through time, the cross-bedding can become very complex (figure 4.17C).

Surface Features

The rhythmic motion of waves or flow of currents also produces surface topography on fine- to medium-grained sediments in the form of **ripple marks** (figure 4.18). The geometry of ripple marks depends on the mode of formation. Ripples that are *symmetric* in cross-section (figure 4.18B) are produced by the alternating back-and-forth motion of water near shore. They are subjected to an equal flow of water from each side, hence the symmetry. Sediments on a lake bottom near shore might show symmetric ripple marks. Ripples produced by a consistent flow of wind or water in one direction will be *asymmetric* in cross-section, flattened on the side from which the current flows, steeper on the downcurrent side. Asymmetric ripples thus not only indicate deposition in the presence of flowing currents but also indicate the direction of flow.

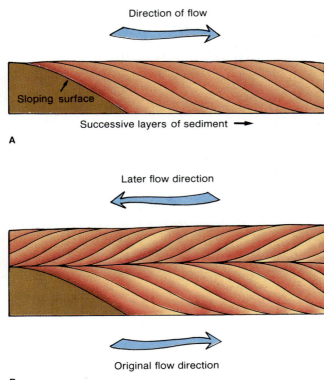

Direction of flow

Sloping surface

Successive layers of sediment →

A

Later flow direction

Original flow direction

B

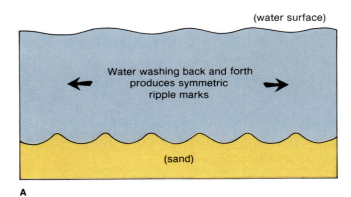

(water surface)

Water washing back and forth produces symmetric ripple marks

(sand)

A

Water flow in one direction produces asymmetric ripples

B

C

Figure 4.17

Cross-bedding. *(A)* Schematic: formation of cross-beds.
(B) Change in current direction results in different orientation of cross-beds. *(C)* Example of cross-bedded sandstone: the Navajo sandstone of the southwestern United States.

C

Figure 4.18

Ripple marks. *(A)* Symmetric ripple marks, formed by water washing back and forth. *(B)* Asymmetric ripple marks, formed by currents flowing consistently in one direction. *(C)* Example of ripple marks in shale.

Figure 4.19
Mud cracks forming in modern clay-rich sediment.

Figure 4.20
Fern fossil preserved as carbon film.
Photo by E. D. McKee, courtesy U.S. Geological Survey.

They are the usual form of ripple found in stream beds. The reasons for their shape will also be explored further in the chapter on wind.

When fine-grained, water-laid sediment dries out, it often shrinks and forms polygonal mud cracks on the surface (figure 4.19). These too may be preserved during lithification. They are formed particularly in finer-grained sediments for several reasons. For one, fine sediments often pack much less tightly during deposition than do sands and gravels (a mud may be 50% to 60% water), so there is a dramatic decrease in volume as the water is removed. Clay minerals are common in fine-grained sediments, too. Many clays have the capacity to absorb water when wet and release it when dried, expanding and contracting in the process. If such clays are deposited and then baked dry by exposure to air and sun, they will shrink and the drying mud will crack as they do so. Mud cracks then indicate two things about the conditions of sedimentation: that the sediments were initially deposited in water and that they were subsequently dried out by exposure to air before lithification. Suitable conditions exist in several environments, including tidal flats and on the bottoms of lakes that dry up as a result of changing climatic conditions.

Figure 4.21
Trilobite with original chitinous skeleton preserved.

Other Features of Sedimentary Rocks

From sedimentary rocks comes nearly all our knowledge of earlier life forms, through fossils. Fossils are the remains or evidence of ancient life. The fossil record is an incomplete one, for not all organisms, or all parts of each organism, are equally well preserved. Soft tissues are subject to decay before lithification or destruction during diagenesis. Usually, they are best preserved as carbon films on a rock surface (figure 4.20). Hard parts like shells, skeletons, and teeth fare better. They may be preserved intact (figure 4.21).

A

Figure 4.23
Examples of molds.

B

Figure 4.22
Fossils preserved by replacement. *(A)* Mineralized logs formed by silica replacement: Petrified Forest National Monument. *(B)* An ancient relative of the scallop, with shell replaced by pyrite (FeS_2).

A

B

Figure 4.24
Oolites. *(A)* Schematic cross section. *(B)* Photograph of oolitic limestone.

More often, they undergo recrystallization during diagenesis (which may destroy very fine structural details), or **replacement**, in which minerals introduced in solution in pore waters replace original organic material as it dissolves or decays away (figure 4.22). If the sediment begins to lithify around the remains before they decay or are crushed by overlying sediment, a **mold** of the object may be formed, which can survive even if the original organic matter then decays or dissolves (figure 4.23). If another sediment later fills the mold and itself lithifies, a matching **cast** is formed. Actual organisms are not the only fossils

preserved in this way. It is the most common mode of preservation of many **trace fossils**—for example, animal tracks and worm burrows.

A much less common but quite distinctive feature of some carbonate sediments is **oolites** (figure 4.24). These are spheroidal objects, typically the size of coarse sand grains, formed by precipitation of layer after layer of calcium carbonate around a core of mineral or shell fragment. Their formation is somewhat analogous to that of pearls, but oolites are completely inorganic. The spheroidal shape is a result of the setting in which they form, on shallow-water carbonate platforms (see next section) where they are rolled around by the water and constantly stirred by waves and currents, becoming evenly coated on all sides. Where present, they are valuable indicators of sedimentary environment.

Sedimentary Environments

The characteristics of a sedimentary rock are plainly determined in large measure by the depositional setting of the sediments. For clastic rocks, the nature of and distance from the source of the sediment are also factors. In this section we will survey briefly some of the common sedimentary environments and principal sediment characteristics of each.

Chemical Sediments

Chemical sedimentation inevitably involves water, so all depositional environments of chemical sediments are under, or close to, water. They may either be shallowly submerged or take the form of basins.

Shallow-water environments in warm waters account for most carbonate sedimentation. Unlike most chemicals, calcite dissolves more readily in cold water than in warm; in other words, it is more readily precipitated from warmer waters. Off the coasts of some continents are broad, shallow shelves; in near-equatorial regions, the waters over these shelves are quite warm. In such areas, *carbonate platforms* may be built up, by direct precipitation of carbonate muds or by the accumulation of shell or skeletal fragments of carbonate-secreting organisms that thrive under the same conditions. Reefs, likewise made of calcium carbonate, are also common in this environment.

Restricted warm basins into which seawater can flow and from which it then evaporates provide the appropriate setting for deposition of evaporites. Such basins once existed over much of North America (recall figure 1.10). Drilling into sedimentary rocks under the Mediterranean Sea likewise suggests that at one time water flow into and out of it was much more limited than it now is, so that seawater would flow in, be trapped, warmed, and evaporated, and leave a salt deposit behind. The thickness of many evaporite deposits is far greater than could be accounted for by taking a single basinful of seawater, even one as deep as the modern ocean, and drying it up. Apparently,

GREAT SALT LAKE

thick evaporite beds form through the cumulative effect of repeated cycles of basin filling and evaporation.

Deep, cold basins of the sea floor may be sites of predominantly chemical sedimentation, if they are sufficiently far removed from clastic input from the continents. Deep bottom waters are too cold for the preservation of calcium carbonate, but they are suitable for the preservation of silica. Biochemical silica-rich muds accumulate as the siliceous skeletons of diatoms and radiolarians settle to the sea floor.

Clastic Sedimentary Environments

Deposition of clastic sediment in the absence of significant water occurs principally in deserts. Here the sediment is found in windblown dunes, usually of sand-sized particles. Finer materials are commonly carried away, to be deposited by other means in other settings.

Rivers provide other depositional environments on the continents. Characteristics of stream-deposited sediments will be explored in greater detail in chapter 13. Here we can note that they commonly consist of the coarser clastics, sand and gravel with relatively little mud (the fine size fractions often stay in suspension in the flowing water). Deposits of stream sediments are also characteristically elongated in shape and restricted in areal extent, which helps to distinguish them from similar materials deposited in large basins. Asymmetric ripple marks can also indicate the direction of stream flow.

Where land and sea meet, several kinds of sedimentation are possible. Streams deposit at their mouths large fan-shaped *deltas* of continentally derived sands and muds that interfinger or mix with marine sediments. The high-energy *beach* environment, with active waves and tides, is characterized by well-sorted and generally well-rounded sands and gravels: the vigorous water action abrades and rounds the sediment grains while washing away the finest size fractions. The more placid setting of a *tidal flat* may provide a place for fine silt and clay to settle out in muddy deposits, which may develop mud cracks.

Further offshore from beach and tidal flat but still close to the continent, the character of sediment is rather like that of a stream delta—a mix of fine- to medium-grained continental clastics with marine sediment, perhaps including carbonate or shells of marine organisms—but it is spread by currents flowing along the coast into a more sheetlike deposit. Only the finest clastics are carried far offshore, either in the water by currents or from the continents by winds. Clastic sediments of the deep sea floor are finely laminated clays that accumulate very slowly. Marine sedimentation will be discussed further in chapter 12.

Box 4.1

Some Unusual Sedimentary Environments: Dripstone and Its Relatives

While chemical sedimentary rocks are, in terms of volume, overwhelmingly marine in origin, deposition of dissolved minerals from fresh water is very common also. Much of it occurs during diagenesis and cementation of sediments and is therefore not very obvious. It does not, in that case, produce entire new rocks. Only occasionally do freshwater chemical sediments achieve prominence.

One example is the case of _dripstone_ (figure 1). Where abundant subsurface water seeps through limestone rocks, large underground caverns may be dissolved out of those rocks. Later, calcite-bearing solutions dripping from cracks in the roofs of such caverns, depositing their dissolved mineral load, build the spectacular stalactites and stalagmites that make Carlsbad and other caverns famous tourist attractions.

Other chemical sedimentary rocks are formed by the hot waters of areas having hot springs and geysers. The two most abundant dissolved mineral materials in those waters are calcite and silica. Deposited little by little as the warmed waters escape and evaporate at the surface, they can build fantastic structures (figures 2, 3).

Figure 1
Formations of dripstone.

Figure 2
Mammoth Terraces in Yellowstone National Park, built of calcite.

Figure 3
Castle Geyser, Yellowstone Park, with structure built of silica.

Figure 4.25

Grain size decreases with increasing distance from shore in clastic sediments.

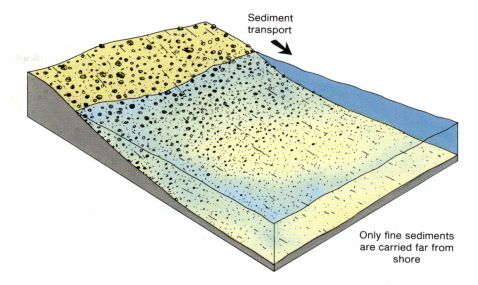

Sediment transport

Only fine sediments are carried far from shore

Other Factors Influencing Sediment Character

We have noted the effect of water temperature on the abundance of carbonate in marine sediments. There are additional factors that affect the composition and/or texture of sediments, particularly clastics.

Nature (Composition) of Source Rock

The following point pertains to clastic sediments and is perhaps self-evident: the underlying control on the composition of a clastic sediment is the composition of the material from which it is derived. If the source region consists of granitic rock, there will not be olivine grains or fragments of limestone in the resultant sediment. If the source region is basaltic, the sediment will not be rich in quartz grains. The same basic constraint will hold true for soil (chapter 18).

Distance from Source

The distance from the source is relevant to water-transported clastic sediments; it influences both the texture and the composition of the sediment. In terms of grain size, all else being equal, the farther from the source the rock and mineral fragments have traveled, the finer the fragments (figure 4.25). The longer the distance traveled, the more opportunity for grains to be partially dissolved, broken, or abraded, and therefore the smaller they get.

Distance from source will also influence the **chemical** or **mineralogical maturity** of the sediment. Not all minerals dissolve or undergo chemical reactions equally readily. The longer the distance traveled, the greater the proportion of the easily dissolved or easily altered material that will be eliminated. The residue will become more and more enriched in so-called resistant minerals. For example, most ferromagnesians weather easily; quartz, in most climates, is fairly resistant to both chemical and physical breakdown. A quartz-bearing sediment transported a long way in a stream channel is likely to become more and more quartz-rich along the way.

These principles are far less valid for clastics transported by wind or by ice. Wind generally moves only very fine material anyway, so further breakup in transit is not readily detectable; and wind does not attack minerals chemically. Materials frozen into ice are subject to far less abrasion or contact with other rock fragments than they would be traveling in water, so further physical breakup is minimized. Ice, too, tends not to react chemically with minerals.

Effect of Climate

As will be explored further in chapter 18, both temperature and moisture influence the extent of chemical reactions that break down rocks at the surface. Briefly, most such reactions involve water, whether the mineral dissolves completely or just reacts with other chemicals already dissolved. Therefore the more rainfall, the more chemical breakdown. Most chemical reactions also proceed more rapidly at high temperatures than at low, so warm climates favor rapid chemical breakdown. Sediments in warm, moist climates will approach maturity—in which the sediment is rich in resistant minerals like quartz—more rapidly than sediment under cold, dry conditions. In the latter case, even easily weathered minerals like olivine and calcic plagioclase may persist for some time.

Figure 4.26

Facies changes at a shoreline (map view).

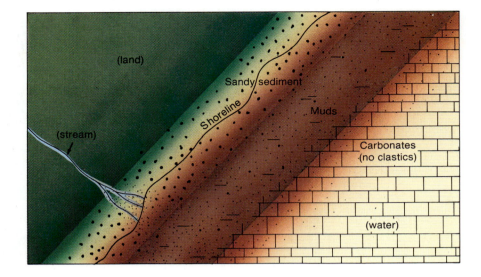

Effect of Topography

Topography principally affects the relative rates of sediment transport and chemical breakdown. That is, the steeper or more rugged the terrain, the more gravity will work with wind, water, or ice to move sediment quickly, allowing less time for chemical breakdown to occur. The transportation itself may also be more vigorous, increasing mechanical breakup of rock and mineral fragments. In very flat terrain, sediment transport is likely to be much slower, unless there are unusually strong winds or fast-flowing streams. The sediment is thus exposed to prolonged chemical breakdown. Sediments in rugged terrain, then, will be mineralogically immature, preserving even readily weathered minerals, but are more likely to be rapidly broken up mechanically. Sediments in gently sloping terrain may suffer only slow breakup but considerable chemical attack and will tend to be more mature mineralogically.

The Facies Concept: Variability within Rock Units

Implicit in the foregoing discussion is the idea that a single rock unit shows a single set of physical and chemical characteristics. This is in fact not necessarily true, especially for units of large areal extent. A single mass of sedimentary material deposited during one time interval may be spatially continuous over a broad area but vary greatly in character (composition, texture) locally.

Sedimentary Facies

Consider the example shown in figure 4.26, which illustrates deposition along a shoreline to which sediment is supplied by streams. Near the beach, the sediment is mainly sandy. Finer, muddy sediments are carried further offshore before they settle to the bottom. The quantity of continentally derived sediment generally decreases with increasing distance from shore, until it becomes insignificant, and sedimentation takes the form of calcite precipitation from seawater. After diagenesis and lithification, these sediments would form several quite distinct rock types.

Yet they are all, in a sense, part of the same sedimentary package. The original sediments were deposited simultaneously in a continuum of sedimentary environments. There will be no sharp demarcation between the limestone, mudstone, and sandstone formed—each will grade into the next. For instance, if these rocks were later uplifted and exposed, and one walked from the mudstone to the limestone, sampling along the way, the nature of the rocks would change, gradually, from a mudstone with a little carbonate, through mixed rocks with similar proportions of each, to limestone with a little silt or clay in it, to perhaps nearly pure limestone.

The different rock types above represent different *facies* of the same basic unit. The term **facies** is used to describe collectively a set of conditions that lead to a particular type of rock (or sediment), or to distinguish that portion of the rock itself that represents a distinct set of conditions. (For clarity, the term **lithofacies,** literally "rock facies," is sometimes used when it is the rocks corresponding to a particular facies, or set of conditions, that are meant.) The hypothetical sedimentary rock unit described could be said to have a sandstone facies, a shale or mudstone facies, and a limestone facies. The sediments that went

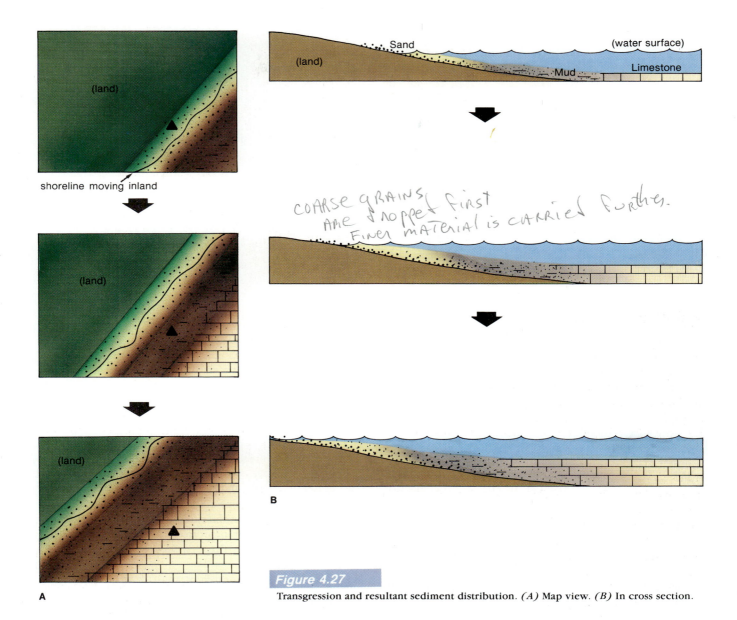

Figure 4.27

Transgression and resultant sediment distribution. *(A)* Map view. *(B)* In cross section.

[handwritten note in figure: COARSE GRAINS ARE DROPPED FIRST. FINER MATERIAL IS CARRIED FURTHER.]

into the rocks of the shale facies would be those deposited in a setting that favors deposition of muds—here, one with some clastic input but far enough from source and shore to exclude sandy or coarser grains, with water currents gentle enough not to sweep the muds away, and an abundant contribution of clastics relative to precipitated carbonate.

The facies phenomenon makes interpretation of the rock record more difficult. Where rocks are very well exposed at the surface, and one can walk through the facies changes, it is fairly easy to recognize what is, or was, going on. Where rocks are poorly exposed, so that one finds here an outcrop of shale, there another of limestone, it is more difficult to know whether the two are different facies of the same unit, equivalent in age and deposited as part of the same overall depositional setting, or wholly unrelated rocks.

Facies and Shifting Shorelines

Most shoreline sedimentation is characterized by some sequence of sedimentary facies that change in the direction perpendicular to the shoreline. Therefore if the position of the shoreline changes, later deposition may lay down sediments of different facies atop earlier sediments in a given spot. Various events, including a worldwide rise and fall of sea level or localized rise and fall of the edge of a continent, can cause the shoreline to shift landward or seaward. For example, if worldwide sea level rises (as is happening now as polar ice melts), the effect is that shorelines move landward.

The impact on distribution of sedimentary facies in map view and in crosssection is shown in figure 4.27. As the sea encroaches on the land, each facies effectively moves inland too: beach sands are deposited at the new shore, which previously was dry land; muds overlie the earlier beach; limestone is deposited over older muds. The process of the ocean advancing over what was dry land is termed a **transgression** of the sea.

Now imagine this whole region buried by later events and the sediments lithified. If one were to drill a hole through the resultant rock at point ▲ in figure 4.27, the sequence of rocks would appear as in figure 4.28: sandstone on the bottom, grading upward into mud and then limestone. This collection of rock types is a typical *transgressive sequence* reflecting deepening water and changing sedimentation patterns at ▲ as the shoreline migrates inland. The reverse process, **regression,** is reflected in a correspondingly inverted sequence of rock types.

Paleogeographic Reconstruction

In chapter 1, we looked at several maps that presented the distribution of land and sea across North America at various times over the past half-billion years (figure 1.10). Such **paleogeographic maps,** maps of ancient geography, are constructed predominantly through the interpretation of sedimentary rocks, because it is the sedimentary rocks, formed at the surface from materials exposed at the surface, that supply the most relevant information. Frequently the paleogeography of a region will bear little resemblance to its present geography: sandstone made from desert sands may be found in the rock record of a region now temperate and well vegetated; remains of ancient tropical soils may be found at now-cold latitudes; deep-water marine sediments can be found high in mountain ranges. Interpreting a single rock unit and variations in its character over a large area can provide considerable information about sources and depositional settings of sediment.

Paleocurrent Directions

We have noted above that the grain size of clastic sediments tends to become finer with increasing distance from the sediments' source. Another commonly used interpretive tool is the orientation of **paleocurrents** (ancient currents, those prevailing at the time the

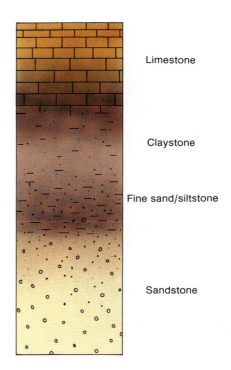

Limestone

Claystone

Fine sand/siltstone

Sandstone

Figure 4.28

Transgressive sequence in a rock column.

sediments were deposited). Paleocurrent directions are deduced from sedimentary structures, including asymmetric ripple marks and crossbeds. Since local eddies in wind or water can cause anomalous orientations in single samples, many measurements of apparent paleocurrent directions are made in a single rock unit, and often the results are interpreted with the aid of statistics.

A sample paleocurrent map is shown in figure 4.29. The rocks are a varied sequence of sandstone, shale, and limestone of broadly similar age. In general, the apparent direction of current flow is away from the central area of the map where rocks of this age are absent (solid shading). The absence of these sedimentary rocks, by itself, could mean either that no deposition was occurring in the shaded area at the time, or that sediments were deposited but at some later time they (or the rocks formed from them) were eroded away. Taken together with paleocurrent evidence, the pattern of occurrence of these sedimentary rocks is consistent with the idea that the shaded areas were topographic highs at the time of sediment deposition. The paleocurrents could indicate that the sediments were deposited on slopes slanting away from those highs.

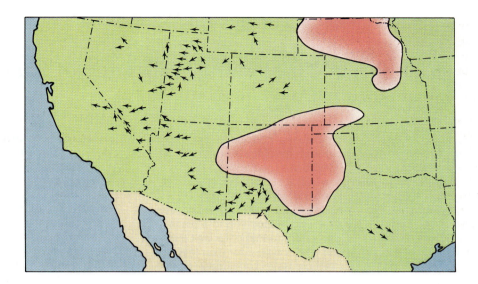

Figure 4.29

Paleocurrent directions in some sedimentary rocks from the southwestern United States. Arrow shows direction of flow.

Source: After Stewart, et al., 1984, U.S. Geological Survey Professional Paper 1309.

Determining Source Regions by Composition

We have already noted that the nature of the rocks of the source region is a fundamental control on the composition of clastic sediments. The principle can be turned around: the composition of clastic sediments may be used to infer the source rocks, within limits, or to differentiate among several possible source regions. If a sedimentary basin is adjacent to several quite different possible source regions—for example, one consisting mainly of granite, one of basalt, and one of limestones and shales—it may be possible to decide which one(s) contributed to the sediments. Abundant coarse quartz and feldspar grains would suggest the granite as source; many fragments of limestone in the sediments would suggest the sedimentary source area; and so forth. If the possible source areas are quite similar compositionally—for instance, all basically composed of granitic rock—then it would be necessary to look at more subtle details of the chemistry and mineralogy of the sediments and possible sources, perhaps in combination with other evidence like paleocurrent indicators.

Paleoclimate

The **paleoclimate,** or ancient climate, can frequently be deduced at least in a general way from sediment character. A chemically mature clastic sediment in which fragments are still quite angular (which have therefore not been transported far) would suggest the rapid chemical weathering of a warm, wet climate. Conversely, a very fine-grained sediment with well-rounded fragments that still contained easily weathered minerals would indicate a colder, drier climate. Carbonate platform sediments suggest warm, shallow seas. Latitude is a primary control on earth surface temperatures, so climatic inferences may be used to deduce the approximate latitudes at which the sediments were originally deposited. They also aid in the recognition of global warming and cooling trends. The presence and types of fossils, if any, may also indicate climate and even, in the case of water-laid sediments, water depth.

Box 4.2

What's in a Rock Pile? Names and Significance of Some Sedimentary Rocks in the Grand Canyon.

Especially when the geology of a region is dominated by layer upon layer of sedimentary rock (as is the Grand Canyon, for example—see figure 1), it is convenient to represent the sequence of layers diagrammatically in a **column**, or **section**. A portion of the Grand Canyon section is shown in figure 2. By convention, different rock types are represented by different symbols: horizontal dashes for shale, dots for sandstone, bricks for limestone, and so on. These distinctive, recognizable units are then named.

The basic unit is the **formation,** a rock unit representing deposition under a uniform set of conditions and at one time, producing a recognizable, mappable unit. A formation need not all be one rock type, but it should reflect a predominance of some particular depositional conditions. Boundaries between formations are drawn where definite changes in depositional conditions can be identified. Formation names as presently assigned have two parts: a locality name, corresponding to a place where that formation is well exposed, and a rock type—for example, "Temple Butte Limestone." (In the past, a prominent characteristic of the rock might have been substituted for a locality—for example, "Redwall Limestone," named for its color and tendency to form vertical cliffs—but this is not correct modern practice.) If the rock types are so varied within a formation that no one type dominates in the section, the word "Formation" may be substituted, as in "Supai Formation."

Within a single formation, varying rock types that can themselves be recognized and mapped may be distinguished. These rock units are termed **members.** Several members are identified in the Redwall Limestone: the Whitmore Wash member (dolomite), the Thunder Springs member (interlayered chert and dolomite), and the Horseshoe Mesa member (limestone with chert). Note the use of localities in the naming of members also. The boundary between the Redwall Limestone and the Supai Formation is put at the change in rock type from carbonate to mostly shale, which reflects an increase in clastic input.

Sometimes, but not always, formations genetically related in some way will be described together as a **group,** but many formations belong to no higher-order unit.

Figure 1
The layered sedimentary rocks of the Grand Canyon.

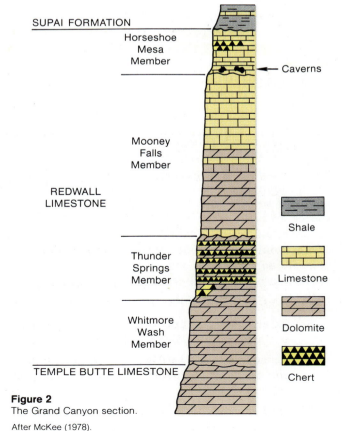

Figure 2
The Grand Canyon section.
After McKee (1978).

Summary

Sedimentary rocks are those formed at low temperatures, near the earth's surface, from sediments. Sediments may be either clastic, formed from mechanical breakup of pre-existing rocks, or chemical, precipitated from solution. The clastic sedimentary rocks are classified principally on the basis of grain size. Chemical sedimentary rocks are subdivided according to composition. Sediments are lithified through various diagenetic processes, including compaction, cementation, and recrystallization. Clastic sedimentary rocks, in particular, may show structures that reflect the conditions of sedimentation—graded bedding, cross-bedding, ripple marks, mud cracks, and the like. Sediments of all kinds may contain the fossilized remains of ancient creatures, preserved intact, recrystallized, replaced, or found as molds and casts.

Different sedimentary environments are conducive to deposition of sediments of different composition, texture, and structure. Individual sedimentary environments are not necessarily sharply bounded but often grade laterally into one another. A single rock unit deposited at one time but over a large area may thus show several different facies, different characteristics indicative of the varying depositional settings from place to place. Reconstruction of the paleogeography of a region may be based on inferences about the depositional environment(s), composition of the sources of clastic sediments, distance from source as indicated by grain size and rounding, paleocurrents, and paleoclimate.

Terms to Remember

bedding	cross-bedding	lithification	paleogeographic maps
biological sediments	diagenesis	lithofacies	regression
breccia	dolomite	matrix	replacement
cast	evaporite	member	ripple marks
cementation	facies	mineralogical maturity	sandstone
chemical maturity	formation	mold	section
chemical sediment	fossils	mud cracks	sediment
clastic	graded bedding	mudstone	shale
coal	group	oolites	sorting
column	lamination	paleoclimate	trace fossils
compaction	limestone	paleocurrents	transgression
conglomerate			

Questions for Review

1. On what basis are clastic sediments subdivided and named?

2. What is meant by saying that a sediment is well sorted? Well rounded?

3. Name and describe two kinds of chemical sediments.

4. Discuss how compaction and cementation contribute to lithification of sediments.

5. Explain how the porosity and permeability of sediments are modified as they are lithified.

6. Describe *graded bedding* and *cross-bedding* and briefly explain one way in which each can arise.

7. Mud cracks form most readily in finer-grained sediments. Why?

8. Give two examples of trace fossils.

9. Compare the kinds of sediments you would expect in a beach environment and in a tidal flat.

10. How are sediments modified as they are transported farther and farther from their source region?

11. What is a *facies*? How does the existence of sedimentary facies complicate the identification or mapping of individual rock units?

12. What is a *transgression*? Describe how this might be recognized in the rock record.

13. Cite and briefly describe two possible paleocurrent indicators.

For Further Thought

1. Many organisms that we now know only as fossils lived over a limited span of geologic time. Consider how this fact might be used to recognize different facies of the same rock unit in an area in which the rocks are generally not well or continuously exposed.

2. In which of the following sediments might you hope to find dinosaur footprints? Explain. (a) a coarse sandstone with pronounced crossbeds; (b) a mudstone with symmetric ripple marks; (c) an oolitic limestone.

Suggestions for Further Reading

Blatt, H.; Middleton, G.; and Murray, R. *Origin of Sedimentary Rocks.* 2d ed. New York: Prentice-Hall, 1980.

Collinson, J. D. *Sedimentary Structures.* Boston: George Allen and Unwin, 1982.

Folk, R. L. *Petrology of Sedimentary Rocks.* Austin, Tex.: Hemphill, 1974.

Freeman, T. J. *Field Guide to Layered Rocks.* Boston, Mass.: Houghton-Mifflin, 1971.

Garrels, R. M., and Mackenzie, F. T. *Evolution of Sedimentary Rocks.* New York: W. W. Norton and Co. 1971.

Laporte, L. F. *Ancient Environments.* Englewood Cliffs, N.J.: Prentice-Hall, 1968.

McKee, E. D. "Paleozoic Rocks of the Grand Canyon." In *Geology of the Grand Canyon,* edited by W. J. Breed and E. Roat, 42–64. Prescott, Ariz.: Classic Printers, 1978.

Reading, H. G., ed. *Sedimentary Environments and Facies.* New York: Elsevier, 1978.

Reineck, H. E., and Singh, I. B. *Depositional Sedimentary Environments.* New York: Springer-Verlag, 1975.

Selley, R. C. *An Introduction to Sedimentology.* New York: Academic Press, 1982.

Stewart, J. H.; McMenamin, M. A. S.; and Morales-Ramirez, J. M. *Upper Proterozoic and Cambrian Rocks in the Cabora Region, Sonora, Mexico.* U.S. Geological Survey Professional Paper 1309, 1984.

Metamorphic Rocks

Introduction

The term **metamorphism** is derived from the Latin for "change of form." It has parallels in other sciences. Biologists use *metamorphosis* to describe collectively the complex changes by which larvae become butterflies and moths. Geologic metamorphism is likewise a process of change—but in rocks: change in the composition, mineralogy, texture, or structure of a rock by which it is transformed into a distinctly different rock. In this chapter we will survey some of the causes and consequences of metamorphism, describe situations that lead to metamorphism, and define some common metamorphic rock types.

Factors in Metamorphism

Rocks are changed when the physical or chemical conditions to which they are subjected change. The two most common agents of metamorphism are heat and pressure. Each mineral is stable only within certain limits of pressure and temperature. Held for long enough under conditions in which it is unstable, a mineral will change to a more stable form. How long is long enough? That depends on the particular mineral and physical conditions. See also box 5.1.

Temperature in Metamorphism

Metamorphic temperatures are limited at the low end by diagenesis, at the high end by melting. We have already noted that there is no sharp division between diagenesis and metamorphism (geologists disagree on how to define the boundary), but it generally corresponds to temperatures in the range of 100° to 200° C (200° to 400° F). The variability at the other end of the metamorphic temperature range is much greater, because the temperature at which a rock melts depends on many factors, including its composition, the prevailing pressure, and the presence or absence of fluids. Typical maximum metamorphic temperatures that most rocks can sustain without melting are in the 700° to 800° C (1300° to 1500° F) range, but dry mafic rocks rich in high-melting-temperature ferromagnesians can be heated to more than 1000° C (1850° F) without melting. Overall, the prevailing temperature influences not only the stability of various minerals but also the rates of chemical reactions, which generally proceed faster at high temperatures.

There are two principal sources of heat for metamorphism. One is the normal increase in temperature with depth already described, the **geothermal gradient.** This is a moderate effect in most

places. Typical geothermal gradients are about 30° C per kilometer, or close to 100° F per mile of depth. Local geothermal gradients can be increased by plutonism, emplacement of hot magma into the crust. Also, rocks very close to a pluton can be raised to near-magmatic temperatures directly. Plutonic activity, then, is a second source of heat for metamorphism.

Pressure in Metamorphism

Pressure takes two forms. Rocks in the earth are all subject to a **confining pressure,** an ambient pressure equal in all directions that is imposed by the surrounding rocks, all of which are under pressure from other rocks above. Confining pressure generally increases with depth. In addition, rocks may or may not be subject to **directed stress,** which, as the name implies, is not uniform in all directions. The compressive, tensile, and shear stresses described in chapter 2 are all examples of directed stresses.

The distinction can be seen by analogy with a person standing on the earth's surface. The body experiences a confining pressure from the surrounding atmosphere (about 14.7 pounds per square inch of body surface at sea level). Atmospheric pressure acts equally all around the body. (If it did not, the body would be compressed in the directions from which higher pressures were exerted, stretched in others.) The feet are subjected in addition to a vertical compressive stress applied by the downward pull of gravity on the mass of the body, and typically the feet are somewhat flattened in response.

Figure 5.1

Metamorphic rocks exposed in a quarry in central Wisconsin.

Box 5.1

Are Diamonds Forever?

We can use metamorphic rocks to investigate pressure and temperature conditions that existed in the crust when the rocks were metamorphosed. Many metamorphic minerals are, in fact, not truly stable at surface conditions. Nor are many other minerals, like diamond, which are formed at the still higher pressures and temperatures of the mantle. Carbon's stable mineralogical form at low pressure is graphite. When rocks containing such minerals are brought to the surface, therefore, why don't the now unstable minerals all break down or change into low-temperature, low-pressure minerals?

The main reason is lack of time. Both chemical reactions and structural changes like recrystallization proceed in solid materials quickly at high temperatures, very much more slowly at low temperatures. Even over millions of years, some technically unstable minerals will be preserved as a result of slow reaction rates. We can see the influence of time in the case of volcanic glasses. Glass is inherently unstable; the orderly atomic arrangements of crystals are more stable. Over periods of thousands or

Figure 1
The irregular whitish patches in this sample of volcanic glass are slowly crystallizing with time.

millions of years, volcanic glasses will slowly *devitrify,* or become nonglassy (crystalline) (see figure 1). We find no glasses at all among rocks billions of years old.

So no, diamonds are not forever. But neither will the diamonds in our watches and jewelry turn to graphite for many many more human generations.

Directed stress in rocks frequently arises in connection with mountain-building processes, when rocks are squeezed, crumpled, and stretched. Details of these *tectonic* processes will be explored further in chapters 7 and 11. Directed stress can also develop on a smaller scale, as for example when magma is forcibly injected into cracks in rocks during igneous intrusion. The common result of directed stress is deformation (figure 5.1).

Chemistry and Metamorphism

A third type of factor that enters into some metamorphic situations is a change in chemical conditions in a rock. If rocks are fairly permeable, fluids will flow readily through them, and either the fluids themselves, or one or more elements dissolved in them, may react with existing rock to produce new minerals. An example of this is when warm fluid associated with magma migrates away from the silicate melt, carrying dissolved gases and other elements. **Hydrothermal activity, involving the action of hot water**, is common in the country rock around igneous intrusions. Gases may pass through or escape from permeable rocks. Even in the absence of fluids, chemical elements can migrate through rocks, but they do so much more slowly.

The Role of Fluids

As noted above, the presence of fluids facilitates migration of chemical elements through rocks and thus makes possible changes in bulk rock chemistry and mineralogy. For the same reason it also facilitates mineralogical changes even if the rock as a whole behaves as a closed system, with no gain or loss of elements. Elements can migrate, or diffuse, through rock more readily in the presence of fluid, so atoms can more readily rearrange themselves into new minerals in the presence of fluids. The hydroxide minerals and hydrous silicates, those containing (OH^-) groups, require the presence of some water (H_2O) even to form or remain stable. Finally, as noted in chapter 3, wet rocks melt at lower temperatures than dry rocks, so the presence of fluid has some bearing on how high temperatures can get before the line between metamorphism and melting is crossed.

Effects of Metamorphism

The details of the changes caused by metamorphism depend both on the chemical composition and mineralogy of the **parent rock,** the starting material, and on the specific pressure, temperature, and other conditions of metamorphism, such as whether fluids or dissolved chemicals are entering or leaving the system. However, some generalizations are possible.

Effects of Increased Temperature

As temperatures are increased, minerals that are stable at low temperatures react or break down to form high-temperature minerals. A common kind of reaction is one in which a hydrous mineral breaks down. Recall from Bowen's reaction series that the common hydrous silicates, amphiboles and micas, form at moderate to low temperatures. At higher temperatures,

these minerals become unstable. For example, amphibole will break down to form pyroxene and release water; quartz is also produced, balancing the chemical equation:

$$(Mg,Fe)_7Si_8O_{22}(OH)_2 = 7(Mg,Fe)SiO_3 + SiO_2 + H_2O$$
amphibole   pyroxene   quartz   water

At even lower temperatures, the hydrous clay minerals in sedimentary rocks will break down to form more stable micas. If temperatures are increased further, the micas will in turn break down (to feldspar and other minerals, plus water). Certain nonsilicate mineral groups may break down in analogous fashion, and they are not all hydrous minerals. For instance, calcite, a carbonate, may break down with release of carbon dioxide gas (CO_2). Examples of some possible metamorphic reactions are shown in table 5.1.

Effects of Increased Pressure

The effect of increased pressure depends in part on whether or not the stress is directed. In general, increased confining pressure favors denser minerals, those that pack more mass into less space. Garnet, for example, is a very common metamorphic mineral in moderate- to high-pressure situations (figure 5.2). Iron-magnesium garnets are dense even by comparison with other ferromagnesian silicates. While the specific gravities of various pyroxenes and amphiboles are 3.2 to 3.9 and 2.0 to 3.4 respectively, ferromagnesian garnets range from 3.8 to 4.3. Another

Figure 5.2

Garnet-bearing schist, formed at moderately high pressures.

Table 5.1

Some common types of metamorphic reactions

Reactions in silicate rocks

$(Mg,Fe)_7Si_8O_{22}(OH)_2$ = 7 $(Mg,Fe)SiO_3$ + SiO_2 + H_2O
amphibole　　　　　　　　　　pyroxene　　　　quartz　　　　water

$(Mg,Fe)_3Si_2O_5(OH)_4$ + $KAlSi_3O_8$ =
chlorite　　　　　　　　potassium feldspar

$K(Mg,Fe)_3AlSi_3O_{10}(OH)_2$ + 2 SiO_2 + H_2O
biotite　　　　　　　　　　quartz　　　　water

3 $(Mg,Fe)_3(Fe,Al)_2Si_3O_{12}$ + 6 H_2O + $\frac{1}{2}O_2$
garnet　　　　　　　　　water　　　　　oxygen

$KFe_3AlSi_3O_{10}(OH)_2$ + $\frac{1}{2}O_2$ = $KAlSi_3O_8$ + Fe_3O_4 + H_2O
iron-rich biotite　　　oxygen　　potassium feldspar　magnetite　water

Reactions in carbonate rocks

$CaCO_3$ + SiO_2 = $CaSiO_3$ + CO_2
calcite　　　quartz　　wollastonite*　　carbon dioxide

$CaMg(CO_3)_2$ + 2 SiO_2 = $CaMgSi_2O_6$ + 2 CO_2
dolomite　　　quartz　　pyroxene　　carbon dioxide

*Silicate similar to pyroxenes

In all of the above reactions, the water is involved as water vapor; carbon dioxide and oxygen are gases; all other species are solids.
Many more reactions are possible, including various polymorphic transitions in which a low-temperature or low-pressure polymorph is converted to a higher-pressure or higher-temperature form (potassium feldspar, for example, has several polymorphs).
The reactions above should be regarded only as examples; actual reactions may be still more complicated by the effects of solid solution.

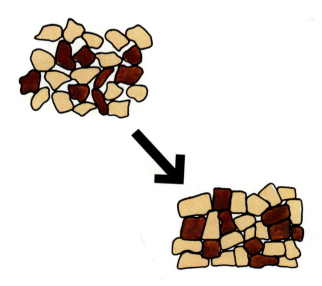

Figure 5.3

Schematic diagram of the effects of recrystallization during metamorphism. The grains become tightly interlocking and porosity decreases.

effect of increased confining pressure that is particularly evident in metamorphosed sedimentary rocks is a tendency for the rock's texture to become much more compact. The minerals recrystallize and in so doing decrease the porosity in the rock, increasing the density of the rock (figure 5.3).

When directed stress is applied, the result is typically a rock with some compositional or textural banding. This can take several forms. If there are already flat, platy minerals such as clays and micas in the rock, they tend to become aligned parallel to each other: parallel to a tensile stress, perpendicular to a compressive stress, parallel to the direction of a shear (figure 5.4). Even when the grains are too small to see with the naked eye, the rock often shows a tendency to split parallel to the aligned plates (rather than through or across the plates). This is **rock cleavage.** It is analogous to mineral cleavage, but rock cleavage is a tendency to break between planar mineral grains rather than between planes of atoms in a crystal. Rock

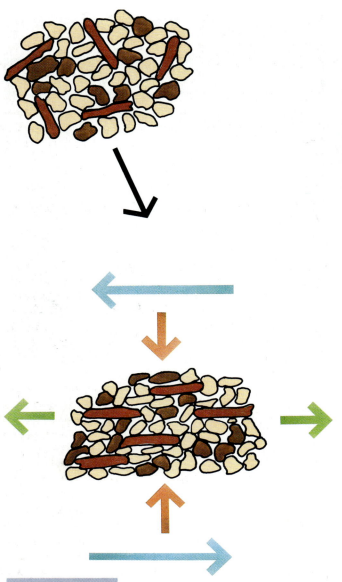

Figure 5.5

Slate, showing characteristic slaty cleavage.

COMPRESSIVE STRESS

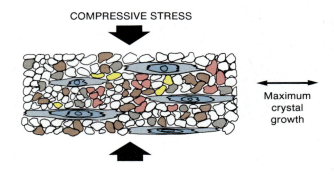

Maximum crystal growth

Figure 5.6

Growing micas grow most readily in the plane perpendicular to the applied compressive stress, leading to development of schistosity.

Figure 5.4

Development of foliation in the presence of directed stress. Elongated or platy minerals are reoriented, becoming aligned in parallel planes.

cleavage is demonstrated by slate, which splits readily into flat slabs from which tiles and flagstones can be made (figure 5.5). In fact, rock cleavage developed in such a fine-grained rock is often called slaty cleavage from this common example of the phenomenon.

As metamorphism intensifies at higher temperatures, crystals of various metamorphic minerals may grow progressively coarser (larger). Many metamorphic minerals have platelike or elongated, needle-shaped crystals (micas and amphiboles, respectively). In the presence of directed stress, these crystals will grow in similar preferred orientation.

Intuitively it makes sense, for example, that flakes of mica tend to grow sideways to, rather than directly opposed to, a compressive stress (figure 5.6). The resultant rock texture, in which coarse-grained platy minerals show preferred orientation in parallel planes, is described as schistosity (figure 5.7). Slaty cleavage in fine-grained rocks and schistosity in coarser-grained ones are both examples of foliation, from the Latin for "leaf" (think of the parallel leaves, or pages, of a closed book).

Another texture sometimes regarded as a form of foliation is compositional, rather than textural, layering. In some metamorphic rocks, especially those subjected to strong metamorphism, recrystallizing minerals in the rock segregate into bands of differing composition or texture. Often the result is a rock of striped appearance, with light-colored bands rich in quartz and feldspar alternating with dark-colored bands rich in ferromagnesian minerals (figure 5.8).

Figure 5.7

Examples of schistose rocks.

Figure 5.8

Gneiss, a compositionally banded metamorphic rock.

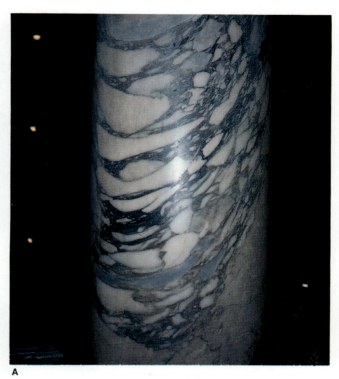

A

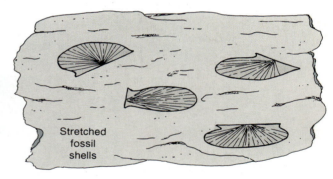

Stretched
fossil
shells

Undeformed
fossil
shape

B

Figure 5.9

Examples of rocks deformed by stress. *(A)* A metamorphosed marble breccia, showing strained fragments. *(B)* Sketch of deformation of fossils as host rock is deformed.

This mineralogical or compositional banding is *gneissic* (pronounced "nice-ic") texture. Although gneissic texture most commonly reflects compositional differences between layers, the term can also be applied to banded rocks with alternating schistose and equigranular layers.

On a megascopic scale, even rock fragments and fossils can be deformed or reoriented by directed stress. Squeeze a tennis ball. Notice that it is flattened in the direction of compression and broadened in the plane perpendicular to it. Stretch a piece of foam rubber. It is elongated in the direction of tensile stress, thinned perpendicular to it. Some sedimentary rocks contain objects whose original shape we know, such as spheroidal quartz pebbles in conglomerate, oolites, or symmetric fossils. Strain during metamorphism may deform these objects in such a way that we can determine the orientation of the deforming stress (figure 5.9).

Box 5.2

Stress and Strain under the Microscope

Strain or deformation on a gross scale is often visible as folds or other physical changes in outcrop (figures 5.1, 5.2). Even when a rock does not obviously show such features, the effects of stress can often be seen in thin section. Compare figures 1 and 2. Figure 1 is a photomicrograph of a thin section of a granite, made up mostly of large clear crystals of quartz (white to dark grey, featureless), coarse feldspar (flecked), and biotite mica (brownish laths). Figure 2 is a metamorphic rock (schist) of similar mineralogy, but the biotite laths are bent and even the quartz grains have a mottled appearance due to small, stress-induced irregularities in their crystal structures.

Figure 1
Thin section of granite. The darker, lathlike grains are biotite mica; the grey and white grains are quartz and potassium feldspar.

Figure 2
Thin section of biotite schist. Note the kinking or bending of biotite, and the mottled appearance of quartz grains caused by stress.

Effects of Chemical Changes

It is more difficult to generalize about the consequences of chemical changes during metamorphism. Briefly, the addition of a particular element will tend to stabilize minerals rich in that element and may cause more such minerals to form. For example, if migrating fluids carried dissolved potassium ions into a silicic rock, more potassium feldspars and micas might be formed. Such an introduction of ions in fluids is termed **metasomatism.** It is not uncommon around plutons, from which fluids may escape into the surrounding rocks.

Loss of a specific chemical constitutent will, conversely, have a destabilizing effect. For instance, if water is driven out of a permeable rock during metamorphism, any remaining hydrous minerals (clays, micas, amphiboles, and so forth) will tend to break down more readily.

Types of Metamorphic Rocks

Metamorphic rocks are subdivided into *foliated* and *nonfoliated* rocks on the basis of the presence or absence of preferred orientation of elongated or platy minerals. Further subdivisions within the foliated rocks are typically based on texture. Most unfoliated rocks are named on the basis of composition.

Foliated Rocks

Many of the common foliated rock types have names directly related to their characteristic textures. As noted previously, for example, **slate** (figure 5.5) is a metamorphic rock exhibiting slaty cleavage. Typically the individual mineral grains in slate are too fine to be seen with the naked eye. With progressive metamorphism, as mica crystals grow, they may become coarse enough to begin to reflect light strongly, and the cleavage planes in the rock take on a shiny appearance in consequence. Such a rock is called **phyllite** (figure 5.10). The individual mica crystals in a phyllite may be barely distinguishable with the unaided eye, but the rock is still basically fine grained.

Further progressive metamorphism with growth of coarser crystals of mica in the presence of directed stress leads to the development of obvious schistosity, and the corresponding rock is termed a **schist** (recall figure 5.7). A compositionally layered **gneiss**—a rock with gneissic texture, as shown in figure 5.8—may form either through still further metamorphism of a schist, with breakdown of some of the micas at higher pressures and temperatures, or during strong metamorphism of other rock types, such as granite.

Figure 5.10

Phyllite, showing glossy surface due to parallel mica flakes.

Note that the rock names of the foliated metamorphic rocks have few compositional implications. Parallel alignment of mica flakes creates slaty cleavage and schistosity, so slates and schists necessarily contain micas. Even so, there are many different compositions of micas, and other minerals may be present as well. The term *gneiss* indicates the existence of compositional layering without specifying anything at all about the minerals in the rock. To supply more information about a specific rock, one might add compositional terms to the rock name. For example, a rock might be described as a "garnet-biotite schist" if those two minerals were prominent in it; a "granitic gneiss" would be a metamorphic rock showing gneissic texture, but having a mineralogic composition like that of granite, rich in quartz and feldspar.

Nonfoliated Rocks

Rocks that consist predominantly of equidimensional grains will not show foliation. If they consist mainly of a single mineral, they may be named on the basis of composition. When a quartz-rich sandstone is metamorphosed, for example, the quartz grains are recrystallized and the rock compacted into a denser product with tightly interlocking grains. The resulting metamorphic rock is termed **quartzite** (figure 5.11). In hand specimen, its appearance will typically differ from that of unmetamorphosed sandstone in that the recrystallized quartz grains in quartzite will have a shinier or more glittery appearance than the original abraded grains in the sandstone. Sometimes the quartzite will have what is described as a sugary

Figure 5.11

Quartzite, metamorphosed quartz sandstone.

Figure 5.12

Marble, metamorphosed limestone.

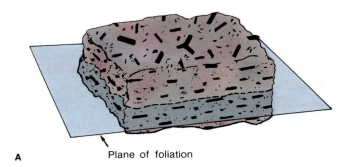

A Plane of foliation

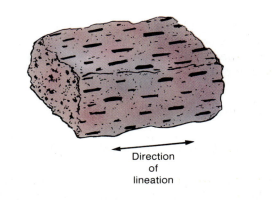

B Direction of lineation

Figure 5.13

Orientation of elongated amphibole crystals in amphibolite.
(A) Foliated amphibolite: crystals fall in parallel planes.
(B) Lineated amphibolite: crystals all line up in the same direction.

Although quartzite and marble do not typically show foliation, if they contain impurities of elongated or platy minerals, such as mica, these minor minerals may show a preferred orientation. Also, because they have formed from sedimentary rocks that may have been bedded or compositionally layered, some quartzites and marbles do show banding unrelated to the presence of directed stress.

appearance. Quartzite is usually also denser than sandstone as a result of the compaction under pressure. A metamorphosed limestone similarly recrystallized is **marble** (figure 5.12). Usually a marble is assumed to consist predominantly of calcite. If in fact it consists mainly of the mineral dolomite, it may, for clarity's sake, be called a dolomitic marble.

An **amphibolite** is a metamorphic rock rich in amphiboles. Strictly speaking, all amphibolites are not necessarily unfoliated rocks, for amphiboles commonly form in elongated or needlelike crystals that may take on a preferred orientation in the presence of directed stress. A textural term may then be inserted to indicate this: a foliated amphibolite is one in which the grains lie in parallel planes, while a lineated amphibolite is one in which they are all aligned in the same direction (figure 5.13).

Sometimes a metamorphic rock neither shows a distinctive texture nor has a sufficiently simple composition to justify applying one of the above terms. In such a case, if the nature of the parent rock can be recognized, the prefix *meta-* is simply added to the parent rock name to describe the present rock. Examples of such terminology are *metaconglomerate* or *metabasalt* to describe, more concisely, a metamorphosed conglomerate or basalt.

Environments of Metamorphism

A variety of geologic settings and events lead to metamorphism. Broadly, metamorphic processes can be subdivided into regional, contact, and fault-zone metamorphism.

Regional Metamorphism

Regional metamorphism is, as its name implies, metamorphism on a grand scale, a regional event. Regional metamorphism is commonly associated with mountain-building events, when large areas are uplifted and downwarped. Rock pushed to greater depths will be subjected to increased pressures and temperatures, and metamorphosed. Regional metamorphism involving changes in both pressure and temperature is also called **dynamothermal metamorphism.** The emplacement of large batholiths, which may or may not be associated with mountain-building, will raise crustal temperatures over broad areas as heat is released from cooling plutons. So batholith formation may likewise result in metamorphism on a regional scale.

Contact Metamorphism

Contact metamorphism is named for the setting in which it occurs, near the contact of a pluton. The pluton, emplaced from greater depths, will be hotter than the country rock, and if it is significantly hotter, the adjacent country rocks will be metamorphosed. The pluton will then be surrounded by a zone of metamorphic rock, also known as a contact **aureole,** or halo. (The term *aureole* comes from the Latin for "golden," as a crown or halo might be.) Higher-temperature metamorphic minerals will be found close to the contact, lower-temperature minerals farther away. Metasomatism may or may not occur in the contact aureole.

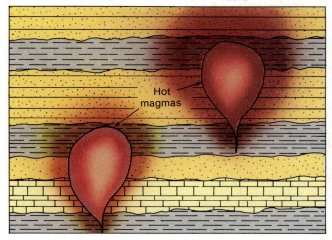

Figure 5.14

Schematic diagram of contact aureoles around plutons emplaced at deep and shallow crustal levels.

Contact metamorphism is a more localized phenomenon than regional metamorphism. It is confined to the immediate environs of the responsible pluton. Contact-metamorphic effects also tend to be most marked around plutons emplaced at shallow depths in the crust. Look at figure 5.14, keeping in mind the general increase of temperature with depth in the crust, and this will become clearer. Consider a mass of magma at some specific temperature. If it crystallizes deep in the crust, the country rock will already be at somewhat elevated temperatures and should already contain moderately high-temperature minerals. The modest increase in temperature around the pluton may not be enough to destabilize these minerals. Close to the earth's surface, the country rocks will be much cooler, perhaps only 100° to 200° C. Emplacement of a magma at 700° to 800° C or more will dramatically change the temperature of the adjacent country rocks, and is more likely to cause the breakdown of minerals stable only at low temperatures and the crystallization of higher-temperature ones. Similarly, emplacement of a mafic magma at a given depth will tend to result in a more pronounced, and generally larger, contact aureole than emplacement of a silicic magma at the same depth, because of the greater temperature contrast between the country rock and the hotter mafic magma.

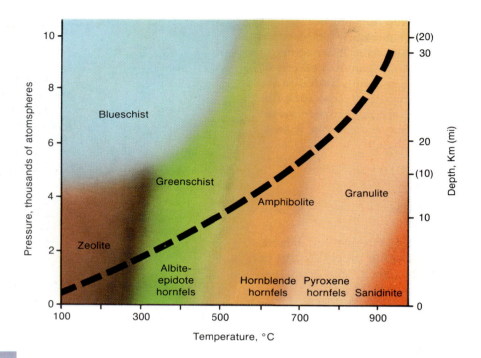

Figure 5.15

Metamorphic facies as functions of temperature and pressure
(depth). Dashed line is average continental geothermal gradient.

Fault-Zone Metamorphism

Fault-zone metamorphism is another localized effect,
of relatively minor importance volumetrically. A *fault*
is a planar break in rocks along which the rocks on
one side have moved relative to the other. Right along
the fault zone, as movement occurs, the rocks may be
severely stressed, crushed, and fragmented. There is
some increase in temperature due to frictional heating
but, except in the case of very large faults, it is typically much less than that associated either with deep
burial or with magma emplacement. The principal effects result from the stress applied. A rock formed from
the angular fragments of rocks broken up in the fault
zone is a **fault breccia.** It may contain minerals that
crystallize only at high pressures, or in the presence
of strong directed stress but at low temperatures.

Metamorphism and the Facies Concept

The concept of *facies* has already been introduced in
the context of sedimentary rocks. An analogous concept can be applied to metamorphic rocks. Rocks of
a particular metamorphic facies contain one or more
minerals characteristic of a particular, restricted range

of pressures and temperatures. Figure 5.15 shows
many of the principal metamorphic facies arranged in
the appropriate positions on a pressure-temperature
diagram. Note that the pressures are equated to depths
on the assumption that these pressures result from the
weight of overlying crustal rock of average density.
The boundaries between facies are not sharp, as the
stabilities of minerals depend in part upon factors
other than pressure and temperature (such as rock
composition and fluids present).

Contact-Metamorphic Facies

Contact-metamorphic facies are the facies of low
pressure and a range of temperatures. The **sanidinite**
facies is characterized by the presence of sanidine, a
very high-temperature polymorph of potassium feldspar. Sanidinite-facies conditions are not commonly
achieved except very close to extremely high-temperature plutons, as of ultramafic rock. The term
hornfels is applied to a variety of fine-grained, contact-metamorphic rocks formed at intermediate temperatures. They can be subdivided into several facies
of more restricted temperature range. For example,
the **pyroxene-hornfels** facies represents temperatures just below those of the sanidinite facies; the
hornblende-hornfels facies is a lower-temperature
facies, since the hydrous hornblende (an amphibole)

is stable at lower temperatures than are most pyroxenes. The zeolites are a group of hydrous silicates, stable only at low pressure and temperature, that give their name to the **zeolite** facies. A given contact aureole may span several facies, grading outward from a high-temperature facies, or mineral assemblage, closest to the pluton, through one or more lower-temperature facies, to unmetamorphosed country rock farther away.

Regional-Metamorphic Facies

The regional-metamorphic facies are characterized by elevated pressure and temperature. The **greenschist** facies is so named because greenschist-facies rocks commonly contain one or both of the greenish silicates chlorite (a mica-like sheet silicate) and epidote. Many but not all greenschist-facies rocks are also texturally true schists. Many amphiboles are stable over the range of conditions represented by the **amphibolite** facies. Garnets are also common under these conditions. Rocks of the **granulite** facies, which are often gneissic in texture, are characterized by various high-temperature, high-pressure mineral assemblages. The hydrous minerals have commonly broken down by the time granulite-facies conditions are reached.

The dashed curve in figure 5.15 represents a typical continental geothermal gradient. Following that curve indicates what will happen to a rock subjected to progressive metamorphism resulting from deeper and deeper burial in the crust—for example, during regional deformation. It will pass progressively through conditions of the greenschist, amphibolite, and ultimately granulite facies, with appropriate changes in mineralogy occurring as low-temperature, low-pressure minerals break down and are replaced by minerals stable at higher pressures and temperatures. Under extreme conditions of metamorphism, at the upper limit of the granulite facies, the minerals with the lowest melting temperatures (generally quartz and potassium feldspar) indeed begin to melt as the boundary between metamorphism and magmatism is reached. If this partly melted rock is then cooled and fully solidified again, the resulting **migmatite,** or "mixed rock," will have a mix of igneous and metamorphic characteristics.

Note also that the contact-metamorphic facies are characterized by lower pressures for a given temperature than the corresponding regional-metamorphic facies. Attainment of high temperatures without concurrent increase in pressure is achieved by emplacement of hot magma at shallow depths rather than by burial.

Facies of Unusually High Pressure

The **blueschist** facies is characterized by high pressure but low temperature. Its name derives from several bluish silicates that form under these conditions. Again, blueschist-facies conditions are not encountered in normal continental crust; unusual pressures or stresses are required. The precise geologic situation under which blueschist-facies rocks form is not well understood. One setting in which we might expect the necessary conditions will be examined further in chapter 7.

Index Minerals and Metamorphic Grade

The **metamorphic grade** of a rock is a rough description of the intensity of metamorphism to which that rock was subjected. That is, a high-grade metamorphic rock is one that shows textural or mineralogic evidence of having been subjected to high temperatures and/or pressures; a low-grade rock, the reverse. Blueschists and rocks of the granulite facies are examples of high-grade rocks, while zeolite-facies rocks are low-grade rocks. In a contact-metamorphic aureole, metamorphic grade increases with proximity to the contact with the pluton. In a regional context, metamorphic grade will increase with deeper burial.

Where regional-metamorphic rocks spanning a range of metamorphic grades are exposed at the surface, the direction of increasing or decreasing grade may be determined using a sequence of **index minerals,** each of which is stable over a restricted temperature range (figure 5.16). The occurrence of each of the index minerals present can be mapped, and lines can be drawn where a particular mineral appears or disappears (figure 5.17). The directional order of these lines, or **isograds** (lines joining points of equal metamorphic grade), will show the regional trend in metamorphic grade. In the example in figure 5.17, the metamorphic grade increases to the northeast. The event causing this regional metamorphism may have been more intense to the northeast. Alternatively, perhaps the rocks now found exposed in the northeast were originally the more deeply buried, and have been uplifted and eroded more since metamorphism, so that higher-grade rocks are now exposed there. Such information is useful in reconstructing past geologic events.

Note also that not all minerals are equally useful as index minerals. Quartz and potassium feldspar, for example, are stable over the whole range of typical

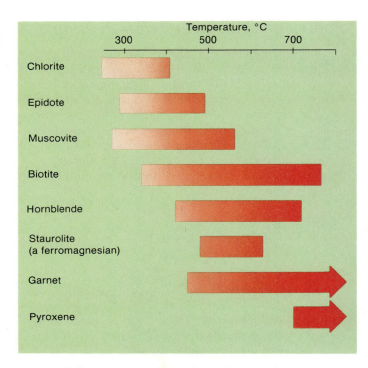

Figure 5.16

Approximate temperature ranges over which some representative index minerals are stable.

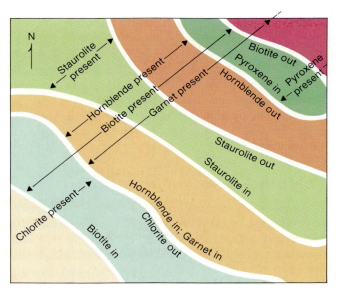

Figure 5.17

Isograds on a map show regional trends in metamorphic grade.

regional-metamorphic conditions. Their presence does not restrict the possible pressures or temperatures under which the rocks formed. Most of the ferromagnesian silicates, on the other hand, are stable over a narrow range of conditions, and many of these are useful index minerals.

Summary

Metamorphism is change in rocks (short of complete melting) brought about by changes in temperature, pressure, or chemical conditions. With progressive metamorphism, existing minerals are commonly recrystallized, the crystals often growing larger in the process. Minerals stable at low temperatures and pressures may break down, to be replaced by other minerals stable at higher-grade conditions. Directed stress can also lead to formation of foliated rocks, in which elongated or platy minerals assume a preferred orientation. The foliated rocks are named on the basis of the particular kind of texture they show. Nonfoliated rocks are often named on the basis of mineralogic composition. Most metamorphism is either contact metamorphism or regional metamorphism. Contact metamorphism occurs in country rocks close to the contacts of invading plutons and is characterized by relatively low-pressure mineral assemblages. Regional metamorphism is caused principally by tectonic activity and/or emplacement of large batholiths and is characterized by elevated pressure as well as elevated temperature. Metamorphic rocks are assigned to a particular facies, corresponding to a specific range of pressures and temperatures, on the basis of the mineral assemblages they contain. Index minerals are useful in assessing the general metamorphic grade of a rock and in determining regional trends in metamorphic grade.

Terms to Remember

amphibolite
aureole
blueschist
confining pressure
contact metamorphism
directed stress
dynamothermal
  metamorphism
facies
fault breccia

foliation
geothermal gradient
gneiss
granulite
greenschist
hornblende-hornfels
  facies
hornfels
hydrothermal
index mineral

isograd
marble
metamorphic grade
metamorphism
metasomatism
migmatite
parent rock
phyllite
pyroxene-hornfels facies

quartzite
regional metamorphism
rock cleavage
sanidinite facies
schist
schistosity
slate
slaty cleavage
zeolite facies

Questions for Review

1. What is metamorphism? What limits the maximum temperatures possible in metamorphism?

2. Describe two sources of heat for causing metamorphism.

3. Explain the distinction between *confining pressure* and *directed stress*. Which leads to such deformation as folding and development of foliation?

4. Describe two ways in which the presence of fluids can influence metamorphism.

5. How do rock cleavage and schistosity develop?

6. Define the following: *phyllite; schist; gneiss; amphibolite.*

7. What is metasomatism? Describe one environment in which it might commonly occur.

8. Explain the difference between contact and regional metamorphism.

9. Outline the progressive metamorphism of rocks buried deeper and deeper in the crust. Describe the facies through which they might pass and some of the changes to be expected.

10. What is a migmatite?

11. Explain what characteristics make a mineral useful as a metamorphic index mineral.

For Further Thought

1. It was pointed out in box 5.1 that many reactions are so slow at near-surface temperatures that minerals unstable at those temperatures can nevertheless be preserved. Consider why this fact allows geologists to gather much more information than if all mineral assemblages constantly adjusted to new pressure, temperature, or other conditions.

2. Sometimes during cooling from peak metamorphic temperatures, rocks do undergo *retrograde* reactions in which lower-grade minerals form again from higher-grade ones. Considering some of the metamorphic reactions in table 5.1, how important would you expect fluids to be in retrograde metamorphism? Explain.

Suggestions for Further Reading

Best, M. G. *Igneous and Metamorphic Petrology.* San Francisco: W. H. Freeman and Co., 1982.

Hyndman, D. W. *Petrology of Igneous and Metamorphic Rocks.* 2d ed. New York: McGraw-Hill, 1985.

Mason, R. *Petrology of the Metamorphic Rocks.* Boston, Mass.: George Allen and Unwin, 1978.

Turner, F. J. *Metamorphic Petrology: Mineralogical, Field, and Tectonic Aspects.* 2d ed. New York: McGraw-Hill, 1981.

Winkler, H. G. F. *Petrogenesis of Metamorphic Rocks.* 5th ed. New York: Springer-Verlag, 1979.

Geologic Time

Introduction

A student once wrote on an examination: "Geologic time is much slower and made up of very large units of time." This answer, while not quite the expected one, contains a large element of truth. Time passes no more slowly for geologists than for anyone else, but they do necessarily deal with the very long span of the earth's history. Consequently they often do work with very large units of time, millions or even billions of years long.

Much of our understanding of geologic processes, including the rates at which they occur and therefore the kinds of impacts they may have on human activities, is made possible through the development of various means of measuring ages and time spans in geologic systems. In this chapter we will explore some of these methods. A final section will examine several applications of geologic age determinations to the study of process rates.

Relative Dating

Before any techniques for establishing numerical ages of rocks or geologic events were known, it was sometimes at least possible to place a sequence of events in the proper order. This is known as **relative dating**.

Arranging Events in Order

Among the earliest efforts in this area were those of Nicolaus Steno. In 1669 he set forth two very basic principles that could be applied to sedimentary rocks. The first, the **Principle of Superposition,** pointed out that in an undisturbed pile of sediments (or sedimentary rocks), those on the bottom were deposited first, followed in succession by the layers above them, ending with the youngest on top (figure 6.1). Today this idea may seem so obvious as not to be worth stating, but at the time it represented a real step forward in thinking logically about rocks. The second, the **Principle of Original Horizontality,** was based on the observation that sediments are commonly deposited in approximately horizontal, flat-lying layers (figure 6.2). Therefore if one comes upon sedimentary rocks in which the layers are folded or dipping steeply, they must have been displaced or deformed after deposition and solidification into rock (figure 6.3).

Time-breaks in the sedimentary record also exist; they may or may not be easily recognized. An **unconformity** is a surface within a sedimentary sequence

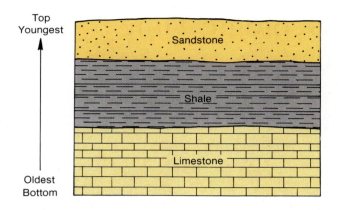

Figure 6.1

The Principle of Superposition. In an undisturbed sedimentary sequence, the rocks on the bottom were deposited first, and the depositional ages become younger higher in the pile.

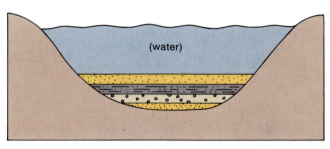

A

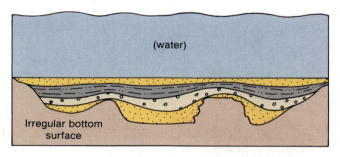

B

Figure 6.2

The Principle of Original Horizontality. *(A)* Sediments tend to be deposited in horizontal layers. *(B)* Even where the sediments are draped over an irregular surface, they tend toward the horizontal.

on which there has been a lack of sediment deposition, or even active erosion, for some period of time. When the rocks are later examined, that time span will not be represented in the record and, if the unconformity is erosional, some of the record once present will have been lost. The most difficult kind of unconformity to recognize is a **disconformity,** an unconformity at which the sedimentary rock layers above

Figure 6.3

Tilted sedimentary rocks, deformed after deposition. Contrast with figure 6.2B.

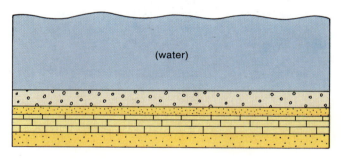

(i) Sediment deposition in water

(ii) Water now absent, deposition ceases

Figure 6.4

Development of a disconformity. Sediment deposition is interrupted for a period of time.

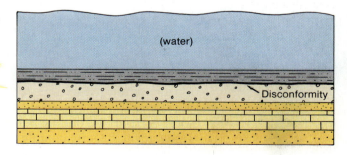

(iii) Deposition resumes after some lapse of time; disconformity established

and below it are parallel (figure 6.4). More obvious is the **angular unconformity** (figure 6.5), at which the bedding planes in rock layers above and below the unconformity are not parallel. The presence of an angular unconformity usually implies a significant period of erosion to produce the new depositional surface for the younger rocks.

In later centuries, reasoning about the relative ages of rocks was extended to include igneous rocks in rock sequences (figure 6.6). If an igneous dike or pluton cuts across layers of sedimentary rocks, the sedimentary rocks must have been there first, the igneous rock introduced later. Often there is a further

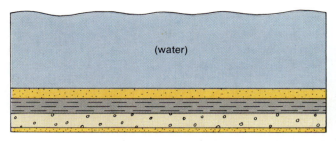

(i) Deposition

Erosional surface

(ii) Rocks tilted, eroded

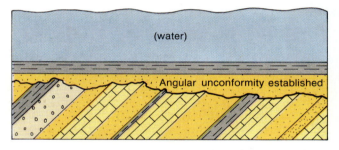

(water)

Angular unconformity established

(iii) Subsequent deposition

A

B

Figure 6.5

Angular unconformity. (*A*) Development involves some deformation and erosion before sedimentation is resumed. (*B*) Angular unconformity in the Grand Canyon.

clue to the right sequence: the hotter igneous rock may have baked, or metamorphosed, the country rocks immediately adjacent to it; so again the igneous rock must have come second. (This is a particular help when the pluton is concordant.) Such geologic common sense can be applied to many rock associations. If a strongly metamorphosed sedimentary rock is overlain by a completely unmetamorphosed one, for instance, the metamorphism must have occurred after the first sedimentary rock formed but before the sediments of the second were deposited. Quite complex sequences of geologic events can be unraveled by taking into account such simple underlying principles. See figure 6.7 for an example.

Correlation

Fossils play a role in determining relative ages too. The concept that fossils could be the remains of older life forms dates back at least to the ancient Greeks, but for some time the idea fell out of favor. It was seriously revived in the eighteenth century, and around the year 1800 William Smith put forth the **Law of Faunal Succession.** Its basic principle is that life forms change through time. Old ones disappear from the fossil record and new ones appear, but the same form is never exactly duplicated independently at two different times in history. This principle in turn implies that when one finds exactly the same type of fossil organism preserved in two rocks, even if they are quite different compositionally and geographically widely

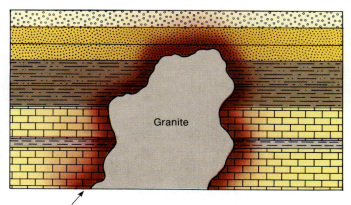

Rocks adjacent to intruding magma may also be metamorphosed by its heat

Figure 6.6

A pluton cutting across older rocks.

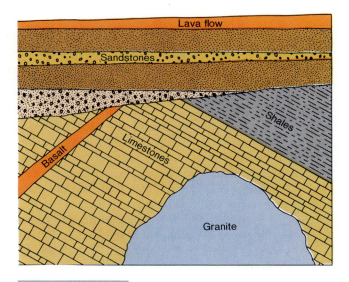

Figure 6.7

Deciphering a complex rock sequence. The limestones must be oldest (Law of Superposition), followed by the shales. The granite pluton and basalt must both be younger than the limestone they crosscut (note the metamorphosed zone around the granite also). It is not possible to tell whether the igneous rocks pre-date or post-date the shales, or to determine whether the sedimentary rocks were tilted before or after the igneous rocks were emplaced. After the limestones and shales were tilted, they were eroded and then the sandstones were deposited on top. Finally the lava flow covered the entire sequence.

separated, those rocks should be the same age. Smith's law thus allowed age **correlation** between rock units exposed in different places (figure 6.8). A limitation on its usefulness, however, is that it can be applied only to rocks in which fossils are well preserved, which are almost exclusively sedimentary rocks. Note that a theoretical basis for the Law of Faunal Succession was later provided by Charles Darwin's theories of evolution and natural selection.

The foregoing ideas were all useful in clarifying relative age relationships among rock units. They did not, however, help answer questions like "How old is this granite?" "How long did it take to deposit this limestone?" "How recently, and over how long a period, did this apparently extinct volcano erupt?" "Has this fault been active in modern times?" Questions requiring numerical replies went unanswered until the twentieth century.

How Old Is the Earth?

How old is the earth? Geologists and nongeologists alike have been fascinated for centuries with this very basic question in geologic time. Many have attempted to answer it, but with little success until the last few decades.

Early Efforts

One of the earliest widely publicized determinations was the seventeenth-century work of Archbishop Ussher of Ireland. He painstakingly counted up the generations in the Bible and arrived at a date of 4004 B.C. for the formation of the earth. This very young age was hard for many geologists to accept. The complex geology of the modern earth seemed to require far longer to develop.

Even less satisfactory from that point of view was the estimate of the philosopher Immanuel Kant. He tried to find a maximum possible age for the earth by assuming that the sun had always shone down on earth and that the sun's tremendous energy output was due to the burning of some sort of conventional fuel. But a mass of fuel the size of the sun would burn up in only one thousand years, given the rate at which the energy is being released. This was plainly an impossible result in view of several millennia of recorded human history. Kant, of course, knew nothing of the nuclear fusion processes by which the sun actually generates its energy.

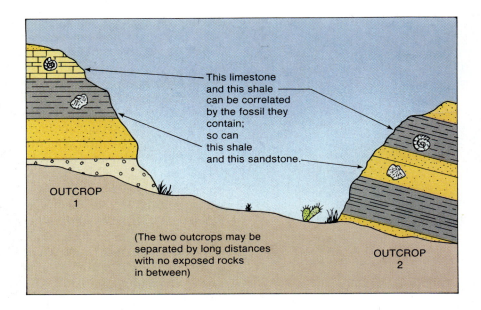

This limestone
and this shale
can be correlated
by the fossil they
contain;
so can
this shale
and this sandstone.

OUTCROP
1

(The two outcrops may be
separated by long distances
with no exposed rocks
in between)

OUTCROP
2

Figure 6.8

Similarity of fossils suggests similarity of ages even in different
rocks widely separated in space.

The Nineteenth-Century View

About 1800, Georges L. L. de Buffon attacked the
question from another angle. He assumed that the
earth was initially molten, modeled it as a ball of iron,
and calculated how long it would take this quantity
of iron to cool to present surface temperatures. The
result was about 75,000 years. To most geologists, this
still wasn't nearly long enough.

Around 1850, the physicist Hermann L. F. von
Helmholtz took the approach of supposing that the
sun's luminosity was due to infall of matter into its
center, converting gravitational potential energy to
heat and light. This gave a maximum age for the sun
(and presumably for the earth) of 20 to 40 million
years. Again, his assumptions were wrong, so his answer was also incorrect. In the late 1800s, Lord
William T. Kelvin reworked Buffon's calculations,
modeling the earth more realistically in terms of rock
properties rather than those of metallic iron. Interestingly, he too arrived at an age of 20 to 40 million
years for the earth. What he did not take into account,
because natural radioactivity was then unknown, was
that some heat is continually being *added* to the earth's
interior through radioactive decay, so it has taken
longer to cool down to its present temperature regime.

The calculations went on, and there was something wrong with each. U.S. Geological Survey geologist Charles D. Walcott tried to compute the total

thickness of the sedimentary rock record throughout
geologic history and, dividing by typical modern sedimentation rates, to estimate how long that pile of sediments would have taken to accumulate. His answer
was 75 million years. Walcott was hampered in his efforts by several factors, including the fact that there
are gaps in many sedimentary rock sequences, other
periods during which no sedimentation occurred, and
periods during which some sediments were eroded
away. He necessarily assumed sedimentation rates
comparable to present ones, but he in fact had no evidence to prove that ancient sedimentation rates were
not quite different. He also had no good idea of the
total thickness of sediments deposited in the time before organisms capable of preservation as fossils became widespread, which turns out to be most of the
earth's history.

In 1899, physicist John Joly published calculations based on the salinity of the oceans. Taking the
total dissolved load of salts delivered to the seas by
rivers, and assuming the ocean was initially pure water,
he determined that it would take about 100 million
years for the present concentrations of salts to be
reached. (This method, incidentally, had been proposed two centuries before by Sir Edmund Halley, of
Halley's comet fame.) Aside from the fact that Joly was
assuming constant rates of weathering and salt input
into the oceans throughout earth's history, he did not
consider the fact that the buildup of salts is slowed
by the removal of some of the dissolved material—
for example, as chemical sediments.

Box 6.1

Tools of the Trade: Mass Spectrometry

The development of radiometric dating required not only recognition of the phenomenon of radioactive decay but also the construction of sophisticated analytical equipment capable of counting or measuring the quantities of different isotopes of an element: the **mass spectrometer** (figure 1).

Prior to analysis, the sample is dissolved in strong acid and chemically processed to separate individual elements of interest. A drop of solution containing such an element is loaded onto a metal filament which is heated in the spectrometer under vacuum. Ions of the element are vaporized off the filament and moved along a tube through a magnetic field to a collector/counter. The magnetic field deflects the stream of ions, isotopes of different masses being deflected slightly differently, allowing each to be counted separately. The relative numbers of different isotopes are indicated by the counter. On a modern spectrometer the data are

Figure 1
A mass spectrometer.
Photo by J. N. Lytwyn.

processed by a computer, which also operates the instrument wholly or partially.

Refinements in analytical techniques have gradually reduced the amount of material needed for analysis. Less than a tenth of a gram (0.0035 ounce) of powdered rock may be enough for rubidium-strontium analysis. Some methods for uranium-lead analysis of zircons can work on a single zircon crystal the size of a small sand grain.

So the debate, frequently a heated one, continued until the discovery of radioactivity provided a much more powerful and accurate tool with which to attempt a solution.

Radiometric Dating

Henri Becquerel didn't set out to solve any geologic problems, or even to discover radioactivity. He was interested in a curious property of some uranium salts: if exposed to light, they would continue to emit light (phosphoresce) for a while afterwards even in a dark room. He had put a vial of uranium salts away in a drawer on top of some photographic plates well wrapped in black paper. When he next examined the photographic plates, they were fogged with a faint

image of the vial, as if they had somehow been exposed to light right through the paper. The uranium was emitting something that could pass through light-opaque materials. We now know that uranium isotopes are among several dozen natural isotopes that are radioactive, that undergo spontaneous decay. Once the phenomenon of radioactive decay was reasonably well understood, physicists and geoscientists began to realize that it could serve as a very useful tool for investigating earth's history.

Radioactive Decay and Dating

A **radioactive** isotope is one with an unstable nuclear configuration. That is, the particular combination of protons and neutrons is not truly stable, and therefore in time the nucleus decays, or changes into a different

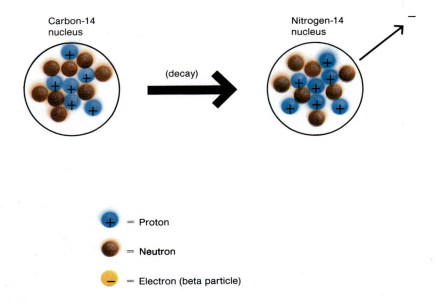

Carbon-14 nucleus

(decay)

Nitrogen-14 nucleus

+ = Proton

= Neutron

− = Electron (beta particle)

Figure 6.9

Schematic representation of the decay of carbon-14 nucleus to nitrogen-14 nucleus. (Electrons not involved; omitted for simplicity.)

nucleus. Up to the instant at which decay occurs, the nucleus of a radioactive isotope behaves no differently from a stable nucleus. The decay, when it does occur, involves emission of radiation, typically consisting of one or more particles plus energy. Principal kinds of radiation emitted are *alpha particles* (helium nuclei, consisting of 2 protons and 2 neutrons), *beta particles* (electrons or their positively charged antiparticles, positrons), or *gamma rays* (electromagnetic radiation, analogous to X rays but more penetrating). The decaying nucleus is by convention called the **parent** nucleus, while the nucleus into which it decays is called the **daughter** nucleus. By way of example, consider the radioactive isotope carbon-14. Its nucleus contains the 6 protons characteristic of carbon and 8 neutrons. When the parent carbon-14 decays (figure 6.9), a neutron is converted to a proton plus an electron. The proton remains in the nucleus, while the electron (beta particle) is ejected. Energy is also released during the decay. The resulting daughter nucleus is nitrogen-14 (having 7 protons and 7 neutrons).

If decay of unstable nuclei were instantaneous, or occurred at altogether random times, the phenomenon would have no application to dating. Indeed, the exact moment at which any single nucleus of a radioisotope will decay cannot be predicted. However, given a large number of atoms of the same radioisotope, it can be predicted that a certain fraction of them will decay over a given period of time. Radioactive decay is thus a statistical phenomenon, one that obeys the laws of probability. In that sense, it is like flipping a coin. One cannot know whether any single flip of an (unbiased) coin will come up heads or tails, but statistically, over a large number of flips, half should come up heads, half tails.

Each radioisotope can be characterized by a parameter called its **half-life,** defined as the length of time required for half of a given initial number of atoms of that isotope to decay. Not only is the half-life of each radioisotope characteristic and measurable in the laboratory; it is also constant. That is, a given radioisotope will decay at the same rate regardless of its chemical or physical state, the compound in which it may occur, or the temperature or pressure to which it is subjected. The half-life of carbon-14, for example, is 5730 years. If we start with a billion atoms of carbon-14, after 5730 years we will have half a billion atoms of carbon-14 left, whether we have the carbon in the form of diamond, coal, organic compounds in human tissues, carbon dioxide gas, or whatever.

Human activities do not affect the rate of decay of unstable atoms. An unstable nucleus cannot be made more or less stable—we cannot speed up, slow down, or stop the decay of a radioisotope. This is part of the problem with radioactive waste materials. They cannot be treated to make them nonradioactive. Instead, if they pose a radiation hazard, they must be isolated for as long as it takes them to decay away enough that the quantity left is harmless.

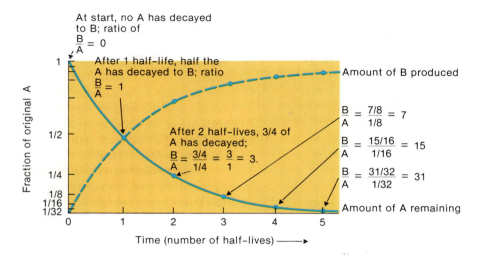

At start, no A has decayed
to B; ratio of
$\frac{B}{A} = 0$

After 1 half-life, half the
A has decayed to B; ratio
$\frac{B}{A} = 1$

Amount of B produced

After 2 half-lives, 3/4 of
A has decayed;
$\frac{B}{A} = \frac{3/4}{1/4} = \frac{3}{1} = 3$.

$\frac{B}{A} = \frac{7/8}{1/8} = 7$

$\frac{B}{A} = \frac{15/16}{1/16} = 15$

$\frac{B}{A} = \frac{31/32}{1/32} = 31$

Amount of A remaining

Fraction of original A

1
1/2
1/4
1/8
1/16
1/32

0   1   2   3   4   5

Time (number of half-lives) ——▶

Figure 6.10

Radioactive decay of parent isotope A proceeds at a regular rate,
with corresponding accumulation of daughter isotope B.

Given the constancy of radioactive decay rates,
one can in principle use the relative amounts of a
parent isotope and the product daughter isotope into
which it decays to find the age of a sample. Figure 6.10
illustrates the fundamental idea. Suppose that parent
isotope A decays to daughter B with a half-life of one
million years. In a rock that contained some A and no
B when it formed, A will gradually decay and B will
accumulate. After one million years (one half-life of
A), half the A initially present will have decayed to
yield an equal number of atoms of B. After another
half-life, half of the remaining half will have decayed,
so three-fourths of the initial amount of A will have
been converted to B and one-fourth will remain. The
ratio of B to A will be 3:1. After another million years,
there will be 7 atoms of B for every atom of A; and so
on. If we find a sample that contains 24,000 atoms of
B and 8,000 atoms of A, then, assuming no B present
when the rock formed, and noting the 3:1 ratio of B
to A, we would conclude that the sample was two half-
lives of A old (in this case, 2 million years).

The foregoing is a considerable oversimplifica-
tion of the dating of natural geologic samples. Many
samples do contain some of the daughter isotope as
well as the parent at the time of formation, and correction for this must be made. Various methods exist
for making these corrections. Other samples have not
remained chemically closed throughout their histories and have gained or lost atoms of the parent or
daughter isotope of interest, which will make the calculated date incorrect. The trained specialist can recognize rock or mineral samples in which this has
occurred.

"Absolute" Ages

Dates calculated on the basis of decay of radioisotopes are sometimes in older texts inaccurately termed
absolute ages. What is really meant by the term is a
numerical age as distinguished from the relative ages
described earlier. The more correct modern term for
a numerical age measured using the phenomenon of
radioactive decay is **radiometric age.** The geologist
who specializes in radiometric dating is a *geochronologist.*

Choice of Isotopic System

Several conditions must be satisfied by an isotopic
system in order for it to be used for dating geologic
materials. The parent isotope chosen must be abundant enough in the sample for its quantity to be measurable. Either the daughter isotope must not normally
be incorporated into the sample initially, or there must
be a means of correcting for the amount initially
present. The half-life of the parent must be appropriate to the age of the event being dated: long enough
that some parent atoms are still present but short
enough that some appreciable decay and accumulation of daughter atoms has occurred. A radioisotope
with a half-life of ten days would be of no use in dating
a million-year-old rock. The parent isotope would have
decayed away completely long ago and there would
be no way to tell *how* long ago. Conversely, a radioisotope with a half-life of 10 billion years would be

Table 6.1

Some radiometric decay schemes commonly used in geology

Parent isotope	Half-life (years)	Daughter isotope	Parent abundant in
potassium-40	1.3 billion	argon-40	potassium-rich minerals (including feldspar, micas)
rubidium-87	48.8 billion	strontium-87	potassium-rich minerals
thorium-232	14 billion	lead-208	zircon and other minor minerals
uranium-235	704 million	lead-207	uranium ores; zircon and other minor minerals
uranium-238	4.5 billion	lead-206	(same as uranium-235)
carbon-14*	5730	nitrogen-14	organic matter; atmospheric CO_2; dissolved carbonate

the ARE STAble, they bReAk down, consequently they can be used for dating

*This method works somewhat differently from the other methods listed. With its very short half-life, carbon-14 cannot be used to date samples more than 50,000 to 100,000 years old. It is used principally for dating wood, cloth, paper, bones, and so on from archaeological sites.

useless for dating a fresh lava flow, because the atoms of the daughter isotope that would have accumulated in the rock would be too few to be measurable. Only about half a dozen isotopes are widely used in dating geologic samples. These are listed in table 6.1, along with the materials in which the parent isotope in each case is readily concentrated.

What Is Being Dated?

Radiometric dates are not the ages of the elements in the rock or the age of the earth. Where valid dates can be obtained, they pertain to some aspect of the history of that particular rock. What kinds of events are dateable can be explored by taking the potassium-argon scheme as an example.

Potassium-40 decays to argon-40. Argon is one of the inert gases, and therefore does not readily bond chemically with other atoms or ions in a crystal. When a potassium-bearing mineral is crystallizing, whether from a melt or a solution, argon is not ordinarily incorporated into the growing crystal. Once the crystal is formed, however, any argon-40 produced by decay of potassium-40 will be trapped within the rigid

crystal structure. After a period of time, a quantity of argon-40 will have accumulated that is proportional to the amount of potassium-40 in the crystal and to the length of time since it formed. Measuring the quantities of potassium-40 and argon-40, and knowing the half-life of potassium-40, we can determine the time since the mineral crystallized. The date of crystallization of a potassium-rich igneous mineral (or rock) or of a chemical sedimentary mineral (rock) can thereby be determined.

Metamorphism tends to blur the results. When a crystal is heated, and the atoms in the structure begin to vibrate more energetically, trapped argon can more readily escape. In general, the higher the temperature, the more easily the argon is lost. The effect of losing some of the argon from a sample is to make it appear too young. When it is later analyzed, it will have too little argon for its true original age and potassium content. Its apparent age will thus have no exact significance in the history of that rock. The date will fall somewhere between the original age of the rock and the time of metamorphism. On the other hand, in the case of very intense, high-grade metamorphism, *all* the accumulated argon can be lost, especially if the minerals are recrystallized during metamorphism. The isotopic clock is then effectively *reset* to the time of metamorphism. If the rock remains undisturbed thereafter, the date it yields when analyzed will be the correct date of the strong metamorphism.

Physical breakup typically is not accompanied by extensive loss of accumulated daughter isotopes. The formation of clastic sediments is therefore not an event that tends to reset the isotopic systems of the constituent minerals. Clastic sediments are said to "inherit" a portion of their daughter isotopes from their source rocks. When analyzed long after deposition, they will tend to yield meaningless ages older than the time of deposition, for they will contain not only the daughter isotopes accumulated as a result of radioactive decay following deposition but also some inherited daughter isotopes. If the grains have lost none of the daughter isotopes at all during breakup, transport, and deposition, the sediment may even preserve the age of the source rocks, although this is rare. In any case, isotopic methods do not generally yield the time of deposition of a clastic sediment, and therefore radiometric methods are not generally applied to such materials.

Each isotopic system behaves a bit differently because of differences in the chemical behavior of the parent and daughter isotopes in each case. Some systems are more easily reset than others. Because different elements are concentrated in different minerals and rock types, not all isotopic systems work equally well on a given rock. As a very general rule, however, it can be said that the time of crystallization of an igneous rock, the date of a very high-grade metamorphic event, or the time of precipitation of a chemical sediment from solution can often be determined quite reliably using an appropriate isotopic system. Where weak metamorphism has been superimposed on a pre-existing rock, radiometric ages obtained are often geologically meaningless (in the sense that they fall somewhere between the primary rock age and the date of metamorphism). Sometimes the age of the source rocks of clastic sediments can be measured, but the time of sediment deposition cannot be.

The foregoing discussion has noted some of the limitations of radiometric dates. Recognition of those limitations is important to realizing when a date obtained is credible, as well as whether a given material can usefully be dated. Several decades of research into radiometric dating of geologic materials, and of laboratory experiments on the behavior of geologic and chemical systems, have increased our understanding to the point that the geochronologist can readily determine which dates are indeed reliable. Some criteria are geologic, relating to the nature of the rock and its thermal history, as deduced from rock type and mineralogy. Other criteria are chronologic. As an example, it was noted above that different isotopic systems behave differently in nature. When several chemically different methods (rubidium-strontium, potassium-argon, and uranium-lead, for example) give the same date on a sample, that date is particularly well constrained, for it would be virtually impossible for such agreement to arise by coincidence. In other words, there are good and bad radiometric dates, but it is now perfectly possible to distinguish which is which. One avoids dating inappropriate materials and applies stringent tests to those dates obtained to assess their value. The result is that the vast majority of radiometric dates are highly reliable, and the availability of these powerful methods has added immensely to our understanding of the earth and of the solar system.

Radiometric and Relative Ages Combined

Because there are many rocks and fossils for which it is not possible to determine accurate radiometric dates, so-called absolute and relative dating methods are often used in conjunction to constrain the ages of units not easily dated. Undateable sedimentary rocks crosscut and perhaps metamorphosed by a dike or pluton must be older than the invading igneous rock, which may be dateable. This will at least put a lower (younger) limit on the age of the sedimentary rock.

A fossil found in a rock sandwiched between two dateable lava flows has its age bracketed by the ages of the volcanic rocks. An unmetamorphosed sedimentary rock directly in contact with a high-grade metamorphic rock must postdate the metamorphism, which puts an upper limit on the time of sedimentation. Following such reasoning, it is often possible to work out a complex sequence of events from a single exposure of rock, and to put age limits, if not exact ages, on the various rock units (see figure 6.11).

The Geologic Time Scale

Particularly in the days before radiometric dating was developed, it was necessary to establish some subdivisions of earth history by which particular periods of time could be indicated. This was initially done using those rocks with abundant fossil remains. The resultant time scale was subsequently refined with the aid of radiometric dates.

The Phanerozoic

The Law of Faunal Succession and the appearance and disappearance of particular fossils in the sedimentary record were first used to mark the boundaries of the time units (table 6.2). The principal divisions were the eras—Paleozoic, Mesozoic, and Cenozoic, meaning respectively, "ancient life," "intermediate life," and "recent life." The eras were subdivided into periods and the periods into epochs. The names of these smaller time units were assigned in various ways. Often they were named for the location of the *type section,* the place where rocks of that age were well exposed and that time unit was first defined. For example, rocks of the Cambrian period are well exposed in Wales, and the type section is there. *Cambria* is a Latin form of the native Welsh name for Wales. The Jurassic period is named for the Jura Mountains of France. Other periods are named on the basis of some characteristic of the type rocks. The Cretaceous derives its name from the Latin *creta,* "chalk," for the chalky strata of that age in southern England and northern Europe. Rocks of Carboniferous age commonly include coal beds. All of the relatively unfossiliferous rocks that seemed to be older (lower in the sequence) than the Cambrian were lumped together as pre-Cambrian, later formally named **Precambrian.**

With the advent of radiometric dating, it became possible to attach numbers to the units and boundaries of the time scale. It became apparent that, indeed, geologic history spanned long periods of time.

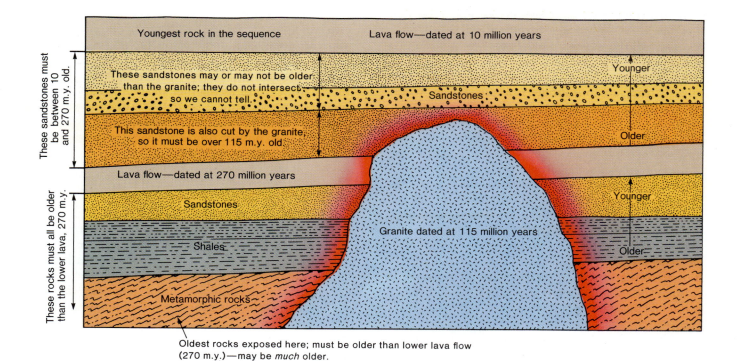

Youngest rock in the sequence   Lava flow—dated at 10 million years

These sandstones must be between 10 and 270 m.y. old.

These sandstones may or may not be older than the granite; they do not intersect so we cannot tell.

Younger

Sandstones

This sandstone is also cut by the granite, so it must be over 115 m.y. old.

Older

Lava flow—dated at 270 million years

These rocks must all be older than the lower lava, 270 m.y.

Sandstones

Younger

Granite dated at 115 million years

Shales

Older

Metamorphic rocks

Oldest rocks exposed here; must be older than lower lava flow (270 m.y.)—may be *much* older.

Figure 6.11

The ability to date some units in an outcrop allows the ages of units not directly dateable to be constrained.

Table 6.2

The geologic time scale (circa 1900)

Era	Period	Epoch	Distinctive life forms
Cenozoic	Quaternary	recent	modern humans
		Pleistocene	stone-age humans
	Tertiary	Pliocene	
		Miocene	flowering plants common
		Oligocene	ancestral pigs, apes
		Eocene	ancestral horses, cattle
		Paleocene	
Mesozoic	Cretaceous		dinosaurs become extinct; flowering plants appear
	Jurassic		birds, mammals appear
	Triassic		dinosaurs, first modern corals appear
Paleozoic	Permian		rise of reptiles, amphibians
	Carboniferous		coal forests; first reptiles, winged insects
	Devonian		first amphibians, trees
	Silurian		first land plants, coral reefs
	Ordovician		first fishlike vertebrates
	Cambrian		first widespread fossils
Pre-Cambrian			scant invertebrate fossils

It further became apparent that the most detailed part of the scale, the **Phanerozoic** (Cambrian and later), was by far the shortest, comprising less than 15% of earth history.

Radiometric Dating and the Time Scale

There was naturally a great deal of interest in assigning very precise dates to the fine subdivisions of the Phanerozoic, but here a difficulty arose due to the nature of those subdivisions and of radiometric dating. It will be apparent from previous discussion that the time of deposition of most sedimentary rocks cannot readily be dated. Very few fossils can be dated directly either, and in any case fossilization may occur long after the organism is dead and buried. Yet the eras, periods, and epochs of the Phanerozoic were defined almost entirely on the basis of the sedimentary/fossil record. In practice, then, it was often necessary to approach the ages of the units by indirect dating and determination of age limits for sedimentary rocks (where possible), as described earlier, then making correlations with rocks in the type sections.

A further complication with respect to the finest subdivisions is that, like any measurements, most radiometric dates have some inherent uncertainty associated with them, arising out of geologic disturbance of the samples, laboratory and statistical uncertainties, and so on. In the best cases, these uncertainties may be much less than 1% of the date determined; in unfavorable cases, they may amount to 10% or more of the age. That is, one would not be able to date a rock at exactly 174,692,361 years; the age might instead be reported as 174 ± 3 million years, meaning that the rock has been determined to be between 171,000,000 and 177,000,000 years old. On a rock billions of years old, the uncertainty may be tens of millions of years; on a million-year-old sample, it might be only tens of thousands of years. Then, too, analytical methods and laboratory instrumentation have been constantly improving, and radiometric dates have been refined. Re-dating of previously dated units has yielded new, more precise ages that may differ slightly from the less precise ages determined earlier.

With the advent of radiometric dating, the approximate time framework of the Phanerozoic was readily established. But the various technical limitations have caused persistent small uncertainties in the exact dates in question. The student should not be surprised, therefore, to find that texts published over the last few decades may differ somewhat in the exact ages shown. A recent revision of the Phanerozoic time scale, with radiometric dates, is shown in table 6.3.

Table 6.3

Modern geologic time scale for the Phanerozoic, with radiometric ages

Era	Period	Epoch[1]	Date at start of time interval (millions of years ago)[2]
		Holocene	0.1
	Quaternary	Pleistocene	2
		Pliocene	5
		Miocene	24
		Oligocene	37
		Eocene	58
Cenozoic	Tertiary	Paleocene	66
	Cretaceous		144
	Jurassic		208
Mesozoic	Triassic		245
	Permian		286
	Carboniferous[3]		
	Pennsylvanian		320
	Mississippian		360
	Devonian		408
	Silurian		438
	Ordovician		505
Paleozoic	Cambrian		570

1. Pre-Cenozoic periods are also subdivided into epochs, but there is little uniformity of nomenclature for these worldwide.
2. Dates used from compilation of Geological Society of America for Decade of North American Geology.
3. Not generally subdivided outside the United States.

The Precambrian

The Precambrian has continued to pose something of a problem. A unit that spans four billion years of time seems to demand subdivision—but, in the virtual absence of fossils, on what basis? The first division into the *Archean* ("ancient") and *Proterozoic* ("pre-life") at least split the Precambrian into two nearly equal halves, but there was considerable disagreement as to how to define the boundary geologically. Although the time around 2 1/2 billion years ago was a time of widespread igneous activity, metamorphism, and mountain building, no single event of global impact could be identified to mark the boundary. There was still less agreement on how the Archean and Proterozoic should be further divided. Eventually, an international commission convened to address these and

Table 6.4

Subdivisions of the Precambrian proposed by the North American Commission on Stratigraphic Nomenclature

Eon	Era	Date (millions of years) at start of unit
(Phanerozoic)		(570)
Proterozoic	Late	900
	Middle	1600
	Early	2500
Archean	Late	3000
	Middle	3400
	Early	3800(?—age of oldest terrestrial rocks)

Note that periods and finer subdivisions have not been established.

other related problems and recommended adoption of the scheme shown in table 6.4. The dates of the boundaries are arbitrary in the sense that they do not represent well-defined events, like the extinction of the dinosaurs. Still, the existence of these divisions allows one to place events in a time framework by referring to them as "middle Proterozoic," "late Archean," or whatever, without resorting to a more cumbersome phrase such as "falling between 2100 million years and 1500 million years in age."

How Old Is the Earth?— A Better Answer

Because the earth is geologically still very active, no rocks have been preserved unchanged since its formation. We cannot therefore determine the age of the earth directly, even by isotopic methods. There is, however, good evidence to indicate that the earth, moon, and meteorites all formed at the same time, during the formation of the solar system. Fortunately, some of these other materials have had quite different subsequent histories. The moon is much smaller than the earth, cooled more rapidly after accretion, and has been geologically inactive for billions of years. The oldest of the samples returned by the manned Apollo lunar missions are approximately 4.6 billion years old.

We also have numerous samples of *meteorites,* extraterrestrial fragments of rock or metal that have fallen to earth. Some of these have been disturbed after formation by collisions in space or by other processes, but the majority of meteorites yield ages between 4.5 and 4.6 billion years. Strong chemical similarities between the earth and meteorites (see chapter 10) support the idea that they formed from the same materials (solar nebula), presumably at the same time.

On the basis of the foregoing and other evidence, the earth is inferred to have formed at about 4.55 billion years ago, and this date is typically taken as the beginning of Precambrian time. The oldest terrestrial rocks that have been accurately dated are actually less than 4 billion years old. Rocks close to this age are found on nearly every continent. The oldest rocks on each continent are generally between 3.6 and 3.9 billion years old.

Dating and Geologic Process Rates

The rates at which geologic processes occur can be estimated in a variety of ways, many of which rely on radiometric dating techniques.

Examples of Rate Determination

As will be seen in the next chapter, the continents and sea floor move over the earth's surface. There are a few approximately fixed reference points (the magnetic poles, for example) by which we can determine how far they have moved. If one can determine the radiometric ages of rocks formed at one time, when a continent was at point A, and those formed at another time when it was at point B, and measure the distance between A and B, the average rate of movement during that period can be calculated. Suppose that the continent is shown to have moved 400 kilometers (about 250 miles) in 10 million years. The average rate of movement is then

$$\frac{400 \text{ km}}{10,000,000 \text{ years}} \times \frac{1000 \text{ m}}{1 \text{ km}} \times \frac{100 \text{ cm}}{1 \text{ m}} = \frac{4 \text{ cm}}{\text{year}}$$

or 4 centimeters/year. This is, of course, only an average rate for the whole 10 million years. The continent may have moved 10 centimeters/year or more at some times, 1 centimeter/year or less at others.

By dating the volcanic rocks from one or a group of volcanoes, we can judge how long volcanic areas remain active and how frequently volcanoes erupt. Such information can be very useful in assessing volcanic hazards. A typical small volcano may erupt intermittently over a period of 100,000 years; a major volcanic center may stay active for 1 to 10 million years. The building of a major mountain range, with all the attendant igneous and metamorphic activity, may take 100 million years.

Some rates can only be broadly constrained. The minimum rate of uplift of rocks in a mountain range might be estimated from the age of marine sedimentary rocks in the mountains, which were once deposited underwater and are now found high above sea level. Such uplift rates are typically 1 centimeter/year or less, often much less. Beaches formed on Scandinavian coastlines during the last Ice Age have been rising since the ice sheets melted and their enormous mass was removed from the land. From the beach deposits' ages and present elevation above sea level, uplift rates can be approximated. Typical rates of this postglacial rebound are of the order of 1 centimeter/year.

Finally, the development of isotopic dating methods has made possible a much better understanding of the rates of organic evolution and the whole history of life on earth.

The Dangers of Extrapolation

One must be somewhat cautious about extrapolating present (modern) process rates too far into the past or future, and one must not assume that rates measured over a short time are representative of a long period. This point is admirably illustrated by the following passage from Mark Twain's *Life on the Mississippi*. Rivers that have developed contorted, twisted channels full of bends (*meanders*) may change course during times of high water flow. The meanders are cut off, or bypassed, as the flowing water seeks a straighter, shorter, more direct path. Here Twain speculates on the implications of the shortening of the Mississippi River by this process:

> In the space of one hundred and seventy-six years the lower Mississippi has shortened itself two hundred and forty-two miles. This is an average of a trifle over one mile and a third per year. Therefore, any calm person, who is not blind or idiotic, can see that in the Old Oolitic Silurian Period just a million years ago next November, the Lower Mississippi was upwards of one million three hundred thousand miles long, and stuck out over the Gulf of Mexico like a fishing rod. And by the same token any person can see that seven hundred and forty-two years from now the Lower Mississippi will be only a mile and three-quarters long, and Cairo and New Orleans will have joined their streets together, and be plodding along comfortably under a single mayor and a mutual board of aldermen. There is something fascinating about science. One gets such wholesale returns of conjecture out of such a trifling investment of fact.

Summary

Before the discovery of natural radioactivity, only relative age determinations for rocks and geologic events were possible. Field relationships and fossils were the principal tools used for this purpose. Rocks could be placed in relative age sequence, or, with the aid of fossils, correlated from place to place.

Radiometric dating has made possible quantitative age measurements, although not all geologic materials can be so dated. Most of the commonly used isotopic dating methods work best for igneous and high-grade metamorphic rocks. It is rarely possible to determine the time of deposition of sedimentary rocks. This has led to difficulties in assigning precise ages to the units of the geologic time scale.

Radiometric ages can be used to explore the rates at which different kinds of geologic processes occur, and have considerably advanced our understanding of earth's development. However, while observations of present geologic processes can be used to understand past geologic history, it cannot be assumed that those processes have always proceeded at rates comparable to those presently observed.

Terms to Remember

absolute age	half-life	Phanerozoic	radioactive
angular unconformity	Law of Faunal Succession	Precambrian	radiometric age
Cenozoic	mass spectrometer	Principle of Original	relative dating
correlation	Mesozoic	Horizontality	unconformity
daughter	Paleozoic	Principle of	
disconformity	parent	Superposition	
era			

Questions for Review

1. Describe the significance of the Principle of Superposition and Principle of Original Horizontality to relative dating of sedimentary sequences.

2. What is the distinction between a disconformity and an angular unconformity? What do they have in common?

3. Explain two ways in which you might determine the relative ages of a pluton and surrounding sedimentary rocks.

4. How is the correlation of rock units made easier by the concept of faunal succession? What is a limitation on its use?

5. Why is it important to radiometric dating that radioactive elements have constant half-lives?

6. Describe any three requirements that must be satisfied in order for a radiometric decay scheme to be useful in dating geological materials.

7. It has proven somewhat difficult to establish radiometric dates for the units of the Phanerozoic time scale because the subdivisions were defined using sedimentary rocks. Explain. How do geologists address this problem?

8. When the geologic time scale was first established, the Precambrian was not subdivided. Why?

9. Why is it not possible to determine the age of the earth directly by radiometric methods? On what basis is its age estimated?

For Further Thought

1. Fossil forms used for correlation are sometimes termed *index fossils*. The best index fossils are those found widely distributed over the earth and which derive from organisms that existed only for geologically short periods of time. Consider why these two criteria are important. What sorts of organisms can you suggest that might satisfy the first criterion especially well?

2. Each of the various radioactive isotopes has a distinct and unique half-life. What do you suppose would be the impact on radiometric dating if all radioisotopes had the same half-life? If there were only one naturally occurring radioisotope?

Suggestions for Further Reading

Dott, R. H., Jr., and Batten, R. L. *Evolution of the Earth.* 3d ed. New York: McGraw-Hill, 1981.

Eicher, D. L. *Geologic Time.* 2d ed. Englewood Cliffs, N.J.: Prentice-Hall, 1968.

Faure, G. *Principles of Isotope Geology.* 2d ed. New York: John Wiley and Sons, 1986.

Harland, W. B. "Geochronologic Scales." In American Association of Petroleum Geologists Studies in Geology #6, pp. 9–32, 1978.

Hurley, P. M. *How Old Is the Earth?* New York: Doubleday and Co., 1959. This classic, written in the early days of radiometric dating of geological materials, is particularly aimed at the nonspecialist reader.

Kummel, B. *History of the Earth.* 2d ed. San Francisco: W. H. Freeman and Co., 1970.

McLaren, D. J. "Dating and Correlation, a Review." In American Association of Petroleum Geologists Studies in Geology #6, pp. 1–7, 1978.

Moorbath, S. "The Oldest Rocks and the Growth of Continents." *Scientific American* 236 (March 1977): 92–104.

Part Two

The "internal processes" are so named not because they affect only the earth's interior but because they are ultimately driven primarily by the internal heat and stress within the earth. Plate tectonics, the subject of chapter 7, is the unifying concept that relates many of these processes as they affect the earth's crust and uppermost mantle. It can be shown that the brittle outer shell of the earth is broken up into a series of slabs or plates that shift in position over the earth's surface. The existence and movement of these plates explains much about why volcanic and earthquake activity occurs where it does, and it accounts for many other processes, such as mountain-building.

Chapters 8 and 9 deal in more detail with earthquakes and volcanic phenomena, respectively—what they are, where they happen, and why they happen. Each chapter also includes some discussion of the hazards posed by these natural geologic phenomena, and looks at the extent to which we can anticipate the events and minimize the negative impacts on humans.

Chapter 10 assembles the various kinds of evidence upon which geologists base their descriptions of the largely inaccessible interior of the earth. We have, in fact, a considerable body of relevant data that allows us to constrain estimates of the chemistry, physical state, pressure, and temperature of the earth's interior and to identify distinct zones within it. The part of the interior that we know best, however, is the crust, much of which is available for direct sampling, observation, and measurement. By way of transition from the deep interior to the surface process of Part Three, chapters 11 and 12 examine, respectively, the continental crust and the ocean basins. Included in these chapters is discussion of crustal structures, which can provide important clues to the kinds of processes by which the crust is formed or deformed.

Plate Tectonics

Introduction

More than a century ago, observers looking at world maps noticed the similarity in outline between the eastern coast of South America and the western coast of Africa (figure 7.1). In 1855, Antonio Snider went so far as to publish a sketch showing the two continents fitted together, jigsaw-puzzle fashion. Such reconstruction gave rise to the bold suggestion that perhaps these continents *had* once been part of the same landmass, which had later broken up.

This concept of **continental drift,** the idea that individual continents could shift position on the globe, had an especially vocal champion, Alfred Wegener, in the early part of this century. He began to publish on this subject in 1912 and continued to do so for nearly two decades. Several other prominent scientists found the idea plausible. However, most people, scientists and nonscientists alike, had considerable difficulty in conceiving how something as huge and massive as a continent could possibly "drift" around on a solid earth, or why it should do so. The majority of reputable scientists scoffed at the idea, or at best politely ignored it.

As it happens, most of the supporting evidence was simply undiscovered or unrecognized at the time. Beginning in the 1960s, data of many different kinds began to accumulate that indicated that the continents have indeed moved. Continental drift turns out to be one aspect of a broader theory known as **plate tectonics** that has evolved over the last two decades. **Tectonics** is the study of large-scale movement and deformation of the earth's outer layers. *Plate tectonics* relates such deformation to the existence and the movement of rigid plates of rock over a weak or plastic layer in the upper mantle.

The Early Concept of Continental Drift

Alfred Wegener had based his proposition of continental drift on several lines of evidence, of which the fit of the edges of the continents was only one.

Fossil Evidence

Wegener had become aware, for example, that paleontologists studying fossil remains had found some plant and animal forms that occurred in the rock record over limited areas of several continents now widely separated in space. Leaves of a fossil plant, *Glossopteris,* were found in southern Africa, Australia, South America, India, and even Antarctica. Certain dinosaur and other vertebrate remains were

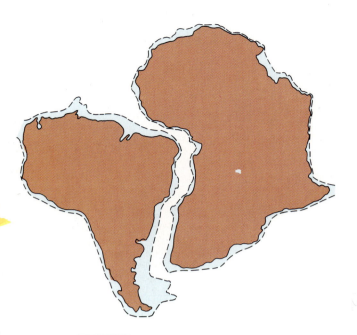

Figure 7.1

The jigsaw-puzzle fit of South America and Africa suggested that they might once have been joined together and were subsequently separated by continental drift.

similarly distributed. It was extremely difficult to imagine the same life form developing, identically and simultaneously, on those five continents, spaced as they are on the globe.

Some paleontologists postulated long-distance transport of seeds and spores by wind or water over the intervening oceans. For the larger animals, the past existence of land bridges between continents, now vanished, was proposed. Wegener suggested instead that at the time these organisms flourished, the continents had been together. They had split and drifted apart after the plant and animal remains had been entombed in the rocks.

Climatic Evidence

Climatic considerations seemed to provide more support. Many factors determine a region's climate, but a dominant one is latitude. In general, equatorial regions tend to be warmest, polar regions coldest, with more moderate temperatures in between. Sedimentary rocks, formed at the earth's surface, may preserve evidence of the climatic conditions under which they formed. Fossil remains of plants known to thrive in heat imply a tropical climate; sandstones in which windblown desert dunes are preserved suggest dry conditions; some sediments can be identified as having been deposited by glaciers. (See chapter 16 for some of the criteria by which glacial deposits are recognized.) In many cases, the ancient climate appears to

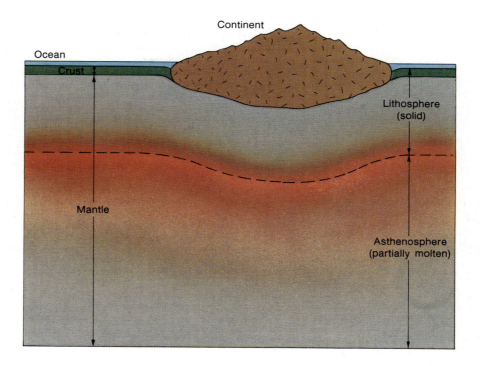

Continent

Ocean

Crust

Lithosphere
(solid)

Mantle

Asthenosphere
(partially molten)

Figure 7.2

The outer zones of the earth (not to scale). The terms *crust* and
mantle have compositional implications; *lithosphere* and
asthenosphere describe physical properties. The lithosphere
includes the crust and uppermost mantle. The asthenosphere lies
entirely within the upper mantle.

have differed drastically from the present climate of
the same region. There is, for example, evidence of
widespread past glaciation over India, where the
present climate is hardly conducive to formation of
large quantities of ice. Moreover, one cannot account
for such discrepancies simply in terms of global climatic changes, for the rock record does not show the
same warming or cooling trends simultaneously on all
continents. Wegener's interpretation was that rocks
indicating cold climatic conditions were indeed deposited when a continent was near a pole, and the
converse for rocks suggesting warm temperatures. Apparent changes in temperature regime were the result
of the continents' drifting from one latitude to another.

Limitations of the Early Model

Wegener saw the continental blocks, then, as plowing
through the ocean floor, piling up mountains at their
leading edges as they went. But there was no evidence
of disruption of the sea floor such as would be expected from this process. Nor could Wegener identify

a sufficiently powerful driving force that could move
continents without seriously violating basic physical
principles. Wegener died decades before his basic vi-
sion would be validated.

Plates and Drifting Continents

The problem of imagining solid continents moving
over solid earth can be eliminated with the realization
that the earth really is not completely solid from the
surface to the center of the core. This can be demonstrated by geophysical methods (see chapters 8 and
10). In fact, there is a plastic, partly molten zone relatively close to the surface. So in a sense, a thin, solid,
rigid skin of rock floats on a weaker, semisolid layer
below. The situation is illustrated schematically in
figure 7.2.

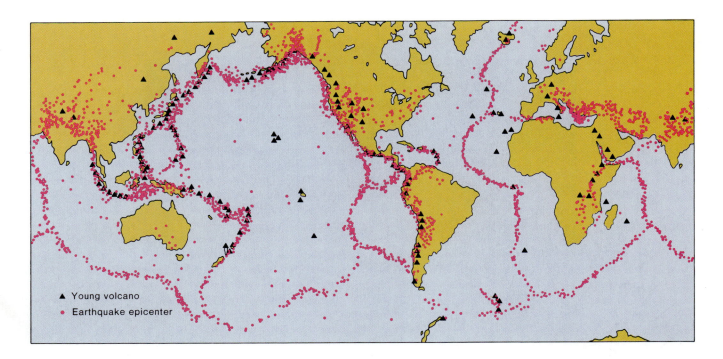

▲ Young volcano
● Earthquake epicenter

Figure 7.3

Locations of modern volcanoes and earthquakes around the world.

Source: Map plotted by the Environmental Data and Information Service of the National Oceanic and Atmospheric Administration; earthquakes from U.S. Coast and Geodetic Survey.

Lithosphere and Asthenosphere

The crust and uppermost mantle are quite solid. The outer solid layer is termed the **lithosphere**, from the Greek *lithos*, meaning "rock." The lithosphere varies in thickness from place to place on the earth. It is thinnest underneath the oceans, where it extends to a depth of about 50 kilometers (about 30 miles). The lithosphere under the continents is thicker, extending in places to depths of over 100 kilometers (60 miles). The next layer below the lithosphere, the **asthenosphere,** derives its name from the Greek for "without strength." This asthenosphere, which lies entirely within the upper mantle, extends to an average depth of about 500 kilometers (300 miles).

The lack of strength, or lack of rigidity, results in part from melting in the upper asthenosphere. This zone is not all molten: where melt is present, only a small percentage of magma exists in otherwise solid rock. Such a small amount of melt is nevertheless enough to cause the asthenosphere to behave plastically rather than rigidly. Even where melting has not actually occurred, temperatures are very close to melting temperatures. Given also the high confining pressures in the asthenosphere (tens of thousands of times atmospheric pressure), the rocks behave plastically under stress at these temperatures.

The presence of the asthenosphere makes continental drift more plausible. The continents need not drag across or plow through solid rock. Instead they can be pictured as sliding over a softened or slightly mushy layer underneath.

Locating Plate Boundaries

If one looks at the distribution of earthquakes and volcanic eruptions on a map (figure 7.3), it can be seen that these phenomena are far from uniformly distributed over the earth. They are for the most part concentrated in belts or linear chains. This pattern suggested that the rigid shell of lithosphere is cracked in places, broken up into pieces. The volcanoes and earthquakes are concentrated at the boundaries of these lithospheric plates, where plates jostle or scrape against one another. (The effect is somewhat like ice floes floating on an arctic sea: most of the grinding and crushing of ice and the spurting up of water from below occur at the edges of the blocks of ice, while their thick, solid central portions are relatively undisturbed.) About half a dozen very large lithospheric plates and many smaller ones have now been identified (figure 7.4).

how do they know?

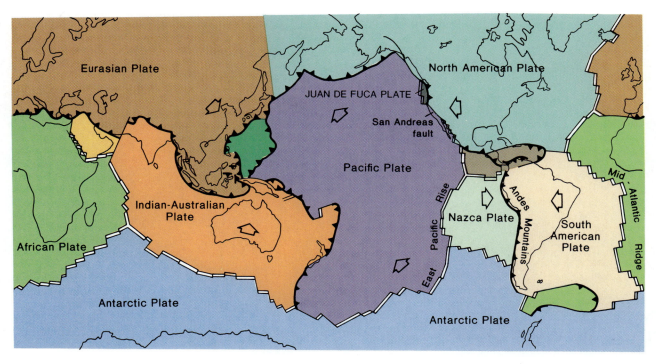

Labels on map: Eurasian Plate, North American Plate, JUAN DE FUCA PLATE, San Andreas fault, Pacific Plate, Indian-Australian Plate, African Plate, East Pacific Rise, Nazca Plate, Andes Mountains, South American Plate, Mid-Atlantic Ridge, Antarctic Plate, Antarctic Plate

Figure 7.4

Principal world lithospheric plates, inferred from information such as that shown in figure 7.3 and other data.

Source: After W. Hamilton, U.S. Geological Survey.

[handwritten note: 6 plates, 5 continents, 1 ocean]

Plate Movements— Accumulating Evidence

Recognition of the existence of the asthenosphere made plate motions more plausible but did not *prove* that they occur. Likewise, the apparent existence of discrete plates of rigid lithosphere near the earth's surface did not prove that those plates had ever moved. Much additional information had to be accumulated and explanations tested before the majority of scientists would accept the concept of plate tectonics. In time it even became possible to document the rates and directions of plate movements.

If South America and Africa have really moved apart, one might expect to see some evidence of this between them, some feature or features on the sea floor to indicate the continents' passage. A topographic map of the floor of the Atlantic Ocean shows an obvious ridge running north-south about halfway in between those continents (figure 7.5). This midocean ridge might be the seam from which they have moved apart. Similar ridges are found on the floors of other oceans. The details of their structures will be explored in

chapter 12. It remained for scientists studying the ages and magnetic properties of seafloor rocks to demonstrate the significance of the ocean ridges to plate tectonics.

Magnetism in Rocks—General

The earth possesses a magnetic field that can, at a first approximation, be described as similar to what would be expected from a huge bar magnet buried at the earth's center (figure 7.6). Magnetic lines of force run from the south magnetic pole to the north; the magnetic poles lie close to the geographic (rotational) poles, though they are not identical. A compass contains a needle that aligns itself parallel to the lines of the magnetic field, north-south, and points to magnetic north. A dip needle, a magnetic needle suspended in such a way that it can pivot vertically but not horizontally, will likewise dip along the trend of the magnetic field lines at that latitude (figure 7.7). Magnetic dip varies with latitude: the dip needle will be horizontal near the equator, vertical at either pole, and at some intermediate angle at latitudes in between.

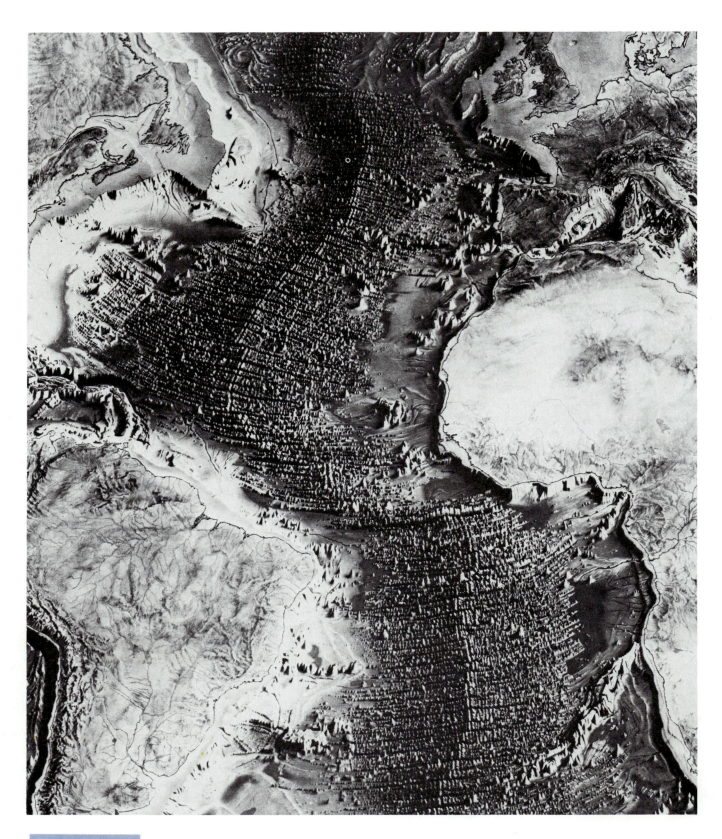

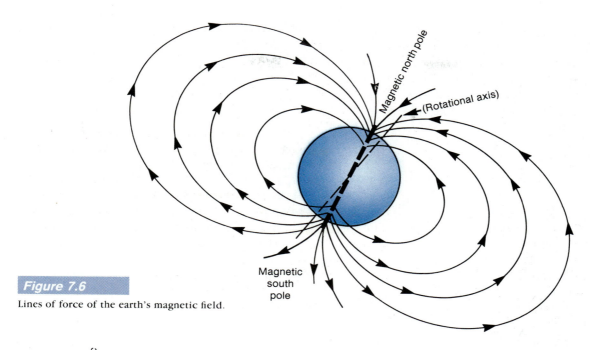

Figure 7.6

Lines of force of the earth's magnetic field.

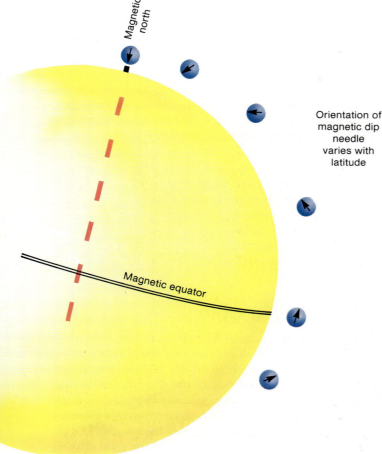

Orientation of magnetic dip needle varies with latitude

Figure 7.7

A dip needle will vary in orientation with latitude, lying horizontally at the magnetic equator, pointing vertically at the poles.

Most iron-bearing minerals are at least weakly magnetic at surface temperatures. Each magnetic mineral has a **Curie temperature** above which it loses its magnetic properties. The Curie temperature varies from mineral to mineral, but it is always below a mineral's melting temperature. A hot magma is therefore not magnetic, but as it cools and solidifies, and ferromagnesian silicates and other iron-bearing minerals crystallize from it, those magnetic minerals tend to line up in the same direction. Like tiny compass or dip needles, but free to move in three dimensions, they align themselves parallel to the lines of force of the earth's magnetic field. They point to magnetic north, and they indicate a magnetic dip consistent with their magnetic latitude. They retain their internal magnetic orientation unless they are heated again. This is the basis for the study of **paleomagnetism**, "fossil magnetism" in rocks.

Magnetic Reversals

However, magnetic north has not always coincided with its present position. In the early 1900s, scientists investigating the direction of magnetization of a sequence of young volcanic rocks in France discovered that some of the earlier flows appeared to be

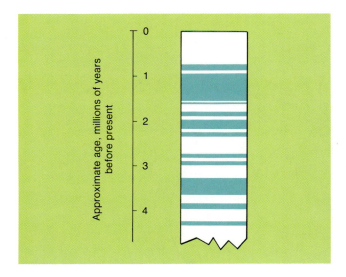

Approximate age, millions of years before present

0

1

2

3

4

Figure 7.8

A portion of the reversal history of the earth's magnetic field for the relatively recent geologic past.

magnetized in the opposite direction from the rest, their magnetic minerals pointing south instead of north. Confirmation of this discovery in many places around the world led to the suggestion, in the late 1920s, that the earth's magnetic field at some past time had flipped, or reversed polarity, with north and south magnetic poles switching places. Although we still do not know just why or how this occurs, the phenomenon of magnetic reversals has by now been well documented with the aid of radiometric dating.

Rocks crystallizing at times when the earth's field has been in its present orientation are said to be *normally magnetized;* rocks crystallizing when the field was oriented the opposite way are described as *reversely magnetized.* Over the history of the earth, the magnetic field has reversed many times, but rocks of a given age show a consistent polarity. Through the combined use of magnetic measurements and age determinations on the same rocks, it has been possible to reconstruct the reversal history of the earth's magnetic field in some detail. A portion of the recent reversal history is shown in figure 7.8, in which times of reversed magnetic orientation are shaded in green.

Paleomagnetism and Seafloor Spreading

The ocean floor is made up largely of basalt, rich in ferromagnesian minerals. During the 1950s, the first large-scale surveys of the magnetic properties of the sea floor were made, and they produced an entirely unexpected result. The floor of the ocean was found to consist of alternating stripes or bands of normally and reversely magnetized rocks, symmetrically arranged around the ocean ridges. At the time, this seemed so incredible that at first it was assumed that the instruments or measurements were faulty. However, other studies consistently obtained the same results. For several years, geoscientists were baffled.

Then, in 1963, an elegant explanation was proposed by the team of F. J. Vine and D. H. Matthews and, independently, by L. W. Morley. A few years previously, geophysicist Professor Harry Hess of Princeton University had suggested the possibility of **seafloor spreading,** the idea being that the sea floor had split and spread away from the ridges. Seafloor spreading could account very simply for the magnetic stripes on the sea floor.

If the oceanic lithosphere splits and plates move apart, this will begin to open a 50-kilometer-deep crack through the lithosphere. But the rift does not stay open, because as it begins to form it provides a path for the escape of some of the magma from the asthenosphere. The magma rises, cools, and solidifies to form new basaltic rock. It will become magnetized in the prevailing direction of the earth's magnetic field. If the plates continue to move apart, the new rock will also split and part, making way for more magma to form still younger rock, and so on.

If during the course of the seafloor spreading, the polarity of the earth's magnetic field reverses, the rocks formed after the reversal will be polarized oppositely from those formed before it. On the ocean floor, we have a continuous sequence of basalts formed over tens or hundreds of millions of years, during which time there have been dozens of polarity reversals. The basalts of the sea floor have acted as a sort of magnetic tape recorder throughout that time, preserving a record of polarity reversals in the alternating bands of normally and reversely magnetized rocks. The process is illustrated schematically in figure 7.9.

Age of the Sea Floor

If any geologic model is correct, it ought to be possible to use it to make predictions about other kinds of data before the corresponding measurements are made (the geologic version of the scientific method). The seafloor spreading model, in particular, implies that rocks of the sea floor should be younger close to the ridges and progressively older further away.

Specially designed research ships can sample sediment from the deep-sea floor and drill into the basalt beneath. The time at which an igneous rock like basalt crystallized from its magma can be determined radiometrically, as described in the last chapter. When

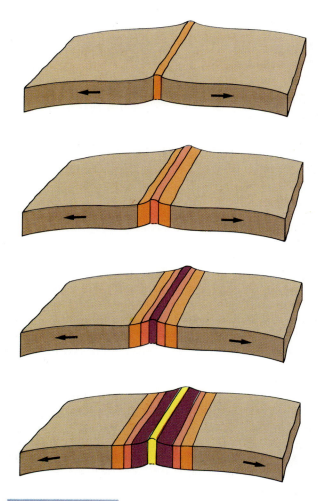

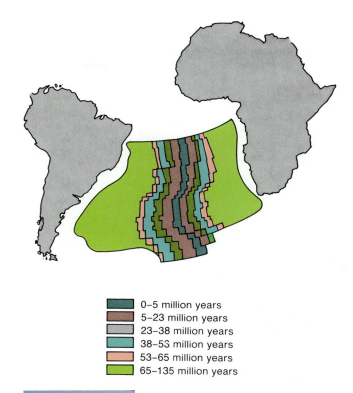

▮	0–5 million years
▮	5–23 million years
▮	23–38 million years
▮	38–53 million years
▮	53–65 million years
▮	65–135 million years

Figure 7.10

The pattern of seafloor ages on either side of the Mid-Atlantic Ridge reflects seafloor-spreading activity. Younger rocks are closer to the ridge.

Source: From a map by W. C. Pitman III, R. L. Larson, and E. M. Herron, 1974, *Geological Society of America.*

Figure 7.9

Formation of magnetic stripes on the sea floor. As each new piece of sea floor forms at the ridge, it becomes magnetized in a direction dependent on the orientation of the earth's field at that time. The magnetism is "frozen in" or preserved in the rock thereafter. Past reversals of the field are reflected in alternating bands of normally and reversely magnetized rocks.

this is done for many samples of seafloor basalt, the predicted pattern indeed emerges. The rocks of the sea floor, and the bottommost sediments deposited on them, are youngest close to the ocean ridges. They become progressively older farther away from the ridges on either side (see, for example, figure 7.10). Like the magnetic striping, the age pattern is also symmetric across each ridge, again as one would predict. This was powerful confirmation of the seafloor-spreading hypothesis. As spreading progresses, previously formed rocks are apparently continually spread apart and moved further from the ridge, while fresh magma rises from the asthenosphere to form new, younger lithosphere at the ridge. The oldest rocks recovered from the sea floor, well away from the ridges, turn out to be about 200 million years old.

Polar-Wander Curves

Evidence for plate movements does not come only from the sea floor. For reasons outlined in the next section, much older rocks are preserved on the continents than beneath the oceans, so longer periods of earth history can be investigated through continental rocks.

Studies of paleomagnetism in continental rocks can span many hundreds of millions of years and yield quite complex data. Magnetized rocks of different ages on a single continent may point to very different apparent magnetic pole positions. The apparent ancient magnetic north and south poles may not simply be reversed but may be rotated from the present magnetic north-south. When the directions of magnetization of many rocks of various ages from a single continent are determined and plotted on a map, it appears that the magnetic poles have meandered far over the surface of the earth, if the position of the continent is assumed to have been fixed on the earth throughout time. The resulting curve, showing the apparent

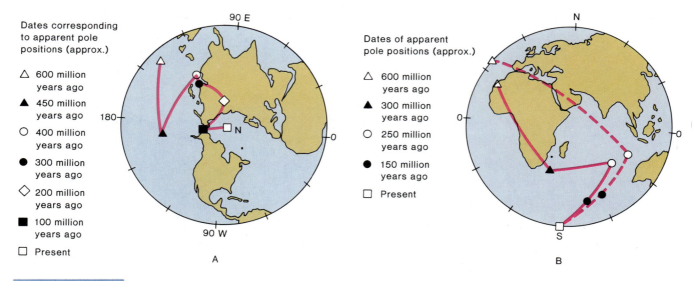

Figure 7.11

Examples of polar-wander curves. The apparent position of the magnetic pole relative to the continent is plotted for rocks of different ages, and these data points are connected to form the curve. The present positions of the continents are shown for reference. *(A)* Polar-wander curve for North America for the last 600 million years, as viewed from the North Pole. *(B)* Polar-wander curves for Africa (solid line) and Saudi Arabia (dashed) suggest that these landmasses moved quite independently up until about 250 million years ago, when the polar-wander curves converged.

From McElhinny, M. W., *Paleomagnetism and Plate Tectonics*. Copyright © 1973 Cambridge University Press. Reprinted by permission.

movement of the magnetic poles relative to one continent or region as a function of time, is the polar-wander curve for that landmass (see figure 7.11).

The very discovery and construction of polar-wander curves was initially troublesome, in that there are good geophysical reasons to believe that the earth's magnetic poles should remain close to the geographic (rotational) poles, as they now are. Worse yet, polar-wander curves for different continents do not match. Rocks of exactly the same age from two different continents may seem to point to two entirely different sets of magnetic poles!

This confusion can be eliminated, however, by a change in perspective. Suppose that in fact the magnetic poles *have* always remained close to the geographic poles, but the *continents* have moved and rotated. The polar-wander curves then provide a way to map the directions in which the continents have moved through time, relative to the approximately stationary magnetic poles and relative to each other.

The Jigsaw Puzzle Refined

We noted early in this chapter that the apparent similarity of the coastlines of Africa and South America triggered early speculation about the possibility of continental drift. Actually, the pieces of this jigsaw puzzle fit even better if we look not at coastlines but at the true edges of the continents (dashed lines in figure 7.1). These are the outer edges of the continental shelves, beyond which the depth of the ocean

increases rapidly to typical ocean-basin depths. Computer-aided manipulation of the puzzle pieces has been used to help determine the best possible physical fit of the pieces. The results are imperfect, perhaps because not every bit of continent has been perfectly preserved in the tens or hundreds of millions of years since continental breakup. However, overall the fit is remarkably good.

Continental reconstructions can be aided also by using details of continental geology—rock types, rock ages, evidence of glaciation, fossils, ore deposits, mountain ranges, and so on. If two now-separate continents were once part of the same landmass, then the geologic features presently found at the margin of one should have counterparts at the corresponding edge of the other. In short, the geology should match when the puzzle pieces are reassembled. An example of such matching is shown in figure 7.12.

Pangaea

Efforts to reconstruct the ancient locations and arrangements of the continents have been relatively successful, at least for the not-too-distant geologic past. It has been shown, for example, that a little more than 200 million years ago there was a single great supercontinent, which has been named **Pangaea** (from the Greek for "all lands"; see figure 7.13). The present seafloor spreading ridges are the lithospheric scars of the breakup of Pangaea. However, they are not the only kind of boundary found between plates.

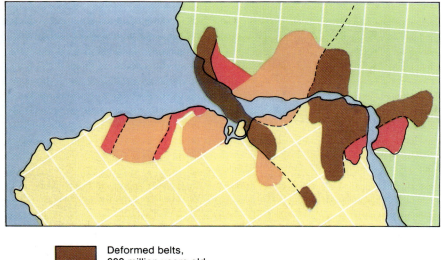

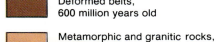
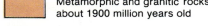
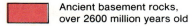

Deformed belts,
600 million years old

Metamorphic and granitic rocks,
about 1900 million years old

Ancient basement rocks,
over 2600 million years old

Figure 7.12

Geologic provinces of distinctive rock types and different ages can be correlated between western Africa and eastern South America.

After P. M. Hurley and J. R. Rand in *The Ocean Basins and Margins*, v. I, Plenum Publishing.

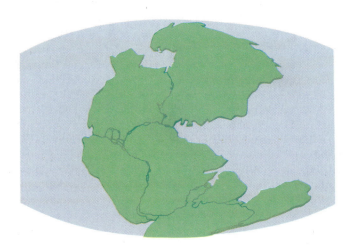

Figure 7.13

Pangaea, the ancient supercontinent of about 200 million years ago.

From R. S. Dietz and J. C. Holden, *Journal of Geophysical Research*, Vol. 75, pp. 4, 939–956. Copyright © 1970 by the American Geophysical Union.

Types of Plate Boundaries

Different things happen at the boundaries between lithospheric plates, depending on the relative motions of the plates and on whether continental or oceanic lithosphere is at the edge of each plate.

Divergent Plate Boundaries

We have already seen what happens at a **divergent boundary**, such as a midocean spreading ridge (figure 7.14). Lithospheric plates move apart, magma wells up from the asthenosphere, and new lithosphere is created. A great deal of volcanic activity thus occurs at spreading ridges. The pulling apart of the plates of the lithosphere results, in addition, in earthquakes along these ridge plate boundaries (see also chapter 8).

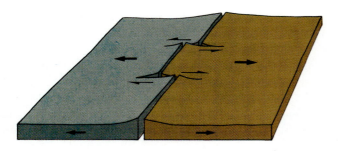

Figure 7.14

A spreading ridge, a divergent plate boundary. Arrows indicate direction of plate motions. Note transform faults between offset ridge segments (see figure 7.16 for detail).

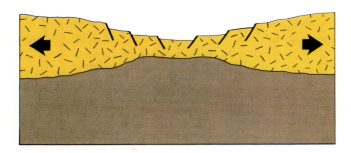

Figure 7.15

Continental rifting. The continental crust is thinned and fractured, and in time a new ocean basin is formed.

Continents can be rifted apart too (figure 7.15). This is less common, perhaps because continental lithosphere is so much thicker than oceanic lithosphere. In the early stages of continental rifting, volcanoes may erupt along the rift, or great flows of basaltic lava may pour out through fissures in the continent. As the continental crust is stretched and thinned, a shallow inland sea may inundate the rift zone. If the rifting continues, a new ocean basin floored with basalt will eventually form between the torn-apart pieces of continent. The Red Sea formed from such a rift in continental lithosphere. Further rifting is slowly ripping apart East Africa: the easternmost strip of the continent is being rifted apart from the rest, and another ocean may one day separate the resulting pieces.

Transform Faults

Actually, the structure of a seafloor spreading ridge is more complex than a single straight crack. A close look at a midocean spreading ridge (figure 7.5) reveals that it is not a continuous break thousands of kilometers

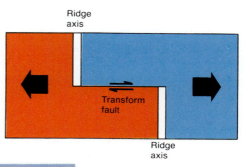

Figure 7.16

Transform fault between offset segments of a spreading ridge (map view). Arrows indicate direction of plate movement. Along the transform fault between ridge segments, plates move in opposite directions.

From Plummer, Charles C., and David McGeary, *Physical Geology*, Third Edition. © 1979, 1982, 1985 Wm. C. Brown Publishers, Dubuque, Iowa. All Rights Reserved. Reprinted by permission.

long. Rather, ridges consist of many short segments slightly offset from one another. The offsets are a special kind of fault, or break in the lithosphere, known as a **transform fault** (figure 7.16). The opposite sides of a transform fault belong to two different plates, which are moving apart in opposite directions. As the plates scrape past each other, earthquakes occur along the transform fault between ridge segments. The lack of earthquake activity along the fracture extensions beyond the spreading ridges is a consequence of the fact that in these regions, both sides of the fracture belong to the same plate and move in the same direction.

The famous San Andreas fault in California is an example of a transform fault that slices a continent sitting along a spreading ridge. The East Pacific Rise, a seafloor spreading ridge off the northwestern coast of North America, disappears under the edge of the continent, to reappear further south in the Gulf of California (see figure 7.4). The San Andreas is the transform fault between these segments of spreading ridge. Most of North America is part of the North American plate. The thin strip of California on the west side of the San Andreas fault, however, is moving northwest with the Pacific plate.

Convergent Boundaries

At a **convergent** plate boundary, as the name indicates, plates are moving toward each other. Just what happens depends on what sort of lithosphere is at the leading edge of each plate. Continental lithosphere is relatively low in density and buoyant, so it tends to float on the asthenosphere. Oceanic lithosphere is closer in density to the underlying asthenosphere, so it is more easily forced down into the asthenosphere as the plates move together.

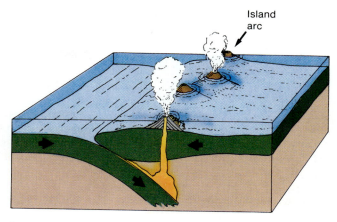

Island
arc

A

Figure 7.17

Continent-continent collision at a convergent boundary. Rocks are deformed and some lithospheric thickening occurs, but neither plate is subducted to any great extent.

In a continent-continent collision, the two land-masses come together, crumple, and deform (figure 7.17). One may partially override the other, but neither sinks into the mantle, and a very large thickness of continent may result. Earthquakes are also frequent during active collision, as a result of the large stresses involved in the process. The extreme height of the Himalaya Mountains is attributed to just this sort of continent-continent collision. India was not always a part of the Asian continent. Paleomagnetic evidence indicates that it drifted northward over tens of millions of years until it ran into Asia. The Himalayas were built up in this collision.

More commonly, oceanic lithosphere is at the leading edge of one or both of the converging plates. One plate of oceanic lithosphere may be pushed under the other plate and descend into the asthenosphere. This type of plate boundary, where one plate is carried down below (*subducted* beneath) another, is called a **subduction zone** (figure 7.18). It is the subduction zones of the world that balance the seafloor equation: if new oceanic lithosphere is constantly being created at spreading ridges, an equal amount must be destroyed somewhere or the earth would simply keep getting bigger. This excess sea floor is consumed in subduction zones. The subducted plate is heated by the hot asthenosphere and in time becomes hot enough to melt. Some of the melt rises to form volcanoes on the overriding plate. Some of the melt may eventually migrate to and rise again at a spreading ridge, to make new sea floor. So in a way, the oceanic lithosphere is constantly being recycled through this process. This explains why no very ancient seafloor rocks are known. The buoyant conti-

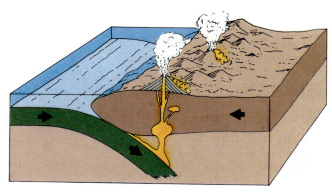

B

Figure 7.18

Subduction zone formed at a convergent boundary when a slab of oceanic lithosphere is forced into the mantle. *(A)* Ocean-ocean convergence: an island arc is formed by volcanic activity above the subducted slab. *(B)* Ocean-continent convergence: the magmatic activity accompanying subduction produces volcanic mountains on the continent.

nents are not so easily reworked; very old rocks may thus be preserved on the continents.

Subduction zones are geologically very active places. Sediments eroded from the continents may accumulate in the trench formed by the downgoing plate. Some of these sediments are caught in the fractured oceanic crust and carried down into the asthenosphere to be melted along with the sinking lithosphere. Volcanoes form where the melted material rises up through the overlying plate to the surface. Where the overriding plate also consists of oceanic lithosphere, the volcanoes may form a string of islands known as an **island arc** (figure 7.18A). Where continent overrides ocean, the volcanoes will be built up as mountains on the continent (figure

Box 7.1

Plate Tectonics and Radioactive Waste: New Solutions?

When plate-tectonic theory was being developed and the existence and nature of subduction zones discovered, some began to see subduction zones as the ultimate in waste disposal. In particular, they were suggested as possible disposal sites for canisters of radioactive wastes.

The appeal of the idea is the image of these potentially hazardous materials being carried deep into the earth, there to vanish, effectively, forever (at least from a human perspective). The principal drawbacks are twofold. First, given the extremely slow rates of plate motion, complete subduction of wastes would take thousands of years at least. Unless the waste canisters were somehow emplaced well into the subducted plate, they would sit exposed to interaction with seawater in the meantime. Seawater, especially seawater warmed by decaying radioactive wastes, is highly corrosive and might breach the canisters, allowing leakage of the wastes. Secondly, a plate undergoing subduction does not slip quietly under the plate above like a spatula under a pancake. Subduction is accompanied by earthquakes and deformation of the plates and the sediments carried on them. Some of the overlying sediments are not in fact subducted but are scraped off onto the overriding plate. It would be difficult to be certain that waste canisters would not be ruptured or caught up in the scraped-off sedimentary pile.

The popularity of this scheme has waned. However, now that the significance of plate tectonics to the formation and destruction of the sea floor has been recognized, there is interest in using the stable interiors of large oceanic plates as possible waste-disposal sites. An international committee is presently studying this idea.

7.18B). The great stresses involved in the convergence and subduction give rise to numerous earthquakes. Parts of the world near or above subduction zones, and therefore prone to both volcanic and earthquake activity, include the Andes region of South America, western Central America, parts of the northwestern United States and Canada, the Aleutian Islands, China, and Japan, and much of the rim of the Pacific Ocean basin (see figure 7.4). Further details of the nature of volcanism and earthquake distribution at subduction zones is included in the next two chapters.

How Far, How Fast, How Long, How Come?

Rates and directions of plate movement can be determined in a variety of ways. Radiometric dating also contributes to answering the question of how long plate-tectonic processes have been active.

Past Motions, Present Velocities

We have already looked at the use of polar-wander curves from continental rocks to determine the directions in which the continents have drifted. The direction of seafloor spreading is usually obvious: away from the ridge. Rates of seafloor spreading can be found very simply by dating rocks at different distances from the spreading ridge and dividing the distance moved by the rock's age (the time it has taken to move that distance from the ridge at which it formed), as illustrated in the last chapter.

Yet another way to monitor rates and directions of plate movement is by using mantle hot spots. These are isolated areas of volcanic activity, usually not associated with plate boundaries. It is believed that the existence of these volcanoes reflects unusual mantle conditions beneath the hot spots, where melting of the asthenosphere is more extensive. Possible causes of hot spots will be explored further in chapter 9. If we assume that mantle hot spots remain fixed in position while the lithospheric plates move over them, the result should be a string of volcanoes of differing ages, the youngest closest to the hot spot.

Emperor Seamounts

Midway 25

Necker 11.3

Hawaiian Chain   Nihoa 7.5

Oahu 2.3–3.3

Hawaii Now Active

Figure 7.19

The Hawaiian islands and other volcanoes in a chain formed over
a hot spot. Numbers indicate the dates of volcanic eruptions, in
millions of years. Movement of the Pacific plate has carried the
older volcanoes far from the hot spot, now under the active
volcanic island of Hawaii.

A portion of the "World Ocean Floor Panorama" by Bruce C. Heezen and
Marie Tharp. Copyright 1977 © Marie Tharp.

A good example can be seen in the north Pacific
ocean (see figure 7.19). A topographic map shows an
L-shaped chain of volcanic islands and submerged
volcanoes. When rocks from these volcanoes are dated
radiometrically, they show a progression of ages, from
about 75 million years at the northwestern end of the
chain to about 40 million years at the bend, through
progressively younger islands to the still active volcanoes of the island of Hawaii at the eastern end of
the Hawaiian Island group. The latter now sit over the
hot spot responsible for the whole chain. The age progression is reflected not only in the radiometric dates
but in topography and surface features: the farther

west one goes in the chain, in general, the more extensively weathered and eroded the island and the
lower its relief above the surrounding ocean.

From the distances and age differences between
pairs of points in the chain, we can again determine
the rate of plate motion. For instance, Midway Island
and Hawaii are about 2700 kilometers apart. The volcanoes of Midway were active about 25 million years
ago. Over the last 25 million years, then, the Pacific
plate has moved over the mantle hot spot at an average rate of 2700 km/25,000,000 yr, or about 11
centimeters per year. The orientation of the volcanic
chain shows the direction of plate movement, most

recently west-northwest. From the kink in the chain at about 40 million years ago, it appears that the Pacific plate's direction of movement changed at that time.

Hot spots occur under continents as well as beneath ocean basins. They are, however, somewhat easier to detect in oceanic regions, perhaps because the associated magmas can more readily work their way up through the thinner oceanic lithosphere.

Average Rates of Motion

Looking at many such determinations from all over the world, geologists find that average rates of plate motion are 2 to 3 centimeters (about 1 inch) per year. In a few places, movement at a rate of up to about 18 centimeters per year is observed, and elsewhere rates may be slower, but a few centimeters per year is typical. This seemingly trivial amount of motion does add up through geologic time. A movement of 2 centimeters per year for 100 million years means a shift of 2000 kilometers, or about 1250 miles! When the motions are extrapolated back into the past we can reconstruct the breakup of Pangaea (figure 7.20).

Why Do Plates Move?

Why does all this happen? A driving force for plate tectonics has not been definitely identified. Several possibilities are consistent with existing data. Probably the most widely accepted explanation, first suggested by Hess in connection with his seafloor-spreading model, is that the plastic asthenosphere is slowly churning in large **convection cells** (figure 7.21). According to this scenario, hot mantle material rises at the spreading ridges; some escapes as magma to form new lithosphere, but most does not. The rest spreads out sideways beneath the lithosphere, slowly cooling in the process. As it flows outward it drags the overlying lithosphere outward with it, thus continuing to open the ridges. When it has cooled sufficiently, the flowing material becomes dense enough to sink back deeper into the mantle. This may be happening under subduction zones.

There is continuing debate about the vertical dimensions of the convection cells. They are not necessarily confined to the asthenosphere. There is geophysical evidence that oceanic lithosphere can be subducted to depths of about 700 kilometers; perhaps convection cells operate down to those depths. Some have proposed convection spanning virtually the whole mantle, from the base of the lithosphere to the core-mantle boundary, 2900 kilometers down. There is no definite evidence on which to choose between these possibilities.

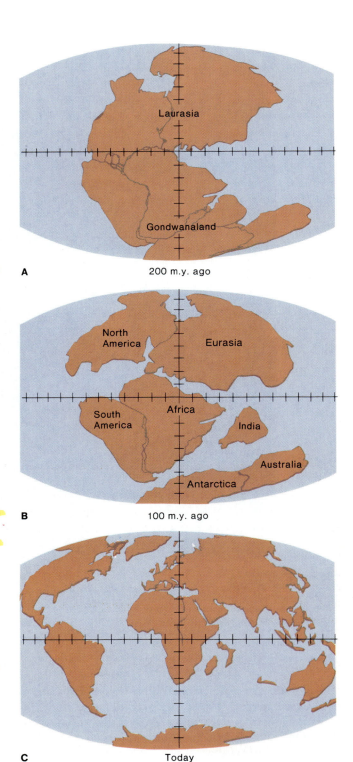

A   200 m.y. ago

B   100 m.y. ago

C   Today

Figure 7.20

Reconstructed plate movements during the last 200 million years: the breakup of Pangaea. Although these changes occur very slowly on the time scale of a human lifetime, they illustrate the magnitude of the natural forces to which we must adjust.

From R. S. Dietz and J. C. Holden, *Journal of Geophysical Research*, Vol. 75, pp. 4, 939–956. Copyright © 1970 by the American Geophysical Union.

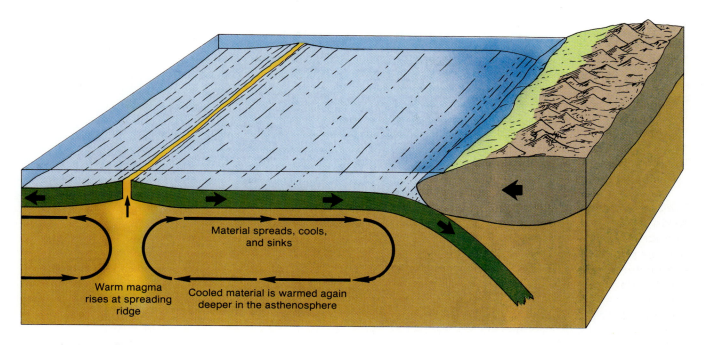

Material spreads, cools, and sinks

Warm magma rises at spreading ridge

Cooled material is warmed again deeper in the asthenosphere

Figure 7.21

A possible driving force behind plate tectonics is slow convection in the weak, partly molten asthenosphere.

An alternative explanation for plate motions is that the weight of the dense, cold, downgoing slab of lithosphere in the subduction zone pulls the rest of the trailing plate along with it, opening up the spreading ridges so magma can ooze upward. In this case friction between the plate and asthenosphere below would help to drive convection as the plates move, rather than the reverse. The full answer may be a combination of these mechanisms, and other mechanisms that have not yet been considered may contribute.

Antiquity of Plate Tectonics

It is not entirely clear how long plate-tectonic processes have been active. Certainly we observe the magnetic stripes characteristic of seafloor spreading over even the oldest, 200-million-year-old ocean floor. From continental rocks, we can reconstruct apparent polar-wander curves going back more than a billion years, although the relative scarcity of undisturbed ancient rocks makes such efforts more difficult for the earth's earliest history. It seems clear that the continents have been shifting in position over the earth's surface for at least 1 to 2 billion years, though not necessarily at the same rates or with exactly the same results as at present. For instance, if mantle convection is responsible, it may well have been more rapid

in the past when the earth's interior was hotter. Plate-tectonic processes of some sort have certainly long played a major role in shaping the earth. They are likely to continue doing so for the foreseeable future.

Plate Tectonics and the Rock Cycle

In chapter 2, we noted that all rocks may be considered related by the concept of the rock cycle. We can also look at the rock cycle in a plate-tectonic context, as illustrated in simplified form in figure 7.22.

New igneous rocks form from magmas rising out of the asthenosphere, at spreading ridges or over subduction zones. The heat radiated by the cooling magmas can cause metamorphism in the continental crust, with recrystallization at elevated temperature changing the texture and/or the mineralogy of the surrounding rocks. Some of these surrounding rocks may themselves melt to form new igneous rocks. The forces of plate collision at convergent margins contribute to metamorphism also, by increasing the pressures and directed stresses acting on the rocks.

Weathering and erosion on the continents wear down pre-existing rocks of all kinds into sediment. Much of this sediment is eventually transported to the edges of the continents, where it is deposited in deep basins and trenches. Through burial under more layers

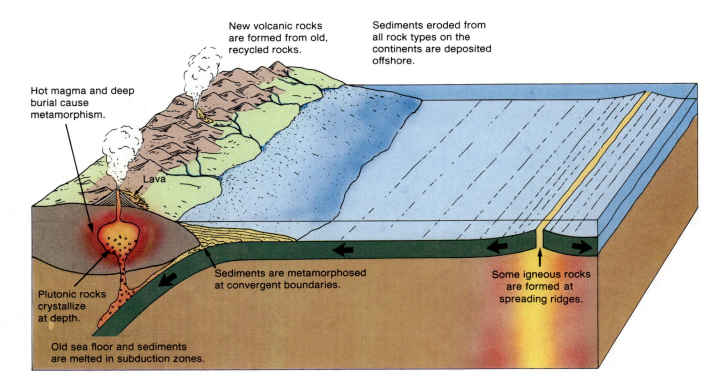

New volcanic rocks are formed from old, recycled rocks.

Sediments eroded from all rock types on the continents are deposited offshore.

Hot magma and deep burial cause metamorphism.

Lava

Plutonic rocks crystallize at depth.

Old sea floor and sediments are melted in subduction zones.

Sediments are metamorphosed at convergent boundaries.

Some igneous rocks are formed at spreading ridges.

Figure 7.22

The rock cycle, interpreted in plate-tectonic terms.

of sediment, it may become lithified into sedimentary rock. Sedimentary rocks in turn may be metamorphosed or even melted by the stresses and the igneous activity at the plate margins. Some of these sedimentary or metamorphic materials may also be carried down with subducted oceanic lithosphere, to be melted and eventually recycled as igneous rock. Plate-tectonic activity thus plays a large role in the process of formation of new rocks from old that is constantly underway on the earth.

Summary

The outermost solid layer of the earth is the 50- to 100-kilometer-thick lithosphere, which is broken up into a series of rigid plates. The lithosphere is underlain by a plastic, partly molten layer of the mantle, the asthenosphere, over which the plates can move. This plate motion gives rise to earthquakes and volcanic activity at the plate boundaries. At seafloor spreading ridges, which are divergent boundaries, new sea floor is created from magma rising from the asthenosphere. The sea floor moves in conveyor-belt fashion, ultimately to be destroyed in subduction zones, a type of convergent plate boundary, where it is carried down into the asthenosphere and eventually remelted. Convergence of continents forms high mountain ranges.

Evidence for seafloor spreading includes the distribution of ages of seafloor rocks, and magnetic stripes on the ocean floor. Continental drift can be demonstrated by such means as polar-wander curves and evidence of ancient climates as revealed in the rock record. Past "supercontinents" can be reconstructed by fitting together modern continental margins and matching up similar geologic features and fossil deposits from continent to continent.

Present rates of plate movement average a few centimeters a year. A mechanism for moving the plates has not been proven definitively. The most likely driving force is slow convection in the asthenosphere (and perhaps in the deeper mantle). Although plate motions are less readily determined in ancient rocks, plate-tectonic processes have probably been more or less active for much of the earth's history. They play an integral part in the rock cycle. We will see in more detail in later chapters how plate tectonics can be related to earthquake occurrence, volcanic activity, and formation of mineral deposits.

Box 7.2

Paired Metamorphic Belts and Plate Tectonics

Along many ancient convergent margins we find features known as **paired metamorphic belts.** These consist of an elongated zone of low-temperature, high-pressure metamorphism (typically blueschist facies) parallel to a belt of high-temperature, low- to moderate-pressure regional-metamorphic rocks (commonly amphibolite grade). Before the development of plate tectonics, these paired belts were puzzling. Now they can be understood in terms of what takes place at a subduction zone (figure 1).

Rocks are sufficiently poor heat conductors that a downgoing slab of oceanic lithosphere stays cold relative to the surrounding mantle for some time. As it is subducted, sediments and perhaps some seafloor basalts are scraped off against the overriding plate at the trench, forming an **accretionary prism** or wedge of material added to the overriding plate. The sediments of the accretionary prism likewise stay relatively cool, insulated by the downgoing slab, but they are subject to intense stresses. Thus the characteristic blueschist-facies metamorphism.

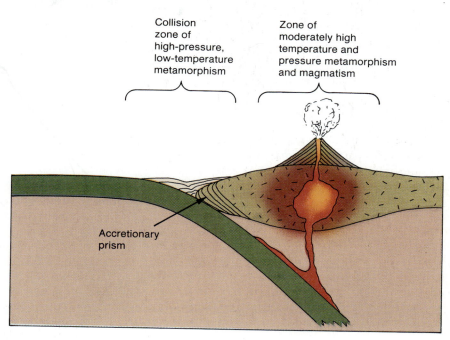

Figure 1
Paired metamorphic belts, a consequence of plate tectonics.

Further into the overriding slab, magmas formed by melting of subducted material and overlying mantle rise into the crust. This pervasive magmatism, coupled with convergence-related stress, produces the higher-temperature metamorphism of the parallel metamorphic belt.

Terms to Remember

accretionary prism
asthenosphere
continental drift
convection cells
convergent boundary

Curie temperature
divergent boundary
hot spots
island arc
lithosphere

paired metamorphic belts
paleomagnetism
Pangaea
plate tectonics
polar-wander curve

seafloor spreading
subduction zone
tectonics
transform fault

Questions for Review

1. Are the terms *lithosphere* and *asthenosphere* equivalent, respectively, to *crust* and *mantle*? Explain.

2. What property of the asthenosphere gives it its name? Why does it behave in this way?

3. How are major plate boundaries identified?

4. What is a Curie temperature, and what does it have to do with paleomagnetic studies?

5. Describe the origin of the magnetic stripes on the sea floor.

6. The ages of seafloor rocks and sediments show a regular pattern around a spreading ridge. Describe it.

7. What is a polar-wander curve? Is the name an accurate one?

8. What is a transform fault, and where are such faults found?

9. Contrast what occurs at a convergent plate boundary when the advancing edge of each plate is continental lithosphere with what occurs when one plate is oceanic lithosphere. How does this help to account for the relative youth of the sea floor?

10. What is an island arc, and where and how would one form?

11. Describe two means of determining rates of plate motion.

12. Convection cells in the asthenosphere may drive plate motions. Explain.

13. Briefly explain the rock cycle in the context of plate tectonics.

For Further Thought

1. It has been proposed (not by geologists) that the phenomenon of apparent polar wander can be explained by a flipping of the earth's whole crust as a single unit, rather than by plate tectonics and independent movements of different continents. Can you suggest any evidence that is inconsistent with that proposal?

2. The moon, though smaller than the earth, has a much thicker lithosphere: the moon's radius is only 1740 kilometers, yet its lithosphere is about 1000 kilometers thick. Would you expect plate-tectonic activity and subduction to occur on the moon as they do on earth? Why or why not?

Suggestions for Further Reading

Bird, J. M., ed. *Plate Tectonics.* 2d ed. Washington, D.C.: American Geophysical Union, 1980.

Burke, K., and Wilson, J. T. "Hot Spots on the Earth's Surface." *Scientific American* 235 (August 1976): 46–57.

Condie, K. *Plate Tectonics and Crustal Evolution.* 2d ed. New York: Pergamon Press, 1982.

Dewey, J. F. "Plate Tectonics." *Scientific American* 226 (May 1972): 56–68.

Dietz, R. S., and Holden, J. C. "The Breakup of Pangaea." *Scientific American* 223 (April 1970): 30–41.

Hurley, P. M. "The Confirmation of Continental Drift." *Scientific American* 218 (April 1968): 52–64.

Marvin, U. B. *Continental Drift: The Evolution of a Concept.* Washington, D.C.: Smithsonian Institution Press, 1973.

McElhinny, M. W. *Paleomagnetism and Plate Tectonics.* Cambridge, U.K.: Cambridge University Press, 1973.

Molnar, P., and Tapponier, P. "The Collision between India and Eurasia." *Scientific American* 236 (April 1977): 30–41.

Wegener, A. *The Origin of Continents and Oceans.* London: Methuen, 1924.

Weyman, D. *Tectonic Processes.* London: George Allen and Unwin, 1981.

Earthquakes

Introduction

Averaged over millennia, plate motions are slow, but they are not always so on a shorter time scale. When the strength of the lithosphere fails and it snaps or shifts suddenly in response to built-up stress, the result is the phenomenon we call an earthquake. In this section, we will survey the nature of earthquakes and their causes and effects. We will also look at some approaches to minimizing the potential damage from earthquakes and the current status of earthquake prediction efforts.

Basic Terms and Principles

Major earthquakes are among the more dramatic demonstrations of the fact that the earth is a dynamic, changing system. They occur along **faults**, planar breaks in rock along which there is displacement of one side relative to the other.

Creep

When movement along existing faults occurs gradually and smoothly, it is termed **creep.** Creep causes broken curbstones, offset fences, and the like (figure 8.1). In buildings built directly across a fault, walls can be stressed and deformed to the point of failure over a period of time. Damage is very localized, however, and lives are rarely lost as a consequence of creep.

Figure 8.1

Curbstone broken by creep along the Hayward fault, Hollister, California.

Photo courtesy U.S. Geological Survey.

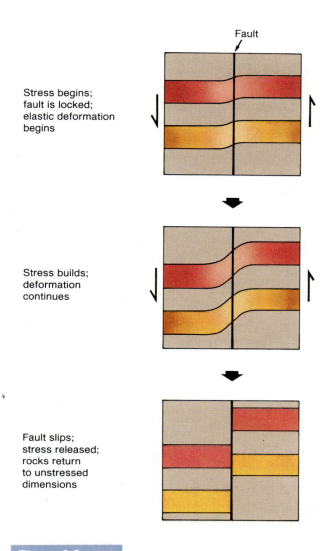

Stress begins; fault is locked; elastic deformation begins

Stress builds; deformation continues

Fault slips; stress released; rocks return to unstressed dimensions

Fault

Figure 8.2

The phenomenon of elastic rebound along fault zones. The rocks deform elastically under stress until failure, then snap back to their original, undeformed condition after the earthquake.

Earthquakes and Elastic Rebound

When friction between rocks on either side of a fault is such as to prevent them from slipping easily, or when the rock under stress is not already fractured, some elastic deformation will occur before failure (figure 8.2). When the stress at last exceeds the rupture strength of the rock (or the friction between rocks along an existing fault), sudden movement occurs along the fault: an **earthquake.** The stressed rocks, released by the rupture, snap back elastically to their previous dimensions, a phenomenon known as **elastic rebound.** The occurrence of movement and stress release is reflected in the displacement of the rocks on either side of the fault following the earthquake.

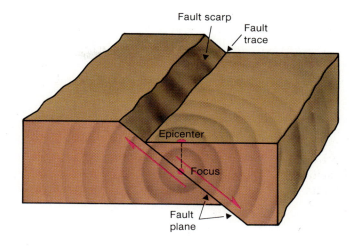

Figure 8.3

Simplified diagram of a fault, illustrating component parts and associated earthquake terminology.

Faults and fractures come in all sizes, from microscopically small to thousands of kilometers long. Likewise, earthquakes come in all sizes, from tremors so small that even sensitive instruments can barely detect them, to massive shocks that can level cities.

Earthquake Terminology

The point on a fault at which the first movement or break occurs during an earthquake is called its **focus,** or *hypocenter* (figure 8.3). In the case of a large earthquake, a section of fault many kilometers long may slip, but there is always a point at which the first movement occurs, and this is identified as the focus. The point on the earth's surface directly above the focus is called the **epicenter.** When news accounts tell where an earthquake happened, it is the location of the epicenter that they report. The line along which the fault plane intersects the earth's surface is the fault trace. If there is vertical movement along the fault, the cliff thus formed is called the fault **scarp.** Note that subsequent erosion may cause the scarp to be eroded back from its original position along the fault trace proper.

Earthquake Locations

If we look at a map showing the locations of major earthquake epicenters over nearly a decade (figure 8.4A), we see that they are concentrated in linear belts corresponding to plate boundaries. Not all earthquakes occur at plate boundaries, but most do. These areas represent the places where plates are jostling

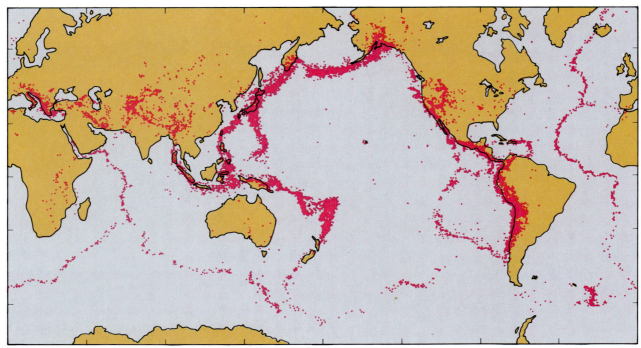

A

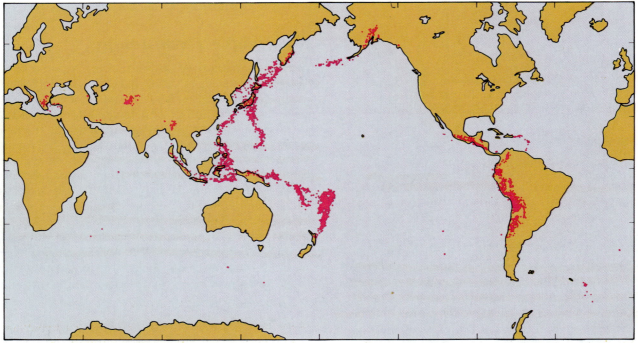

B

Figure 8.4

World earthquake epicenters, 1961–67, from U.S. Coast and
Geodetic Survey. *(A)* All earthquakes. *(B)* Earthquakes with focal
depths greater than 100 kilometers.

From Barazangi and Dorman, *Bulletin of Seismological Society of America*, 1969. Reprinted by permission.

each other, colliding, sliding past each other, places where very large stresses due to plate movements may be built up, and places where major faults or breaks already exist on which further movement can take place.

A map showing only deeper-focus earthquakes (figure 8.4B), earthquakes with focal depths of over 100 kilometers, looks somewhat different: the spreading ridges have disappeared, while subduction zones are still shown by their frequent earthquakes. The explanation is that earthquakes occur in the lithosphere, where rocks are more rigid and brittle and therefore capable of breaking or slipping suddenly. In the plastic, partly molten asthenosphere, material flows, rather than snaps, under stress. Therefore the deep-focus earthquakes are concentrated in subduction zones, where brittle lithosphere is pushed deep into the mantle. Note also that these earthquakes cannot be the result of friction between plates, for they occur within the single subducted plate as stresses are abruptly released.

Seismic Waves and Earthquake Location

When an earthquake occurs, the stored-up energy is released in the form of **seismic waves** that travel away from the focus. There are several types of seismic waves. **Body waves** (*P-waves* and *S-waves,* described below) travel through the interior of the earth; **surface waves,** as their name suggests, travel along the surface.

Types of Body Waves

P-waves are compressional waves. As they travel through matter, it is alternately compressed and expanded. P-waves travel through the earth, then, much as sound waves travel through air. A Slinky Toy can illustrate a compressional type of wave. Stretch the coil out to a length of several feet along a smooth surface, hold one end in place, push the other end in suddenly, then hold that end still also. A pulse of compressed coil will travel away from the end moved (figure 8.5A).

S-waves are shear waves, involving a side-to-side sliding motion of material. The same toy can be used to demonstrate shear-type waves. Stretch the coil out as before, but this time twitch one end of the coil sideways (perpendicular to its length). As the wave moves along the length of the Slinky, the loops of coil move sideways relative to each other, not closer together and farther apart as with the compressional wave (figure 8.5B).

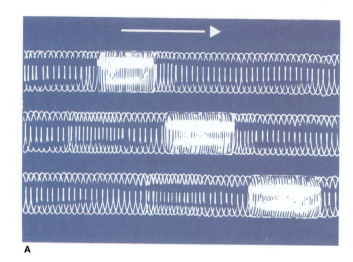

A

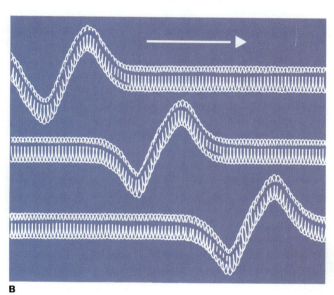

B

Figure 8.5

Schematic diagram of seismic body waves, illustrated using a Slinky Toy. In both cases, the time sequence is from top of figure to bottom; arrows indicate direction in which wave is traveling. *(A)* P-wave (compressional). *(B)* S-wave (shear).

Locating the Epicenter

Seismic body waves give us a way to locate earthquake epicenters. P-waves travel faster through rocks than do S-waves. Therefore at points some distance from the scene of an earthquake, the first P-waves will arrive somewhat before the first S-waves. Both cause ground motions detectable using a **seismograph** (figure 8.6). A trained seismologist can distinguish these first arrivals of P- and S-waves from other ground

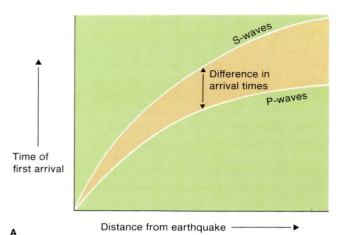

A

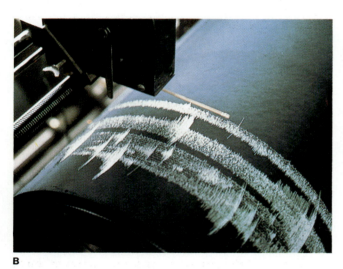

B

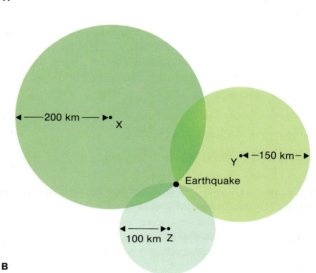

B

Figure 8.6

A seismograph, used for detecting and recording ground motions.
(A) Several seismographs, recording data from several sites.
(B) The larger the ground motion, the greater the oscillation of
the pen.

(A) Photo by F. W. Osterwald, courtesy U.S. Geological Survey. *(B)* Photo
courtesy of U.S. Geological Survey.

Figure 8.7

Use of seismic waves in locating earthquakes. *(A)* Difference in
times of first arrivals of P-waves and S-waves is a function of the
distance from the focus. *(B)* Triangulation using data from
several seismograph stations allows location of the earthquake.

displacements (including surface waves and background noise like passing traffic). The difference in
arrival times of the first P- and S-waves is a function
of distance to the earthquake.

The effect can be illustrated by considering a pedestrian and a bicyclist traveling the same route,
starting at the same time. If the bicyclist can travel
faster, he or she will arrive at the destination first. The
longer the route to be traveled, the greater the difference in time between the arrival of the cyclist and the
later arrival of the pedestrian. Likewise, the farther the

receiving seismograph is from the earthquake epicenter, the greater the time lag between the first arrivals of P-waves and S-waves. The principle is
illustrated graphically in figure 8.7A.

Once several recording stations have determined their distances from the epicenter in this way,
it can be located on a map (figure 8.7B). If the epicenter is 200 kilometers from station X, it is located
somewhere on a circle with a 200-kilometer radius
around point X. If it is also found to be 150 kilometers from station Y, the epicenter must fall at either of
the two points that are both 200 kilometers from X

Table 8.1

Frequency of earthquakes of various magnitudes

Description	Magnitude	Number per year	Approximate energy released (ergs)
great earthquake	over 8	1 to 2	over 5.8×10^{23}
major earthquake	7–7.9	18	2–42×10^{22}
destructive earthquake	6–6.9	120	8–150×10^{20}
damaging earthquake	5–5.9	800	3–55×10^{19}
minor earthquake	4–4.9	6,200	1–20×10^{18}
smallest usually felt	3–3.9	49,000	4–72×10^{16}
detected but not felt	2–2.9	300,000	1–26×10^{15}

For every unit increase in Richter magnitude, ground displacement increases by a factor of 10 while energy release increases by a factor of 30. Therefore, most of the energy released by earthquakes each year is released not by the hundreds of thousands of small tremors, but by the handful of earthquakes of magnitude 7 or larger.

Data from B. Gutenberg, in *Earth*, by Press, F., and R. Seiver. W. H. Freeman and Company. Copyright © 1978. Reprinted by permission.

and 150 kilometers from Y. Knowing its distance from a third station, Z, should resolve the position of the epicenter to one point.

In practice, precise location of epicenters typically requires data on distances from more than three seismograph stations, because geologic inhomogeneities in the earth (differences in rock type or density, for example) cause local distortions in the arrival times at individual stations. Computers are often used to refine the location of the epicenter further. However, the general approach is as outlined above.

Size or Severity of Earthquakes

All seismic waves represent means of energy release and transmission. They cause the ground shaking that people associate with earthquakes. Most structural damage is done by the surface waves. There are various ways of describing the size of an earthquake. The two parameters most commonly used are *magnitude* and *intensity.*

Magnitude

The amount of ground shaking (amount of vertical motion) is related to the **magnitude** of the earthquake. Earthquake magnitude is most often reported using the *Richter magnitude scale,* named after geophysicist Professor Charles Richter, who developed it.

A magnitude number is assigned to an earthquake on the basis of the amount of ground displacement or shaking that it produces, as measured by a seismograph. The reading is adjusted for the distance of the instrument from the earthquake (because ground motion naturally decreases with increasing distance from the site of the earthquake as the energy is dissipated), so that measuring stations in different places will arrive at approximately the same magnitude value. The Richter scale is a logarithmic one, meaning that an earthquake of magnitude 4 causes ten times as much ground movement as one of magnitude 3, 100 times as much as one of magnitude 2, and so on. The amount of energy released rises even faster with increased magnitude—by about a factor of 30 for each unit of magnitude. An earthquake of magnitude 4 releases approximately 30 times as much energy as one of magnitude 3, and 900 times as much as one of magnitude 2.

There is no upper limit to the Richter scale. The largest recorded earthquakes have had magnitudes of about 8.9. Earthquakes of this size have occurred in Japan and in Chile. Although we only hear of the very severe, damaging earthquakes, there are in fact hundreds of thousands of earthquakes of all sizes each year. Table 8.1 summarizes the frequency and effects of earthquakes in different magnitude ranges.

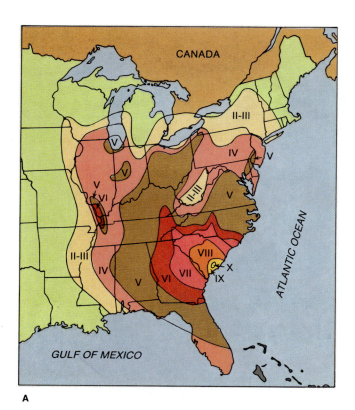

A

Intensity

An alternative way of describing the size of an earthquake is by its **intensity.** Intensity is a measure of the effects of the earthquake on humans and on surface features. It is not a unique, precisely defined characteristic of an earthquake. The surface effects produced by an earthquake of given magnitude will vary considerably as a function of such factors as local geologic conditions, quality of construction, and distance from the epicenter. A single earthquake, then, will produce effects of many different intensities in different places, though it will have one magnitude assigned to it (figure 8.8). Also, the extent of the area experiencing a given intensity of damage will vary with local geology, even for earthquakes of the same magnitude.

Intensity is a somewhat subjective measure in that it is based on direct observation by individuals rather than on instrumental measurements. Different observers in the same spot may assign different intensity values to a single earthquake. On the other hand, intensity is a more direct indication of the human impact of a particular seismic event in a given place than

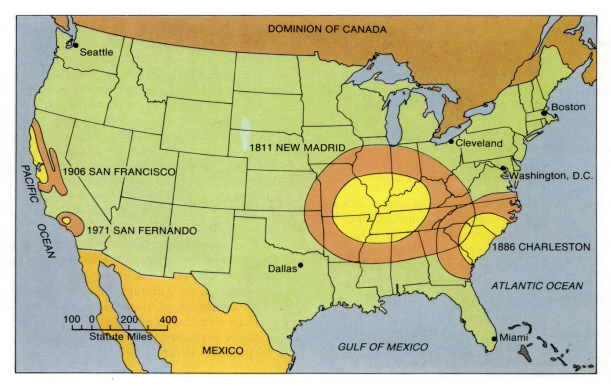

B

Figure 8.8

Regional variations in earthquake intensity. *(A)* Zones of different intensity for a single earthquake: the Charleston earthquake of 1886. *(B)* Areas of equal-intensity damage for several major earthquakes. The outer zone in each case experiences Mercalli intensity VI, the inner zone intensity VI or above.

(A) Data from U.S. Geological Survey. *(B)* After U.S. Geological Survey Professional Paper 1240–B.

is magnitude. There are several dozen intensity scales in use worldwide. That most widely applied in the United States is the Modified Mercalli scale, a modern version of which is summarized in table 8.2.

Types of Earthquake-Related Hazards and Their Mitigation

Earthquakes can have a variety of harmful effects, some obvious, some more subtle. Earthquakes of the same magnitude occurring in two different places can cause very different amounts of damage, depending on many variables including the nature of the local geology, whether the affected area is near the coast, or whether the terrain is steep or flat. We can take action to minimize some of the hazards; some can only be avoided by avoiding the dangerous area.

Ground Motion

Ground shaking and *displacement along the fault* are obvious hazards. Ground shaking by seismic waves causes damage to and sometimes complete failure of buildings (figure 8.9). Sudden shifts of even a few tens of centimeters can be devastating, especially to weak materials such as adobe. Offset between rocks on opposite sides of the fault can also break power lines, pipelines, buildings, roads, bridges, and other structures that cross the fault (figure 8.10). In the 1906 San Francisco earthquake, maximum relative horizontal displacement across the San Andreas fault was over 6 meters (nearly 20 feet). Such effects will of course be most severe on or very close to the fault, so the simplest strategy would be not to build near fault zones. However, many cities have already developed near major faults. Sometimes cities are rebuilt many times in such places. The ancient city of Constantinople—now Istanbul—has been leveled by earthquakes repeatedly since ancient times. Yet there it sits, still vulnerable.

Short of moving whole towns, can anything be done in such places? Power lines and pipelines can be built with extra slack where they cross a fault zone, or designed with other features to allow some "give" as the fault slips and stretches them. Such considerations had to be taken into account when the Trans-Alaska Pipeline was built, for it crosses several major known faults along its route.

Table 8.2

Modified Mercalli intensity scale (abridged)

Intensity	Description
I	Not felt.
II	Felt by persons at rest, on upper floors.
III	Felt indoors; hanging objects swing. Vibration like passing of light trucks.
IV	Vibration like passing of heavy trucks. Standing automobiles rock. Windows, dishes, doors rattle; wooden walls or frame may creak.
V	Felt outdoors. Sleepers wakened. Liquids disturbed, some spilled; small objects may be moved or upset; doors swing; shutters and pictures move.
VI	Felt by all; many frightened. People walk unsteadily. Windows, dishes broken; objects knocked off shelves, pictures off walls. Furniture moved or overturned. Weak plaster cracked. Small bells ring. Trees, bushes shaken.
VII	Difficult to stand. Furniture broken. Damage to weak materials, such as adobe; some cracking of ordinary masonry. Fall of plaster, loose bricks, tile. Waves on ponds; water muddy; small slides along sand or gravel banks. Large bells ring.
VIII	Steering of automobiles affected. Damage, partial collapse of ordinary masonry. Fall of chimneys, towers. Frame houses moved on foundations if not bolted down. Changes in flow of springs and wells.
IX	General panic. Frame structures shifted off foundations if not bolted down; frames cracked. Serious damage even to partially reinforced masonry. Underground pipes broken; reservoirs damaged. Conspicuous cracks in ground.
X	Most masonry and frame structures destroyed with their foundations. Serious damage to dams, dikes; large landslides. Rails bent slightly.
XI	Rails bent greatly. Underground pipelines out of service.
XII	Damage nearly total. Large rock masses shifted; objects thrown into the air.

Modified *Elementary Seismology*, by C. Richter. W. H. Freeman and Company. Copyright © 1958. Reprinted by permission.

A

B

Figure 8.9

Building failure from 1985 earthquake, Mexico City.
(A) Pancake-style collapse of 15-story, reinforced-concrete
structure. *(B)* Identical structures may not respond in the same
way to an earthquake. Only one of these four apartment towers
collapsed.

Photos by M. Celebi, courtesy U.S. Geological Survey.

Fourth Avenue landslide, Anchorage, Alaska, 1964. Note how far the shops and street at right have dropped relative to left side of street.

Photo courtesy U.S. Geological Survey.

Designing so-called earthquake-resistant buildings is a greater challenge and is a relatively new idea that has developed mainly in the last few decades. Engineers have studied how different types of buildings have fared, and often failed, in real earthquakes. Scientists can conduct laboratory experiments on scale models of skyscrapers and other buildings, subjecting them to small-scale shaking designed to simulate the kinds of ground movement during an earthquake. In such ways, they can see how differently designed structures respond to these forces and on the basis of their findings, begin to develop special building codes for earthquake-prone regions. However, this approach has many complications. For one thing, we have very few reliable records of just how the ground does move in a severe earthquake. In order to have such records, sensitive instruments must be in place near the fault zone beforehand, and those instruments must survive the earthquake.

Even with good records from an actual earthquake, it is difficult to be sure that the laboratory replicas are accurate enough that the model buildings will respond to model earthquakes in the laboratory in the same ways that real skyscrapers or other structures respond to real earthquakes. Such uncertainties are a major factor in concerns about the safety of nuclear power plants near active faults. Some attempts to circumvent the limitations inherent in scale modeling have recently been made. In 1979, the United States and Japan set up a cooperative program to test earthquake-resistant building designs. The tests have included experiments on full-sized, seven-story reinforced concrete structures, in which ground shaking was simulated using hydraulic jacks. Results were not entirely as anticipated on the basis of earlier modeling studies, and suggest that full-scale experiments and observations may be critical in designing optimum building codes in earthquake-prone regions.

Box 8.1

The Infamous San Francisco Earthquake, 1906

The first thing I was aware of was being wakened sharply to see my bureau lunging solemnly at me across the width of the room. . . . Then I remember standing in the doorway to see the great barred leaves of the entrance on the second floor part quietly as under an unseen hand . . . and suddenly an eruption of nightgowned figures crying out that it was only an earthquake. . . . Almost before the dust of ruined walls had ceased rising, smoke began to go up against the sun. . . . South of Market, in the district known as the Mission, there were cheap man-traps folded in like pasteboard, and from these, before the rip of the flames blotted out the sound, arose the thin, long scream of mortal agony. . . . In the park were the refugees huddled on the damp sod with insufficient bedding and less food and no water. . . . Hot, stifling smoke billowed down upon them, cinders pattered like hail. . . . I came out . . . and saw a man I knew hurrying down toward the gutted district. . . . "Bob," I said, "it looks like the day of judgment." He cast

Figure 1
San Francisco in flames, 1906.
Photo by T. L. Youd, courtesy U.S. Geological Survey.

back at me over his shoulder unveiled disgust at the inadequacy of my terms. "Aw!" he said, "it looks like hell!" (Mary Austin, as quoted in Rhodes and Stone, 1981)

Terrifying as that San Francisco earthquake was, it caused only an estimated 700 deaths and $4 million in property damage (figure 1). History records other earthquake disasters claiming nearly a million lives.

Reprinted with permission from "The Temblor—A Personal Narration," by Mary Austin, in *Language of the Earth*, by F. H. T. Rhodes and O. Stone. Copyright © 1981, Pergamon Press.

A further complication is that the same building codes cannot be applied everywhere. Not all earthquakes of given magnitude produce the same patterns of ground motion. It is important to consider not only how structures are built but what they are built *on*. Buildings built on solid bedrock seem to suffer far less damage than those built on deep soil. In the 1906 San Francisco earthquake, buildings built on filled land reclaimed from San Francisco Bay suffered up to four times as much damage from ground shaking as those built on bedrock. Mexico City is underlain by thick layers of weak volcanic ash and clay; most smaller and older buildings lack the deep foundations needed to reach more stable sand layers at depth. This is one reason why damage from the 1985 earthquakes there was so extensive, and why so many buildings completely collapsed. By contrast, Acapulco—much closer to the epicenter—suffered far less damage. It stands firmly on bedrock.

The characteristics of the earthquakes in a particular region must be taken into account, too. Severe earthquakes are generally followed by many **aftershocks, further earthquakes weaker than the principal tremor.** It is usually the main shock that causes the most damage, but where aftershocks are many and nearly as strong as the main shock, they may also cause serious destruction. One day after the magnitude 8.1

Figure 8.11

Landslide in Turnagain Heights area, Anchorage, 1964. Close inspection will reveal a number of houses amid the jumble of downdropped blocks in the foreground.

Photo courtesy U.S. Geological Survey.

main shock of the 1985 Mexico City earthquake, another major shock struck, with magnitude close to 7.3, leveling more buildings, deepening the rubble, and hampering rescue efforts.

The duration of a shock will affect how well a building survives it. In reinforced concrete, ground shaking leads to formation of hairline cracks, which then widen and lengthen as shaking continues. A concrete building that can withstand a one-minute main shock might collapse in an earthquake in which the main shock lasts three minutes. Many of the California building codes, used as models around the world, require that buildings be designed to withstand a 25-second main shock; but earthquakes can last ten times that long.

Finally, a major problem is that even the best building codes are typically applied only to new construction. Where a large city is located near a fault zone, thousands of vulnerable older buildings may already have been built in high-risk areas without special design features. The costs to redesign, rebuild, or even modify all those buildings would be staggering. Most legislative bodies are reluctant to require such efforts—indeed, many do nothing even about municipal buildings built in fault zones.

Fire

A secondary hazard of earthquakes is *fire,* which may be more devastating than the results of ground movement. In the 1906 San Francisco earthquake, 70% of the damage was due to fire, not simple building failure. As it was, the flames were confined to a ten-square-kilometer area only by dynamiting rows of buildings around the burning section. Fires start as fuel lines and tanks and power lines are broken, touching off flames and fueling them. At the same time, water lines are broken, leaving no way to fight the fires effectively. Putting in numerous valves in all water and fuel pipeline systems helps to combat these problems, because breaks in pipes can then be isolated before too much pressure or liquid is lost.

Ground Failure

Landslides can be a serious additional danger in hilly areas (figure 8.11). Earthquakes are one of the possible triggering mechanisms of the sliding of unstable slopes. The best solution is not to build in such areas. Even if a whole region is hilly, it may be possible to

Figure 8.12

Effects of soil liquefaction during an earthquake: Niigata, Japan, 1964. The buildings themselves were designed to be earthquake resistant, and they toppled over intact.

Photo courtesy National Geophysical Data Center.

avoid the most dangerous sites by making detailed engineering studies of rock and soil properties and slope stability. There may also be visible evidence of past landslides that would indicate especially dangerous areas.

Ground shaking may cause a further problem in areas where the ground is very wet—in filled land, near the coast, or in places with a high water table. This problem is **liquefaction.** When wet soil is shaken by an earthquake, the soil particles may be jarred apart, allowing water to seep in between them. This greatly reduces the friction between soil particles that gives the soil strength, and it causes the ground to become somewhat like quicksand. When this happens, buildings can just topple over or partially sink into the liquefied soil—the soil has no strength to support them. Dramatic examples of the effects of liquefaction were seen after a major earthquake in Niigata, Japan, in 1964. One multi-story apartment building tipped over

to settle at an angle of 30° to the ground—while the structure remained intact! (See figure 8.12.) It may be possible in some areas prone to liquefaction to install improved underground drainage systems to try to keep the soil drier, but otherwise there is little that can be done about this hazard, beyond avoiding areas at risk. It should also be noted that not all areas with wet soils are subject to liquefaction. The nature of the soil or fill will play a large role in the extent of the danger.

Tsunamis

Coastal areas, especially around the Pacific Ocean basin where so many large earthquakes occur, may also be vulnerable to **tsunamis.** They are seismic sea waves, sometimes improperly called "tidal waves." (They have nothing to do with tides.) When an undersea or nearshore earthquake occurs, sudden movement of the sea floor may set up waves traveling away from that spot, like ripples on a pond caused by a dropped pebble.

Figure 8.13

Boats washed into the heart of Kodiak, Alaska, by tsunami in 1964.

Photo courtesy U.S. Geological Survey.

Contrary to modern movie fiction, a tsunami is not seen as a huge breaker in the open ocean, toppling ocean liners in one sweep. In the open sea the tsunami will be only an unusually broad swell on the water surface. Like all waves, tsunamis only develop into breakers as they approach shore and the undulating waters touch bottom (see chapter 15). The breakers associated with tsunamis can easily be over 15 meters (45 feet) high and may reach up to 65 meters (close to 200 feet) in the case of larger earthquakes. Their effects can be correspondingly dramatic (figure 8.13). Several such breakers may crash over the coast in succession; between waves, the water may be pulled swiftly seaward, emptying a harbor or bay and perhaps pulling unwary onlookers along. Tsunamis can travel very quickly—speeds of 1000 kilometers per hour (600 MPH) are not uncommon—and a tsunami set off on one side of the Pacific may still cause noticeable effects on the other side of the ocean.

A tsunami caused by a 1960 earthquake in Chile was still vigorous enough to make 7-meter-high breakers when it reached Hawaii some 15 hours later, and 25 hours after the earthquake, the tsunami was detected in Japan.

Given the speeds at which tsunamis travel, there is little that can be done to warn those near the earthquake epicenter, but people living some distance away can be warned in time to evacuate, saving lives if not property. In 1948, two years after a devastating tsunami hit Hawaii, the U.S. Coast and Geodetic Survey established a Tsunami Early Warning System based in Hawaii. Whenever a major earthquake occurs in the Pacific region, tidal (sea-level) data are collected from a series of monitoring stations around the Pacific. If a tsunami is detected, data on its source, speed, and estimated time of arrival can be relayed to areas in danger and people evacuated as necessary.

The Tsunami Early Warning System does not, admittedly, always work perfectly. In the 1964 Alaskan earthquake, disrupted communications prevented warnings from reaching some areas quickly enough. Also, the response of residents to the warnings was variable, with some ignoring them and some even going closer to shore to watch the waves—often with tragic consequences. Two tsunami warnings were issued in early 1986 following earthquakes in the Aleutian islands. These were largely ignored. Fortunately, in this case, the feared tsunamis did not materialize.

Earthquake Control?

It seems clear that if earthquakes are ultimately caused by forces strong enough to move continents, there is no way that human efforts can stop them from occurring. However, it may be possible to moderate some of the most severe effects.

Locked Faults and Seismic Gap Theory

As noted earlier, friction between rocks along existing faults may prevent movement until sufficient stress has built up to overcome the friction. If friction is relatively small, slippage occurs when little stress has accumulated, and creep or rather small earthquakes result. The more stress built up before failure occurs, the stronger the eventual earthquake.

Many of the worst earthquakes have happened along sections of major faults that had for a time become locked, or immobile. If we map the locations of earthquake epicenters along major faults, we find that there are sections that show little or no seismic activity, while small or moderate earthquakes continue along other sections of the same fault zone. Such quiescent sections of otherwise active fault zones are termed seismic gaps. These areas may be sites of future serious earthquakes. On either side of a locked section, stresses are being released by creep and/or earthquakes. In the seismically quiet locked sections, friction is apparently sufficient to prevent the fault from slipping, so the stresses are simply building up. The fear, of course, is that the stresses will build up so far that when that locked section of fault finally does slip again, a very large earthquake will result.

At one time it was actually suggested that carefully placed nuclear explosions could be used to unstick locked faults. This idea is not being considered very seriously any more, partly because of the concern about radiation release and partly because the sudden, poorly controlled jolt of a nuclear explosion in a locked section of fault with a great deal of built-up stress could itself cause the feared large earthquake. What is needed is some alternative way to release locked faults gently, in a more controlled way.

Fluid Injection

In the mid-1960s, the city of Denver began to experience small earthquakes. They were not particularly damaging, but they were puzzling. Denver had not previously been earthquake-prone. In time a connection with an Army liquid-waste-disposal well at the nearby Rocky Mountain Arsenal was suggested. The Army denied any possible link, but a comparison of the timing of the earthquakes with the quantities of liquid pumped into the well at different times showed a very strong correlation (figure 8.14). The earthquake foci were also concentrated near the arsenal well. It seemed likely that the liquid, by increasing the fluid pressures in rocks along old faults, was allowing them to slip by decreasing the resistance to shearing. Experiments later conducted at an abandoned oil field near Rangely, Colorado, suggested similar possibilities. When the fluid pressure of liquids pumped into the ground exceeded a certain level, old faults were reactivated. Existing stresses in the rocks were greater than the fluid-reduced shear strength, and earthquakes occurred as the faulted rocks slipped. Other observations and experiments around the world have supported the concept that fluids in fault zones may facilitate movement along a fault.

Such observations have prompted speculation by many scientists that fluid injection might be used along locked sections of major faults to allow the release of built-up stress. Unfortunately, we are presently far from sure of the results to be expected from injecting fluid (probably water) along large locked faults. There is no guarantee that only small earthquakes would be produced. Indeed, injecting fluid in an area where a fault had been locked for a long time could lead to release of all the stress at once, in a major, damaging earthquake, just as might happen if a nuclear explosion jarred it loose. There is great concern not only for the possible casualties and damage that could result, but also about the legal and political consequences of a serious earthquake induced by human activities. There is also some question about whether it is even possible to release large amounts of energy through small earthquakes, considering how little energy is released in low-magnitude earthquakes (recall table 8.1).

It would probably be safer to use the fluid injection technique in areas of major fault zones that have not been seismically quiet for long, where less stress

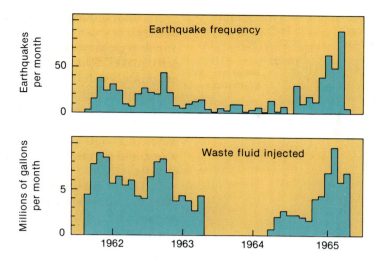

Figure 8.14

Correlation between waste disposal at the Rocky Mountain Arsenal (lower diagram) and frequency of earthquakes in the Denver area.

From David Evans, "Man-made Earthquakes in Denver," in *Geotimes*, Vol. 10. Reprinted by permission of the author.

has built up, to prevent that stress from continuing to build. The first attempts are also likely to be made in areas of low population density, to minimize the potential for inadvertently causing serious harm. Certainly much more research is needed before this method can be applied safely and reliably on a large scale. It is, however, an intriguing possibility for the future.

Earthquake Prediction?

It might be best if no one lived near major fault zones, but of course millions already do. Therefore, it would be very useful to be able to predict major earthquakes, so that people could at least be evacuated and lives saved. Some progress has been made in this direction.

Earthquake Precursors

Earthquake prediction is based on the study of earthquake **precursor phenomena,** things that happen or rock properties that change prior to an earthquake. We may be able to identify warning signs of an earthquake. Many different, possibly useful precursor phenomena are being studied. For example, the ground surface may be uplifted and tilted prior to an earthquake. P-wave velocities in rocks near the fault (measured using artificially caused shocks, such as those from small explosions) will drop and then rise before an earthquake. The same pattern is seen in the ratio

of P-wave to S-wave velocity. Electrical resistivity (the resistance of rocks to the flow of electric current through them) increases, then decreases, before an earthquake. The time scale over which precursory changes occur varies: it may be on the order of weeks, months, years, even decades. In a general way, there seems to be a correlation between the time over which precursory changes occur and the size of the eventual earthquake: the longer the cycle, the larger the earthquake.

Various precursor phenomena can be related to a model of what happens in the rocks on a microscopic scale that is known as **the dilatancy** model. The theory holds that as the rocks are subjected to stress, many small pores and cracks open up in them. With continued stress, the cracking becomes so extensive that water can seep into the cracks and pores. This may in turn increase pore fluid pressures, effectively lubricating the rocks. They eventually slip, releasing the built-up stress through an earthquake and snapping back elastically to their original condition afterwards, with the cracks and pores closing up again.

Figure 8.15 demonstrates how some of the precursor phenomena mentioned might be explained in terms of the dilatancy model. For instance, as cracks and pores open in the rocks, the rocks' volume will increase. The effect is to cause uplift and/or tilting of the ground surface. Or consider electrical conductivity. Empty (air-filled) holes in rocks will not conduct electricity well. As cracks and pores open up, the

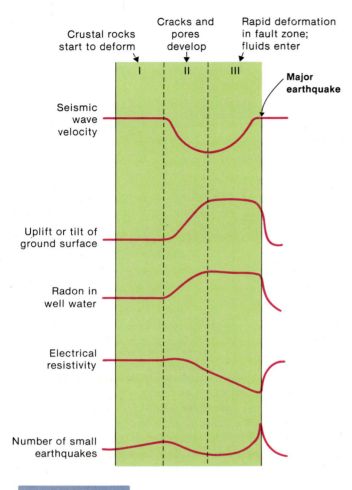

overall ability of the rock to conduct electricity decreases—in other words, resistivity increases. But as fluids seep into the cracks, they improve the ability of the rock to conduct electricity. Water, especially with dissolved minerals in it, conducts electricity far better than air. So resistivity goes down as water seeps in. In the case of seismic-wave velocities, seismic waves travel faster in denser material: faster in rock than in water, faster in water than in air. As cracks and pores open, then, the seismic waves are slowed down; as cracks fill with water, seismic-wave velocities begin to increase. In all cases, after the earthquake, when the rocks have snapped back to their unstressed dimensions and the cracks have all closed up, rock properties should return to what they were before this whole process began.

Other observations may also be explained by dilatancy. For example, an increase in the concentration of radon gas in well water has been observed prior to earthquakes. Radon is a chemically inert, radioactive gas produced during the decay of the small quantities of uranium naturally present in most rocks. As cracks open and water seeps through rocks, it would become easier for radon formed in the rocks to escape into well waters. Changes in other chemical properties of well water, and in water levels, also have been used as precursors.

These and other less well documented precursors (like unusual animal behavior before large earthquakes) have all been considered in efforts to predict earthquakes. Such efforts have been complicated by the fact that not all earthquakes have shown the same patterns of precursor events. The dilatancy model itself is not universally accepted. It appears to account for many commonly observed precursory changes, but it has not been proven. It has not yet been possible to design instruments that can be placed deep in the lithosphere to scrutinize stressed rocks in a fault zone on a microscopic scale before and during earthquakes. A consequence of the present imperfect understanding of earthquakes is that while there have been some spectacular successes in earthquake prediction, there have been equally conspicuous failures.

Current Status of Earthquake Prediction

In the People's Republic of China, some 10,000 scientists and technicians, and 100,000 part-time amateur observers, work on earthquake prediction. Their motivation is great: from 1966 to 1976 alone, 11 earthquakes of magnitude 6.8 or greater occurred in China. In February 1975, after months of smaller earthquakes, radon anomalies, and increases in ground tilt followed by a rapid increase in both tilt and microearthquake frequency, the scientists predicted an imminent earthquake near Haicheng in northeastern China. The government ordered several million people out of their homes, into the open. Nine and one-half hours later a major earthquake struck, and many lives were saved because people were not crushed by collapsing buildings.

Over the next two years, the earthquake scientists successfully predicted four large earthquakes in the Hebei district. They also concluded that a major earthquake could be expected near T'ang Shan, about 150 kilometers southeast of Peking. In the latter case, however, they could only say that the event was likely to occur sometime during the following two months.

Box 8.2

Seismic Risks and Public Response

Assuming that routine earthquake prediction will become reality, some planners have begun to look beyond the scientific questions to possible social or legal complications of such predictions. For example, some people have speculated that if a major earthquake were predicted several years ahead for a particular area, property values would plummet because no one would want to live there. A counterargument to that, perhaps, is San Francisco: despite widespread acceptance of the idea that another large earthquake will strike, the city continues to thrive.

There is concern about the logistics of evacuating a large urban area on short notice if a near-term prediction were made. Many urban areas have hopelessly snarled traffic every rush hour; what happens if everyone in the city wants to leave at once? Some critics say that a rapid evacuation might involve more casualties than the eventual earthquake. And what if a city is evacuated, people inconvenienced or hurt, property perhaps damaged by vandals and looters—and then the predicted earthquake never comes? Will the issuers of the warning be sued? Will future warnings be ignored?

In the People's Republic of China, vigorous public-education programs (and several recent major earthquakes) have made earthquakes a well-recognized hazard. Strong government support for earthquake-prediction efforts, and the involvement of large numbers of lay people in those efforts, have given them a high level of visibility and widespread community support. "Earthquake drills," which stress orderly response to an earthquake warning, are held in Japan on the anniversary of the 1923 Tokyo earthquake in which more than 100,000 people died. By contrast, in the United States, surveys have repeatedly shown that even people living in such high-risk areas as along the San Andreas fault are often unaware of the very existence of the earthquake hazard. Among those who are aware, there is widespread belief that the risks are not very great. Many survey respondents indicate that they would not take any special action even if a specific earthquake warning were issued. Past mixed response to tsunami warnings underscores doubts about public response to earthquake predictions.

In principle, prediction seems desirable. In practice, it may be fraught with complications.

When the earthquake—magnitude 8.2 with aftershocks up to magnitude 7.9—did happen, there was no immediate warning, no sudden change in the precursor phenomena, and over 650,000 people died.

At present, only four nations—Japan, the Soviet Union, the People's Republic of China, and the United States—have government-sponsored earthquake prediction programs. Such programs typically involve intensified monitoring of active fault zones to expand the observational data base, coupled with laboratory experiments designed to increase understanding of precursor phenomena and the behavior of rocks under stress. Even with these active research programs, scientists can generally forecast only that a major earthquake will occur within a certain period of time—sometimes months or years—in the future. Sometimes they are right, sometimes wrong. Often they are hampered by lack of personnel or equipment in critical areas; they cannot monitor everywhere at once.

Also, as noted earlier, we do not completely understand precursors yet. There may or may not be any distinctive warning signals shortly before a given event.

The predictions have not been regarded as reliable or precise enough to justify actions such as large-scale evacuations in this country. It may some day be possible to predict the timing and size of major earthquakes accurately enough that orderly evacuations of populated areas can be undertaken routinely when serious earthquakes are imminent, thereby saving a great many lives. Property damage is inevitable as long as people persist in living and building in earthquake-prone areas. In any case, most seismologists feel that consistently reliable earthquake predictions are probably still more than a decade in the future.

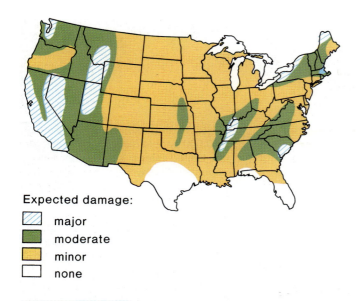

Expected damage:

- ⧄ major
- ▨ moderate
- ▨ minor
- ☐ none

Figure 8.16

Seismic risk map for the contiguous United States.

Source: National Oceanic and Atmospheric Administration.

Seismic Risks in the United States

Figure 8.16 is a seismic risk map prepared by the U.S. Geological Survey. Several of the shaded high-risk areas are not normally associated with earthquake activity—the ones near the center of the country, for instance, or those in the southeastern coastal region. This map is based partly on the distribution of known faults, partly on historical records of earthquakes. It reflects not just the frequency of past earthquakes, but also their severity, which is taken to indicate the possible severity of future earthquakes in terms of ground motion.

Areas of Well-Recognized Risks

When the possiblity of another large earthquake in San Francisco is raised, most geologists debate not "whether?" but "when?" At present, that section of the San Andreas fault has been locked for some time. The last major earthquake there was in 1906. At that time, movement occurred along at least 300 kilometers, and perhaps 450 kilometers, of the fault. The 1971 San Fernando earthquake was only of magnitude 6.6, yet it cost 64 lives and an estimated $1 billion in property damage (figure 8.17). The death toll would

have been far higher had it not taken place early in the morning. The 1983 Coalinga, California, earthquake, a magnitude 6.5 event that occurred in what had been a seismic gap along the San Andreas fault, caused $31 million in property damage even though it happened in a relatively undeveloped area.

If another earthquake of the size of the 1906 event—estimated at Richter magnitude 7.9—were to strike the San Francisco area during a busy work day, there would be an estimated 50,000 or more casualties, including the number of people at risk of drowning in case of dam failures in the area. There also would be an estimated $24 billion in property damage from ground shaking alone. Scientists working on earthquake prediction hope very much that they will at least be able to give fairly precise warning of the next major Californian tremor by the time it occurs. At present, the precursor phenomena, insofar as they are understood, do not indicate that a great earthquake (magnitude 8 or greater) will occur near San Francisco for at least several more years. The more time seismologists have to improve their methods, the better the eventual warning will be.

On Good Friday, March 27, 1964, an earthquake with a magnitude estimated at approximately 8.5 struck southern Alaska. The main shock, which lasted three to four minutes, was felt more than 1200 kilometers from the epicenter. About 12,000 aftershocks, many over magnitude 6, occurred during the next two months, and more continued intermittently through the following year. Structures were damaged over more than 100,000 square kilometers; ice on rivers and lakes cracked over an area of 250,000 square kilometers. Many areas suffered permanent uplift or depression. The uplifts—up to 12 meters on land, and over 15 meters in places on the sea floor—left some harbors high and dry and destroyed the habitats of marine organisms. Downwarping of 2 meters or more flooded other coastal communities. Tsunamis following the main shock destroyed four villages and seriously damaged many coastal cities; they were responsible for about 90% of the 115 deaths. Some of these waves traveled as far as Antarctica and accounted for 16 deaths in Oregon and California. Landslides—both submarine and above ground—were widespread and were responsible for most of the remaining casualties. Levels of water in wells shifted abruptly as far away as South Africa. Total damage was estimated at $300 million. Southern Alaska sits above a subduction zone. The area can certainly expect more earthquakes, some of which could be severe. The coastal regions there will also continue to be vulnerable to tsunamis.

A

B

Figure 8.17

Damage resulting from the 1971 earthquake in San Fernando, California. *(A)* Remains of freeway interchange, Los Angeles County. *(B)* Lower Van Norman dam. The reservoir level had been lowered considerably before photo was taken. Had the dam failed, an area in which 80,000 people lived would have been flooded.

Photos courtesy U.S. Geological Survey.

Table 8.3

Earthquake safety rules from the National Oceanic and Atmospheric Administration

During the Shaking

1. *Don't panic.*

2. If you are indoors, stay there. Seek protection under a table or desk, or in a doorway. Stay away from glass. Don't use matches, candles, or any open flame; douse all fires.

3. If you are outside, move away from buildings and power lines, and stay in the open. Don't run through or near buildings.

4. If you are in a moving car, bring it to a stop as quickly as possible, but stay in it. The car's springs will absorb some of the shaking and it will offer you protection.

After the Shaking

1. Check, but do *not* turn on, utilities. If you smell gas, open windows, shut off the main valve, and leave the building. Report the leak to the utility and don't reenter the building until it has been checked out. If water mains are damaged, shut off the main valve. If electrical wiring is shorting, close the switch at the main meter box.

2. Turn on radio or television (if possible) for emergency bulletins.

3. Stay off the telephone except to report an emergency.

4. Stay out of severely damaged buildings that could collapse in aftershocks.

5. Don't go sightseeing; you will only hamper the efforts of emergency personnel and repair crews.

Other Potential Problem Areas

Perhaps even more dangerous is the situation in hazardous areas that people are *not* conditioned to regard as earthquake prone, for example in the central United States. Though most Americans think "California" when they hear "earthquake," in fact the strongest and probably the most devastating series of earthquakes in the United States happened in the vicinity of New Madrid, Missouri, in 1811–12. The three strongest shocks are estimated to have had Richter magnitudes of 8.6, 8.4, and 8.7. They were spaced out over almost two months, from December 16, 1811, to February 7, 1812. A total of 1,874 shocks were recorded by one resident engineer in the period from the December 16 earthquake through the following March 15—and this without the aid of sensitive modern seismographs. Some towns were leveled, others drowned by flooding rivers. The most vigorous tremors were felt from Quebec to New Orleans, and along the East Coast from New England to Georgia. Lakes were uplifted, "blown up," and emptied, while elsewhere the ground sank to make new ones. Boats were flung out of rivers; an estimated 150,000 acres of timberland were destroyed. Aftershocks continued for more than *ten years*. Total damage and casualties have never been accurately estimated.

The potential damage from even one earthquake as severe as the worst of the New Madrid shocks is enormous: twelve million people now live in the immediate area. Yet very few of those twelve million are probably aware of the risk. The big earthquakes there were a long time ago, beyond the memory of anyone now living. Unfortunately, the danger of more such serious earthquakes is real. Beneath the midcontinent is a failed rift zone, where the continent began to rift apart, then stopped. Long and deep faults in the lithosphere there remain a zone of weakness in the continent and a probable site of more major tremors. It is only a matter of time.

National Geologic Hazard Warning System

Since 1976, the Director of the U.S. Geological Survey has had the authority to issue warnings of impending earthquakes and other potentially hazardous geologic events (volcanic eruptions, landslides, etc.). An Earthquake Prediction Panel has been established to review scientific evidence that might indicate an earthquake threat and to make recommendations to the Director regarding the issuance of appropriate public statements. These could range in detail and immediacy from a general notice to residents in a fault zone of the existence and nature of earthquake hazards there, to a specific warning of the anticipated time, location, and severity of an imminent earthquake. There have been no very large earthquakes in the United States since this warning system was established, so its effectiveness has yet to be fully tested, as does public response to formal, official earthquake predictions in this country.

In early 1985, the panel made its first endorsement of an earthquake prediction. They concurred that an earthquake of approximately Richter magnitude 6 can be expected on the San Andreas fault near Parkside, California between 1985 and 1993. The prediction is based in large part on cyclic patterns of seismicity in the area: on average, major earthquakes occur there every 22 years, and the last was in 1966. Local residents have expressed neither surprise nor concern over the prediction.

Box 8.3

Earthquake Focal Mechanisms and Plate Tectonics

Much of our understanding of relative plate motions has come from **first-motion studies** of earthquakes along faults occurring at plate boundaries. It is possible to tell from a seismogram whether the first ground motion caused by seismic waves from a given earthquake was up or down at the particular receiving station. This in turn indicates whether the lithosphere in that direction from the epicenter was first subjected to compression or dilatation (stretching) when the rocks shifted. By plotting the regions subject to compression or dilatation around a given earthquake epicenter, one can deduce which way the lithosphere moved.

This was especially helpful in the correct interpretation of transform faults (figure 1A). In this example, note that compression was observed in the northeast and southwest quadrants of the map, dilatation in the other two quadrants. This suggests that the lithosphere shifted east above the transform fault, west below. Note that if the situation were simply a case of a passive mountain ridge being split apart and offset by horizontal movement along an ordinary fault (figure 1B), the pattern of compression and dilatation would be the opposite of that observed around the transform fault.

First-motion studies are also one way of detecting underground nuclear explosions. The first motion from an explosion should be compressional in all directions and thus differs from the pattern produced by an earthquake along a planar fault. Other differences between patterns of seismicity resulting from explosions and from faulting include complexity of seismic-wave pattern and efficiency of generation of certain S-waves.

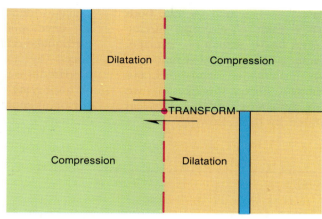

A

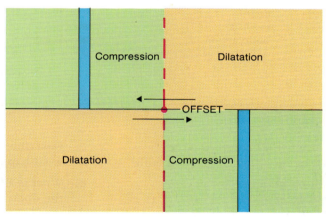

B

Figure 1
First-motion studies show the direction of movement of a transform fault (map view). *(A)* Actual distribution indicates movement away from spreading ridges, and illustrates the nature of the fault. *(B)* If the ridge segments were passive and simply being offset by the fault, the distribution of first motions should be the reverse of what is observed.

Indeed, much of the governmental support for research in seismology has been motivated, at least in part, by interest in developing reliable methods for identifying nuclear explosions.

Summary

Earthquakes result from sudden slippage or failure of rocks along fault zones in response to stress. Most earthquakes occur at plate boundaries and are related to plate-tectonic processes. The pent-up energy is released through seismic waves, which include both compressional and shear body waves, plus surface waves, which cause the most structural damage. Earthquake hazards include damage from ground rupture and shaking, fire, liquefaction, landslides, and tsunamis. We cannot hope to stop earthquakes, but we can try to limit their destructive effects. Physical damage could be limited by the following: seeking ways to cause locked faults to slip gradually and harmlessly, perhaps by using fluid injection to reduce frictional resistance to shear; designing structures in active fault zones to be more resistant to earthquake damage; identifying and, wherever possible, avoiding development in areas at particular risk from earthquake-related hazards. Casualties could be reduced by increasing public awareness of and preparedness for earthquakes in threatened areas, and by improving our understanding of earthquake precursor phenomena so that accurate and timely predictions of earthquake occurrence can be made.

Terms to Remember

aftershocks	fault	locked fault	seismic waves
body waves	first-motion studies	magnitude	seismograph
creep	fluid injection	precursor phenomena	surface waves
dilatancy	focus	P-waves	S-waves
earthquake	intensity	scarp	trace
elastic rebound	liquefaction	seismic gap	tsunami
epicenter			

Questions for Review

1. What determines whether creep or large earthquakes will occur along an existing fault?

2. What is the distinction between an earthquake's epicenter and its focus?

3. Deep-focus earthquakes are confined to subduction zones. Why?

4. Name the two kinds of seismic body waves and briefly describe them.

5. A given earthquake will have one magnitude but a range of intensities. Explain.

6. Cite and discuss any three factors that limit the extent to which we can design building codes for earthquake-resistant structures.

7. Describe the phenomenon of liquefaction and what happens to structures when it occurs.

8. What is a tsunami? Are tsunamis hazardous in the open ocean, near shore, or both?

9. Earthquakes in Denver indirectly suggested a possible future means of reducing the danger of major earthquakes along locked faults. Explain the concept involved.

10. Choose any two earthquake precursor phenomena and relate them to the changes in stressed rocks postulated according to the dilatancy model.

11. A seismic risk map of the United States shows an area of high risk in the middle of the country. Why?

For Further Thought

1. Investigate the history of any modern seismic activity in your area. What is the geologic explanation for this activity, and what is the likelihood of future activity believed to be? (You might consult the U.S. or state Geological Survey for information.) If a significant hazard is perceived, what plans are in place for responding to it?

2. Research one or more past earthquake predictions made in the United States or elsewhere. How specific were the predictions? How accurate? What was the public response, if any?

3. If you live near a major urban area, collect data on the population in the city on a workday and the rate at which commuters can be moved in or out of the city during rush hour. Speculate on the implications of your findings for the impact of short-term earthquake predictions.

Suggestions for Further Reading

Asada, T., ed. *Earthquake Prediction Techniques.* Tokyo: University of Tokyo Press, 1982.

Austin, M. "No More Wooden Towers for San Francisco, 1906." In *Language of the Earth,* edited by F. H. T. Rhodes, and R. O. Stone, 64–67. New York: Pergamon Press, 1981.

Berlin, G. L. *Earthquakes and the Urban Environment.* Boca Raton, Fla.: CRC Press, 1980.

Eiby, G. A. *Earthquakes.* New York: Van Nostrand Reinhold, 1980.

Hays, W. W., ed. *Facing Geologic and Hydrologic Hazards.* U.S. Geological Survey Professional Paper 1240-B, 1981. See especially section Z, "Hazards from Earthquakes," 6–38.

Lomnitz, C. *Global Tectonics and Earthquake Risk.* Amsterdam: Elsevier Scientific Publishing Co., 1974.

Penick, J. L., Jr. *The New Madrid Earthquakes.* 2d ed. Columbia: University of Missouri Press, 1981.

Press, F. "Earthquake Prediction." *Scientific American* 232 (May 1975): 14–23.

Tank, R., ed. *Environmental Geology.* 3d ed. New York: Oxford University Press, 1983. This collection of readings includes six chapters relating to earthquakes.

Volcanoes

Introduction

We have already surveyed some aspects of the composition, texture, and classification of volcanic rocks (chapter 3). Here we focus particularly on the often dramatic events accompanying their formation: volcanic eruptions. Volcanoes vary considerably in character, in the kinds of materials they emit, and in the extent to which they pose hazards to people and their works. Differences in volcanic character are directly related to magma composition, and often to the plate-tectonic setting in which the volcano forms.

Before 1980, most Americans probably regarded volcanoes as remote phenomena, of no immediate relevance to their own lives. Then Mount St. Helens exploded with a roar in a cloud of ash (box 9.1), and volcanoes suddenly and forcefully demanded our attention. Increased scrutiny of possible volcanic hazards in the United States has drawn attention to a number of areas in this country that may be threatened by eruptions in the not-too-distant future.

Box 9.1

Mount St. Helens

On the morning of May 18, 1980, a thirty-year-old volcanologist with the U.S. Geological Survey was watching the instruments monitoring Mount St. Helens in Washington state. David Johnston was one of many scientists keeping a wary eye on the volcano. They expected some kind of eruption, perhaps a violent one, but were uncertain of its probable size or of what kind of warning signs might precede it. Johnston's observation post was more than 9 kilometers (over 5 miles) from the mountain's peak. Suddenly he radioed to the control center, "Vancouver, Vancouver, this is it!"

Indeed it was. Seconds later the north side of the mountain blew out in a massive blast that would ultimately cost nearly $1 billion in damages and 25 lives, with another 37 persons missing and presumed dead. The mountain's elevation was reduced by more than 400 meters (figure 1). David Johnston was among the casualties. His death illustrates the fact that even the experts still have much to learn of the ways of volcanoes.

Figure 1
Aftermath of Mount St. Helens eruption, May 18, 1980.
Photo by Austin Post, courtesy U.S. Geological Survey.

Kinds and Locations of Volcanic Activity

Much volcanic activity is a consequence of eruption of lava, magma that reaches the earth's surface. Lava is only one kind of product of volcanic eruptions, however. Nor are all volcanic rocks produced out of cone-shaped objects on the continents.

In chapter 3, a volcanic rock was defined as one formed from a magma—a silicate melt—at or close to the earth's surface. By this definition, the largest volume of volcanic rock is actually produced at the seafloor spreading ridges. They spread at the rate of only a few centimeters per year, but there are some 50,000 kilometers (about 30,000 miles) of these ridges presently active in the world. Altogether, that adds up to an immense volume of volcanic rock. Because it is out of sight under the oceans, it is an often-forgotten form of volcanism.

Still, the toll might have been far higher. Fortunately, officials had taken the threat of Mount St. Helens seriously after its first few signs of reawakening. The immediate area near the volcano was cleared of all but a few essential scientists and law-enforcement personnel. Access to hundreds of thousands of square kilometers of terrain around this was severly restricted to both residents and workers (the latter mainly loggers). Others with no particular business in the area were banned altogether. Many people grumbled at being forced from their homes. A few flatly refused to go. Numerous tourists and reporters were frustrated in their efforts to get a close look at the action. When the major explosive eruption came, there were far fewer casualties than there might have been. On just a normal spring weekend, 200 or more people would have been on the mountain, with more living and camping nearby. Considering the influx of the curious once eruptions began, and the numbers of people turned away at roadblocks leading into the restricted zone, it is possible that with free access to the area tens of thousands could have died (see figure 2).

Figure 2
Fawn Lake (foreground) and flattened, ash-covered forest after the same eruption.
Photo by N. P. Dion, courtesy U.S. Geological Survey.

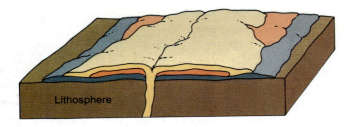

Lithosphere

Figure 9.1

Schematic diagram of a fissure eruption.

Fissure Eruptions

The outpouring of magma forming volcanic rock at spreading ridges is an example of **fissure eruption,** the eruption of magma out of a long crack in the lithosphere, rather than from a single pipe or vent (figure 9.1). There are also examples of fissure eruptions on the continents, in which many layers of lava are erupted in succession. One example in the United States is the Columbia Plateau, an area of about 50,000 square kilometers (20,000 square miles) in Washington, Oregon, and Idaho, covered by layer upon layer of basalt, over one and a half kilometers deep in places (figure 9.2). This area may represent another ancient beginning of a rift in the continent that ultimately stopped spreading. It is no longer active, but it serves as a reminder of how large a volume of magma can come welling up from the asthenosphere where zones of weakness or fractures in the lithosphere provide suitable openings. Even larger examples of such *flood basalts* on continents, covering up to 750,000 square kilometers, are found in India and Brazil.

Extensive lava flows often develop a distinctive structure upon cooling. Like most materials, lava and volcanic rock contract as they cool. This contraction can fracture the flow into a mass of polygonal columns (figure 9.3). Such **columnar jointing** is especially common in basaltic flows, perhaps because they start at higher temperatures and therefore undergo more contraction. Ireland's famous Devil's Causeway is formed of columnar-jointed basalt.

Individual Volcanic Cones— Locations

When most people think of volcanoes, they visualize eruptions from a central vent of a mountainlike object. Figure 9.4 is a map of the locations of such volcanoes presently or recently active on the continents.

A

B

Figure 9.2

Multiple lava flows from fissure eruption. *(A)* The Columbia River flood basalts (map extent). *(B)* At least five separate lava flows, one atop another, can be seen in this photograph.

As can be seen from this figure and by comparison with figure 7.4, most of this volcanic activity occurs at or close to plate boundaries. In fact, most of the volcanoes shown in figure 9.4 are located over subduction zones. The so-called Ring of Fire, the collection of volcanoes rimming the Pacific Ocean, is really a ring of subduction zones. As with earthquakes, there are a few volcanic anomalies not associated with plate boundaries, the *hot spots* described first in chapter 7. Hawaii is one such hot spot. What accounts for hot spots in the mantle is not clear. The usual cause proposed is an isolated **plume** of rising hot magma at the

A   B

Figure 9.3

Columnar jointing. *(A)* Formation of fractures in a cooling lava
flow by contraction. *(B)* Photo of columnar jointing in a flow in
Yellowstone Park.

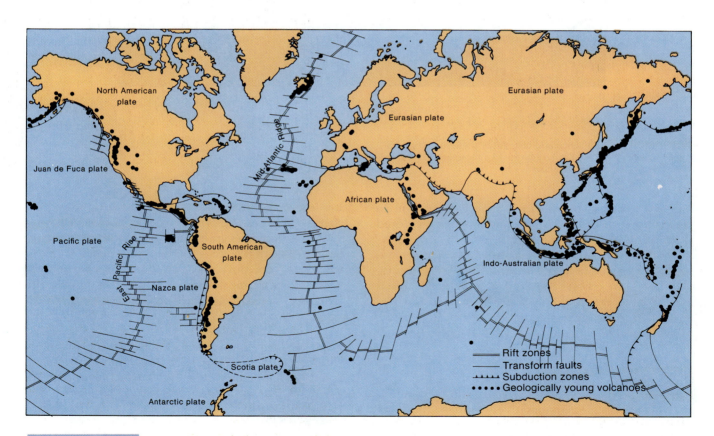

Figure 9.4

Volcanic areas of the modern world (excluding spreading ridges
and oceanic islands).

After Decker and Decker (1981).

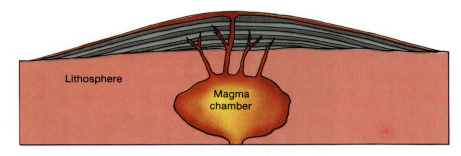

A

B

Figure 9.5

Shield volcanoes. (A) Schematic diagram of a shield volcano in cross section. (B) Very thin lava flows, like these on Kilauea, are characteristic of shield volcanoes.

hot spot. The relationship of magma generation to plate tectonics and mantle plumes will be explored later in this chapter.

The volcanoes of figure 9.4 can be divided into several types based on the kind of structure they build, which in turn reflects the kind of volcanic material that has been erupted. The nature of the volcanic material also indicates the probable kind(s) and extent of hazards each volcano poses.

Shield Volcanoes

Basaltic lavas, relatively low in silica and high in iron and magnesium, are comparatively fluid, so they flow very freely and far when erupted. Consequently, the kind of volcano they build is a very flat one in shape, low in relation to its diameter. This low, shieldlike shape has led to use of the term **shield volcano** for such a structure (figure 9.5). Though the individual lava flows may be thin— perhaps less than a meter in thickness—the buildup of hundreds or thousands of flows through time can produce quite large volcanic structures. The Hawaiian islands are all shield volcanoes. Mauna Loa, the largest peak on the still-active island of Hawaii, rises 3 1/2 kilometers (about 2 1/2 miles) above sea level. If measured properly from its true base, the sea floor, it is still more impressive: about 10 kilometers high, and 100 kilometers in diameter at its base. With such broad, flat shapes, the Hawaiian islands do not necessarily look like volcanoes from sea level, but seen from above their character is clear (figure 9.6). Hot-spot-produced volcanoes are often shield volcanoes.

A

Figure 9.6

Mauna Loa, an example of a shield volcano. *(A)* Viewed from low altitudes. Note flat shape, and fresh lava flows visible at left. *(B)* Bird's-eye view of the island of Hawaii, taken by Landsat satellite, shows its volcanic character more clearly. Large peak is Mauna Loa; smaller one, Mauna Kea.

(A) Photo by G. A. MacDonald, courtesy U.S. Geological Survey. *(B)* © NASA

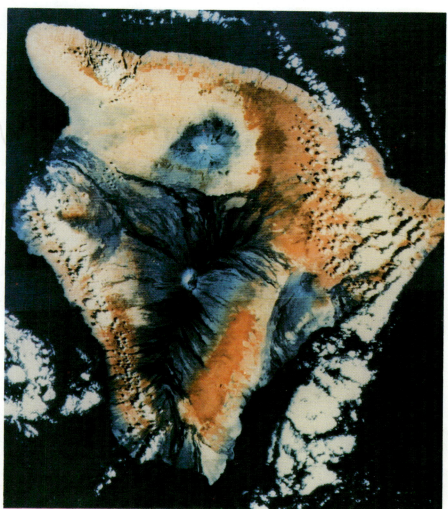

B

Even with limited variation in magma composition, there can be obvious variation in the appearance of the resultant lava flow. Especially fluid lavas that form a smooth, quenched, hardened skin as they cool develop a ropy appearance as they flow (figure 9.7A); this ropy lava is termed **pahoehoe** (pronounced "pahoy-hoy"). Other lavas flow less readily and produce jumbled, blocky flows (figure 9.7B); this material is called **aa** ("ah-ah"), reputedly so named from the pained cries of persons attempting to walk barefoot across its jagged surface. When lavas are extruded under water, rapid quenching of the hot flow surfaces leads to development of bulbous flow forms resembling pillows. In these **pillow lavas,** each pillow has a glassy rind and a coarser-grained, more slowly cooled interior (figure 9.7C).

Volcanic Domes

The less mafic, more silicic lavas, andesitic and rhyolitic in composition, tend to be more viscous, thicker, stiffer, flowing less readily. They ooze out at the surface like thick toothpaste from an upright tube, piling up close to the volcanic vent rather than spreading freely. The resulting structure is a more compact and steep-sided **volcanic dome.** Modern eruptions of Mount St. Helens are characterized by this kind of stiff, viscous lava, and a volcanic dome has formed in the crater left by its 1980 explosion (figure 9.8). Such thick, slowly flowing lavas also seem to solidify and stop up the vent from which they are erupted before much material has emerged. Volcanic domes, then, tend to be relatively small in areal extent compared to shield volcanoes, although through repeated eruptions over time, such volcanoes can build quite high peaks.

Explosive Eruptions; Cinder Cones

Magmas, as noted previously, not only consist of melted silicates but also contain dissolved water and gases, which are trapped under great pressures while the magma is deep in the lithosphere. As the magma wells up toward the surface, the pressure on it is reduced and the gases try to bubble out of it and escape. The effect is much like popping the cap off a soda bottle: the soda contains carbon dioxide gas under pressure, and when the pressure is released by removal of the cap, the gas comes bubbling out.

A

B

C

Figure 9.7

Types of lava flows. *(A)* Pahoehoe, with a smooth, ropy surface. *(B)* Aa, rough and blocky. *(C)* Ancient pillow lavas. The dark bands are the glassy rinds of individual pillows.

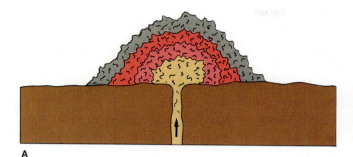

A

A

B

Figure 9.8

Volcanic domes. *(A)* Formation of a dome (schematic). *(B)* Dome built in the crater of Mount St. Helens.

(B) Photo by R. E. Wallace, courtesy U.S. Geological Survey.

B

Figure 9.9

Pyroclastics. *(A)* Cinders from Sunset Crater, Arizona. *(B)* Volcanic bombs.

Sometimes the built-up gas pressure in a rising magma is released suddenly and forcefully by an explosion. Bits of magma and rock are flung out of the volcano. The magma may freeze into solid pieces before falling to earth. The bits of violently erupted volcanic material are described collectively as **pyroclastics,** from the Greek words for "fire" (*pyros*) and "broken" (*klastos*) (see figure 9.9). The most energetic pyroclastic eruptions are more typical of volcanoes with the more viscous andesitic or rhyolitic lavas, because the thicker lavas tend to trap more gases. Gas usually escapes more readily and quietly from the fluid basaltic lavas, though even basaltic volcanoes may emit quantities of finer pyroclastics.

The fragments of pyroclastic material can vary considerably in size. The very finest make a flourlike dust. Coarser, gritty volcanic **ash** and **cinders** range up to golf-ball size. The largest chunks, the volcanic **blocks,** can be the size of a house. Block-sized blobs of liquid lava may also be thrown from a volcano; these volcanic **bombs** commonly develop a streamlined shape as they deform in flight before solidifying completely.

A

Figure 9.10

Cinder cones. *(A)* Night eruption of Stromboli (foreground) shows ejection of hot pyroclastics, piling up in a cinder cone. *(B)* Small cinder cones in the summit crater of Haleakala, a dormant volcano on Maui in the Hawaiian islands. Note layered lava flows on far rim of crater, small flow from flank of cinder cone in foreground, and low profile of nearby Mauna Loa on the horizon.

(A) Photo courtesy U.S. Geological Survey.

When pyroclastics fall close to the vent from which they were thrown, they may pile up into a very symmetric cone-shaped heap known as a **cinder cone** (figure 9.10). A rock formed predominantly from ash-sized pyroclastic fragments is a **tuff**; a coarser rock containing large angular blocks is a **volcanic breccia** (figure 9.11).

Composite Volcanoes

Many volcanoes, andesitic ones especially, erupt different materials at different times. They may emit some pyroclastics, then some lava, then more pyroclastics, and so on. Volcanoes built up in this layer-cake fashion are called **stratovolcanoes** or, alternatively, **composite volcanoes**, because they are built up of more than one kind of material (figure 9.12). Most of the potentially dangerous volcanoes of the western United

B

A

B

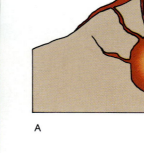

Figure 9.11

Rocks formed from pyroclastic material. *(A)* The Bandelier Tuff (New Mexico). *(B)* Volcanic breccia on flanks of Mount Washburn, Yellowstone Park.

Figure 9.12

Formation of a composite volcano.

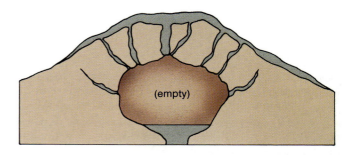

States, in the Cascade Range, are composite volcanoes—as in fact are most volcanoes formed over subduction zones. They typically have fairly stiff, gas-charged lavas, which sometimes flow and sometimes trap enough gases to erupt explosively with a rain of pyroclastic material. The combination of pyroclastics and viscous lavas tends to produce rather steep-sided cones of relatively symmetric shape.

Calderas

An eruption from a volcanic vent is fed from a magma chamber below, sometimes a very large one. When much of the magma has erupted, or perhaps has drained back down to deeper levels, the volcano is left partially unsupported. If the rocks are too weak they may collapse into the hole, forming a depressed **caldera** much larger than the original summit crater from which the lava emerged (figure 9.13). Alternatively, a caldera may be formed by a violent explosion, like that of Mount St. Helens, which greatly

A

(magma)

B

(empty)

C

New caldera width

Old crater width

Old volcano profile

Figure 9.13

Caldera formation by collapse. *(A)* Filled magma chamber helps support weight of rocks above. *(B)* Extraction of magma leaves a void and reduces support for the roof of the magma chamber. *(C)* Caldera collapse. The roof falls into the old chamber, producing a large depression.

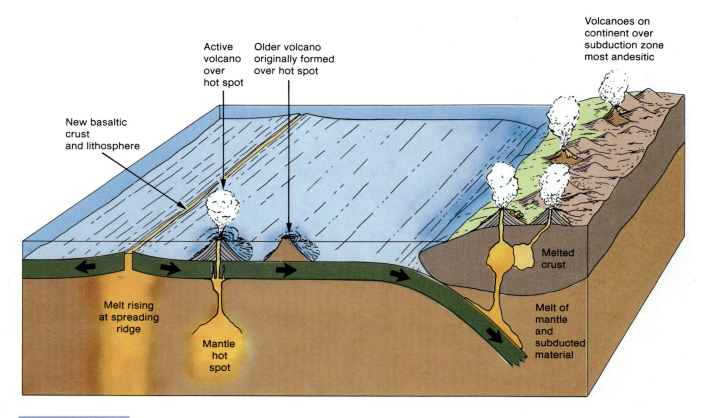

Figure 9.14

Volcanism and plate tectonics.

enlarges the vent. Major calderas can be very large features, covering tens or even hundreds of square kilometers. The best known example in the United States is the misnamed Crater Lake. The lake fills the depression formed by caldera collapse of the ancient volcano Mount Mazama.

Volcanism and Tectonics

Magmas typically form in one of three settings (figure 9.14): at divergent plate boundaries, in subduction zones, and at hot spots (*intraplate volcanism*). Each environment tends to produce certain compositions of magma, and thus is characterized by certain kinds of volcanoes.

Magmatism at Divergent Plate Boundaries

Most divergent boundaries are seafloor spreading ridges. The magma produced there is derived by partial melting of mantle material in the asthenosphere. The melting may be facilitated by the reduction in pressure from overlying lithosphere associated with faulting and spreading (figure 9.15). Given the ultramafic character of the upper mantle, and the fact that it must be fairly extensively melted to produce the volume of magma needed to make the amount of new rock generated at spreading ridges, we would expect the resulting melt to be rather mafic, and indeed it is. The dominant rock type is basalt, which forms the sea floor.

Much of the volcanism along continental rift zones is also basaltic, with voluminous flows formed from mantle melting as at seafloor spreading ridges. In addition, rising hot basaltic magma may warm the more granitic crustal rocks sufficiently to cause melting (figure 9.16). Some silicic volcanism is thus also found in continental rift zones, though volumetrically it is usually less important than the basaltic volcanism.

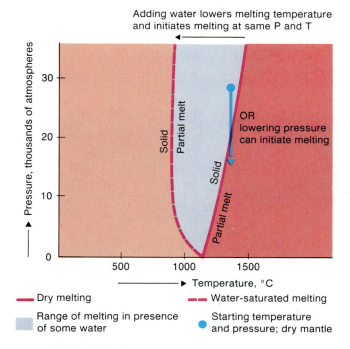

Adding water lowers melting temperature and initiates melting at same P and T

OR lowering pressure can initiate melting

━━━ Dry melting
━ ━ Water-saturated melting
▨ Range of melting in presence of some water
● Starting temperature and pressure; dry mantle

Figure 9.15

These melting curves for ultramafic material similar in composition to the mantle show that the addition of water lowers its melting temperature. Reducing the pressure on the material can also cause melting even with no change in temperature.

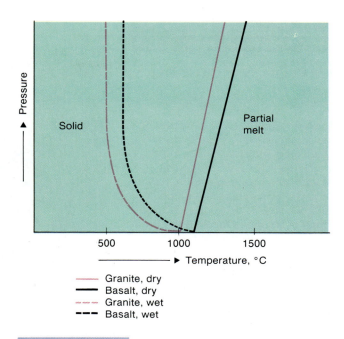

─── Granite, dry
━━━ Basalt, dry
─ ─ ─ Granite, wet
━ ━ ━ Basalt, wet

Figure 9.16

Granitic material has lower melting temperatures than basalt. This explains why basaltic magmas can sometimes cause melting of the continental crust through which they rise toward the surface.

Subduction-Zone Volcanism

The complexity of a subduction zone leads to a variety of possible magma sources.

The downgoing slab typically comprises seafloor basalt, mafic to ultramafic upper-mantle rocks of the associated deeper lithosphere, and some seafloor sediments. Any or all of these can melt, wholly or in part. Where the overriding plate is continental, the subducted sediments are in part material weathered from the continents. Those sediments will be relatively more silicic and tend to melt (producing small quantities of granitic melt) before appreciable melting of the more mafic rocks occurs. Ultimately, at least the basaltic crustal rocks melt also.

The downgoing sediments, and to a lesser extent the underlying basalt, contain water. During subduction, **dewatering** occurs: this water is released and rises into the portion of the upper mantle between the subducted and overriding plates. The water lowers the melting temperature of the mantle material and induces magma formation. The composition of the resulting magma is andesitic to basaltic, depending on the extent of melting. This is the characteristic volcanism of island arcs.

Interaction with and assimilation or melting of the overriding plate by rising magmas can further modify the composition of the magma before it reaches the surface. Where the overriding plate is continental, assimilation of granitic or granodioritic crustal material by mafic magma often produces an intermediate-composition, andesitic melt. Hence the common observation that volcanoes on continental edges above subduction zones are andesitic (and, often, explosive in eruptive style as well).

Intraplate Volcanism

There is not complete agreement among geologists about the nature of hot spots. Isolated hot-spot volcanoes are usually attributed to the presence of mantle plumes, columns of magma rising in the asthenosphere, spreading laterally when they reach the base of the lithosphere. If the lithosphere is sufficiently weak, some of the magma breaks through to form a volcano.

The composition of the material erupted depends largely on the composition of the lithosphere. That is, the melt in the plume would be expected to be basaltic, like the magma at a spreading ridge. If it

Life on a Volcano's Flanks

The so-called East Rift of Kilauea has been quite active in recent years. There were major eruptions in this area in the 1970s. Less than ten years later a new subdivision, Royal Gardens, was begun in the area. In 1982-83, lava flows from a renewed round of eruptions reached down the slopes of the volcano and quietly obliterated new roads and several houses. These photos illustrate some of the results (figure 1). The eruptions have continued intermittently since.

When asked why anyone would have built in that particular spot, where the risks were so obvious, one native shrugged and replied, "Well, the land was cheap."

It will probably remain so, too—a continuing temptation to unwise development.

A

B

Figure 1
Lavas invade the Royal Gardens subdivision on Hawaii. *(A)* Flows from March through July of 1983 covered 330 lots and destroyed 16 homes. *(B)* A superfluous stop sign. A lava flow several meters thick blocks a road so new that no lines had yet been painted on it.

(A) Photo by J.D. Griggs, courtesy U.S. Geological Survey.

rises up through the (mafic) sea floor, it will still be basaltic when it erupts, whether or not some sea floor is assimilated. If the magma must make its way up through continental crust, there is more potential for assimilation of granitic material and production of a more silicic final magma.

What causes plumes is not known for certain. They may form over regions of locally high concentrations of (heat-producing) radioactive elements in the mantle, with the extra heat causing extra melting. Some have suggested that plumes rise over anomalous regions of the outer core. If plumes do originate below the asthenosphere, it is difficult to explain why the rising magma columns should not be disrupted by the convection believed to occur in the asthenosphere. Whatever their cause, hot spots are long-lived features: recall, for example, that we can trace the path of the Pacific plate over the now-Hawaiian hot spot over the last 70 million years.

Hazards Relating to Volcanoes and How to Combat Them

Lava

At least until the eruption of Mount St. Helens, most people in this country, if they thought about volcanoes at all, probably thought of lava as the principal hazard. Actually lava is not generally a very serious threat, at least to life. Most lava flows advance at speeds of only a few kilometers an hour, or less, so one can evade the advancing lava readily even on foot. The lava will, of course, destroy or bury any property over which it flows. Lava temperatures are typically over 500° C (over 950° F) and may be over 1400° C (2550° F). Combustible materials—houses, forests—will burn at such temperatures. Other property may simply be engulfed in lava, which then solidifies into solid rock (figure 9.17 and box 9.2).

Lavas, like all liquids, flow downhill, so one way to protect property is simply not to build close to the slopes of the volcano. However, throughout history, people *have* built on or near volcanoes, for many reasons. They may simply not expect the volcano to erupt again. Soil formed from weathering of volcanic rock may form only slowly, but it is often very fertile. The Romans cultivated the slopes of Vesuvius and other volcanoes for that reason. Sometimes, too, a volcano is the only land available, as in the Hawaiian islands or Iceland. Where there has been development in the eventual path of the lava, particularly if the structures threatened are important economically or culturally, action may be necessary when eruptions occur.

Figure 9.17

Lava flow over road in Hawaii. The distance that can be covered by the fluid lavas of shield volcanoes is illustrated by the distance this 1974 flow has come from its source, Mauna Ulu (dimly visible on horizon).

Iceland sits astride the Mid-Atlantic Ridge. The land is wholly volcanic in origin. The fact that Iceland rises above the water surface may indicate that it is on a hot spot as well as a ridge. Certainly it is a particularly active volcanic area. Off the southwestern coast of Iceland are several smaller volcanic islands, which are also part of that country. One of these tiny islands, Heimaey, accounts for the processing of about 20% of Iceland's major export, fish.

Heimaey's economic importance stems from an excellent harbor. In 1973 the island was split by rifting, and several months' eruption followed. The residents were evacuated within hours of the start of the event. In the ensuing months, homes, businesses, and farms were set afire by falling hot pyroclastics or buried under pyroclastic material or lava. Many buildings were saved only by repeated shoveling of heavy loads of hot pyroclastic fragments off their roofs. When the harbor itself, and therefore the island's livelihood, was threatened by encroaching lava, officials took the unusual step of deciding to fight back.

The key to the harbor's defense was the fact that as lava cools it becomes thicker, more viscous, and flows more slowly even before it solidifies completely. When a large mass of partly solidified lava has

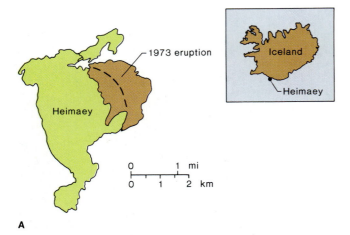

A

Figure 9.18

The 1973 eruption of Heimaey, Iceland. *(A)* Map showing extent of lava filling of harbor. *(B)* Lava flow is controlled by quenching the hot melt.

(B) Photo courtesy U.S. Geological Survey.

B

accumulated at the advancing edge of a flow, it begins to act as a natural dam to stop or slow the progress of the flow. How could a lava flow be cooled more quickly? The answer surrounded the island: water. Using pumps on boats and barges in the harbor, streams of water were directed at the lava's edge. Plastic pipes, prevented from melting by the cool water running through them, were carried across the solid crust that had started to form on the flow and positioned to bring water to places that could not be reached directly from the harbor. And it worked: part of the harbor was filled in, but much of it was saved, along with the fishing industry on Heimaey (see figure 9.18).

This bold scheme succeeded in part because the needed cooling water was abundantly available, and in part because the lava was moving slowly enough already that there was time to undertake flow-quenching operations before the harbor was filled. Also, the economic importance of the harbor justified the expense of the effort. Similar efforts in other areas have not always been equally successful.

Where it is not practical to arrest the flow altogether, it may be possible to divert it—from a course along which a great deal of damage may be done, to an area where less valuable property is at risk. Sometimes the motion of a lava flow is slowed or halted temporarily during an eruption, because the volcano's output has lessened or because the flow has encountered a natural or artificial barrier. The magma contained within the solid crust of the flow remains molten for days, weeks, or months thereafter. If a hole is then punched in this crust by explosives, the remaining fluid magma inside can flow out and away. Careful placement of the explosives can divert the flow in a chosen direction.

This was tried in Italy in early 1983 when Mount Etna began another in an intermittent series of eruptions. By punching a hole in a natural dam of old volcanic rock, officials hoped to divert the latest lavas to a broad, hollow, uninhabited area. This would have served the dual purpose of directing the lava away from populated areas and providing it with a wide area in which to spread out, which would ideally cause the lava to cool more rapidly and thus to slow down. Unfortunately, the effort was only briefly successful. Part of the flow was deflected, but within four days the lava had abandoned the planned alternate channel and resumed its original flow path. Later, new flows threatened further destruction of inhabited areas.

Lava flows may be hazardous, but in one sense they are at least predictable: like other fluids, lavas flow downhill. Their possible flow paths can be anticipated, and once they have flowed into a relatively flat area, they tend to stop. Other kinds of volcanic hazards can be more challenging to deal with and affect much broader areas.

Pyroclastics

Pyroclastics are often more dangerous than lava flows. They may erupt more suddenly, explosively, and spread faster and farther. The largest blocks and volcanic bombs present an obvious danger because of their size and weight. For the same reason, however, they usually fall quite close to the volcanic vent, and so they affect a relatively small area.

The sheer volume of the finer ash and dust particles can make them as severe a problem, and they can be carried over a much larger area. Also, ashfalls are not confined to valleys and low places. Instead, like snow, they blanket the countryside. The May 18, 1980, eruption of Mount St. Helens was by no means the largest such eruption ever recorded, but the ash from it blackened the midday skies more than 150 kilometers away, and measurable ashfall was detected halfway across the United States. Even in areas where only a few millimeters of ash fell, transportation ground to a halt as drivers skidded on the slippery roads and engines choked on the dust in the air. Homes, cars, and land were buried under the hot ash. Volcanic ash is also a health hazard that makes breathing both uncomfortable and difficult. The cleanup effort required to clear the debris strewn about by Mount St. Helens was enormous. It has been estimated that almost 3 cubic kilometers of material (about two-thirds of a cubic mile)—some fresh pyroclastics, some older bits of the volcano—were blown out in the major eruption of Mount St. Helens; 600,000 tons of ash landed on the city of Yakima, Washington, more than 100 kilometers away.

Past explosive eruptions of other violent volcanoes have been equally devastating. When the city of Pompeii was destroyed by Mount Vesuvius in A.D. 79, it was buried not by lava but by ash. (This is why extensive excavation of the ruins has been possible; figure 9.19.) Contemporary accounts suggest that most of the people had ample time to escape as the ash fell (though many chose to stay, and died), but the town was ultimately obliterated. Its very existence had been forgotten until some ruins were discovered in the late 1600s. In 1912, Mount Katmai in Alaska erupted violently, emitting an estimated 20 cubic kilometers (about 5 cubic miles) of pyroclastics, nearly eight times what Mount St. Helens cast out. Fortunately, there were no densely inhabited areas nearby. Even that event is dwarfed by comparison with the 1815 explosion of Tambora, in Indonesia, which ejected an estimated 175 cubic kilometers of debris!

Figure 9.19
Portion of the excavated ruins of Pompeii.

Lahars

Pyroclastic materials present a special hazard with snow-capped volcanoes like Mount St. Helens. The heat of the falling ash melts the snow and ice on the mountain, producing a mudflow of meltwater and volcanic ash. When solidified, the resulting deposit is called a lahar. Such mudflows, like lava flows, also flow downhill. They may follow stream channels, choking them with mud and causing floods of stream waters. Flooding produced in this way was a major source of damage near Mount St. Helens (figure 9.20). In A.D. 79, Herculaneum, closer to Vesuvius than was Pompeii, was partially invaded by volcanic mudflows.

A

B

Figure 9.20

Mudflow and flood damage from Mount St. Helens eruption,
1980. *(A)* Damaged homes along the north fork of the Toutle
River. *(B)* Vehicle wrapped around tree by the force of mudflow.

Photos by R. L. Schuster, courtesy U.S. Geological Survey.

Figure 9.21

Nevado del Ruiz, before the disastrous volcanic mudflows of 1985. Note the steepness of the slopes, the snow cover, and the small slide (dark) at right.

Photo courtesy U.S. Geological Survey.

This made possible the preservation, as mud casts, of the bodies of some of the volcano's victims. A very recent example of the devastation possible from such mudflows was provided by the 1985 eruption of Nevado del Ruiz in Colombia (figure 9.21). The swift and sudden mudflows triggered as its snowy cap melted were the major cause of the more than 20,000 deaths in towns below the volcano.

Nuées Ardentes

Another special kind of pyroclastic outburst is a deadly, denser-than-air mixture of hot gases and fine ash known as a **nuée ardente,** from the French for "glowing cloud." A nuée ardente is very hot—temperatures can be over 1000° C in the interior—and

Figure 9.22

Nuée ardente from Mount St. Helens, May 18, 1980.

Photo by P. W. Lipman, courtesy U.S. Geological Survey.

rushes down the slopes of the volcano at more than 100 kilometers per hour (60 mph), charring everything in its path and flattening trees and weak buildings. A nuée ardente accompanied Mount St. Helens's major eruption (figure 9.22).

Perhaps the most famous such event in recent history took place during the 1902 eruption of Mont Pelée, on the Caribbean island of Martinique. The volcano had begun erupting weeks before, emitting both ash and lava, but was believed by many to pose no imminent threat to surrounding towns. Then, on the morning of May 8th, with no immediate advance warning, a nuée ardente emerged from that volcano and swept through the nearby town of St. Pierre and its harbor. In a period of about three minutes, an estimated 25,000 to 40,000 people died or were fatally injured, burned, and suffocated. The single reported survivor in the town was a convicted murderer who had been imprisoned in the town dungeon, where he was shielded from the intense heat. He spent four terrifying days buried alive without food or water before rescue workers dug him out. Figure 9.23 gives some idea of the devastation.

Figure 9.23

Devastation from a nuée ardente: St. Pierre destroyed by Mont Pelée, 1902.

Photo by I. C. Russell, courtesy U.S. Geological Survey.

Just as andesitic volcanoes more often have a history of explosive eruptions, so many of them have a history of eruption of nuées ardentes. Again the composition of the lava is linked to the likelihood of such an eruption. While the emergence of a nuée ardente may be sudden and unheralded by special warning signs, it is not generally the very first sign of activity shown by the volcano during an eruptive stage. Steam had issued from Mont Pelée for several weeks before the day St. Pierre was destroyed, and lava had been flowing out for over a week. This suggests one possible strategy for avoiding the dangers of a nuée ardente: when a volcano known or believed to be capable of such eruptions shows signs of activity, leave.

Sometimes human curiosity overcomes both fear and common sense even when danger is recognized. A few days before the destruction of St. Pierre, the wife of the U.S. consul there wrote to her sister: "This morning the whole population of the city is on the alert and every eye is directed toward Mont Pelée, an extinct volcano. Everybody is afraid that the volcano has taken into its heart to burst forth and destroy the whole island. Fifty years ago Mont Pelée burst forth with terrific force and destroyed everything within a radius of several miles. For several days the mountain has been bursting forth in flame and immense quantities of lava are flowing down its sides. *All the inhabitants are going up to see it. . . .*" (Italics added. From Garesche, 1902).

Evacuations can themselves be disruptive. If in retrospect they turn out to have been false alarms, people may be less willing to heed a later call to clear the area. There are two volcanoes named La Soufrière in the Caribbean, both similar in character to Mont Pelée, both potentially deadly. In 1976, La Soufrière on Guadeloupe began its most recent eruption. Some 70,000 people were evacuated and remained displaced for months, but only a few small explosions occurred. Government officials and volcanologists were blamed for disruption of people's lives, needless as it turned out to be. When La Soufrière on St. Vincent began to erupt in 1979, officials were much less anxious to start issuing evacuation orders, despite recognition that here also was the potential for many casualties: a 1902 eruption of this volcano had killed 1600 people. Luckily, on this occasion no great harm resulted from failure to evacuate.

Toxic Gases

In addition to lava and pyroclastics, volcanoes emit a variety of gases. Many of these, such as water vapor and carbon dioxide, are harmless. Others are poisonous. The latter may include carbon monoxide, various sulfur gases, hydrochloric and hydrofluoric acids, and other deadly substances. People have been killed by volcanic gases even before they realized the danger. During the A.D. 79 eruption of Vesuvius, fumes overcame and killed many unwary observers, including historian Pliny the Elder. The dangers were illustrated very dramatically in late 1986, in Cameroon, when accumulated volcanic gases in a lake were suddenly released, killing more than 1500 people living nearby. Again the best defense against toxic volcanic gases is the commonsense one: to get well away from the erupting volcano or escaping gases as quickly as possible.

Steam Explosions

Some volcanoes are deadly not so much because of any characteristic inherent in the particular volcano but because of where they are located. In the case of a volcanic island, there is the possibility that large quantities of seawater will seep down into the rock, come close to the hot magma below, turn to steam, and blow up the volcano like an overheated steam boiler. This is termed a phreatic eruption. The classic example is Krakatoa, which exploded in this fashion in 1883. The force of its explosion was estimated to have been comparable to that

of 100 million tons of dynamite; the sound was heard 3000 kilometers away in Australia; some of the dust was shot 80 kilometers into the air, causing red sunsets for years afterward; ash was detected over an area of 750,000 square kilometers; and the shock of the explosion generated a tsunami over 40 meters high! Krakatoa itself was an uninhabited island, yet that eruption killed an estimated 36,000 people, mostly in low-lying coastal regions inundated by tsunamis.

In earlier times, when civilization was concentrated in fewer parts of the world, a single eruption such as that of Krakatoa could be even more destructive. During the 14th century B.C., the volcano Santorini on the island of Thera exploded in the Mediterranean Sea, and the resulting tsunami obliterated many coastal cities around the Mediterranean. It has been suggested by some scholars that this event contributed to the fall of the Minoan civilization.

Secondary Effects: Climate

It has long been recognized that a single volcanic eruption can have a global impact on climate. Very explosive eruptions put large quantities of volcanic dust high into the atmosphere. The dust can take years to settle, and in the interim it partially blocks out incoming sunlight, thus causing measurable cooling. After Krakatoa's 1883 eruption, worldwide temperatures dropped nearly half a degree centigrade, and the cooling effects persisted for almost ten years. The larger 1815 eruption of Tambora caused still more dramatic cooling: 1816 became known around the world as the "year without a summer." Such past experience forms the basis for fears of a "nuclear winter" in the event of a nuclear war, for modern nuclear weapons are quite powerful enough to cast volumes of fine dust into the air, and more dust and ash would be generated by ensuing fires.

Moreover, the climatic impacts of volcanoes are not confined to the effects of volcanic dust. The 1982 eruption of the Mexican volcano El Chichón did not produce a particularly large quantity of dust, but it did shoot volumes of unusually sulfur-rich gases into the atmosphere. These gases produced clouds of sulfuric acid droplets that spread around the earth. Not only do the acid droplets block some sunlight, like dust; in time they also settle back to the earth as acid rain.

Box 9.3

Thermal Features in Volcanic Areas

Figure 1
Hot spring in Yellowstone Park. Colors are due to algae. The particular colors present are a function of water temperature.

Figure 2
Geyser eruption, Yellowstone National Park.

Figure 3
Roaring Mountain fumaroles, Yellowstone Park.

In areas of active or recent volcanism, there is a good deal of hot magma or rock close to the surface. Water circulating below the surface is warmed in turn. If the water makes its way back to the surface, it may produce a variety of thermal features. Such warming of subsurface waters is also the basis for geothermal power (chapter 21). **Hot springs** are springs fed by magma-warmed waters (figure 1). They may be so hot that they boil, or just barely warm to the touch. If the return of the heated water to the surface is partially obstructed, the water may be trapped until some of it is heated to steam and the steam blasts it forcibly out. The resulting features are **geysers** (figure 2). After a geyser erupts, there is a lag during which water seeps in to replace that which was erupted, and the replenished water is heated until some is converted to steam again. This accounts for the intermittent eruption pattern of geysers. Finally, some underground water may be converted entirely to steam, which escapes through cracks or vents in rocks, sometimes mixed with volcanic gases. These fuming gas vents are **fumaroles** (figure 3).

Circulating warmed waters are also capable of reacting with and altering rocks. This **hydrothermal (hot-water) alteration** can be both extensive and spectacular (figure 4).

Figure 4
The hydrothermally altered volcanic rocks of the Grand Canyon of the Yellowstone River are particularly colorful.

Problems of Predicting Volcanic Eruptions

In terms of their activity, volcanoes are divided into three categories: **active; dormant,** or "sleeping"; and **extinct,** or "dead." There are unfortunately no precise rules for assigning a particular volcano to one or another category. A volcano is generally considered *active* if it has erupted within recent history. Where there have been no recent eruptions, but the volcano is fresh-looking and not very much eroded or worn down, it is regarded as *dormant,* inactive for the present but with the potential to become active again. Historically, a volcano that not only has no recent eruptive history but that also appears very much eroded has been considered *extinct,* very unlikely ever to erupt again. However, as volcanologists have learned more about the frequency with which volcanoes erupt, it has become clear that these guidelines are too simplistic. Volcanoes differ widely in their normal patterns of activity. Statistically, a "typical" volcano erupts once every 220 years, but 20% of volcanoes erupt less than once every 1000 years, and 2% less than once in 10,000 years. Long quiescence, then, is no guarantee of extinction.

The first step in predicting volcanic eruptions is monitoring, keeping an instrumental eye on the volcano. However, there are not nearly enough personnel or instruments available to monitor every volcano all the time. There are an estimated 300 to 500 active volcanoes in the world (the uncertainty arises from not knowing whether some are truly active or only dormant). Monitoring those alone would be a large task. The dormant volcanoes might return to activity, so they also should be watched. Theoretically, the extinct volcanoes at least can safely be ignored, but that assumes that we can distinguish the extinct from the long-dormant. The detailed eruptive history of many volcanoes is imperfectly known over the longer term.

Volcanic Precursors

There are several advance warnings of volcanic activity. One commonly used one is seismic activity. The rising of a volume of magma and gas up through the lithosphere beneath a volcano to the surface puts stress on the rocks of the lithosphere, and the process may produce months of small (and occasionally large) earthquakes. In August of 1959, earthquakes were detected 55 kilometers below Kilauea, a depth that corresponds to the base of the lithosphere under Hawaii.

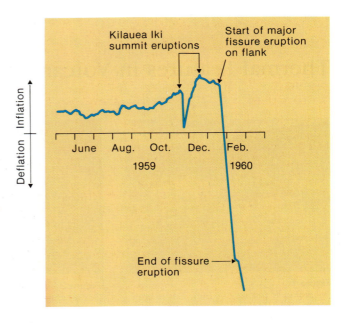

Figure 9.24

Tilt record of Kilauea, 1959–60. Increasing tilt of volcano's slopes reflects bulging due to rising magma below. Note the deflation after eruptions.

From *Volcanoes*, by Decker, R., and B. Decker. W. H. Freeman and Company. Copyright © 1981. Reprinted by permission.

Over the subsequent few months, the earthquakes became shallower and more numerous, as the magma pushed upward. In late November of that year, by which time up to 1000 small earthquakes were being recorded every day, a crack opened on the flank of the volcano and lava poured out.

The eruptions of Mount St. Helens have likewise been preceded by increased frequency and intensity of seismic activity. Sometimes the shock of a larger earthquake may itself unleash the eruption. The major explosion of Mount St. Helens in 1980 is believed to have been set off indirectly in this way: an earthquake shook loose a landslide from the bulging north slope of the volcano, which lessened the weight confining the trapped gases inside it and allowed them to blast forth. Detailed studies of the seismic activity at Mount St. Helens over the past few years have also suggested that its major eruptions may be preceded by characteristic *harmonic tremors,* distinctive rhythmic patterns of seismicity. If so, this could be an additional tool in eruption prediction.

Bulging, tilt, or uplift of the surface of the volcano is also a warning sign. It often indicates the presence of a rising magma mass, the buildup of gas pressure, or both. Figure 9.24 is a record of tilt and of eruptions on Kilauea. Note that eruption is preceded by an inflating of the volcano as magma rises up from the asthenosphere to fill the shallow magma chamber. Unfortunately, it is not possible to know beforehand at just what point the swollen volcano will crack to release its contents. That varies from eruption to eruption with the pressures and stresses involved and the strength of the overlying rocks. This limitation is not unique to Kilauea. Uplift, tilt, and seismic activity may indicate that an eruption is approaching, but we do not yet have the ability to predict its exact timing.

Other possible predictors are being evaluated. Many volcanologists speculate that changes in the mix of gases coming out of a volcano may give clues to impending eruptions. Surveys of ground surface temperatures may reveal especially warm areas where magma is particularly close to the surface and about to break through. As with earthquakes, there have been reports that volcanic eruptions have been anticipated by animals, which have behaved strangely for some hours or days before the event. Perhaps they are sensitive to some changes in the earth that scientists have not thought to measure.

Certainly much more work is needed before the precise timing and nature of major volcanic eruptions can be anticipated consistently. Recognition of impending eruptions of Kilauea and Mount St. Helens has plainly been successful in saving many lives through evacuations and restricting access to the danger zones. On the other hand, the exact moment of the 1980 explosion of Mount St. Helens was not known until seconds beforehand, and it is not currently possible to tell whether or just when a nuée ardente might emerge from a volcano like La Soufrière. In early 1985, for example, scientists detected patterns of harmonic tremors at Mount St. Helens disturbingly similar to those preceding the 1980 event and suggested that another such explosion (not necessarily as large) could be expected within days. It did not materialize. Nor can volcanologists readily predict the volume of lava or pyroclastics to be expected from an eruption, or the length of the eruptive phase.

Response to Eruption Predictions

The safest course, when data indicate an impending eruption that might threaten populated areas, is evacuation until the activity subsides. However, a given volcano may remain more or less dangerous for a long time. An active phase, consisting of a series of intermittent eruptions, may continue for months or years. In these instances, either property must be abandoned altogether for such prolonged periods, or inhabitants must be prepared to move out not once but many times. Given the uncertainty of eruption prediction at present, some precautionary evacuations will continue to be shown, in retrospect, as unnecessary or unnecessarily early.

Accurate prediction and assessment of the hazard is particularly difficult with volcanoes reawakening after long dormancy, where historical records for comparison with current data are sketchy or lacking. Beside the Bay of Naples is an old volcanic area known as the Phlegraean Fields. In the early 1500s, the town of Tripergole rose up some 7 meters as a chamber of gas and magma bulged below; the region was shaken by earthquakes; then suddenly a vent opened and in three days the town was buried under a new, 140-meter-high cinder cone. For the next four centuries, the only thermal activity was steaming fumaroles and hot springs. But within the last decade, the nearby town of Pozzuoli has been uplifted nearly 3 meters. It has been shaken by over 4000 earthquakes in a single year, leaving the town so badly damaged that many of its 80,000 residents have had to sleep in "tent cities" outside the town for months. Business can be conducted in what is left of the town by day, but it must be evacuated at night. The threat, and the disruption of people's lives, have persisted for over a year, but the volcanologists are still unsure what sort of volcanic activity can be expected here, how large in scale, or how soon.

Present and Future Volcanic Hazards of the United States

Hawaii

Several of the Hawaiian volcanoes are active or dormant. It is of course not practical to evacuate the island indefinitely, and eruptions are frequent in some places. A first, commonsense step toward reducing losses from eruptions might be to prevent any new development in recently active areas. Even this is not being done—see box 9.2.

Cascade Range

Mount St. Helens is only one of a set of volcanoes that threaten the western United States and Canada. There is subduction beneath the Pacific Northwest, which is the underlying cause of continuing volcanism there (figure 9.25). Several more of the Cascade Range volcanoes have shown signs of reawakening. Lassen Peak last erupted in 1914–21, not so very long ago geologically. Its products are very similar to those of Mount St. Helens; violent eruptions are certainly possible. Mount Baker (last eruption 1870), Mount Hood

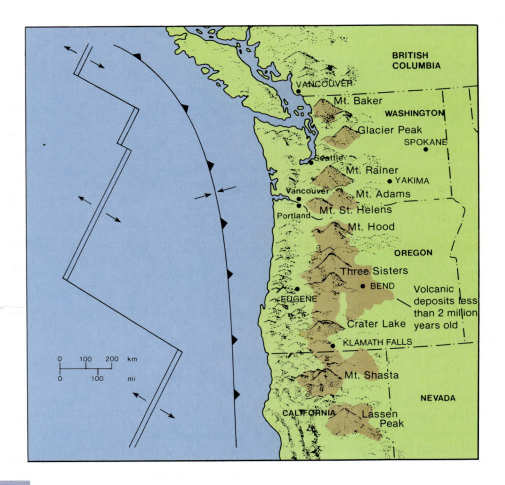

Figure 9.25

The Cascade Range volcanoes and their spatial relationship to the subduction zone and to major cities.

(1865), and Mount Shasta have shown seismic activity, and steam is escaping from these and from Mount Rainier, which last erupted in 1882. In fact, at least nine of the Cascade peaks are presently showing thermal activity of some kind (steam emission, hot springs). The eruption of Mount St. Helens has in one sense been useful, in that it has focused attention on this threat. Scientists are watching many of the Cascade Range volcanoes very closely now. With skill and perhaps some luck, they may be able to recognize when others of these volcanoes are close to eruption.

Mammoth Lakes, Long Valley

In 1980, the Mammoth Lakes area of California suddenly began experiencing earthquakes. Within one 48-hour period in May of 1980, four earthquakes of magnitude 6 rattled the region, interspersed with hundreds of lesser shocks. Mammoth Lakes lies within the Long Valley caldera, a 13-kilometer-long oval depression

formed during violent pyroclastic eruptions 700,000 years ago (figure 9.26). Smaller eruptions occurred around the area as recently as 50,000 years ago. Geophysical studies have shown that a partly molten mass of magma close to 4 kilometers across still lies below the caldera. Furthermore, since 1975 the center of the caldera has bulged upward more than 25 centimeters (10 inches); at least a portion of the magma appears to be rising in the crust. In 1982, seismologists realized that the patterns of earthquake activity in the caldera were strikingly like those associated with eruptions of Mount St. Helens. In May of 1982, the Director of the U.S. Geological Survey issued a formal notice of potential volcanic hazard for the Mammoth Lakes/Long Valley area. A swarm of earthquakes of magnitudes up to 5.6 occurred there in January of 1983. Seismicity is continuing. There has so far been no eruption, but the scientists continue to monitor developments very closely, seeking to understand what is happening below the surface so as to anticipate what volcanic activity may develop, and when.

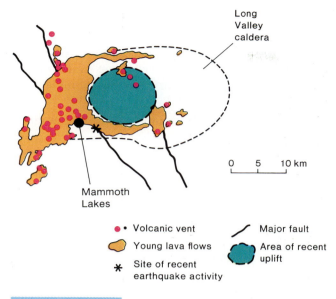

Long Valley caldera

Mammoth Lakes

0   5   10 km

- • Volcanic vent
- Young lava flows
- * Site of recent earthquake activity
- \ Major fault
- Area of recent uplift

Figure 9.26

The Long Valley caldera and Mammoth Lakes—possible future eruption site?

Source: After R. A. Bailey, U.S. Geological Survey, 1983.

Yellowstone Park

Another area of uncertain future is Yellowstone National Park. Yellowstone at present is notable for its geothermal features—geysers, hot springs, fumaroles. All of these features reflect the presence of hot rocks at shallow depths in the crust below the park. Until recently it was believed that the heat was just left over from the last cycle of volcanic activity in the area, which began 600,000 years ago. Recent studies suggest that in fact considerable magma may remain beneath the park, which is also still a seismically active area. One reason for concern is the scale of past eruptions: in the last cycle, 1000 cubic kilometers (over 200 cubic miles) of pyroclastics were ejected! Whether activity on that scale is likely in the foreseeable future, no one presently knows. The research goes on.

Summary

Most volcanic activity is concentrated near plate boundaries. Volcanoes differ widely in eruptive style and thus in the kinds of dangers they represent. Spreading ridges and hot spots are characterized by the more fluid, basaltic lavas. Subduction-zone volcanoes typically produce much more viscous, silica-rich, gas-charged andesitic magma, so in addition to lava they may emit large quantities of pyroclastics and other deadly products like nuées ardentes. Lava is perhaps the least serious hazard associated with volcanoes: it moves slowly, it can sometimes be diverted, and its path can be predicted. The results of explosive eruptions are less predictable, and the eruptions themselves more sudden. One secondary effect of volcanic eruptions, especially explosive ones, includes global cooling, which occurs as a result of dust and gases being thrown into the atmosphere and blocking incoming sunlight.

Early signs of potential volcanic activity include bulging and warming of the ground surface and increased seismic activity. Volcanologists cannot yet predict precisely the exact timing or type of eruption, except insofar as they can anticipate eruptive style on the basis of historic records, the nature of the products of previous eruptions, and tectonic setting.

Terms to Remember

aa	columnar jointing	hot springs	plume
active	composite volcano	hydrothermal	pyroclastics
ash	dewatering	lahar	shield volcano
blocks	dormant	lava	stratovolcano
bombs	extinct	nuée ardente	tuff
caldera	fissure eruption	pahoehoe	volcanic breccia
cinder cone	fumaroles	phreatic eruption	volcanic dome
cinders	geysers	pillow lava	

Questions for Review

1. Describe the nature of fissure eruptions and give an example.

2. What is a shield volcano and from what kind of lava are shield volcanoes usually built?

3. Compare/contrast a volcanic dome with a shield volcano. What causes the differences in form between these two kinds of volcanic structures?

4. What causes the eruption of pyroclastics, and with what type of lava are they more often associated?

5. Describe the processes leading to magma production at subduction zones.

6. Hot-spot volcanoes commonly erupt basaltic magma; volcanoes formed on continents above subduction zones (for example, the Andes) often erupt more silicic lavas. Suggest why this may be so.

7. Lava is usually more of a hazard to property than to lives. Why? Describe one strategy for protecting property from an advancing lava flow.

8. On what kind of volcano is a volcanic mudflow most likely to develop? Name a recent example.

9. What is a nuée ardente? What strategy can be used to minimize the loss of life from eruption of nuées ardentes?

10. What causes a phreatic explosion? Cite an example.

11. Volcanic eruptions may alter climate. How?

12. Describe two kinds of precursors used in the prediction of volcanic eruptions.

13. What is the underlying cause of volcanic activity in the Cascade Range of the western United States?

For Further Thought

1. Look up the projections made prior to the May 1980 eruption of Mount St. Helens of the areas that might be affected by its eruption, including secondary effects such as mudflows. Compare the actual consequences with the predictions.

2. Find out if there is any history of volcanic eruptions in your area (a) in historic times, or (b) within the last 1 to 2 million years. If so, what was the nature of the activity? What is the explanation for it? Is future activity expected?

Suggestions for Further Reading

Axelrod, D. I. *Role of Volcanism in Climate and Evolution*. Geological Society of America Special Paper 185, 1981.

Bailey, R. A. "Mammoth Lakes Earthquakes and Ground Uplift: Precursors to Possible Volcanic Activity?" In U.S. Geological Survey, *Earthquake Information Bulletin,* 15, no. 3 (1983): 88–102.

Bullard, F. M. *Volcanoes of the Earth*. 2d rev. ed. Austin, Tex.: University of Texas Press, 1984.

Decker, R., and Decker, B. *Volcanoes*. San Francisco: W. H. Freeman and Company, 1981.

Foxworthy, B. L., and Hill, M. *Volcanic Eruptions of Mount St. Helens: The First 100 Days*. U.S. Geological Survey Professional Paper 1249, 1982.

Garesche, W. A. *Complete Story of the Martinique and St. Vincent Horrors:* L. G. Stahl, 1902.

Gore, R. "A Prayer for Pozzuoli." *National Geographic* 165 (1984): 614–25.

MacDonald, G. A. *Volcanoes*. 2d ed. New York: Prentice-Hall, 1983.

Rampino, M. R., and Self, S. "The Atmospheric Effects of El Chichón." *Scientific American* 250 (January 1984): 48–57.

Volcanoes and the Earth's Interior. San Francisco: W. H. Freeman and Co., 1983. A selection of readings from *Scientific American,* 1975–82.

Williams, H., and McBirney, A. R. *Volcanology*. San Francisco: Freeman, Cooper and Co., 1979.

Williams, R. S., Jr., and Moore, J. G. "Iceland Chills a Lava Flow." *Geotimes* 18 (August 1973): 14–17.

Chapter 10

The Nature of the Earth's Interior

Introduction

In several previous chapters, reference has been made to various aspects of the physical or chemical properties of the earth's interior. One might legitimately ask just how geoscientists know so much about the interior. After all, we stand here on the surface of a planet nearly 13,000 kilometers in diameter; our deepest wells, drilled to extract natural gas, penetrate only about 10 kilometers into the crust. Yet we discuss even the composition of the core with some confidence. This chapter assembles the various kinds of evidence used to determine the physical and chemical makeup of the interior of the earth.

Chemical Constraints

There is both direct and indirect evidence relating to the earth's internal composition.

Direct Sampling and Chemical Analysis

Analysis of surface samples and drill cores reveals the compositions of a variety of crustal rocks of all types—igneous, metamorphic, and sedimentary. Where tectonic deformation has brought deep plutonic and high-grade metamorphic rocks closer to the surface, we have some samples even of lower-crustal rocks, though these exposures are few. Crustal rocks are extremely varied in composition, but our sampling is sufficiently thorough that there is general agreement on the average composition of the crust (table 10.1).

Deeper samples, from the upper mantle, are conveniently brought to the surface by volcanic rocks. Rising magmas formed by partial melting in the mantle may carry along with them fragments of still-unmelted mantle, or pick up pieces of mantle or crustal rocks above the depth of melting as they ascend. The rock inclusions are termed **xenoliths,** from the Greek *xenos,* meaning "stranger," and *lithos,* "rock" (figure 10.1). It is frequently possible to deduce the depths from which xenoliths come on the basis of the mineral assemblages present in them. In any case, they cannot have come from depths below the source region of the magma containing them. Where several types of xenoliths are present in the same volcanic

rock, we clearly have samples of several different subsurface rock units present below that point, although it may not be possible to know the vertical sequence in which those units actually occur.

The deepest-source magmas for which we have good evidence of the depth of formation are the diamond-bearing volcanic pipes known as **kimberlites,** which originate at depths of approximately 200 kilometers. Thus even our sampling of xenoliths extends only a short distance into the earth, and these deepest xenoliths are few. They do at least include some upper mantle material. Further inferences about the composition of the upper mantle can be drawn from the compositions of the magmas produced from it, combined with laboratory experiments on melting and magmatic crystallization. The nature of the magmas' parent mantle rocks can frequently be deduced from the compositions of these partial melts.

Cosmic Abundances of the Elements

The compositions of stars can be determined from analyses of their spectra: different wavelengths of light correspond to different elements. Most stars are quite similar in composition, at least in terms of the principal elements we can detect; 90% of them are compositionally similar to our own sun. The sun's light spectrum can thus be analyzed in some detail to give us an estimate of the composition of most stars, which in turn encompass most of the mass of the universe.

Table 10.1

Composition of the earth's crust[*]

Oxide	Continental crust	Oceanic crust	Overall average
SiO_2	59.2	48.0	58.1
Al_2O_3	15.4	15.2	15.1
FeO	7.5	10.7	7.4
MgO	4.3	7.7	4.4
CaO	6.0	12.2	6.8
K_2O	2.6	0.6	2.4
Na_2O	2.8	2.6	3.1
TiO_2	1.0	2.2	1.0

[*]All concentrations in weight percent; averages of several independent estimates.

Figure 10.1

A xenolith in basalt.

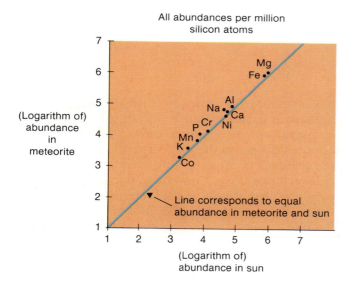

All abundances per million
silicon atoms

(Logarithm of)
abundance
in
meteorite

Line corresponds to equal
abundance in meteorite and sun

(Logarithm of)
abundance in sun

Figure 10.2

The relative abundances of (nonvolatile) elements in the sun and
in primitive meteorites are very similar.

One limitation of this approach is that the presence
and amounts of rarer elements are hard to detect.

Fortunately, there is a remarkable similarity between the major-element composition of the sun (excluding gases) and of primitive meteorites (figure
10.2; see also chapter 22). This suggests that one can
use the relative amounts of the rarer elements in the
same meteorites to estimate their abundances in the
rest of the solar system, and perhaps in the universe.

In such ways, one can assemble the so-called
cosmic abundance curve of the elements (figure
10.3). If we compare it with the relative abundances
of major elements in the earth's crust, again the overall
similarity of pattern is striking. The principal discrepancies are in elements such as the inert gases and
some other volatile elements (like hydrogen, oxygen,
and carbon). These may have failed to condense in the
primitive earth as it formed from the solar nebula.

This in turn suggests that all of the nonvolatile
elements should have condensed in the primitive
earth. The estimated bulk composition of the earth

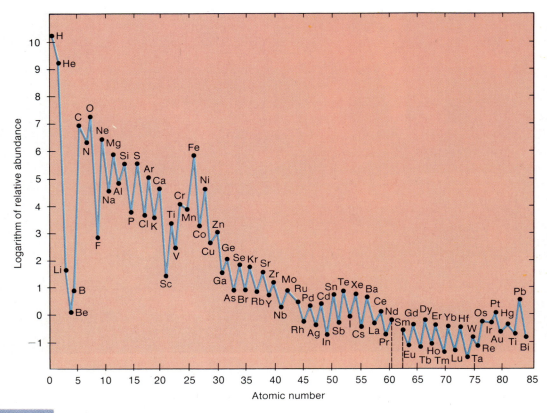

So-called cosmic abundance curve of the elements. Note that the
scale on the vertical axis is a logarithmic one. Abundances are
reported per million (10^6) atoms of silicon.

Table 10.2

Approximate average composition of whole earth*

Element	Percentage in earth
iron (Fe)	33.3
oxygen (O)	29.8
silicon (Si)	15.6
magnesium (Mg)	13.9
nickel (Ni)	2.0
calcium (Ca)	1.8
aluminum (Al)	1.5
sodium (Na)	0.2

*All concentrations in weight percent; averages of several independent
  determinations.

thus determined is shown in table 10.2. If the crust
is relatively depleted in iron, nickel, and magnesium
by comparison, then perhaps these elements are concentrated deeper in the earth. The abundance of ferromagnesians in upper mantle samples is consistent
with such reasoning. Conversely, one cannot realistically postulate that the earth's interior is made up of
something relatively rare, like lead or gold.

Physics and Geophysics

Bulk properties of the earth also restrict what we can
propose for the earth's interior. Its gravitational field
is a function of its mass. Taking mass and size together
yields an average density for the earth of about 5.5
grams per cubic centimeter. This is about twice the
density of continental crustal rocks and also significantly greater than the density of the ultramafic upper-
mantle rocks of which we have samples. Even allowing for the fact that the deep interior is strongly

compressed, it is still necessary to postulate the presence of materials in the interior that are considerably denser than any terrestrial materials that we have actually sampled.

The Magnetic Field

The earth also has a sizeable magnetic field, which could, so far as its orientation is concerned, be represented approximately by a giant bar magnet aligned nearly through the rotational poles. It might be tempting to postulate a large lump of solid iron, or perhaps iron-nickel alloy such as we find in some meteorites, in the middle of the earth. Iron is a relatively abundant element, it is strongly magnetic, and it is dense. But a mass of solid iron in the interior will not, in fact, account for the magnetic field as neatly as it might at first appear to do. This is because of the problem of temperature. Recall from chapter 7 that each magnetic material has a *Curie temperature* above which it loses its magnetic properties. For iron, this Curie temperature is less than 800° C. Yet magmas from the upper mantle frequently have temperatures of 1000° C or more. Temperatures in the deep interior must be hotter still. Any solid iron deep within the earth, then, will be well above its Curie temperature. The magnetic field has still to be accounted for, but the explanation must be more complex.

Seismic Waves as Probes of the Upper Mantle

The seismic body waves provide a unique means of investigating the nature of the interior below the depths from which we can obtain samples. Body waves from large earthquakes can be detected all over the earth, and their paths and travel times through it allow us to deduce properties such as the density and physical state of the materials at depth.

In a general way, body-wave velocities are proportional to the density of the material through which they propagate: the denser the medium, the faster a given type of body wave will travel. If they cross a boundary between two materials of distinctly different densities, they are refracted, or deflected into a different direction (figure 10.4). In a compositionally uniform material that increases smoothly in density with depth as a result of pressure, a body wave will be continuously refracted, and the resulting path is curved (figure 10.5).

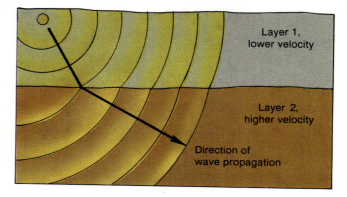

Figure 10.4

Refraction of seismic wave at a boundary between two layers of different density.

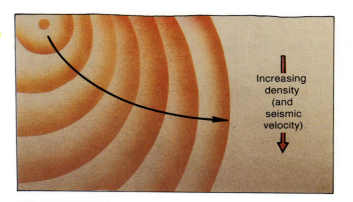

Figure 10.5

When density increases gradually with depth, continuous refraction in small increments produces a curved path.

All kinds of waves can be refracted. Light is similarly deflected on passing from air into water or a transparent mineral. A person spearfishing from a boat must allow for this effect, because the fish is not really where it will appear to be from above the water surface. The effects of the refraction of light waves can also be observed by putting a hand into an aquarium or a pencil into a partially filled glass of water.

The Moho

One of the first internal features to be detected seismically was the crust/mantle boundary. It is marked by an increase in seismic-wave velocities in the mantle relative to the crust, which corresponds to a change in composition and consequent increase in density (confirmed by xenolith evidence). This boundary is the **Mohorovičić discontinuity,** often known as the **Moho** for short, named after the Yugoslavian seismologist who first reported the seismic break by which it is identified. The Moho is most easily recognized beneath the continents, where the greatest compositional contrasts between lower crust and upper mantle exist. Even under the continents there is considerable lateral variation in seismic-wave velocities in crust and mantle, related to compositional variations. Commonly, the Moho in any spot is defined as the depth at which P-wave velocities first reach 7.8 kilometers/second or higher.

The higher seismic velocities in the upper mantle, and the phenomenon of wave refraction, together make possible the measurement of the thickness of the crust. The principle is illustrated in figure 10.6. When an earthquake occurs, the seismic body waves traveling only through the crust reach stations near the site ahead of those following a longer, deeper path to the mantle and back up to the surface. But at some distance away, the higher seismic velocities of the mantle will compensate for the greater distance traveled by waves following the deeper routes, and those waves traveling part of the way in the mantle will arrive ahead of waves traveling entirely in the crust.

The effect can be visualized by means of a human analogy. Consider a person with some distance to travel, who can either walk directly to the destination, or take a (faster) bus, which requires walking a bit out of the way at either end of the trip. For very short trips, the direct walk will be the faster way to go, avoiding the detour to the bus. But for long trips, the time spent getting to and from the bus is more than compensated by the time saved along most of the length of the route.

How long the trip must be before taking the bus becomes time effective depends partly on how much faster the bus travels than the pedestrian does, and partly on how far out of the way the bus route is. Likewise, how far apart earthquake and seismograph must be in order for the body waves going down to the mantle to arrive ahead of those traveling wholly through the crust will depend on the relative seismic velocities in crust and mantle, and on the thickness of the crust. We have samples of crust and upper-mantle

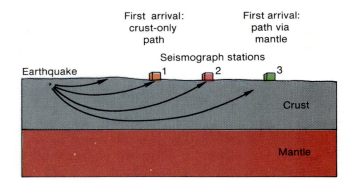

First arrival: crust-only path   First arrival: path via mantle

Seismograph stations

Earthquake   1   2   3

Crust

Mantle

Figure 10.6

The thickness of the crust can be determined using the seismic velocities in crust and upper mantle and the distance from an earthquake source at which waves traveling through the mantle reach a detecting seismograph ahead of waves traveling a shorter path entirely through the crust.

rocks whose seismic velocities can be determined experimentally in the laboratory. Therefore the thickness of the crust can be deduced. In the actual case, the problem is further complicated by seismic-velocity variations in crust and mantle as well as by local variations in crustal thickness.

Fine Structure of the Upper Mantle

The existence of the weak asthenosphere is important to the plausibility of plate tectonics, for it provides a means to explain plate movements. The existence of the asthenosphere was deduced from seismic-wave studies. Body waves passing through this zone showed unusually long travel times, from which a decrease in seismic velocities in this zone was deduced (figure 10.7). This **low-velocity layer** extends from the base of the lithosphere (by definition) to, typically, a depth of 175 to 250 kilometers. It does not encircle the whole earth uniformly, nor does it everywhere include partly molten material. Rocks of the low-velocity layer are all quite close to their melting temperatures, however, and therefore relatively less elastic and more plastic in their behavior. Plastic materials, which tend to flow under prolonged stress, transmit both compression and shearing deformation less efficiently than elastic materials, if at all. Under the short-term stress of a passing seismic wave, the rocks of the low-velocity layer may be sufficiently elastic to transmit the waves, though more slowly than in fully elastic rocks. Under sustained stress these same rocks will flow and deform permanently. Similarly, a ball of wax or putty may bounce when dropped quickly against a hard surface and retain its original shape, but the same material may flow plastically and

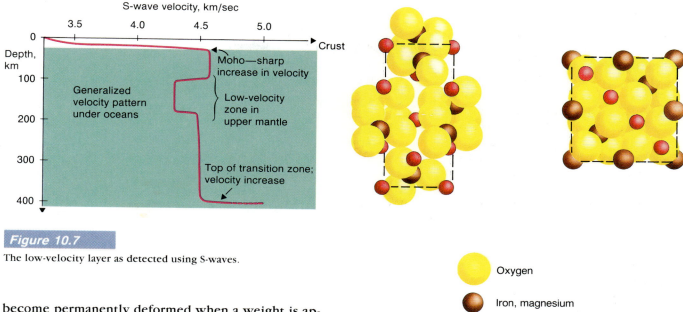

Figure 10.7

The low-velocity layer as detected using S-waves.

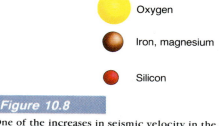

Figure 10.8

One of the increases in seismic velocity in the transition zone is attributed to a change in the crystal structure of olivine, from its low-pressure form (left) to a denser, more compact form (right) at depth.

become permanently deformed when a weight is applied over a long period of time. As can be seen in figure 10.7, the seismic velocities in the low-velocity layer are still substantially higher than crustal velocities, but markedly lower than velocities in the more rigid mantle immediately above and below that layer.

Still deeper in the upper mantle, beneath the low-velocity layer, seismic velocities increase again with depth and density. For the most part the increases are gradual, as if caused simply by the increasing density to be expected as a compositionally uniform mantle becomes more and more compressed with increasing depth. There are, however, several small jumps in velocity within the **transition zone**, between approximately 400 and 700 kilometers' depth. These are attributed to **phase changes**, changes in mineralogy or crystal structure without, necessarily, any change in composition. (This is in contrast to the Moho, which definitely reflects a compositional change.) Several specific phase changes have been proposed to explain the velocity increases. The jump at about 400 kilometers, for example, has been attributed to the collapse of the magnesian olivine abundant in upper-mantle rocks to a more compact, denser structure (figure 10.8). Laboratory experiments on olivine under pressure have suggested that the change will indeed occur at the appropriate depths/pressures. Deeper still, the various silicates of the mantle may break down into simple oxides (MgO, FeO, SiO_2, and so on) capable of achieving very dense crystal structures in response to the extremely high pressures.

Below the transition zone, from about 700 kilometers' depth to the base of the mantle, seismic velocities resume their gradual increase with depth (figure 10.9). The lower mantle, then, is not believed

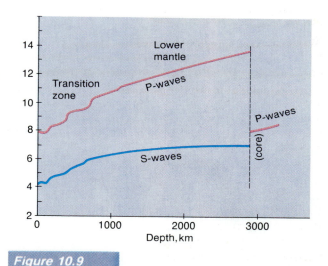

Figure 10.9

Seismic-wave velocity variations with depth in the mantle. Below the transition zone, velocities increase smoothly with increasing depth/pressure.

to include any sharp compositional or phase changes. Continuous changes—for example, a change in the ratio of iron to magnesium with depth—are still possible within the limits of the data.

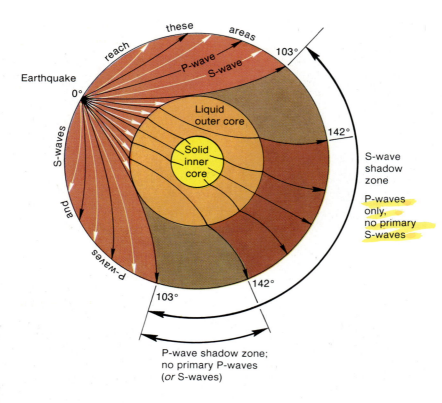

Figure 10.10

Seismic shadow zones for S-waves and P-waves caused by the
liquid outer core.

The Shadow Zone

The strength of seismic waves from a major earthquake is such that the body waves could, in principle,
be expected to reach the opposite side of the earth.
P-waves indeed do so. S-waves do not, and the pattern
of their absence led to recognition of another major
zone of the interior and the determination of its size.

Compressional waves can pass through solids or
liquids. Sound waves can travel through water, and P-
waves can travel through rock or magma, although
their velocities will naturally differ in the different
media. Liquids will not, however, support and propagate a shear wave. A liquid subjected to shear stress
will flow in response, and there is no elasticity to pull
it back again. S-waves will therefore not travel through
a liquid.

When a major earthquake occurs, no direct S-
waves are detected at distances greater than about 103
degrees of arc from the source (figure 10.10). A simple
explanation for this S-wave seismic shadow is that a
large fluid mass is blocking S-wave transmission. This
is the liquid outer core. Its physical state (liquid) is
indicated by its ability to block S-waves; its size is apparent from the size of the seismic shadow that it casts.

Actually, the outer core also casts a conical or ringlike
P-wave shadow (see figure 10.10). This much smaller
shadow zone, extending in a band from 103 degrees
to 142 degrees of arc away from the earthquake
source, is a consequence of the differences in elastic
properties between solids and liquids, and the correspondingly anomalous refraction that P-waves
undergo at the core-mantle boundary.

The Inner Core

Seismic waves can also bounce off, or be reflected
from, the boundaries between layers of different
seismic properties. It is in this way that the presence
of a further zone within the core was detected (figure
10.11). Some P-waves are reflected from the boundary
between the inner and outer core. P-wave velocities
in the outer core can be approximated from the travel
times of those P-waves passing only through the outer
core (and shallower zones). Once velocities in mantle
and outer core are known, the travel times of P-waves
reflected off the inner core can be used to deduce the
depth to the inner/outer core boundary.

Travel times of P-waves traveling straight through
the earth, inner core and all, indicate significantly
higher velocities in the inner core than in the outer.

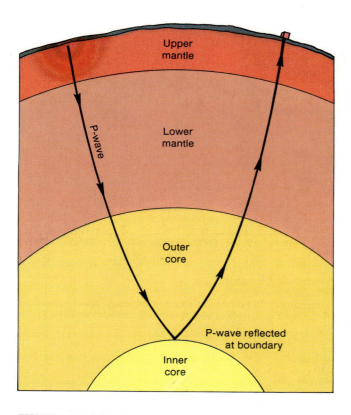

Figure 10.11

The travel times of P-waves that bounce off the inner/outer core boundary allow the depth to that boundary to be determined.

From this it is inferred that the inner core is again solid rather than liquid. Temperatures in the deepest interior are not believed to increase very much with depth, but pressures, especially in the very dense core, certainly do. Therefore, even if the inner and outer core are compositionally identical, it is perfectly plausible that the inner core, at much higher pressures, could be solid while the outer core is molten.

Seismic Data and the Composition of the Core

The results of laboratory experiments on seismic velocities and densities of various materials under very high pressures have been compared with velocities and densities indicated by actual seismic data and show that silicate materials are unlikely candidates for the core. Metallic iron gives a fairly good fit to actual data, but it is somewhat too dense, especially for the outer core. Iron meteorites are typically made of an iron-nickel alloy, so it has been suggested that nickel is present in the earth's core with iron. But nickel is still denser than iron; if iron is a little too dense for the seismic data, an iron-nickel alloy would be a worse

fit. Therefore it is generally accepted that there must be at least one other element present in moderate quantity (perhaps 10%) to "lighten" the core, lowering its density to conform to actual values estimated from seismic data. There are several elements that are both cosmically abundant and miscible with molten iron that have been proposed as minor constituents of the outer core. They include metallic silicon, oxygen, and sulfur.

The Magnetic Field Revisited

The demonstration of the presence of a liquid outer core, and the inference that it consists mostly of iron, together permit an explanation of the earth's magnetic field after all. A magnetic field can be generated by flow in an electrically conducting fluid, even at temperatures above the Curie point of the corresponding solid. Molten metallic iron seems a suitable fluid. Direct proof of this model of generation of the magnetic field is lacking, in large measure because we cannot make appropriate direct measurements of motions in or electrical conductivity of the outer core. It is a credible explanation, however, and consistent with available data. It is also consistent with the observation that the magnetic poles are close to the rotational poles: intuitively, it makes sense that fluid motions in the outer core might be influenced, controlled, or even generated by the earth's rotation. A remaining awkwardness is the phenomenon of magnetic reversals, well documented by paleomagnetic studies. These have yet to be satisfactorily explained by this or any other model of the earth's magnetic field.

Constraining Pressure and Temperature

Just as we cannot sample the deep interior directly, so we must also infer the pressures and temperatures in the deep mantle and core.

Pressure

Estimates of internal pressures are comparatively easy to make. Below the lithosphere, in which additional lateral stresses arise from plate motions, pressures are effectively determined by the thickness and densities of overlying rocks at any point. The densities of various layers can be obtained from seismic body-wave data, as noted earlier. This makes possible the calculation of pressure as a function of depth in the earth,

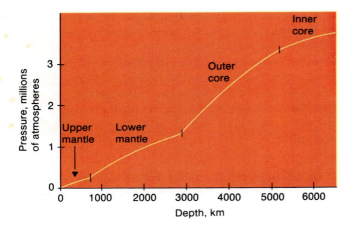

Figure 10.12

Pressure as a function of depth in the earth.

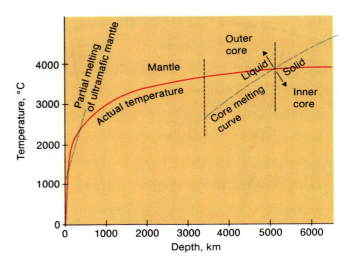

Figure 10.13

Temperature as a function of depth in the earth. Major constraints are the melting curves of ultramafic rock (mantle) and iron (core).

shown graphically in figure 10.12. Maximum pressures, in the inner core, are over 3 1/2 million times atmospheric (surface) pressure.

Temperature

Internal temperatures are somewhat more difficult to deduce, for temperatures do not increase in any predictable way with the thickness of overlying rock. If we approach the problem by simply extrapolating crustal geothermal gradients downward, projected temperatures quickly become impossibly high. Extrapolating at a temperature increase of 30° C per kilometer, for example, yields expected temperatures of about 21,000° C (37,000° F) at the base of the upper mantle, and 86,700° C (156,000° F) at the core-mantle boundary. It would be impossible for rocks to be solid, or even liquid, under such pressure-temperature conditions, so actual temperatures in mantle and core must be substantially lower.

Fortunately, knowledge of the composition of the interior, coupled with the demonstrated variations in physical state of the interior, permit us to put upper or lower limits on the temperatures of various zones. For example, the fact that there is some partial melting in the low-velocity layer indicates that temperatures must be at or above the minimum melting

temperature of the ultramafic silicate rock of the upper mantle. On the other hand, most of the mantle is solid. So, assuming no major compositional changes deeper in the mantle (and remember, there is no seismic evidence for any such changes), temperatures deeper in the mantle must be *below* the melting curve for such material at appropriate pressures.

Similar reasoning may be applied to the core. The core is at least predominantly iron, if not pure iron, so we can approximate core melting properties by those of iron. The outer core is molten. There, temperatures must be above the melting curve for iron at those pressures. The inner core is solid; temperatures there must be below the melting point of iron. (If small amounts of one or more additional elements are also in the core, as indicated by seismic data, their presence will tend to reduce the melting temperature of iron somewhat at a given pressure, but the effect is probably not sufficient to invalidate altogether temperature estimates based on the properties of pure iron.)

Combining temperatures estimated in this way for the mantle and core, we obtain an estimate for the geothermal gradient at depth as shown in figure 10.13.

Summary

Information of many kinds contributes to determining the physical state and chemical composition of the earth's interior, shown in figure 10.14. Chemical and mineralogical data for the crust and uppermost mantle are obtained by direct sampling and analysis, coupled with laboratory studies of magmas and melting. Estimates of the bulk composition of the earth can also be constrained using compositional data from meteorites and stars. Physical parameters (mass, density, and so forth) and observations such as the existence of a magnetic field further limit the possible makeup of the interior.

Detailed studies of the propagation of seismic body waves have provided most of the information on the structure of the interior and the densities and physical states of the zones within it. Seismic data have made possible the identification of the silicate mantle, with the low-velocity layer underlain by several phase transitions in the upper mantle, and the much denser, iron-rich core. The S-wave shadow zone demonstrates that the outer core is liquid and indicates its size. High-pressure laboratory experiments suggest that one or more elements besides iron and nickel are present in the outer core at least, as its density is somewhat too low for pure iron or iron-nickel alloy. Pressures in the interior are estimated straightforwardly using the densities of the various zones in the earth. Temperature estimates are constrained using the melting curves of ultramafic rocks (for the mantle) and of iron (for the core) for pressures corresponding to those prevailing in the interior.

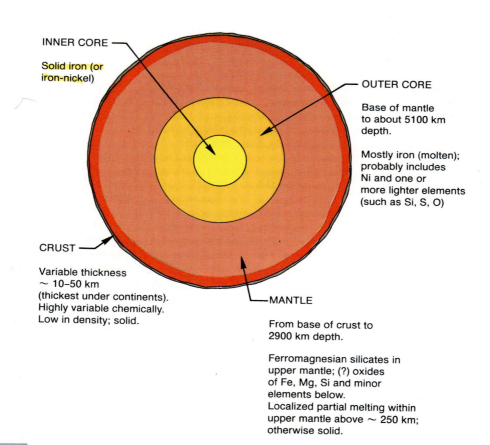

INNER CORE

Solid iron (or iron-nickel)

OUTER CORE

Base of mantle to about 5100 km depth.

Mostly iron (molten); probably includes Ni and one or more lighter elements (such as Si, S, O)

CRUST

Variable thickness ~ 10–50 km (thickest under continents). Highly variable chemically. Low in density; solid.

MANTLE

From base of crust to 2900 km depth.

Ferromagnesian silicates in upper mantle; (?) oxides of Fe, Mg, Si and minor elements below. Localized partial melting within upper mantle above ~ 250 km; otherwise solid.

Figure 10.14

Composite picture of the earth's interior and its various zones.

Box 10.1

Project Mohole

In the late 1950s, a group of earth scientists proposed a revolutionary exercise to explore the earth: drilling a hole through the crust, through the Moho, and into the upper mantle. Such an operation would provide a systematic and complete cross section through the crust at the site chosen, and allow direct sampling of part of the upper mantle, for which scientists previously had to rely on the haphazard sampling of xenoliths in magmas. Scientifically, it was an exciting idea.

Technically, it was forbidding. The engineering technology for drilling through a minimum of 5 kilometers of crust to reach the Moho did not exist. Moreover, in order to keep the costly and difficult drilling to a minimum, it would be necessary to drill through the relatively thin oceanic crust— which meant drilling in about 5 kilometers of water. Compounding the difficulty of deep drilling in rock, then, would be the difficulty of drilling in open ocean, keeping a drilling ship poised for months or years over a hole several kilometers below, despite rough seas and ocean currents.

The initiators of the idea enlisted the backing of the prestigious National Academy of Sciences. The National Science Foundation supplied funds for a feasibility study. Early projections (made in 1958) put the probable cost of "Mohole" at $5 million. More refined feasibility studies resulted in a revised estimate of $14 million by mid-1959. This was still comparable to the cost of a single space shot, and seemed reasonable in the light of the probable scientific gains.

Because deep-ocean drilling techniques were so new and untried, it was felt to be desirable, as an interim step, to drill a small hole in the deep ocean first. In the early 1960s, at a site off the southwestern California coast, the modified drilling barge used for Phase I set a then-world record for deep drilling with a 197-meter hole in the sea floor. Technically this was a great feat and was widely hailed. It was still a long way from the Moho.

Bids were taken for the next stage, which included the hole to the Moho. The contract from the National Science Foundation was awarded to the fifth-lowest bidder, amid loud allegations of political influence.

Thirteen months later, in 1963, the firm revealed its plan for a huge seagoing drilling platform to drill Mohole. They also revealed a projected cost of $67.7 million. Privately, some NSF officials conceded that costs might run to $125 million. Such sharply higher cost estimates gave many people pause.

Mohole was stalled for several months while its reasonableness was debated. One of its strongest Congressional proponents died, and his successor moved to terminate funding for it. Some attempt was made to salvage the project in the Senate. But when it was learned that the contracting firm was making substantial political contributions in an effort to influence the outcome, Congress voted firmly to squelch Project Mohole by cutting off all funding. The year was 1966, nearly a decade after the project's inception.

Drilling technology has improved radically since that time, and some deep continental drilling is now in progress along with shallow drilling of the sea floor. But nothing of the scope of Project Mohole is likely to be proposed again in the foreseeable future.

Terms to Remember

cosmic abundance curve	Mohorovičić discontinuity	refraction	transition zone
kimberlites	phase changes	seismic shadow	xenoliths
low-velocity layer			

Questions for Review

1. By what means do we obtain samples of the deep crust and uppermost mantle below the depths to which we can drill?

2. From what information is the cosmic abundance curve of the elements derived, and how is it used to constrain the composition of the earth's interior?

3. Can the earth's magnetic field be due to a lump of solid iron in the core? Why or why not? How else can the field be explained?

4. What is the Moho, and how is the depth to the Moho determined?

5. What causes the decrease in seismic velocities associated with the low-velocity layer?

6. There are not believed to be significant compositional changes in the lower mantle. Why not?

7. Describe the origin of the S-wave shadow zone. What does it tell us about the earth's interior?

8. In what way is the depth to the inner/outer core boundary determined?

9. The core is believed to be neither pure iron nor iron-nickel alloy. Why?

10. Can temperatures in the earth's interior be determined by extrapolation from geothermal gradients in the crust? How else are internal temperatures constrained? Discuss briefly.

For Further Thought

The relatively high cosmic abundance of iron suggests it as a plausible dense element for the principal component of the core. In the absence of that constraint, consider whether one could demonstrate in some other way that the core is not made of another metal, such as gold or lead, and if so, how.

Suggestions for Further Reading

Garland, G. D. *Introduction to Geophysics: Mantle, Core & Crust.* 2d ed. Toronto: Holt, Rinehart & Winston, Inc., 1979.

Gass, I. G.; Smith, P. J.; and Wilson, R. C. L., eds. *Understanding the Earth.* Cambridge, Mass.: MIT Press, 1971. See especially "Mohole: Geopolitical Fiasco" by D. S. Greenberg, 342–48; "The Composition of the Earth" by P. Harris, 52–69; and "The Earth's Heat and Internal Temperatures" by J. H. Sass, 80–87.

Henderson, P. *Inorganic Geochemistry.* New York: Pergamon Press, 1982.

Scientific American 249 (September 1983). Contains a series of articles on the earth and its various zones.

Volcanoes and the Earth's Interior. San Francisco: W. H. Freeman and Co., 1983. A selection of readings from *Scientific American,* 1975–82.

York, D. *Planet Earth.* New York: McGraw-Hill Book Co., 1975.

The Continental Crust

Introduction

The surfaces of the continents are the part of the earth most accessible for observation, measurement, and sample collection. In favorable cases, we can extrapolate features recognized at the surface some distance into the crust. The first portion of this chapter describes common kinds of structures and other features formed in the continental crust and how they are identified. This is followed by a brief survey of the geophysical techniques—including seismic methods already discussed—that allow probing of the deeper crust. From studies of geologic structures, we learn much about the nature of mountain building and find that there are several different kinds of mountains. Following a discussion of mountain-building processes, the chapter concludes with a look at the continents through time, and what can be said about the longevity and growth of continents.

Geologic Structures

We have already looked at a variety of rock textures produced during the formation of a single rock unit—for example, the foliation of schists, or the porphyritic or vesicular textures of some volcanic rocks. Additional structure can be imposed on rocks after formation, through the application of stress. The type of structure produced varies, depending on the nature of the stress and on whether the rock behavior is more brittle or more plastic. Brittle behavior favors the formation of fractures; plastic behavior, of folds. (For a review of these terms and related rock properties, see chapter 2.)

Describing the Orientation of Structural Features

It is frequently important not only to identify the kind of geologic structure present but to be able to describe its orientation. This allows one to derive information about the directions from which the deforming stresses have come, which may in turn aid in tectonic interpretation. The orientation of lines and planes can be described in terms of two parameters, *strike* and *dip*. **Strike** is a compass direction, measured parallel to the earth's surface. **Dip,** as the name suggests, is displacement downward from the horizontal, measured in degrees.

These concepts are most readily visualized for a linear feature, such as a metamorphic lineation, described in chapter 5 (figure 11.1). The strike of the lineation is the compass direction in which the lineation points, conventionally measured as degrees east or west from north. Values of strike, then, will range from 0 to 90 degrees east or west. For example, the lineation in figure 11.1 would be described as striking N 30° E. The dip of a lineation is the vertical angle between the lineation and a horizontal plane. Dip angles therefore range from 0 degrees (horizontal) to 90 degrees (vertical). For any dip between these limits, one also specifies the direction of the dip. For the example shown in figure 11.1, the dip would be described as 24° NE. Because elongated minerals are rarely perfectly aligned even in a lineated rock, one might make several measurements on a single outcrop in order to obtain a clear idea of the trend and coherence of the lineation, just as one might do in determining a paleocurrent direction (chapter 4).

Strikes and dips can also be assigned to planes, although in this case their definitions are less intuitive (figure 11.2). The strike of a plane is the compass orientation of the line of intersection of that plane with a horizontal plane. Dip is then measured perpendicular to the strike, by definition. Note that for any (nonhorizontal) plane, one could measure a displacement from the horizontal in many different directions, and the measured angle would differ depending on the direction chosen (figure 11.3). The angle measured perpendicular to the strike will be the *maximum* of these values. Specifying that dip is to be measured perpendicular to the strike of the plane ensures consistency in reported data, as well as providing information on the maximum slope of the plane.

The concepts of strike and dip of a plane can be explored by immersing a sheet of stiff cardboard or a thin board partway into a basin of water. The strike of the board is the orientation of the line formed where board meets water surface; dip would be measured at right angles to that direction. In an outcrop, the direction of dip of a sloping surface can be determined most readily by pouring water on the outcrop: the water trickles down the steepest slope, which is also in the direction of dip. The strike is the trend of the horizontal line perpendicular to the dip.

Many geologic features are planar: sedimentary beds, igneous dikes and sills, metamorphic foliation. However, these planes may be deformed into nonplanar shapes that do not maintain the same orientation over broad areas. If a limestone bed is folded, for example, the bed will be curved, and its orientation becomes more complex to describe. Strike and dip can

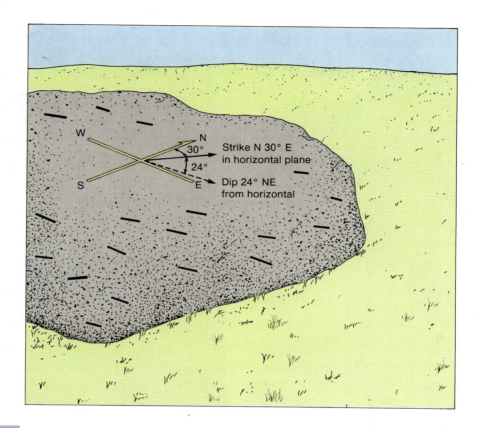

Figure 11.1

Strike and dip of a lineation. Strike is the compass direction in which the lineation points, as it would appear in map view (looking down from above). Dip is the vertical angle between the lineation and the plane of the horizontal.

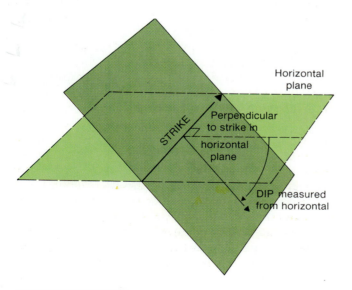

Figure 11.2

Strike and dip of a plane.

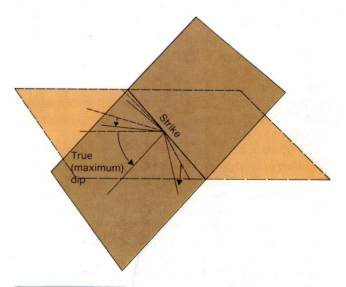

Figure 11.3

The angle between a dipping plane and the horizontal depends on where you measure it.

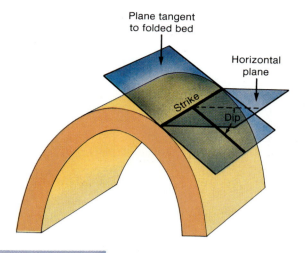

Figure 11.4

Strike and dip of a plane tangent to a folded bed.

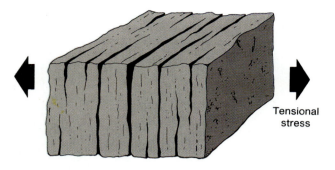

Figure 11.5

A joint set made up of parallel planar fractures.

Tensional stress

still be measured at any single point by considering the orientation of a hypothetical plane tangent to (just touching) the curved surface at that point (figure 11.4). In order to characterize the geometry of the structure fully, it is then necessary to take strike and dip measurements at many points, not just one or two.

Joints and Faults

We have already noted (chapters 2, 8) that brittle rocks fracture in response to stress. If there is no movement of the rocks on one side of the fracture relative to the other, the planar fractures are described as **joints.** The columnar jointing of a basalt flow is one example: the tensional stress created by contraction during cooling causes fracturing of the flow in place, but not displacement or rearrangement of the columns. Jointing can also result from compression. Often, the imposition of stress on a regional scale results in multiple fractures with the same orientation: **joint sets** (figure 11.5).

When there is relative movement between rocks on either side of the fracture, it is a **fault,** as discussed in chapter 8 (figure 11.6). The nature of the movement, or type of faulting, may be more important in interpreting geology and tectonics. Faults, like other planar features, can be assigned strikes and dips. They may be described further in terms of the steepness of the fault plane and the direction of relative movement. The relative movement may be horizontal, vertical, or a combination of the two (figure 11.7).

Figure 11.6

A fault as seen in the field.

Photo courtesy U.S. Geological Survey.

Movement only in the horizontal direction will correspond to movement along the strike of the fault, which is then called a **strike-slip** fault. Transform faults are one kind of strike-slip fault. Simple strike-slip faults also occur along which features originally rectly opposite one another across the fault are separated horizontally by the displacement. If the displacement is such that an observer facing the fault would find the matching features across the fault displaced to the right, the fault is described as *right-lateral*. A *left-lateral* fault shows the opposite movement. (See figure 11.8.) If the fault has cut and

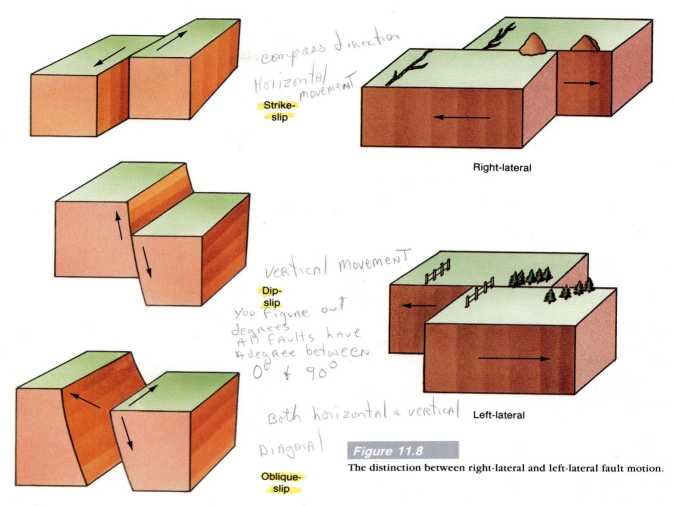

(handwritten annotations)
compass direction
Horizontal movement

Strike-slip

vertical movement

Dip-slip

you Figure out degrees
All FAULTS have A degree between 0° & 90°

Both horizontal & vertical
DIAGONAL

Oblique-slip

Right-lateral

Left-lateral

Figure 11.8

The distinction between right-lateral and left-lateral fault motion.

Figure 11.7

Types of fault displacement: strike-slip, dip-slip, and oblique-slip.

displaced one or more features that can be recognized as having once been continuous across the fault plane, the amount of displacement can easily be determined by the horizontal distance now separating matching features along the fault.

If the relative movement along the fault is entirely vertical, up or down in the direction of dip, the fault is a **dip-slip** fault. Unless the fault plane is perfectly vertical, one can distinguish a *hanging wall* and a *footwall* (figure 11.9). These can be identified by imagining oneself standing within the fault between the rock blocks: the footwall would be underfoot, below the inclined fault plane, with the hanging wall hanging overhead. If the hanging wall moves downward relative to the footwall, the fault is termed a **normal** fault; if the hanging wall moves relatively upward, it is a **reverse** fault. A reverse fault with a shallowly dipping (low-angle) fault plane is also

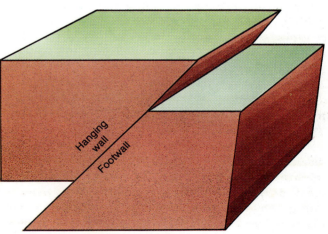

Hanging wall

Footwall

Figure 11.9

Identification of the hanging wall and footwall of a fault.

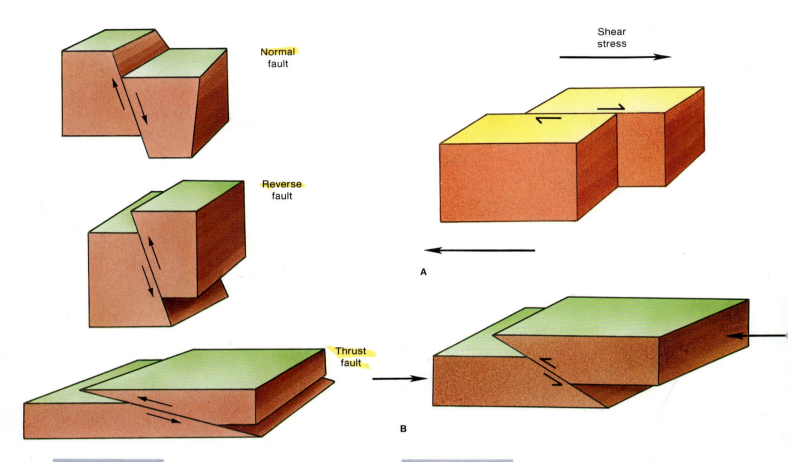

Figure 11.10

Dip-slip faults are subdivided on the basis of the sense of motion and the angle of the fault plane into normal, reverse, and thrust faults.

Figure 11.11

Stress is related to resultant fault movement. *(A)* Horizontal shear produces strike-slip faulting. *(B)* Thrust faulting results from compression.

known as a **thrust fault.** Thrust faulting on a regional scale may have the effect of thickening the crust by piling slices of rock atop each other. The various dip-slip faults are illustrated in figure 11.10. There are also faults along which the relative movement has both a horizontal and a vertical component. Such faults are termed **oblique-slip** faults.

The relative movements of faulted blocks provide information on the orientation of the stresses responsible (figure 11.11). For example, strike-slip faulting indicates shear stress, oriented horizontally. Thrust faulting results from compressive stress, concentrated in the direction of fault movement. Normal faulting is commonly a result of tension. If the crust is bulged up and stretched, for example by a subcontinental hot spot, the tension associated with crustal stretching over the bulge often causes high-angle (steeply dipping) faults to form. Some blocks may settle into the space created by the stretching; others may be pushed up by the buoyancy of the rising magma. Both normal and reverse faulting might be observed in this instance.

Folds

Plastic deformation, particularly when it involves compression or shear, is characterized by the production of folds. Folds, like faults, come in all sizes—microscopic, of handsample size, and on the scale of mountains (figure 11.12). Each fold has several components (figure 11.13). The **nose** of the fold is the most sharply curved part of the fold (just as a person's nose is the most sharply curved part of the face). The **axial plane** divides the two **limbs,** or sides, of the fold. The **axial trace** of the fold is the line along which the axial plane intersects the land surface, and is thus directly analogous to a fault trace (chapter 8). A more frequently used feature, the **axis of the fold,** is the line of intersection of the axial plane with the surface of the fold (or with some particular surface within it, such as a bedding plane). Axial planes and fold axes may be oriented in terms of strike and dip just as any other planar or linear features may be.

A

B

Figure 11.12

Types of folding. *(A)* Small-scale folds in a schist. *(B)* Folding on an outcrop scale.

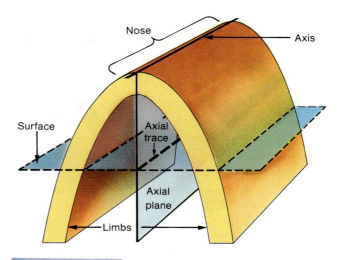

Figure 11.13

Nomenclature of the component parts of a fold.

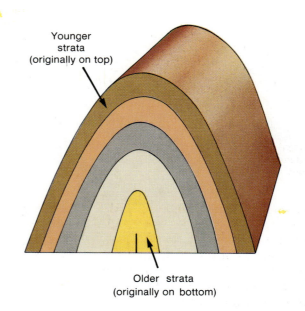

Younger strata (originally on top)

Older strata (originally on bottom)

Figure 11.14

An anticline, or arching fold, in layered sediments. Note that the oldest strata are at the center.

Large-scale folds (of regional scale) are subdivided on the basis of whether they arch upward or are bowed down into trough shapes. When the rocks folded were originally flat-lying sedimentary rocks, the resulting structures differ not only in the orientation of the beds but also in the relative ages of the rocks in various parts of the fold. An arching fold with a horizontal axial trace is an **anticline** (figure 11.14). The limbs of the anticline dip away from the axial trace. If the units folded were originally undeformed sedimentary rocks, the rocks at the core of the anticline

will be the oldest. A trough-shaped fold, a **syncline**, shows the reverse pattern (figure 11.15). The limbs of the fold dip toward the center of it, and if a sedimentary sequence is folded, the youngest units will be folded into the center of the syncline, to be exposed there after erosion.

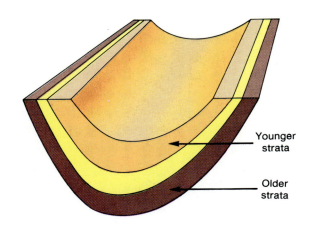

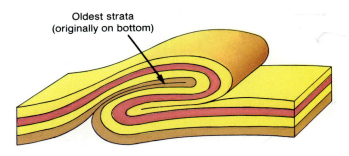

Oldest strata
(originally on bottom)

A

Figure 11.15

A syncline, showing the reverse age pattern.

Younger strata

Older strata

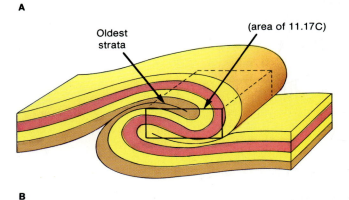

Oldest strata

(area of 11.17C)

B

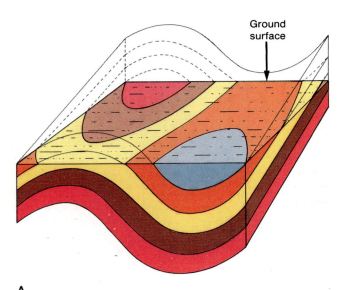

Ground surface

A

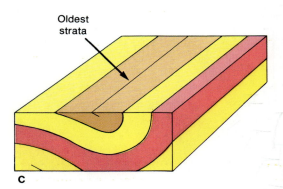

Oldest strata

C

Figure 11.17

Extreme crustal shortening produces recumbent or overturned folds. (See also figure 11.30.) *(A)* A recumbent anticline. *(B)* An overturned anticline. *(C)* Erosion of the overturned anticline reveals a synformal structure, but the relative ages of the rocks are inconsistent with a syncline.

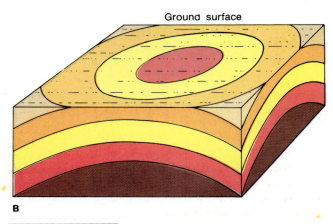

Ground surface

B

Figure 11.16

Plunging folds. *(A)* A simply plunging anticline and syncline. *(B)* A dome.

Strictly speaking, the terms *anticline* and *syncline* are reserved for folds in which the rocks show the age patterns described above. Other folds of similar form or shape, but not showing the appropriate age relationships, are called *antiforms* and *synforms* respectively. However, *anticline* and *syncline* are commonly used by nonspecialists to describe all arch-shaped and trough-shaped folds, and this informal usage will generally be followed here, for simplicity.

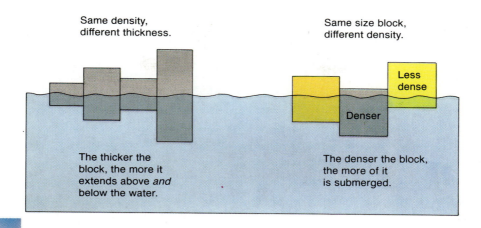

Same density,
different thickness.

Same size block,
different density.

Less dense

Denser

The thicker the
block, the more it
extends above *and*
below the water.

The denser the block,
the more of it
is submerged.

Figure 11.18

Blocks of similar density but different
thickness float at different heights in water.
Density differences affect the proportion of
each block that projects above the water.

Folds need not have either vertical axial planes or horizontal fault traces. A fold with a dipping axial trace is a **plunging** fold (figure 11.16A). The plunge of the fold is described by the orientation of the axial trace. A special case of plunging fold is that in which the fold effectively plunges radially in all directions. The upwarped fold in this case is a **dome** (figure 11.16B); the corresponding downwarped fold is a **basin.**

When the amount of compression is great, the axial plane of the fold may be tilted close to the horizontal as the rocks are doubled back upon themselves. The result is a **recumbent** fold (figure 11.17A). In extreme cases, the fold may be completely **overturned** (figure 11.17B). In such complex cases, erosion can produce unusual patterns of repeated rock units and relative ages. For example, figure 11.17C shows that erosion of this overturned anticline in sedimentary rocks produces a synform, but the rocks at the center of the trough are the oldest in the sequence, as would be appropriate for an anticline.

Note that any kind of rock can be folded, just as rocks of all kinds can be faulted. The emphasis in the foregoing discussion has been on layered sedimentary rocks mainly because fold structures are particularly easy to see and describe in such rocks. Likewise, the age relationships described for anticlines and synclines are observed only for rocks spanning an age progression in an orderly way, such as layered sediments or a sequence of lava flows. In principle a massive rock unit like a granite pluton can be folded or deformed, but the rocks in it are all the same age, and the nature of the folding may be much harder to see.

Continents Afloat: Vertical Movements

The lithosphere is less dense than the asthenosphere, and therefore both continental and oceanic lithosphere float buoyantly on the asthenosphere, as an iceberg floats on water. The overall density contrast is greater for continental lithosphere. The mantle portion of all lithosphere is very similar in density to the underlying asthenosphere: both are about 3.3 grams per cubic centimeter (g/cc). The iron-rich basaltic oceanic crust has a somewhat lower density of about 3 grams/cubic centimeter. The granitic continental crust is still lower in density, about 2.7 grams/cubic centimeter, so continental lithosphere is relatively more buoyant. The effect of density on buoyancy can be demonstrated by comparing the behavior in water of an ice cube, a block of dry wood, a cork, and a Ping-Pong ball. The ice is only slightly less dense than water. Therefore, although ice floats, most of its volume is submerged below the surface. Cork and most woods are less dense than ice, so a higher proportion of each of these objects will project above the water surface. The density of the Ping-Pong ball is not far above the density of the air it encloses, and most of its volume bobs above the water. While much of the difference in elevation between the continents' surfaces and the sea floor can be ascribed to the much greater thickness of the continental lithosphere, the continents also possess additional buoyancy resulting from the lower density of continental crust. Figure 11.18 illustrates the effects of thickness and density differences for a set of floating blocks.

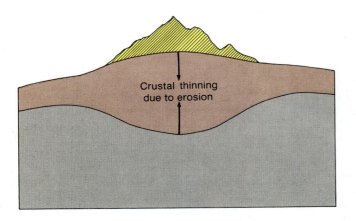

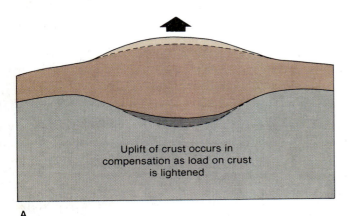

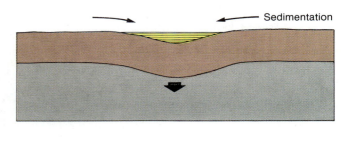

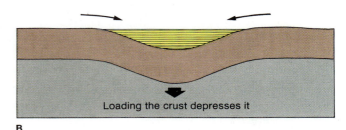

Crustal thinning
due to erosion

Uplift of crust occurs in
compensation as load on crust
is lightened

A

Sedimentation

Loading the crust depresses it

B

Figure 11.19

Isostatic compensation. *(A)* As a mountain erodes and the
continent thins, the decrease in elevation is compensated by
uplift of the whole lithosphere, decreasing the depth of the root
below. *(B)* As a basin is filled, the thickened crust sinks.

Isostasy

The tendency of lithospheric masses to float at elevations consistent with their relative densities is **isostasy,** named from the Greek for "same standing." The
phenomenon of isostasy implies that if the load on a
region changes, the lithosphere will shift vertically in
adjustment (figure 11.19). When the lithosphere and
underlying asthenosphere are in **isostatic equilibrium,** the mass of the overlying column of rock above
a certain depth in the mantle is the same everywhere.
For example, as a mountain erodes, the thickness of
the continental crust is decreased, and the total mass
of the column of lithosphere beneath the mountain's
surface is reduced. The lithosphere there will warp
upward in compensation, allowing additional dense
asthenosphere to flow in beneath the thinned crust.
For a crude illustration of the phenomenon, float a
stack of two or three flat slabs of wood in water; note
the elevation of the bottom slab relative to the water
surface; then remove the top slab and look again.

If lithospheric blocks over the asthenosphere
behaved exactly like blocks of wood bobbing in water,
isostatic equilibrium would readily be maintained.
Two factors slow the restoration of equilibrium in the
lithosphere/asthenosphere system when it has been
disturbed. One is the viscosity of the asthenosphere,
its resistance to flow. Just as a pile of bricks does not
immediately sink into warm asphalt, so the lithosphere does not immediately sink into the asthenosphere as it is loaded. Conversely, as erosion or the
melting of an ice sheet unloads the lithosphere, it does
not immediately bob up in response. The second factor
is the limited flexibility of the lithosphere. The thin,
pliable rubber surface of an inner tube or waterbed
deforms easily when stressed and bounces back
quickly when the stress is removed. Lithosphere tens
of kilometers thick is far more rigid; the relatively cold
crust is particularly stiff. Some time may elapse between loading or unloading and the corresponding
deformation.

There exist some dramatic examples of the
slowness of isostatic adjustment. One of the best
known is the modern rebound of the Scandinavian region. That region was weighted down by thick ice
sheets (about 2 1/2 kilometers, or 1 1/2 miles, thick)
during the last ice age. The ice sheets had completely
melted by about 10,000 years ago. Yet the region is
still rising, or rebounding, from the removal of that
ice load (figure 11.20). Isostatic adjustment is continuing, but slowly. Present uplift rates are one centimeter per year, or less.

Geophysics and Crustal Structure

Just as seismic methods permit investigation of the deep, inaccessible zones of the earth's interior, so a variety of geophysical methods make possible deductions about subsurface crustal features.

Gravity

The force of gravity attracts all matter on earth toward the earth's center. The strength of the gravitational attraction between two objects is a function of the mass of each and of the distance between them. If the earth were a perfectly homogeneous and spherical object, the gravitational force acting on an object at its surface would be the same at any point on the surface. The same would be true if the earth consisted of perfectly spherical concentric shells of materials of different densities, provided that each shell were homogeneous. In either case the same amount of mass, distributed the same way, would be pulling inward from any point on the surface. In practice, because the earth is not so uniform, there are small lateral variations in the measured force of gravity from place to place, which are termed **gravity anomalies**.

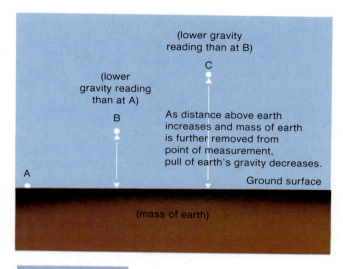

Figure 11.21

The effect of topography on gravity.

One source of variation is topography (figure 11.21). The gravitational attraction between two objects decreases as the distance between them increases. As an object at the surface is raised or lowered, the total gravitational pull on it decreases or increases respectively, as the distance from object to earth is increased or decreased. A gravity measurement made from an airplane over some point on the surface will give a lower value than the corresponding measurement made at the surface. All gravity data are routinely corrected to their equivalent values as if measured at mean sea level, before any possible anomalies due to geology are considered.

Another possible source of local gravity variation is the additional mass represented by topographic variations (figure 11.22). That is, if we assume the continents to be uniform in composition below a certain elevation—sea level, for instance— but the surface is not uniform in elevation, there will be more or less total mass in a vertical column beneath the surface from place to place. Where there is more total mass, gravity should be a little stronger, and vice versa.

If the objective is to detect deep-crustal features, we want to subtract out this kind of effect also. From sampling the surface or near-surface rock types we can choose reasonable densities for the rocks of, for example, a mountain range. Then we subtract the gravity increment due to those rocks, correcting the data over a region to the values that would be measured at sea level if all rocks above sea level were stripped away.

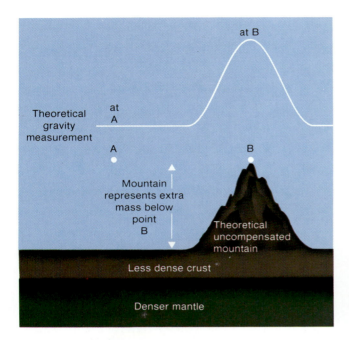

Figure 11.22

If a mountain were uncompensated by a crustal root, the gravitational pull atop a mountain should be greater than the pull at the same altitude where no mountain exists.

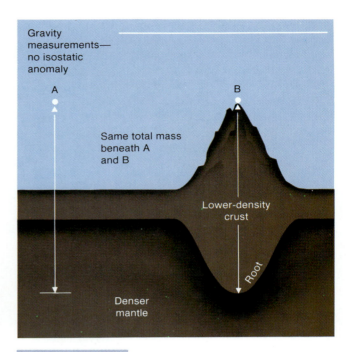

Figure 11.23

Gravity measurements corrected only for topography and for the mass of the mountains above sea level show an apparent mass defect, caused by the crustal root displacing denser mantle.

The results provide strong evidence that, in most cases, the topography of the continental surface is mirrored by the profile of the base of the continental crust (Moho), that mountains are isostatically compensated by "roots" of crustal rock (figure 11.23). Mountain ranges typically show negative gravity anomalies when the raw data are corrected only for topography and for the rock masses above sea level. That is, the pull of gravity is less than would be expected if the column of rock below sea level under the mountains corresponded in densities and thicknesses of layers to the rock column below low-elevation areas. But if the less dense, granitic rocks of the continental crust extend to greater-than-average depths beneath the mountains, displacing denser mantle rocks, the apparent mass deficiency is explained.

Given the densities of crustal and mantle rocks, it is possible to calculate the size of the crustal root needed to compensate for a mountain range of particular size, or for any particular topography. One can then assume the appropriate compensation and make a further correction to the gravity measurements on that basis. Some anomalies may still remain. They can be explained in two principal ways. One is by the presence of localized anomalous masses within the crust. A dense mafic pluton within the crust, for example, may cause a local positive gravity anomaly above it. A second kind of anomaly, of regional scale,

may arise as a consequence of the time lag in isostatic readjustment. For example, in the case of the ice-depressed Fennoscandian region, the ice mass that caused the original downwarping is gone, but the lithosphere has not yet fully rebounded or adjusted isostatically. This results in a negative gravity anomaly, an apparent mass defect, over the region. When isostatic adjustment is complete—when the crust has risen again to an elevation consistent with the new lighter load and denser mantle material has moved in underneath the rising lithosphere—this residual gravity anomaly will be eliminated. Regional gravity anomalies, then, can be used to detect areas that are not in complete isostatic equilibrium.

Magnetics

We have already discussed aspects of the earth's magnetic field (chapters 7 and 10), which is believed to originate from fluid motions in the molten outer core. The spatial distribution of this field determines the broad regional character of the strength and orientation of the magnetic field as measured near the surface. Superimposed on the global pattern are small local variations resulting from the variable magnetic properties of crustal rocks.

We have discussed how magnetic minerals, crystallizing and cooling through their Curie temperatures, may take on a preferred magnetic orientation parallel to the earth's prevailing field at the time. This accounts for the normally and reversely magnetized strips of sea floor, and for similarly polarized igneous rocks on the continents. During precipitation or deposition, magnetic sedimentary minerals may also align themselves with the earth's field as they settle into sedimentary deposits. Magnetic orientation may be changed during intense metamorphism, as rocks are heated close to Curie temperatures of the magnetic minerals in them. On cooling, the minerals' magnetism may be wholly or partially realigned to correspond to the field at the time of metamorphism. Also, different minerals and rock types differ in the intensity of magnetization that they can acquire. Rocks rich in ferromagnesian minerals and especially in magnetite show the strongest magnetization.

These differences in magnetic properties and orientation lead to localized **magnetic anomalies**, somewhat analogous to gravity anomalies. Strong magnetic anomalies are often associated with mafic plutons in the continental crust. A large linear magnetic anomaly, trending southwestward from the Great Lakes region, is associated with the failed rift buried beneath the surface there. This magnetic anomaly is probably due to extensive intrusion of basaltic rocks within the rift zone in the predominantly granitic continent. Magnetic anomalies are also used in prospecting for ore deposits, especially iron ores.

Heat Flow Data

The earth's interior is hotter than the surface; therefore heat flows from the interior outward. The amount of this **heat flow,** though small, is measurable. On the continents, it averages about 60 milliwatts per square meter at the surface.

> To put heat flow in perspective, if the escaping heat were somehow collected over a square meter for a year, it would be sufficient to heat about two gallons of water to the boiling point. Another way to look at it is that to keep one 60-watt light bulb burning constantly would require the energy equivalent of the heat flow from 1000 square meters.

Heat flow is related to the amount of heat supplied to the base of the crust from the interior (still warm from earth's formation), and to the amount of heat production from continuing decay of radioactive elements, most of which are concentrated in the continental crust. Localized areas of high heat flow on the continents are most often associated with recent volcanism, which brings hot magma from the asthenosphere up into the normally cooler crust, closer to the

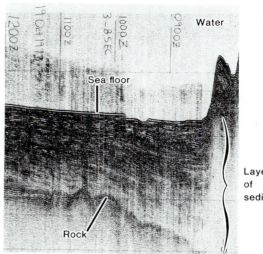

Figure 11.24

Reflection of seismic waves from bedding planes and interfaces between layers of different density reveals subsurface structure.

surface. Regions of very high heat flow are often *geothermal areas* characterized by geysers, hot springs, and other warm-water features. The generation of geothermal energy (see chapter 21) depends in part upon identifying these abnormally warm regions within the crust. Regional heat-flow surveys are thus becoming increasingly important in identifying the most promising regions for geothermal power generation.

Seismic Methods

Some of the same techniques described in chapter 10 can be applied to the problem of elucidating the fine structure of the earth's crust. Seismic-wave refraction occurs between crustal layers of different density, just as it does at the Moho or the mantle-core boundary, although the effects are less pronounced where the density contrasts between layers are smaller. Seismic waves are also reflected off interfaces between layers of different density within the crust, and even bedding planes within sedimentary sequences. Both the presence of layering and the attitude (orientation) of the layers may be determined, which is useful in identifying subsurface folds and faults (figure 11.24). Petroleum exploration, which relies in part upon the location of geologic structures capable of trapping and collecting oil and gas, has been greatly enhanced by the use of seismic methods. Seismic waves are attenuated by passage through partly molten masses within the crust just as they are by passage through the low-velocity zone in the mantle. It is in this way that the presence of magma in the crust beneath Yellowstone National Park and the Long Valley caldera has been confirmed.

Figure 11.25

Volcanic mountains: Mount St. Helens (foreground); Mt. Rainier (rear).

Photo by D. R. Mullineux, courtesy U.S. Geological Survey.

Seismic data are also used in conjunction with other geophysical methods. The combination of seismic and gravity data can be especially powerful. A gravity anomaly of given size or a profile of particular shape does not yield a unique interpretation of the subsurface geology. Many different combinations of subsurface mass/density distribution could produce the same net gravity anomaly pattern as measured from the surface. However, seismic data may constrain the crustal models by providing information on the densities of various subsurface layers that have not been sampled directly, and on the locations of the boundaries between layers.

Mountains and Mountain Belts

The topographic prominence of mountains tends to draw attention to them and to inspire investigation of and speculation about the nature of mountain ranges and how they form. One immediate observation is that mountains *do* tend to occur in ranges or belts rather than in isolation. Forces and processes of sufficient scale to cause the amount of crustal thickening represented by a mountain a thousand meters or more high usually also have a regional impact and create not one mountain but many. The principal exception would be a volcano built up over an isolated hot spot, and even in that case, as the overlying lithospheric plate moves, a trail of additional volcanoes will be formed over the hot spot.

The topography of a mountain range as we now observe it may or may not closely reflect either the nature of the processes by which the mountains formed, or the structures in the rocks in the mountain range. The sculpturing effects of water and ice not only tend to reduce the overall elevation of the land through time but to shape and re-shape the eroded rocks. The height and form of a mountain range is likely to be as much a function of the age of the mountain-building or crust-thickening events (old mountains are typically more severely eroded and of lower overall relief), and of the readiness with which the rocks erode, as of the nature of the mountain-building process.

Types of Mountains

Mountains form in a variety of ways, some by a combination of processes or with a variety of structures. Often a single process or structure will predominate throughout a range, simply because the rock types and the stresses to which the rocks were subjected, or the tectonic setting of the region, were the same throughout the range.

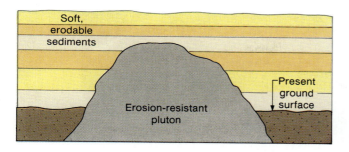

Formation of a mountain by subtraction of erodible rocks above and around it.

The relationship between volcanic activity and formation of mountains is easily visualized and was explored in some detail in chapter 9. Lava or pyroclastics or both pile up around a volcanic vent, creating a topographic high at that spot (figure 11.25). The chains of volcanoes that form above subduction zones contribute to the formation of mountains on continents along convergent plate boundaries, although other processes contribute to crustal thickening there as well (see the following section on mountain building).

Some mountains form, in effect, by subtraction (figure 11.26). Magma invades a sequence of relatively easily weathered rocks and solidifies into one or more plutons. If the region is elevated so that it is subjected to erosion, that erosion will remove the softer country rocks preferentially, and what is left is the durable, resistant igneous core of pluton (figure 11.26). The granite peaks of Yosemite National Park and the White Mountains of New Hampshire are examples of residual mountains formed in this way.

Broad arching or doming on a regional scale may produce a topographic high that ultimately forms a mountain range. The Black Hills of South Dakota (figure 11.27) are an example. They are also cored by ancient igneous and metamorphic rocks, uplifted at the center of the dome, that resisted the erosion that stripped away the overlying blanket of sedimentary rocks.

Tension or compression on a regional scale can lead to the formation of large-scale joint sets and parallel faults. With continued stress and movement along the faults, crustal blocks are shifted up or down relative to one another (figure 11.28). A block dropped down relative to the adjacent blocks is a **graben**; a block uplifted relative to those on either side is a **horst.** If the relative vertical movement associated with the block faulting is significant, fault-block mountains are formed. A common characteristic of fault-block mountains, especially young ones, is that they rise sharply from the surrounding terrain, as they

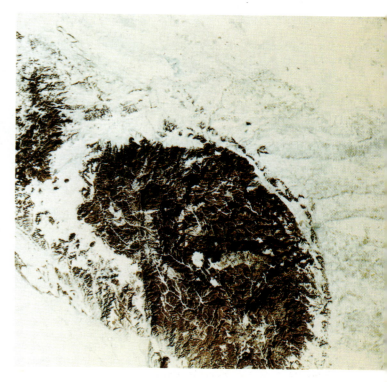

The Black Hills of South Dakota, formed by doming followed by erosion of sedimentary rocks to reveal a granite core (Landsat satellite photo).

© NASA

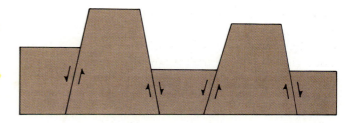

Original sharply bounded
fault blocks softened by erosion
and sedimentation

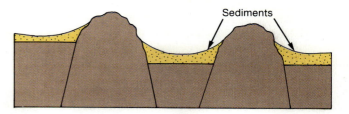

Sediments

Formation of horsts and grabens creates fault-block mountains.

Figure 11.29

The Tetons rise abruptly from the surrounding plains.

are bounded by steeply-dipping fault scarps. The Teton Mountains (figure 11.29) are one of several ranges of fault-block mountains in the western United States.

The great vertical relief associated with major mountain ranges—the Appalachians, the Andes, and many others—requires a degree of crustal thickening that is virtually impossible to achieve without some folding on a large scale. As compression proceeds and folds progress from broad, open folds through tight, upright folds, the crust is shortened horizontally and thickened vertically (figure 11.30). Folding is thus an important feature of all major mountain belts, although magmatic activity and faulting are invariably also present, to some extent. Differential erosion of different rock units within the folds occasionally causes the fold structures to stand out starkly (figure 11.31). Thrust faulting is another mechanism by which net crustal thickening can occur (figure 11.32).

With the development of plate-tectonic theory, it has become apparent that mountain systems hundreds or thousands of kilometers long are associated with plate boundaries. Continental mountain ranges, in particular, develop at the various convergent plate boundaries.

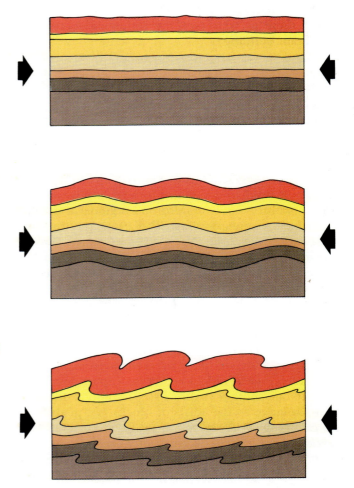

Figure 11.30

Folding as a mechanism of crustal thickening.

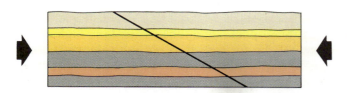

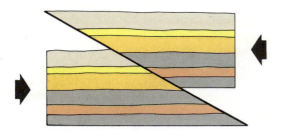

Figure 11.32

Thrust faulting also thickens and shortens the crust.

Mountain Building

The term **orogenesis** is used to describe collectively the set of processes by which mountains are formed. It thus comprises a variety of folding, faulting, volcanic, plutonic, and metamorphic activity. An **orogeny** is the set of events occurring within a specific time period that has led to the formation of a particular mountain system.

All continents have regions that have remained geologically quiet or stable for a long time. ("Long," in this context, would typically mean hundreds of millions of years, or more.) Such a region is called a **craton.** Within many cratons there are **shield** areas consisting of exposed Precambrian igneous and metamorphic rocks, which are flanked by younger stable areas. In North America, the continental shield is exposed principally in central Canada. Much of the central United States is also part of the North American craton, but the upper-crustal rocks are younger and more often sedimentary than igneous or metamorphic. Orogenesis by definition is confined to the margins of a craton. Often this corresponds to the continental margin. Orogenesis is commonly preceded by an accumulation of sediments at the margin, which later become incorporated into the developing mountain belt.

Sedimentation at the Continental Margin

Continental margins are classified as **active** or **passive,** depending on whether there is an active plate boundary at or near the margin.

At a passive, tectonically quiet margin, typical sediments are a mix of relatively mature sands and shales eroded from the stable continental interior, and marine limestones precipitated from seawater. Thick sequences of sediments with well-developed, regular bedding can accumulate at passive margins. As sediments deepen, on the oceanic crust particularly, isostatic adjustment may depress the sedimentary basins further, allowing still greater thicknesses of sediment to accumulate.

At an active margin—most often associated with ocean-continent convergence—the pattern of sedimentation is more complex, involving a greater diversity of sediment (figure 11.33). Pre-existing continental rocks are still eroded and the resulting sediments deposited in adjacent sedimentary basins. The magmatism above the subduction zone adds an additional supply of volcanic and plutonic material to the continent and may create a topographic high, from which sediments are shed both landward toward the craton and seaward toward the trench. Volcanic

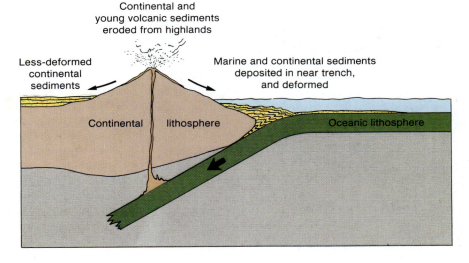

Figure 11.33

Sedimentation at a convergent margin.

(Figure labels)
Continental and young volcanic sediments eroded from highlands

Less-deformed continental sediments

Marine and continental sediments deposited in near trench, and deformed

Continental   lithosphere

Oceanic lithosphere

activity adds a volcanogenic component to the sediment. Erosion may also be more rapid in the higher-relief region of the magmatic belt, and therefore the sediments produced may be less mature chemically, less extensively weathered, as well as more voluminous than the continental sediments on a passive margin. The higher proportion of continental clastics reduces the relative importance of marine limestones in the sequence. The sedimentary basin nearer the continent will be less subject to deformation, while sediments to the seaward side and especially in the trench may be considerably deformed.

Orogenesis in Zones of Convergence

A passive margin will not undergo orogenesis until conditions change so as to transform it into an active margin. For example, plate movements may change in such a way that a subduction zone develops along the margin. The situation then becomes similar to that of the active margin previously described.

As convergence progresses, stress may crumple and fracture the rocks of the pre-existing continent and the magmatic additions associated with subduction. Strong horizontal compression may cause tight folding and development of thrust faults in the continental crust (figure 11.34), contributing to its thickening. The continent may also be extended, as some of the sediments in the trench are wedged against the margin of the overriding plate in the accretionary prism (see chapter 5). Here too compression results in reverse faulting, while metamorphism is associated both with the stress of convergence and with the heat of magma rising into the crust.

Where the margin was passive for some time prior to the initiation of subduction, a considerable period of sedimentation precedes orogenesis. Once

subduction has begun, sedimentation and orogenesis proceed simultaneously thereafter for the duration of subduction.

The situation becomes more complex when the subducting plate carries with it a continent near its leading edge. As sea floor is consumed by subduction, the continent advances toward the subduction zone. Its buoyancy, however, will prevent it from being subducted to any significant extent. Typically, after a period of compressive deformation, the two plates will become sutured together, and subduction will cease.

If a continent lies at the advancing edge of the overriding plate, then, as noted in chapter 7, the result will be continent-continent collision, producing a belt of greatly thickened continental crust at the suture zone (figure 11.35). Isostatic considerations dictate that the thicker crust will also ride higher above the asthenosphere, and thus particularly high mountains will be formed. It is no coincidence that the world's highest peak (Mount Everest) is found in a mountain range formed in this way (the Himalayas). Continent-continent collision is believed to be the cause of major mountain ranges, such as the Urals, now found within cratonic regions. The Appalachian Mountains were formed when Europe and Africa collided with North America, prior to the opening of the Atlantic Ocean.

If the overriding plate is of oceanic lithosphere, with an island arc upon it fed by magma from the subducting plate, a similar suturing process occurs, and the island arc becomes sutured onto the edge of the continent. In this case, however, the thinner oceanic lithosphere may break under continued compressive stress, and the old overriding plate beyond the island arc, being the less buoyant, may start to be subducted (figure 11.36). The same arc may then be supplied with magma derived from the plate of which it was once a part. The net effect of this process is to increase the size of the continent.

Figure 11.34

While oceanic lithosphere is subducted at a convergent margin, folding and faulting on the continent cause crustal thickening.

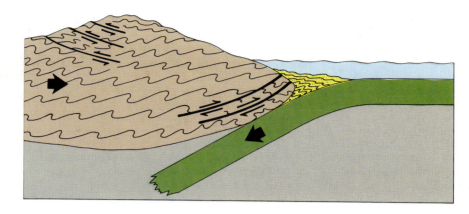

Figure 11.35

A continent-continent collision builds unusually thick lithosphere.

Continent — continent collision, with tight folding and thrust faulting thickens crust

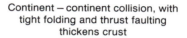

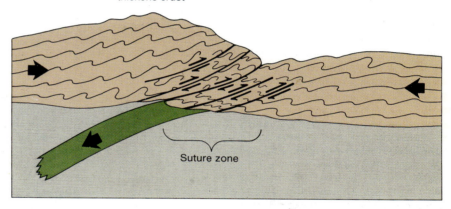

Suture zone

Figure 11.36

The direction of subduction reverses when a continent on the subducting plate arrives at a subduction zone.

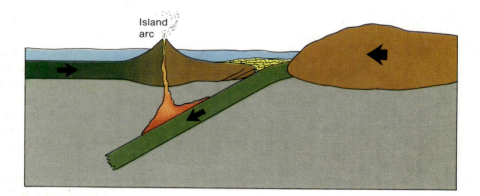

Island arc

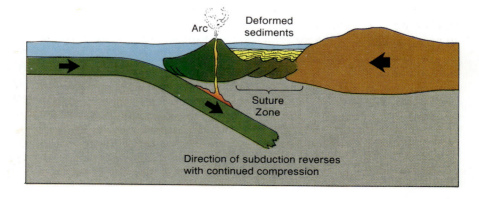

Arc   Deformed sediments

Suture Zone

Direction of subduction reverses with continued compression

After the Orogeny

The rate of collision, orogenic deformation, and crustal thickening may outstrip isostatic adjustment to the new vertical distribution of more-dense and less-dense material. A common sequel to orogeny and the accompanying cessation of compression is slow uplift (and erosion). The uplift may contribute to the formation of fault-block mountains: the upper crust will be stretched as uplift proceeds; it may fracture, and continued upward pressure may then push up horsts from which erosion will carve complex mountain topography. Material eroded from the horsts is often deposited in the adjacent grabens, which contributes to the leveling of the surface through time.

When the surface has been smoothed by erosion and isostatic equilibrium has been restored, the mountain belt has become *cratonized,* incorporated into the stable continental block. By the time cratonization is complete, the mountain belt formed by orogeny has largely been erased topographically. Only the deformed metamorphic and plutonic rocks that formed the core of the belt, now exposed at the surface, testify to the orogenic activity of the past.

Box 11.1

New Theories for Old: Plate Tectonics and Geosynclines

Before the advent of plate-tectonic theory, the prevailing theory used to explain mountain building was the geosynclinal theory. A **geosyncline** is a syncline of regional scale—hundreds or thousands of kilometers long. Geosynclines, acting as sedimentary basins at the margin of a continent, were believed to be the first phase of a mountain-building episode. Later elaboration of the model resulted in subdivision of geosyncline into a *miogeosyncline* and *eugeosyncline*. The miogeosyncline was on the edge of the continent proper, where "clean" limestones, sandstones, and some shales were deposited. The deeper eugeosyncline was to the seaward side and was characterized by the accumulation of a much more diverse or "dirty" sedimentary package, including more poorly weathered rock fragments and volcanic material (figure 1). With progressive deepening of the basin, a point was reached at which melting of the most deeply buried sediments began in the eugeosyncline. The magmas thus formed rose to invade the rocks above; the geosyncline became unstable and began to deform. Ultimately, the sediments of the eugeosyncline were transformed into an igneous/metamorphic complex. Mountains were built, and the crust

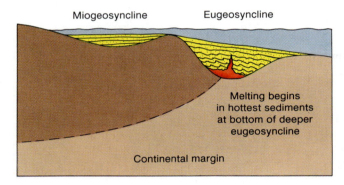

Miogeosyncline   Eugeosyncline

Melting begins in hottest sediments at bottom of deeper eugeosyncline

Continental margin

Figure 1
Eugeosyncline and miogeosyncline, part of an older theory of mountain building.

then stabilized, with the newly formed mountains representing a marginal extension of the continent.

This is a good example of the way in which theories are modified or abandoned in response to new or improved data. When geosynclinal theory was developed, it represented the best, most coherent, most logical explanation of the data then available. At the time, radiometric dating was in its infancy; many of the geophysical techniques we now use routinely were undeveloped. As these more powerful investigative tools were employed to gather more data, new theories had to be developed to accommodate the newer information along with the old. Eventually the geosynclinal theory was essentially discarded, to be replaced by a more sophisticated plate-tectonic model of orogenesis at convergent margins. We may not yet have seen the ultimate orogenic model. The discovery of suspect terranes has suggested that further refinements are necessary. This is the nature of the science. It deals not with a static body of facts and truths but with an ever-growing body of knowledge. Where additional data are consistent with an established theory, confidence in it is increased. Inconsistencies suggest a need for modification. Irreconcilable inconsistencies may require that the old theory be discarded altogether, to be replaced with one that *can* account for all of the available data bearing on the subject.

The Growth of Continents through Time

A subject of considerable interest (and heated debate) is the extent and nature of the continental crust through time. Were the continental masses essentially created completely as a result of the initial melting and differentiation of the primitive earth, and have they always existed in approximately their present forms and extent? With the development of paleomagnetic techniques, it has become apparent that continents have *moved* through time, but that does not address the question of the total mass of continental material present at any given time.

How Large Were the Early Continents?

The oldest reliably dated continental material is 3.8 to 4 billion years old, and some rocks of approximately that age exist on virtually every continent. Early geochronologic studies suggested that continents might show a pattern of an older shield region flanked by successively younger regions (figure 11.37). The chemistry of the younger rocks was also consistent with their derivation, in large measure, from mantle rather than from crustal material. This was interpreted to indicate that continents had grown continuously through time. Such an interpretation is also consistent with the observation that many of the orogenic processes just described have the net effect of adding material at the continental margins.

Other geochemical evidence suggests that the extent of depletion of the mantle in those elements that are concentrated in the crust has *not* increased comparably and continually through time. That would be an argument for a more nearly constant volume of continental material over billions of years.

These lines of evidence can be reconciled through plate tectonics, considering the crustal recycling for which plate-tectonic activity is responsible. While some sediment is scraped off a downgoing slab at a convergent margin, some continentally derived sediments are also subducted along with oceanic lithosphere at an ocean-continent convergence. That material is then remixed with the much larger volume of upper-mantle material. Elements concentrated in the crust are returned to the mantle and at the same time diluted in it. Fresh magma extracted from that mantle will show predominantly mantlelike chemistry. Also, the rocks crystallized from such magma will yield an age that reflects the time of last extraction of the magma from the mantle, even though some of the elements in those rocks previously spent some time in the crust.

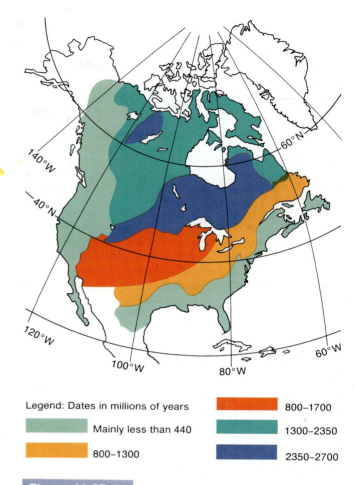

Legend: Dates in millions of years

Mainly less than 440	800–1700
800–1300	1300–2350
	2350–2700

Figure 11.37

Pattern of younger ages toward the margins of North America.

From ''Pre-Drift Continental Nuclei,'' by P. M. Hurley and J. R. Rand, in *Science,* 164, June 1969, p. 1231. Copyright 1969 by the American Association for the Advancement of Science. Reprinted by permission.

The majority view at present, therefore, is that the most extensive creation of continental crust from the mantle occurred in the late Archean, during the period between 3 1/2 and 2 1/2 billion years ago. Thereafter, the total volume of continental crust has probably not increased substantially. Instead, continental material has been recycled repeatedly through the mantle.

Suspect Terranes

An interesting new twist on models of continental growth has been provided by the discovery, within the last decade, of continental regions whose geology does not appear to be consistent or continuous with that of surrounding areas. Geologists use the term **terrane** to describe a region or a group of rock units of a common age, type, or deformational style. Recent studies of the geology of western North America has revealed an apparent jumble of juxtaposed terranes of quite diverse kinds. Their rocks and histories appear unrelated to each other except insofar as they are now together in one place. The question arose whether some of these terranes might have formed far apart, then been added to the continent later. This would account for marked geologic differences between adjacent terranes. The regions for which this was proposed were named **suspect terranes.**

Paleomagnetic and fossil evidence has since confirmed that some of these suspect terranes have indeed traveled long distances to reach their present positions. An example is the terrane called Wrangellia (figure 11.38). The data suggest that before it was added to the western margin of North America, 100 million years ago, Wrangellia traveled nearly 6000 kilometers, probably from somewhere south of the equator. Since its accretion onto the continent, Wrangellia has been further broken up and the fragments displaced by younger strike-slip faulting. As shown in figure 11.38, further data suggest that a broad band of rocks along the western margin of the continent consists of suspect terranes, formed separately from the continent and later accreted onto or attached to it. The full interpretation of the geology of the region is going to be considerably more complex than was formerly believed. Similar mosaics of suspect terranes are now being identified in mountain belts worldwide.

The origin of the small landmasses making up suspect terranes is not yet known. There may be several sources. Some could have originated as large island-arc systems. Others may have been small blocks of continental crust separated by continental rifting or strike-slip faulting from a larger continent, much as a slice of North America west of the San Andreas fault is being propelled northward, and East Africa is being separated from the rest of Africa. There are many oceanic plateaus, some underlain by crustal roots up to 40 kilometers thick, that are clearly fragments of continental crust. These may have formed during the breakup of Pangaea. As plate movements continue, many are destined to collide eventually with larger continental landmasses, perhaps to become future suspect terranes.

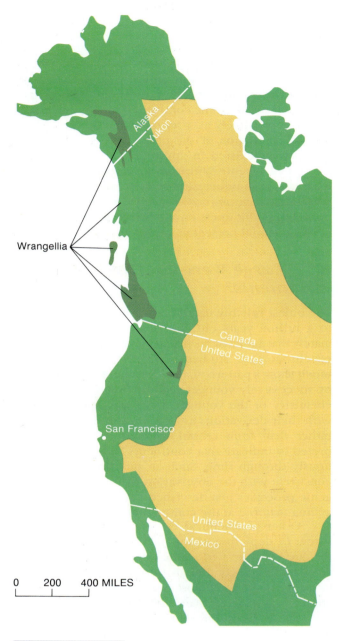

Figure 11.38

Wrangellia.

Source: After U.S. Geological Survey Annual Report.

Summary

Deformation of the continental crust takes the form of folds (plastic deformation) and faults (brittle deformation). The nature and orientation of these structures provides information on the kinds and orientation of the stresses that formed them. Both folds and thrust faults can result in horizontal crustal shortening and in crustal thickening.

The lower density and greater thickness of the continental crust relative to oceanic crust gives continents greater buoyancy, and accounts for their relatively high relief. When a continental block is in isostatic equilibrium, the extra relief above sea level is compensated by a low-density root extending into the mantle. When erosion, orogenesis, glaciation, or other processes change the distribution of crustal mass, isostatic adjustment occurs, with the lithosphere rising or sinking until equilibrium is restored.

Gravity measurements, suitably corrected, can aid in assessing the present extent of isostatic adjustment, and in recognizing anomalous mass distribution within the crust. Other geophysical methods also probe the continental crust. Seismic data permit detection of subsurface structures and determination of rock densities. Magnetic anomalies are used particularly in locating iron-rich rocks and ores within the more granitic continental crust. Regions of high heat flow are associated with the presence of bodies of magma or young, hot plutons within the crust.

Large mountain ranges are formed through orogenesis, which comprises deformation, metamorphism, and magmatic activity. The present topography of individual mountains is, in most cases, due to erosion by water or ice rather than to the nature of the crustal structures involved in their formation. Orogenesis most commonly occurs at convergent plate boundaries, where new material is accreted or sutured onto the margin of a craton. Some of this added material, now found in the so-called suspect terranes, has been transported over considerable distances. The fact that material is added to continental margins by successive orogenic events does not mean that there is necessarily a net increase in the total mass or extent of the continents through time, for continental material is also cycled back into the mantle through subduction.

Terms to Remember

active margin	fault	limbs	recumbent fold
anticline	geosyncline	magnetic anomalies	reverse fault
axial plane	graben	normal fault	shield
axial trace	gravity anomalies	nose	strike
axis	heat flow	oblique-slip fault	strike-slip fault
basin	horst	orogenesis	suspect terrane
craton	isostasy	orogeny	suture
dip	isostatic equilibrium	overturned fold	syncline
dip-slip fault	joint	passive margin	terrane
dome	joint set	plunging fold	thrust fault

Questions for Review

1. Define the strike and dip of a plane.

2. What is the distinction between joints and faults? Between a thrust fault and a normal fault?

3. Describe two processes by which continental crust may be thickened.

4. How are anticlines and synclines distinguished when they occur in sedimentary rock sequences?

5. Under what circumstances do overturned folds develop?

6. Explain the concept of isostatic adjustment, giving an example. Why is such adjustment not instantaneous?

7. How are the crustal roots of continents detected?

8. Describe a geologic situation that would give rise to locally high heat flow.

9. Name and briefly describe three kinds of mountains.

10. Define: (a) *craton*; (b) *shield*; (c) *active margin*; (d) *passive margin*.

11. Summarize the principal orogenic processes at a continental margin with an adjacent subduction zone.

12. What is a *terrane*? A *suspect terrane*?

For Further Thought

1. The identification of suspect terranes is difficult, as demonstrated by their relatively recent discovery. Suppose that you have found a batholith adjacent to a sequence of interlayered sedimentary and volcanic rocks. What features or properties might you look for in order to decide whether or not these are two distinct terranes, formed as different continental blocks, that have become juxtaposed?

2. In general, would you expect shield regions to be more or less likely to be in isostatic equilibrium than continental margins? Why? Would you expect differences in the extent of isostatic equilibrium between active and passive margins? Explain.

3. What geophysical technique(s) would be especially useful in trying to detect and map the extent of each of the following subsurface features? (a) Folds in bedded limestone and shale. (b) An ancient mafic pluton in a section of marble. (c) A young granite pluton intruded into quartz-mica-feldspar schists. (d) A bed of sedimentary iron ore within a sedimentary rock sequence.

Suggestions for Further Reading

Condie, K. *Plate Tectonics and Crustal Evolution.* New York: Pergamon Press, 1976.

Davis, G. H. *Structural Geology of Rocks and Regions.* New York: John Wiley and Sons, 1984.

Hobbs, B. E.; Means, W. D.; and Williams, P. F. *An Outline of Structural Geology.* New York: John Wiley and Sons, 1976.

Miyashiro, A.; Aki, K.; and Sengor, A. M. C. *Orogeny.* New York: John Wiley and Sons, 1982.

Sharma, P. V. *Geophysical Methods in Geology.* New York: Elsevier, 1976.

Tarling, D.H., ed. *Evolution of the Earth's Crust.* New York: Academic Press, 1978.

Weyman, D. *Tectonic Processes.* London: George Allen and Unwin, 1981.

Windley, B. *The Evolving Continents.* 2d ed. New York: John Wiley and Sons, 1984.

The Ocean Basins

Introduction

For many centuries, the deep ocean basins were virtually unknown territory—dark, cold, inaccessible. That has changed, first with the development of echo-sounding and seismic methods for exploring the topography and structure of the sea floor. Later tools included sampling devices capable of collecting cores of sediment from the sea floor and dredging rocks off the bottom, and cameras to photograph the features of the deep. Still more recently, submersible research vessels, like miniature submarines, have allowed scientists to observe, sample, and photograph such features as ridge systems directly. This chapter will review briefly the principal structural features of the ocean basins, the patterns of sediment distribution on the sea floor, and the circulation of ocean currents.

The Face of the Deep: Topographic Regions of the Sea Floor

If the oceans were drained of water, a complex topography would be revealed (figure 12.1). Far from being featureless, as they once were thought to be, the oceans contain mountains and valleys to rival those on the continents. The sea floor can be subdivided into several distinct physiographic, or topographic, regions: the continental margins; the abyssal regions (broadly defined, the "deep ocean" where the bottom lies below 1000 fathoms, which is 6000 feet, or nearly 2 kilometers); and the oceanic ridge systems. (Figure 12.1 reveals additional smaller topographic features—including mountains, canyons, and plateaus—which will be described separately below.) The characteristics of the continental marginal zones differ, depending on whether those margins are active or passive.

Passive Continental Margins

A *passive margin,* as noted in the previous chapter, is seismically and volcanically quiet, not associated with an active plate boundary. A generalized topographic profile across a passive continental margin is shown in figure 12.2. The marginal region can be seen to consist of several zones distinguished, topographically, on the basis of slope.

Immediately offshore from the continent is the **continental shelf,** a nearly flat, shallowly submerged feature. The average slope of the world's continental shelves is about 1/10 of a degree, or ten feet

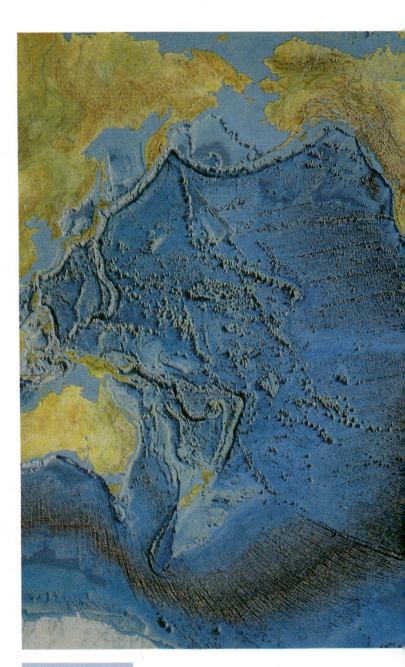

of drop per mile of distance offshore. This accounts in part for the generally shallow waters above continental shelves, the majority of which lie no deeper than 100 fathoms (600 feet) at their outer limits. The average width of all continental shelves is about 70 kilometers (40 miles); the shelves at passive margins are systematically broader than those of active margins, and a few shelves at passive margins extend

Figure 12.2

Cross section of the topography of a passive continental margin, showing continental shelf, slope, and rise.

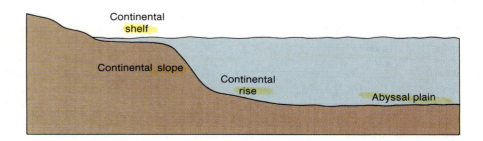

Continental shelf

Continental slope

Continental rise

Abyssal plain

nearly 1000 kilometers (600 miles) offshore from the adjacent continent. The Atlantic margin of North America is an example of a passive margin.

The continental shelves are a natural site for the accumulation of sediment derived from the continent, and many shelves have quite flat surfaces as a consequence of the deposition of layer upon layer of sediment over the underlying bedrock. This is not true of all shelves. During the last ice age, worldwide sea levels were lowered to such an extent that much of the now-submerged continental shelves was dry land. Rivers now terminating at the present shoreline flowed out tens or hundreds of kilometers across the continental shelves, carving valleys and depositing sediments just as they do now on the exposed continents. In regions directly subjected to glaciation, the ice sheets themselves carved up the shelf and left behind deposits of sand, gravel, and boulders that now add topographic relief to the shelves.

In some places—for example, the Gulf of Mexico—further relief on the continental shelf is provided by features originating from below: **salt domes** (figure 12.3). Within the thousands of meters of sediments below the shelf in the Gulf are thick salt beds. Salt is lower in density than most clastic sediments and is therefore somewhat buoyant. It can also flow plastically. When buried in a sedimentary pile, salt may mobilize, and from a bed of salt one or more stocklike plugs of salt will rise, migrating upward through the overlying sediments, often doming the layers above. A practical significance of salt domes is that they can play an important role in localizing oil and gas deposits.

The outer limit of the continental shelf is defined by the abrupt break in slope to the steeper continental slope. The angle of the continental slope varies widely, from nearly vertical to only a few degrees. The continental slope is the principal transition zone from near-continental elevations to deep-ocean depths. At the bottom (outside) edge of the continental slope, the topography again flattens out, the slope decreasing once more to less than 1°. The depth at which this occurs is highly variable, ranging from 1 1/2 to 3 kilometers below sea level. The region beyond this second break in slope is the **continental rise.** The sea bottom continues to deepen across the rise until the deepest, flat *abyssal plains* are reached. (The continental rise in some areas is not well developed, and in these cases the continental slope terminates in the deep ocean plains.)

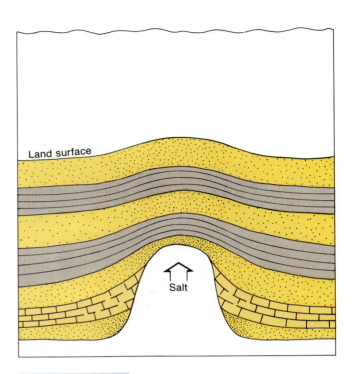

Figure 12.3

A salt dome rising from deep in the sediments may cause a topographic high above.

Formation of Passive Margins

The nature and formation of the passive continental margin can be understood in plate-tectonic terms. Passive margins are believed to have begun as continental rift systems, torn apart by mantle plumes and/or convection systems welling up from underneath (figure 12.4A). The continental lithosphere is stretched and thinned somewhat in the process. It will ultimately fracture along long faults perpendicular to the direction of stretching, and with continued tension, grabens will form along the rift, making a long rift valley. This is now occurring in east Africa, which in time will be split off from the rest of the continent, if rifting continues. Already the floors of some of the graben-formed valleys of the East African rift system lie below sea level. They remain dry only because mountains presently block the ingress of the sea. Lake Tanganyika shows its rift origins in the complementary shapes of its east and west shores (figure 12.5).

Eventually the rifting proceeds to the point that seawater floods the rift valleys (figure 12.4B). While the new seas remain shallow, rapid evaporation may cause deposition of salt in these basins. This is the probable origin of the thick salt beds now buried along many continental margins.

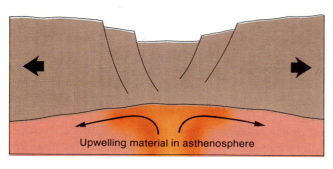

A

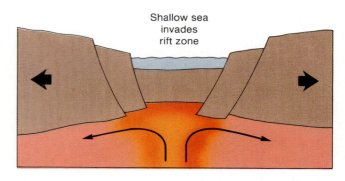

Shallow sea
invades
rift zone

B

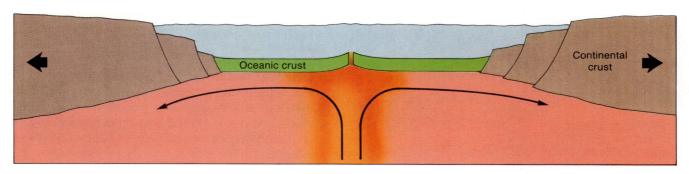

Oceanic crust

Continental
crust

C

Figure 12.4

Formation of a passive margin. *(A)* The continent is stretched
and fractured. *(B)* As rifting progresses, a shallow sea forms.
Evaporation may leave salt deposits in the rift zone. *(C)* When
rifting is complete, oceanic crust is formed between the
fragments of continent.

on the continents it self.

With further tension, the rifting of the continent
is completed (figure 12.4C). The rift valley is now
floored by oceanic crust; an oceanic spreading ridge
divides the fragments of continent; a narrow strip of
true ocean basin lies between the separated continental masses. The Red Sea is in this stage of development. With continued spreading the ocean
continues to widen, as the North Atlantic has done following the breakup of Pangaea.

Throughout this process, the newly formed continental margins are passive in the sense that there is
no subduction and therefore none of the associated
seismic and volcanic activity. Limited volcanism does
occur as magma works its way up through the fractured continent, and earthquakes will certainly be associated with the downdropping of the grabens as well
as with the spreading ridge when it has evolved. However, the active plate boundary in the system is the
spreading ridge, once formed. There are not plate

boundaries where the oceanic and continental lithosphere meet at each margin; on a given side of the
ridge, ocean and continent belong to the same plate.
The passive margin is created by rifting, but once the
intervening ocean is formed, further rifting is restricted to the seafloor spreading ridge. This is the
reason for the subsequent tectonic quiescence of the
continental margin so formed.

The lithospheric thinning and block-faulting together account for the thinning of continental crust
and the deepening of the ocean seaward from the continent. The exact geometry of the faulting on a given
margin will exercise a fundamental control on the topography of the margin. For example, the steep angles of continental slopes at some passive margins are
probably fault scarps formed during the development
of the margin. The topography is then modified by
sedimentation. The extent to which sedimentary

Figure 12.5

Satellite photograph of Lake Tanganyika, in the East African rift.

blankets soften the contours of the margin is principally a function of the quantity of sediment supplied from the continent. If the accumulated sedimentary sequence is a thick one, its weight may continue to depress the margin and deepen the sedimentary basin.

Active Continental Margins

Active margins are those at or near subduction zones. Like passive margins they have continental shelves, but the shelves adjacent to active margins are narrower than those at passive margins. The continental slope of an active margin is characteristically steeper than that of a passive margin, increasing in slope with depth. The slope of an active margin terminates not in a rise but in a trench, a long, deep, steep-walled valley trending approximately parallel to the continental margin. The locations of principal oceanic trenches are shown in figure 12.6 (see also figure 12.1). Note that trenches occur both at active continental margins and at ocean-ocean convergence zones, parallel to island arcs. The trenches represent the greatest ocean depths. The deepest of them plunge to more than 10 1/2 kilometers (6 1/2 miles) below sea

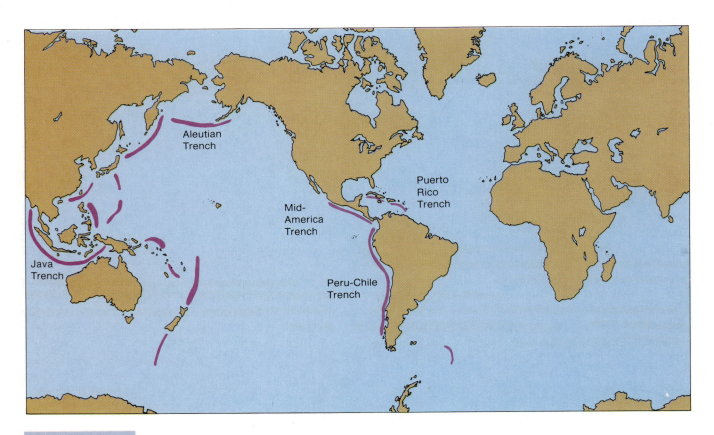

Figure 12.6

Locations of principal trenches on the sea floor.

level, and their depth below the sea floor rivals the elevation of the highest mountains on the continents. Trenches mark the plate boundary at which subduction is occurring, and serve as sites of sediment accumulation where there is adjacent land to serve as a sediment source.

Before the development of plate-tectonic theory, the trenches were a major puzzle to marine geologists. No erosional mechanism for forming them could be imagined. Trenches are characterized by negative gravity anomalies (figure 12.7), which indicates that they are significantly out of isostatic equilibrium. However, it was difficult to picture a large volume of low-density lithosphere below the trench, and there was no evidence for this. (In fact the lithosphere is apparently depressed or held down by the stresses attendant on plate convergence.) Trenches also show abnormally low heat flow, which we now attribute to the presence of the cold descending slab of lithosphere. The existence of that slab is further supported by the pattern of seismic activity at trenches: earthquake epicenters are located not only at the trench proper but along a plane dipping away from the trench into the mantle, termed a **Benioff zone** (figure 12.8). Earthquakes occur there because brittle lithosphere is present there, being subducted under another plate. The existence and orientation of the Benioff zone is further evidence of the subduction that accounts for the presence of the trench.

Certain specialized active margins arise in unusual tectonic settings. Subduction along the western edge of North America has caused a seafloor ridge system to be partially overridden by the continent. A portion of this margin, along the California coast, is now a transform fault, separating offset segments of a spreading ridge off the coast of Oregon and Washington from other spreading segments in the Gulf of California.

Oceanic Ridge Systems

Although the zones of active spreading are conventionally drawn as lines on a map, an oceanic ridge system is a broad feature (recall figure 12.1). The rift and its associated structures span wide areas of the sea floor. A topographic profile across an oceanic ridge system is shown in figure 12.9. Looking beyond the surface irregularities, two principal features may be observed: (1) the sea floor slopes away from the ridge crest symmetrically on either side, and (2) there is a valley at the very center of the ridge.

The valley is a **rift valley,** similar to a continental rift valley and formed in much the same way. The tensional stresses of rifting (spreading) cause

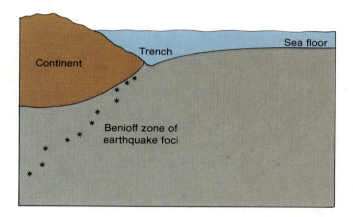

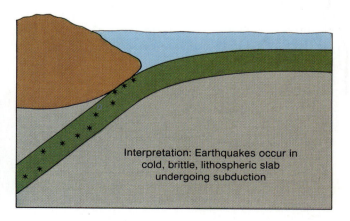

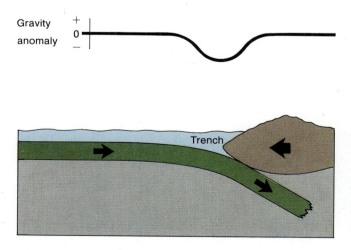

Figure 12.7

Negative gravity anomaly over a trench indicates that it is out of isostatic equilibrium.

Figure 12.8

The Benioff zone of progressively deeper earthquakes dipping away from a trench along the subducted slab.

Figure 12.9

Topography of a spreading ridge in cross section (see also figure 12.1).

Figure 12.10

Fracture zones are the seismically quiet extensions of transform faults.

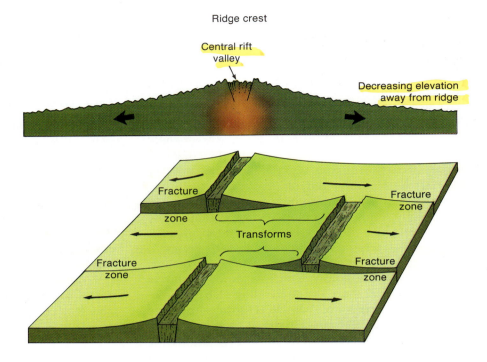

Ridge crest

Central rift valley

Decreasing elevation away from ridge

Fracture zone

Transforms

Fracture zone

Fracture zone

Fracture zone

deep, high-angle faulting, and with spreading, grabens bounded by these faults drop into the gap. The decreasing elevation of the sea floor away from the ridge was once thought to be due to relative uplift of the central portion of the ridge by upwelling mantle material. However, calculations have shown that the profile can be explained by thermal contraction. Newly formed sea floor at the ridge, though solid, is warm. As it cools, it contracts. The cooling and contraction of the oceanic lithosphere as it moves away from the ridge at which it formed accounts for its declining elevation. The fact that the sea floor is somewhat elevated even tens or hundreds of kilometers away from the rift is an indication of the slowness of the cooling, a function of the poor heat-conduction properties of rocks. Recall also from chapter 7 that the production and rifting of new sea floor at the ridge leads to symmetric patterns of magnetic anomalies across the ridge.

We have already described (chapter 7) the distribution of seismicity along a rift system, and the fact that earthquakes are concentrated along the ridge proper and the transform faults between offset ridge segments. The fractures formed between offset segments do not disappear when plate movement has carried them beyond the spreading ridge (figure 12.10). They become seismically quiet where both plates move in the same direction, but the fractures remain, and there may be large scarps across the fractures. The fracture systems may be detectable outward from the rift over most of the region of elevated topography.

Along the ridge crest, seawater seeps into the young, hot lithosphere through fractures in the crust. The water is warmed as it circulates through the warm crust, and reacts with the fresh seafloor rocks. It may escape again at **hydrothermal** (hot-water) **vents along the ridge, rising up into the overlying colder** seawater (figure 12.11). Some hydrothermal vents are crusted with minerals, once dissolved in the circulating waters and later deposited as the temperature and chemistry of the hydrothermal waters change upon reaction with uncirculated seawater. Some of these minerals are possible future mineral resources (see chapter 20).

A particularly surprising discovery was made by the crews of submersible vehicles exploring these hydrothermal vents: life. In some areas, complex communities, including such higher organisms as giant clams and tube worms (of species previously unknown), were found (figure 12.11B). How could they live? Similar organisms in near-surface waters derive their energy ultimately from plants. But no sunlight to stimulate plant growth reaches the ridge vents, and little organic matter rains down from above. Bacteria seem to be a key part of the answer. The vents teem with bacteria, which derive their energy from chemical reactions involving compounds released at the vents (for example, hydrogen sulfide, H_2S). Some of the higher organisms then feed directly on the bacteria; others contain bacterial colonies that produce the organic carbon the animals need to grow.

Figure 12.11

Seawater circulating through new sea floor at a spreading ridge produces hydrothermal vents. *(A)* Schematic diagram of circulation patterns. *(B)* Strange creatures teem in the warm waters.

(B) Photo by W. R. Normark, courtesy U.S. Geological Survey.

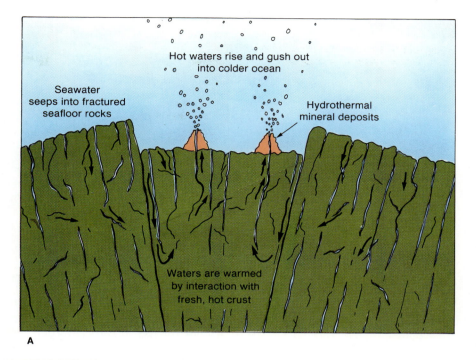

Seawater seeps into fractured seafloor rocks

Hot waters rise and gush out into colder ocean

Hydrothermal mineral deposits

Waters are warmed by interaction with fresh, hot crust

A

B

Figure 12.12

Topography of a submarine canyon.

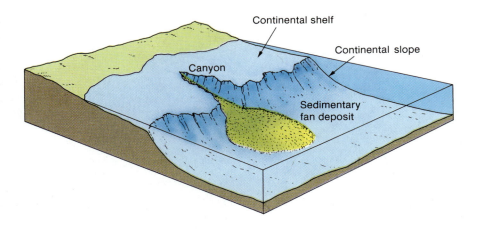

Abyssal Plains

The **abyssal plains** are essentially flat areas occupying what is also called the "deep ocean basin," away from continental margins and ridges. Except for the trenches, the abyssal plains constitute the deepest regions of the sea floor. Although the plains themselves are quite level, they are not featureless (figure 12.1). Often they are dotted with **abyssal hills**, low hills rising several hundred meters above the surrounding plains, or the higher *seamounts*. These features, of volcanic origin, will be described further in the next section, as they are not unique to the abyssal plains. Some abyssal plains are also cut by channels resembling shallow stream channels. They appear to be related to, or to be extensions of, the submarine canyons of many continental margins (see next section). If so, they are probably carved by the same density currents, although it is difficult to understand how sea-bottom currents could continue to flow across nearly flat plains with sufficient velocity to cause very much erosion.

Other Structures of the Ocean Basins

The ocean basins include a variety of structural features that occur in more than one of the regions described in the previous sections. These include submarine canyons (distinct from the trenches), hills of various sizes, and shallowly submerged platforms.

Submarine Canyons

A notable feature of many continental slopes at both active and passive margins is the presence of deep, steep-walled **submarine canyons** that cut across the shelf/slope system (figure 12.12). These canyons are usually V-shaped, sometimes winding; sometimes they have tributaries on the continental shelves. Typically they enlarge and deepen seaward and extend to the base of the continental slope, where they often terminate in a fanlike deposit of sediment resembling an alluvial fan (see next chapter). They are cut into hard rock as well as into soft sediment.

The origin of submarine canyons has long been debated. A few are continuous with, and appear to be extensions of, major stream systems on land. Logically, it might be supposed that they were cut by those stream systems at a time when the continental shelves were exposed by a lowering of sea level. However, the canyons dissect not only the continental shelf, but in many cases the whole length of the continental slope as well. Streams stop downcutting where they flow into the sea. It seems highly unlikely that sea level would ever have been lowered to the base of the continental slope. It is therefore impossible to attribute these canyon systems entirely to rivers. Moreover, many submarine canyons have no corresponding stream systems on land; they seem to have formed independently of stream action. Another hypothesis attributed them to submarine landslides, but this was not borne out by subsequent experiments and measurements; nor would it as easily explain canyons in solid rock.

A possible mechanism of canyon carving was suggested in part by observations following an earthquake off the Grand Banks of Newfoundland in 1929.

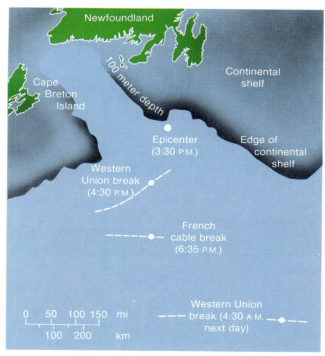

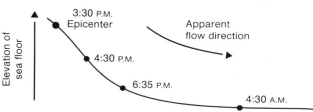

Figure 12.13

Cable breaks following the Grand Banks earthquakes provided some of the strongest indirect evidence for turbidity currents.

After F. P. Shepard, *The Earth Beneath the Sea* (NY: Athaneum, 1971), p. 25.

Several transcontinental telephone cables that had been run along the continental slope were broken. Furthermore, many of the breaks occurred hours after the earthquake, and in apparent progressive sequence with increasing distance from the epicenter (figure 12.13). The overall timespan involved was too great for the breaks to be attributed directly to seismic-wave action. The most plausible explanation is that the breakage was caused by a fast-moving **turbidity current**, a density current of suspended sediment and water rushing down the continental slope. Turbidity currents are analogous to nuées ardentes: just as the latter are denser-than-air ash clouds that flow along the ground as a consequence of their density, so the turbidity current is a denser-than-water mass that flows along the ocean bottom.

Fast-flowing turbidity currents could be forceful enough not only to break underwater cables but to carve canyons in rock or sediment. As the slope grows shallower at the continental rise, the currents would slow down and begin to deposit their sediment load. Indeed, individual beds of sediment at the bases of canyons often show graded bedding fining upward. This would be consistent with their deposition by gradual settling out of suspension, beginning with the coarsest material. Multiple graded beds, one atop another, could reflect successive turbidity flows. A turbidity current could be initiated as sediment is stirred up by an earthquake, or perhaps in conjunction with an undersea landslide. Since sediment is an essential component of turbidity currents, they are most common where there is an abundant sediment supply on the continental shelf. It is not clear that turbidity currents can account for all submarine canyons. One remaining puzzling feature, too, is the rapid velocities—up to 60 kilometers per hour, or over 40 MPH—inferred for some turbidity currents. Such velocities exceed those measured for any undersea currents, though the measurements are admittedly limited in number. (It has not been possible to measure turbidity-current velocities directly; this would require anticipating the time and place of such an event beforehand so as to have instruments in place.) Turbidity currents do appear to be the most plausible mechanism for forming the majority of the canyons. Indeed, those not associated with continental river systems are difficult to account for in any other way.

Seamounts, Guyots, and Coral Reefs

The ridge systems and deep ocean basins are dotted with hills, some tall enough to break the surface and form islands, some not so high. These are generally of volcanic origin, whether associated with hot spots or with volcanism near the spreading ridges. The **seamounts** are those larger hills rising more than 500 fathoms (3000 feet, or about 1 kilometer) above the surrounding ocean floor, but not extending above the sea surface. Seamounts may or may not occur in linear chains (see figure 12.1). Where a row of seamounts exists, age data may indicate formation in succession as the oceanic plate moved over a stationary hot spot, as described in chapter 7.

A distinctive type of seamount is a flat-topped one, called a **guyot** (pronounced "ghee-oh," with a hard g). The simplest explanation for the flat top is that a conical hill is planed off at sea level through

Figure 12.14

Possible modes of formation of guyots. (A) A volcano formed near a ridge sinks as the lithosphere is carried away from the ridge, and meanwhile the volcano is planed off by waves. (B) A volcano, flattened by wave erosion, sinks as isostatic adjustment compensates for the weight of the volcano on the oceanic lithosphere.

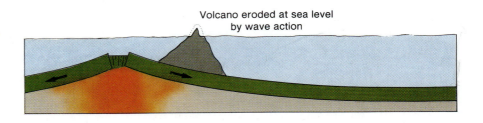

Volcano eroded at sea level
by wave action

With movement away from ridge,
volcano sinks below surface

A

wave action, but the question then becomes, how does the guyot come to be submerged? (Many lie below the lowest sea levels expected during past glacial episodes.) There are two possible mechanisms to account for the submergence of guyots, which may operate separately or in conjunction. The first, which is most relevant to guyots found near spreading ridges, is related to the thermal contraction of sea floor as it moves away from the ridge (figure 12.14A). As the sea floor cools and contracts, it sinks deeper and carries the guyot with it. A second mechanism is isostatic compensation (figure 12.14B): the mass of the guyot loads the lithosphere, which should sink or warp downward locally in response. Hot-spot volcanoes on the sea floor, becoming inactive as they move away from the hot spot, may evolve into guyots by this mechanism.

The formation of ringlike coral atolls is also linked to the existence of oceanic islands (figure 12.15). The corals require a solid foundation on which to build a reef. They also must be near the surface, in the lighted (*photic*) zone of the water where the microscopic plants on which they feed, and with which they coexist, will be abundant. Thus reef construction begins around the rim of an island. Weathering erodes the island. It may also be sinking, for reasons described above, or becoming submerged because sea

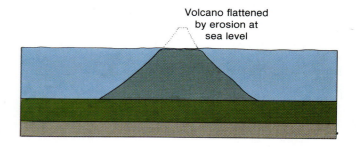

Volcano flattened
by erosion at
sea level

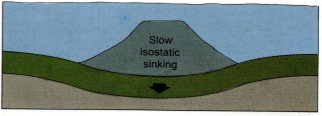

Slow
isostatic
sinking

B

level is rising worldwide. Reef growth continues upward, to keep the live colonies in the photic zone; meanwhile the island's surface moves downward. Eventually a large ring of reef may be all that projects above the water surface. Note that not all oceanic islands are hospitable to corals. Corals also require water warmer than 20° C (68° F), which further restricts their occurrence.

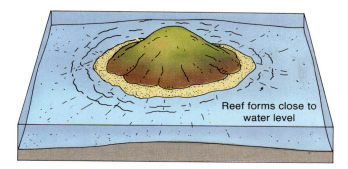

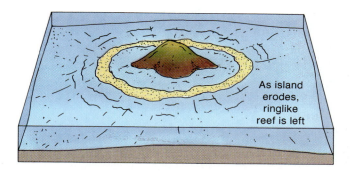

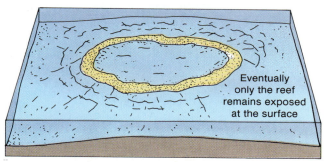

Reef forms close to water level

As island erodes, ringlike reef is left

Eventually only the reef remains exposed at the surface

A

B

Figure 12.15

Coral atolls are closely related to marine volcanoes. *(A)* Schematic formation of an atoll. A reef forms around a volcanic island, which then erodes and/or sinks, leaving a circular atoll. *(B)* Coral atolls in the Pacific (satellite photograph).

(B) © NASA

Plateaus

In addition to ridge systems and seamount/island chains, the ocean basins contain broader **plateaus,** shallowly submerged or projecting partially above sea level. Some of the larger examples are mapped in figure 12.16. Seismic and gravity data have shown that a number of these plateaus consist of continental, not oceanic, crust and are compensated isostatically by roots of continental crust, just as the continents themselves are (figure 12.17). The origin of oceanic plateaus of continental affinity is unclear. The most likely explanation is that they are splinters off larger continents, perhaps split away by continental rifting during the formation of ocean basins.

Considering the longevity (or lack of it) of the sea floor over geologic time, it seems that the plateaus will, sooner or later, be transported via plate movements to a subduction margin at which sea floor is being consumed. The greater buoyancy of continental lithosphere will prevent the plateaus' subduction. Instead, they might be pasted onto overriding continents. Perhaps we have in the plateaus examples of the origins of suspect terranes found in the present continents.

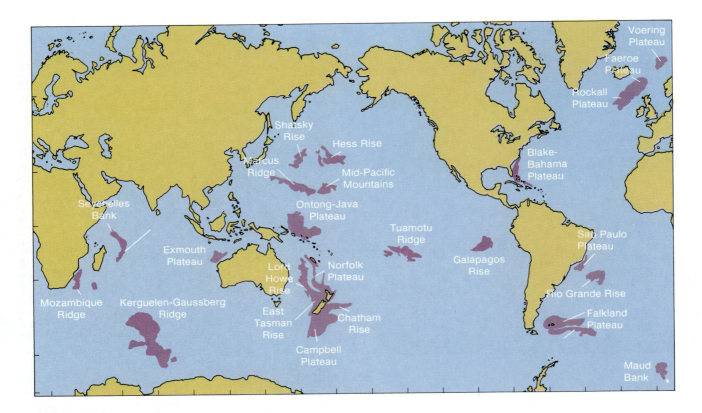

Figure 12.16

Distribution of principal plateaus in the ocean basins.

From "Continental Accretion: From Oceanic Plateaus to Allochthonous Terranes," by Z. Ben-Avraham, et al., in *Science*, 213, 1981, pp. 47–54. Copyright 1981 by the American Association for the Advancement of Science. Reprinted by permission.

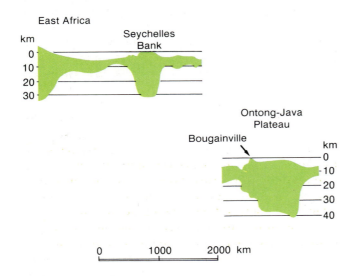

Figure 12.17

Some plateaus have a crustal structure that closely resembles that of continental crust.

After Ben-Avraham, Z., "The movement of continents": *American Scientist* 69 (1981) 285–299.

The Structure of the Sea Floor; Ophiolites

Seismic evidence puts some constraints on the structure of the oceanic lithosphere (figure 12.18A). The oceanic crust is subdivided into three layers. Layer 1, which averages about 450 meters (1500 feet) thick, consists of unconsolidated or unlithified sediment. This is known both from the low measured seismic velocities and from direct core sampling. Layer 2, an average of 1.75 kilometers (1.1 miles) thick, has also been sampled by deep coring and been shown to consist of basalt. The Layer 2 basalts often show the pillowed structure characteristic of submarine extrusion. This is entirely reasonable considering that Layer 2 would be at the top of new oceanic lithosphere being created at a spreading ridge. The thickest layer of the oceanic crust, Layer 3, is about 4.7 kilometers (3 miles) thick. It has not been sampled. However, its seismic velocity is also consistent with a composition of basalt, or its coarse-grained equivalent, gabbro. The base of Layer 3 is the Moho. Below that, seismic velocities jump to values typical of the ultramafic rocks of the mantle (probably peridotite, rich in olivine with some pyroxene).

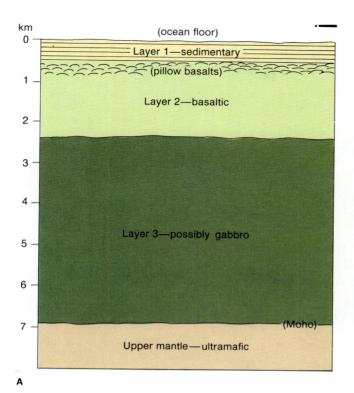

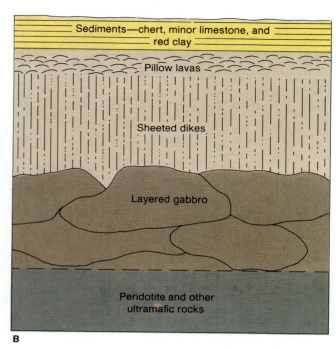

Figure 12.18

Ophiolites may be samples of the oceanic crust and upper mantle. *(A)* Typical cross section of the oceanic lithosphere. *(B)* Typical ophiolite sequence as reconstructed from outcrops on the continents.

As noted in chapter 10, deep coring through the oceanic crust would be both technologically difficult and extremely expensive. However, we may have samples of oceanic crust accessible on land, in the form of **ophiolites** (figure 12.18B). An ophiolite is a particular mafic/ultramafic rock association that is widely believed to be a slice of old oceanic crust, on the basis of similarities between the types and arrangement of rocks in the ophiolites and analogous layers of the sea floor. Ophiolites contain pillow basalts. The pillow basalts are overlain by beds of *chert,* a microcrystalline rock formed from silica-rich sediment; some areas of the sea floor today are sites of accumulation of siliceous sediment. Below the pillow basalt of an ophiolite is a complex of *sheeted dikes* (closely spaced, parallel, vertical dikes, of basaltic composition) and gabbro. The gabbro would be the coarse plutonic rock crystallized at depth during the formation of new sea floor, with the sheeted dikes being the feeder fissures of the pillow-basalt eruptions; abundant high-angle fractures would certainly be expected at a spreading ridge. Below the gabbros,

ophiolites show ultramafic rock, plausible upper-mantle material. This ultramafic rock is typically highly altered by weathering or low-grade metamorphism.

The close correspondence between the layers in ophiolite sequences and what we know of the sea floor strongly suggests the analogy between the two. The correspondence is not necessarily perfect. Not all marine sediments are siliceous by any means, yet chert is the principal component of the sedimentary part of the classic ophiolite sequence. The exact thicknesses of the individual layers in the ophiolite sequence is difficult to know, for ophiolites are commonly highly fractured, but the reconstructed thicknesses seem approximately right for oceanic crust. In the absence of samples of oceanic Layer 3, we cannot be sure that it consists of sheeted dikes and gabbro; however, the seismic velocities are approximately right, and the sheeted dikes make sense tectonically. On balance, it does seem highly likely that the ophiolites are indeed samples of oceanic crust and a bit of upper mantle. It then remains to explain how denser oceanic crust comes to be perched on the continents.

Ophiolites are exposed in mountain ranges. Mountains are commonly formed by convergence. It may be that during convergence, especially in a case in which an ocean is slowly disappearing between two approaching continents, a few bits of ocean floor were

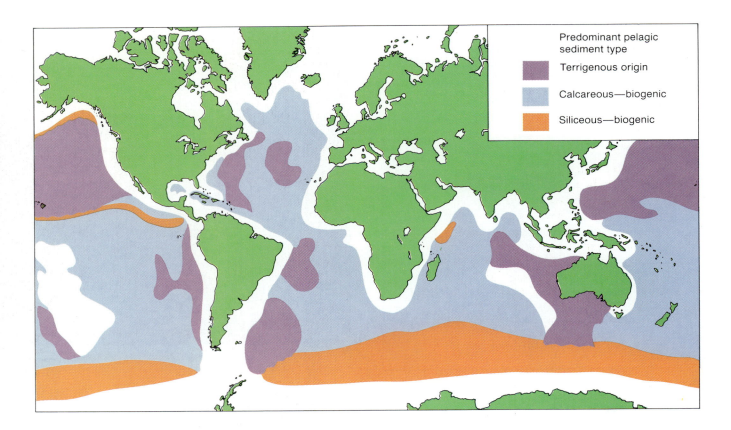

Figure 12.19

Distribution of principal sediment types on the sea floor.

After Arrhenius, G., "Pelagic Sediments," in M. N. Hill, *The Sea*, Vol. 3. Copyright © 1963 by John Wiley and Sons, Inc. Reprinted by permission of John Wiley and Sons, Inc.

caught between converging plates and forced up onto a continent. This would be consistent both with where ophiolites are found and with their typically highly deformed condition. The process by which a slab of oceanic lithosphere is slipped up onto a continent has sometimes been termed **obduction**, to distinguish it from *sub*duction, in which the oceanic plate goes below the continent. On the basis of rock density, one would expect obduction to be relatively rare, and ophiolites are rather rare.

Sedimentation in Ocean Basins

Several sources supply sediment to the ocean basins. The principal sources are the following: the continents, from which the products of weathering and erosion are transported to the sea by wind, water, or ice; seawater, with its content of dissolved chemicals; and the organisms that live in the sea, whose shells and skeletons provide a biogenic component to

oceanic sediment. There is even a small but detectable contribution from the micrometeorites that rain unnoticed through the atmosphere. A generalized map of the distribution of seafloor sediment types is shown in figure 12.19.

Terrigenous Sediments

Continental-marginal sediments are generally dominated by **terrigenous** material. Terrigenous sediment—literally, "earth-derived"—is clastic sediment derived from the continents. We have already noted (chapter 4) a tendency for the coarser sediments to be deposited first, closest to shore, with progressively finer materials transported farther seaward before they settle out of the water.

Sediments at the margins are not static, once deposited. They are redistributed by currents flowing along the continental shelf and slope, and by turbidity currents and submarine landslides that carry them down the continental slope and beyond. The presence of a trench or rise near the margin may limit the extent to which terrigenous sediments, especially coarser ones, are transported away from the continent. The chemical and mineralogical makeup of the terrigenous sediments is as varied as the weathering products of the diverse rock types of the continents. The finest size fraction is typically dominated by clay minerals.

Pelagic Sediments

This finest size fraction stays in suspension most readily, and may be transported beyond topographic barriers on the sea floor to settle into the abyssal plains. It is one component of **pelagic** sediment. Pelagic sediments (named from the Greek *pelagos,* "sea") are the sediments of the open ocean (away from continental margins), regardless of origin. As can be seen from figure 12.19, however, the pelagic sediments of much of the sea floor are dominated by material not of terrigenous origin.

Many of the larger marine creatures with shells or skeletons live in the shallower waters of nearshore regions. They contribute relatively little to pelagic sediment. Volumetrically, the most important biological contributors to pelagic sediment are microorganisms, the *foraminiferans* and *radiolarians* illustrated in chapter 4. The majority of species of these creatures have, respectively, calcareous ($CaCO_3$-rich) and siliceous (SiO_2-rich) hard parts. They live in the near-surface waters; when they die, they settle toward the bottom, the soft organic matter decaying away in the process. Although each shell or skeleton is tiny, the immense numbers of organisms involved collectively produce a significant accumulation of sediment. The result is a fine-grained, water-saturated, calcareous or siliceous **ooze** on the bottom. The distribution of the biogenic oozes on the sea floor is only partially a reflection of the distribution of the corresponding organisms in near-surface waters. Both calcium carbonate and silica are also subject to dissolution in deep-sea water, which selectively removes them from the sediments into solution.

Recall, for example, that calcite dissolves more readily in cold water than in warm. It also dissolves more readily at higher pressure. The ocean bottom waters are very cold and under high pressure from the overlying water column above. Calcite crystallized in warm shallow waters and then carried to the bottom will begin to dissolve under the new pressure/temperature conditions.

Figure 12.20

Where terrigenous sedimentation is limited, manganese nodules may litter the sea floor.

Photo by K. O. Emery, courtesy U.S. Geological Survey.

Other Oceanic Sediments

Another locally important biogenic component of marine sedimentary rock is the calcareous reef, which may be built either by corals or by calcareous algae. We have already noted that reefs must be built on a solid base, but within the photic zone. This, plus the effect of temperature on calcium carbonate solubility, dictates that most reefs will be found in warm, shallow waters, on continental shelves or shallowly submerged platforms.

In relatively warm waters, such as are found on platforms, especially in tropical latitudes, carbonates may be precipitated directly out of seawater to form limestone beds. Some clays are precipitated directly from seawater also, although the bulk of marine clay is probably of terrigenous origin. And over much of the deep-sea floor, where overall sedimentation rates are slow, precipitation of manganese oxides and hydroxides forms the lumpy manganese nodules (figure 12.20) that may be an important mineral resource in the future (see chapter 20).

Box 12.1

Radioactive Waste on the Deep-Sea Floor?

When looking for a disposal site for radioactive wastes, we require that the site have long-term geologic stability. Another requirement is that the wastes should be well isolated from contact with the biosphere. The prospect of putting radioactive wastes into the sea appalls some, who envision contaminated fish and the rapid spread of wastes. Perhaps unexpectedly, areas of the abyssal plains may actually satisfy disposal-site criteria as well as can many proposed sites on the continents.

Abyssal plains far from plate boundaries are tectonically quiet, and are expected to remain so for many millions of years. Dating of sediments in cores taken from these regions confirms that stability. Many areas of the abyssal plains are covered by blankets of finely laminated sediment rich in clay. Clay minerals are an attractive medium for radioactive waste disposal, for they have a great capacity for *ion exchange,* the exchange of ions bonded to the clays for ions in surrounding solution. The highly charged ions of many of the more toxic elements in radioactive waste would tend to be adsorbed preferentially by the clays and held in the sediment. Even in the worst case, in which these elements escaped the sediments into the water, they would be escaping into the deep, cold, slow-moving, poorly mixed, lifeless bottom waters. In time the wastes might dissipate into shallower waters, but by the time they did so they would also have been dispersed and diluted in an immense volume of water.

As an alternative to disposal on land, deep seabed disposal of radioactive waste is presently under serious consideration by an international commission.

Ocean Circulation–An Introduction

Discussion of the details of oceanic circulation patterns is beyond the scope of this chapter, but some general comments and observations about ocean currents and controls on their movement can be made.

The Ocean-Atmosphere Connection

The principal driving force behind the motion of oceanic currents, particularly near-surface currents, is atmospheric circulation. On a superficial level, friction between low-altitude winds and the sea surface generates ripples and waves on the water. Surface water circulation patterns are established largely by atmospheric pressure differences from place to place. From figure 17.2 it can be seen that atmospheric circulation patterns can be described in terms of belts of easterly or westerly winds. These provide the principal drive behind near-surface water transport also. The interaction of the momentum imparted by surface winds and the rotation of the earth results in net water movement oriented somewhat differently from the directions of wind flow. The dominant flow directions of shallow oceanic currents are shown in figure 12.21.

It can be seen from this figure that there are several belts or zones of water **convergence,** where opposing currents bring waters together from different directions. The result is a pileup of water, a measurable mound on the sea surface, up to several meters in height. The principal convergences are at about 30° north and south latitude. The additional water height and mass associated with these zones create a pressure gradient with respect to surrounding waters, which further modifies current flow. Water flows down the pressure gradient away from such a high; a rotational component is added by the earth's rotation. The result is a **gyre, a nearly closed circulation pattern.** See, for example, the North Atlantic circulation in figure 12.21.

Boundary Currents

Circulation patterns are further modified by the presence of continents disrupting the free flow of currents. Many of the surface currents flow predominantly east/west in the open ocean, which keeps them in approximately the same climatic zone along their flow paths. When they are deflected by a continent, these currents begin to move more nearly north/south,

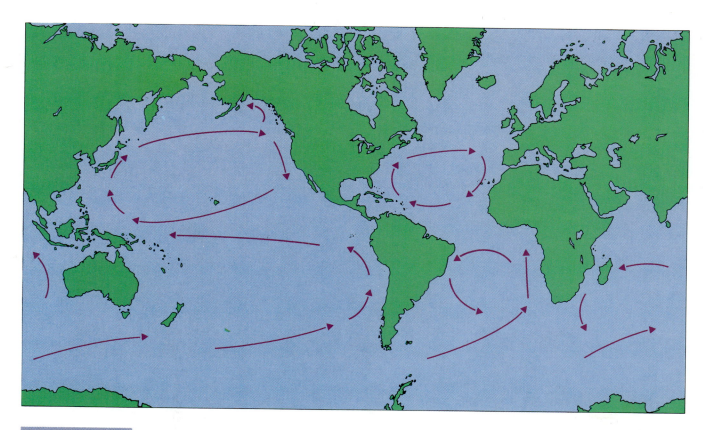

Figure 12.21

Principal near-surface circulation of the world oceans.

Figure 12.22

The Gulf Stream in cross section. Note temperature contrast with adjacent waters.

From Bishop, J. M., *Applied Oceanography.* Copyright © 1984 by John Wiley and Sons, Inc. Reprinted by permission of John Wiley and Sons, Inc.

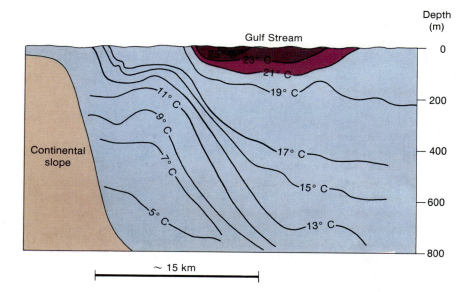

across climatic zones. This is the origin of currents markedly warmer or colder than the general regional surface temperature of the surroundings. The Gulf Stream (figure 12.22) is one example. Formed by the northward deflection of a warm, low-latitude current, it carries that warmer water northward, moderating temperatures along the Atlantic coast of North America. Temperatures in the Gulf Stream may be more than 5° C (10° F) higher than in the surrounding water masses on either side.

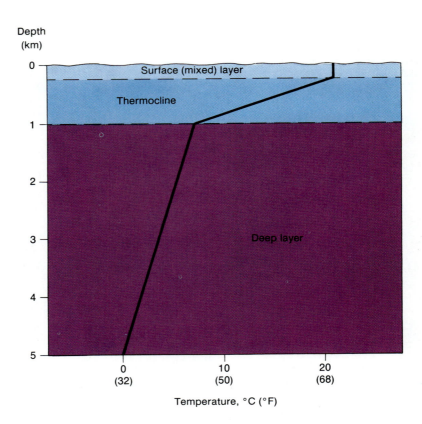

Figure 12.23

Vertical temperature distribution in the oceans.

Water Zonation and the Deep Waters

With the exception of turbidity currents, most of the vigorous circulation of the oceans is confined to the near-surface waters. These waters are also distinct from most of the volume of the oceans in terms of temperature (figure 12.23). Only the shallowest waters, within 100 to 200 meters of the surface, are well mixed, by waves, currents, and winds. The mixing produces a thin layer of fairly uniform temperature and salinity. The photic zone of water lies within this surface layer too, so plant growth is confined to this zone, as are many organisms that require marine plants for food. The average temperature of the surface layer is about 15° C (60° F).

Below the surface layer is the **thermocline,** a zone of rapidly decreasing temperature that extends to 500 to 1000 meters below the surface. Temperatures at the base of the thermocline are typically about 5° C (40° F). It is the interface between the warm mixed layer and the so-called **deep layer** of cold, slow-moving, rather isolated water. Temperatures in the deep layer continue to decrease with depth, but more slowly than in the thermocline. The temperature of the bottommost water is close to freezing, and may even be slightly below freezing (the water prevented from freezing solid by its dissolved salt con-

tent). This cold, deep layer originates largely in the polar regions. Dense, cold polar water masses sink, and then slowly begin to migrate toward the warmer climate of the equatorial region. The deep waters do move, but slowly; they may take centuries to travel from poles to equator.

Upwelling

When winds blow parallel or nearly parallel to a coastline, the resultant currents may tend to cause surface waters to be blown offshore. This, in turn, creates a region of low pressure, and may result in **upwelling** of deep waters to replace the displaced surface waters (figure 12.24). The deep waters are not only relatively cold but relatively enriched in dissolved nutrients, in part because few organisms live in the cold, dark depths to consume those nutrients. When the nutrient-laden waters rise into the photic zone, they can support abundant plant life and, in turn, animal life that feeds on the plants. Many rich fishing grounds are located in zones of coastal upwelling. Given the distribution of continents, winds, and currents, conditions tend to be particularly favorable to the development of upwelling off the western coasts of continents. The west coasts of North and South America and of Africa are subject to especially frequent upwelling events. The productivity of fisheries off the coast of Peru is noteworthy (see also box 12.2).

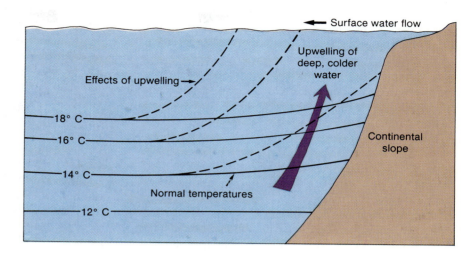

Strong offshore winds
cause low pressures near shore

← Surface water flow

Upwelling of
deep, colder
water

Effects of upwelling →

Continental
slope

18° C

16° C

14° C

Normal temperatures

12° C

Box 12.2

Currents and Climate: El Niño

As noted in the text, zones of upwelling along the western coasts of continents may be particularly productive. The anchoveta fishing grounds of Peru are, most of the time, especially rich. From time to time, however, the upwelling is suppressed for a period of weeks or longer. The reasons are not precisely known, and probably several factors are involved. Abatement of coastal winds would reduce the pressure gradient driving the upwelling. Other wind changes in the western Pacific can cause anomalous west-to-east flows of warm surface water, depressing the thermocline, thickening the warm surface layer in the eastern Pacific. The reduction in upwelling of the fertile cold waters has a catastrophic effect on the Peruvian anchoveta industry. Such an event is called *El Niño* ("the (Christ) Child") by the fishermen, for it commonly occurs in winter, near the Christmas season. The intensity of and interval between El Niño events, however, is quite variable.

Significant El Niño conditions occurred in 1957-58, 1965, 1972-73, and 1982-83. During the course of each of these events, various other meteorological problems arose worldwide—droughts in some places, torrential rains elsewhere. The coincidence in time, together with the high probability that meteorological factors do influence El Niño in some measure, has prompted various investigators to blame El Niño for the droughts and floods also. A great deal more research will be required in order to clarify any causative links among meteorological phenomena and El Niño episodes.

Summary

The sea floor can be subdivided into continental-marginal regions, spreading ridge systems, and abyssal plains. The continental margins are further divided, on the basis of tectonics, into active and passive margins. A passive continental margin, formed during continental rifting and the creation of a new ocean, typically has a broad continental shelf; the angle of the continental slope is variable. Active, convergent margins show narrow shelves, usually steep continental slopes, and trenches at the foot of the slope, formed by subduction. The shelves and slopes of continental margins are often dissected by canyons, most likely cut by turbidity currents. The abyssal plains are the deepest parts of the oceans. They may be dotted with hills, seamounts, and guyots. Oceanic ridge systems possess a central rift valley; the topography slopes away from the ridge crest, and fracture zones, extensions of transform faults, stretch for tens or hundreds of kilometers from the ridge crest. Hydrothermal activity occurs as a result of the circulation of water through the hot rocks along the ridge.

The relatively thin oceanic crust is divided into Layer 1 (sedimentary), Layer 2 (basaltic), and an unsampled Layer 3 that may be gabbroic. Ophiolites may be slices of ancient, obducted oceanic crust. The crust underlying some of the large plateaus in the ocean basins, by contrast, has a density and thickness comparable to continental crust; these plateaus may be future accreted terranes. Terrigenous sedimentation in the oceans is largely confined to the continental margins. Pelagic sediment consists predominantly of fine clays (partially terrigenous) and calcareous and siliceous oozes. The most active oceanic circulation is confined to the well-mixed surface layer, the uppermost 100 to 200 meters. This circulation is predominantly wind-driven. Most of the open-ocean currents flow east/west; where they encounter continents and are deflected into boundary currents, they flow north/south across climatic zones. The deep layer of water below the thermocline is much colder and slower moving. In favorable settings, these deep, nutrient-rich waters rise in upwelling events along continental coastlines, greatly increasing the biological productivity in the near-surface waters of the affected areas.

Terms to Remember

abyssal hills	convergence	obduction	seamount
abyssal plains	coral atoll	ooze	submarine canyon
Benioff zone	deep layer	ophiolite	terrigenous
continental rise	guyot	pelagic	thermocline
continental shelf	gyre	plateau	trench
continental slope	hydrothermal vents	rift valley	turbidity current
	manganese nodules	salt dome	upwelling

Questions for Review

1. Describe the process of ocean-basin creation that leads to formation of a passive margin.

2. What are the principal topographic features of a passive continental margin?

3. Where and how are trenches formed? What evidence supports the plate-tectonic explanation of trenches?

4. Sketch a topographic cross section of a spreading ridge and account for the central valley and the general slope of the topography.

5. What are the abyssal plains? How flat and featureless are they?

6. Describe the nature and probable origin of a turbidity current. What features of the sea floor might be explained by turbidity currents?

7. What is a guyot? Explain one mechanism by which guyots might be formed.

8. How do coral atolls develop?

9. The oceanic plateaus may be the suspect terranes of the future—explain briefly.

10. What is an ophiolite? Why are ophiolites believed to be samples of oceanic lithosphere, and how might they come to be found on the continents?

11. Cite and briefly explain two major components of pelagic sediment.

12. In oceanic circulation, what is a convergence and how does it arise?

13. What is the origin of boundary currents, and how may they modify the climate on the adjacent continent?

14. Describe the way seawater temperature changes with depth. How well mixed are the oceans?

15. Briefly describe the phenomenon of coastal upwelling.

For Further Thought

1. In what ways might you be able to distinguish a sediment sample from the abyssal plain from a sediment sample collected at the base of the continental slope of a passive margin?

2. Suppose, in drilling in the sediments on the continental shelf off New England, you find buried fossil coral reefs. The water there today is too cold for corals. Suggest at least two possible explanations for the presence of the fossil corals.

Suggestions for Further Reading

Bishop, J. M. *Applied Oceanography*. New York: John Wiley and Sons, 1984.

"El Niño." *Oceanus* 27, no.2 (Summer 1984).

Fanning, K. A., and Manheim, F. T., eds. *The Dynamic Environment of the Ocean Floor*. Lexington, Mass.: D.C. Heath and Co., 1982.

Heezen, B. C., and Hollister, C. D. *The Face of the Deep*. New York: Oxford University Press, 1971.

Hill, M. N., ed. *The Sea*. New York: Interscience Publishers, 1962-83. This eight-volume set, though somewhat dated in parts, still assembles in one place a wealth of data and basic information on the oceans and ocean basins.

Kennett, J. P. *Marine Geology*. Englewood Cliffs, N.J.: Prentice-Hall, 1982.

Pickard, G. L. *Descriptive Physical Oceanography*. 3d ed. New York: Pergamon Press, 1979.

Scrutton, R. A., ed. *Dynamics of Passive Margins*. Washington, D.C.: American Geophysical Union, 1982.

Scrutton, R. A., and Talwani, M., eds. *The Ocean Floor*. New York: Wiley-Interscience, 1982.

Seibold, E., and Berger, W. H. *The Sea Floor*. New York: Springer-Verlag, 1982.

Shepard, F. P. *Geological Oceanography*. 3d ed. New York: Crane, Russak, 1977.

Part Three

Part opener photo © Steve McCutcheon

It is at the earth's surface that we see the ceaseless interplay between the internal heat of the earth that builds up the land and creates mountains, and the force of gravity that tends to pull material down again to level the land. The leveling process is greatly accelerated by the action of ice, water, and wind. These agents, in turn, are ultimately powered in large measure by the external heat supplied to the earth by the sun: that heat drives the evaporation stage of the water cycle of which ice and water are both a part, and atmospheric temperature differences drive the wind.

The subjects of chapters 13 and 14, streams and groundwater, are particularly closely linked. These two chapters survey the occurrence, movement, and effects of water above and below the ground surface. Consideration is also given to associated hazards, especially flooding, and to the growing concern of water resources. Land and water may interact especially vigorously at the coast. Chapter 15 looks at characteristic features and processes of coastal areas, as well as at some ways in which human activities may modify them. The actions of glaciers may be slower and more difficult to observe, but a thick mass of moving ice is a powerful erosive agent. Over time, much of the earth's surface has been shaped by glaciers, the subject of chapter 16.

Air is more tenuous than water or ice, and wind is correspondingly a less powerful agent of surface change. Its effects are most noticeable in sparsely vegetated areas, such as deserts. A discussion of deserts and desert features is thus combined with the discussion of wind-related processes and features in chapter 17. A variety of chemical and physical processes contribute to weathering, the breakdown of rocks in place. Soil is a result of weathering; weathering, soil, and soil erosion are considered together in chapter 18. Chapter 19 deals with mass wasting, the downslope movement of material. The focus is particularly on the more rapid movements, landslides—what they are, why they occur, and what can be done to reduce the risks from landslides in areas where they threaten humans or their works.

Chapter 13

Streams

Introduction

When rain falls on the earth's surface, or accumulated snow melts, the water either sinks into the ground (**infiltrates**) or remains at the surface. Subsurface waters will be considered in the next chapter. Surface water tends to flow from higher to lower elevations. It may wash in broad sheets over the surface, but commonly the flowing water is collected into a channel. It has then formed a **stream.** The term *stream* is used to describe any body of flowing water confined within a channel, regardless of its size. In this chapter we will survey streams and their characteristics, including their effects as agents of erosion, sediment transport, and deposition. Later in the chapter we will consider the phenomenon of stream flooding, and the interplay of flooding and human activities.

Drainage Basins and Size of Streams

The geographic area from which a stream system draws water is its **drainage basin** (figure 13.1). The volume of water in a stream at any spot is related in part to the size of the area drained, as well as to other factors such as the amount of precipitation in the area and how readily the water infiltrates into the ground. Typically, only very small drainage basins are drained by a single channel. As small streams flow into larger ones, drainage patterns become more complex.

One way to describe the complexity of drainage is in terms of stream **order** (figure 13.2). A **tributary stream** is one that flows into another stream. A *first-order stream* is one into which no tributaries flow. A *second-order stream* is one that has only first-order streams as tributaries, and so on. In general, the higher the order of the stream, the larger the area drained by

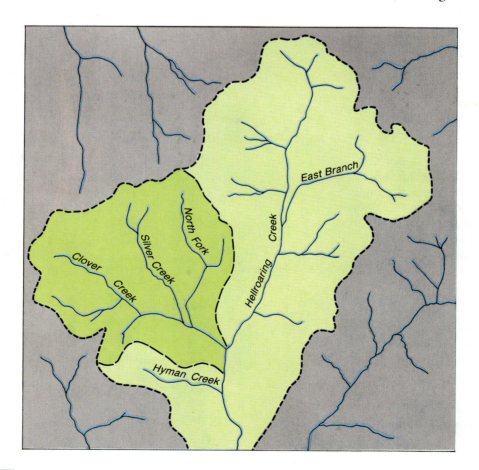

Figure 13.1

A stream system and its drainage basin: Hellroaring Creek in the Gallatin National Forest. The shaded area is a sub-basin drained by North Fork, Silver Creek, and Clover Creek. The boundaries of drainage basins are local topographic highs—in this case, mountain ridges.

the stream and its tributaries. Deciding when a trickle of surface water in a channel properly constitutes a stream, and therefore determining the order of streams further down in a given drainage system, is in part a matter of judgment and the scale of map being used. The U.S. Geological Survey identifies between 1 1/2 and 2 million streams in the United States. The more than 1 1/2 million first-order streams have an average drainage area of 2 1/2 square kilometers each; the 4200 fifth-order streams average about 1300 square kilometers in drainage area; the Mississippi River, the single tenth-order stream, drains 320,000,000 square kilometers.

Drainage Geometry

The majority of stream systems exhibit a branching drainage pattern described as **dendritic,** from the Greek *dendros,* "tree." In a dendritic drainage pattern stream channels are irregular, and tributaries join larger streams at a variety of angles. The system in figure 13.1 is an example.

Local geology may cause the drainage network to assume a more regular geometry. Where sets of parallel fractures cause zones of weakness and/or topographic lows, streams may establish a rectilinear **trellis drainage** pattern in which tributaries join the main stream at right angles (figure 13.3A). Trellis drainage may also develop where the topography is

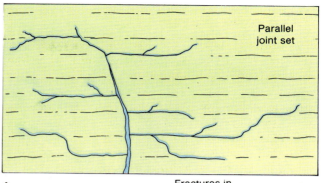

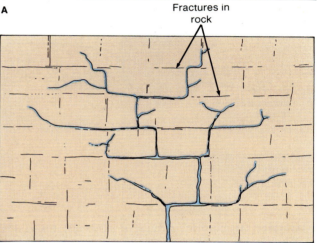

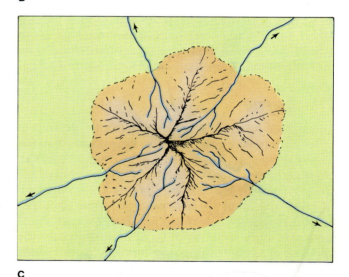

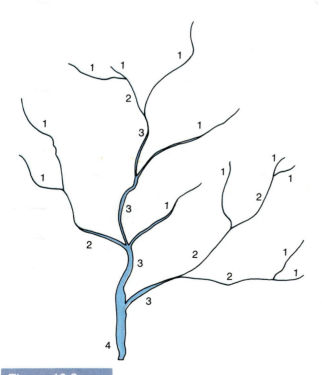

Figure 13.2

Stream order is determined by the number of levels of tributaries flowing into the stream.

Figure 13.3

Specialized geometry in drainage networks arises from special geologic conditions. *(A)* Fracture-controlled trellis drainage. *(B)* Rectangular drainage controlled by intersecting joint sets. *(C)* Radial drainage around a conical hill.

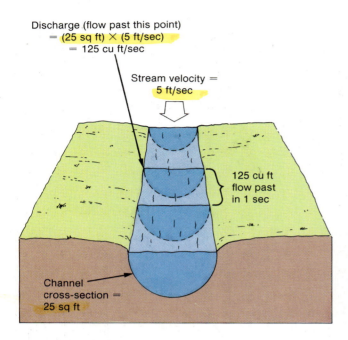

Discharge (flow past this point)
= (25 sq ft) × (5 ft/sec)
= 125 cu ft/sec

Stream velocity =
5 ft/sec

125 cu ft
flow past
in 1 sec

Channel
cross-section =
25 sq ft

Figure 13.4

Discharge = area × velocity.

Figure 13.5

Boulder scoured to a concave shape by stream erosion.

dominated by parallel ridges of resistant rock, and streams tend to flow between those ridges. Intersecting joint sets may create **rectangular drainage with right-angle bends in all streams** (figure 13.3B). Where there exists a conical topographic high, such as a volcano, **radial drainage** carries water away from the center of the high in all directions (figure 13.3C).

Discharge

The size of a stream at any point can be defined in terms of its **discharge,** the volume of water flowing past that point in a given period of time. It is calculated as the product of the cross-sectional area of the stream times the flow velocity (figure 13.4). Conventionally, discharge is measured in cubic feet per second or cubic meters per second. As a practical matter, discharge can be difficult to measure, for the velocity of the stream is not everywhere the same. Flow is generally slower along the channel bed and sides, where there is friction between water and channel, and faster near the center of the channel and surface of the stream. Discharge is a function of many factors, including climate (precipitation, availability of water), flow velocity, characteristics of the surrounding soil that influence water infiltration, and the overall size of the drainage basin above that point, from which the stream flow is drawn. In turn, a stream's discharge has a bearing on its effectiveness in erosion and sediment transport.

Streams as Agents of Erosion

Moving water is the principal erosive agent at the earth's surface. Most effective in eroding unconsolidated sediments, water and water-borne sediment can even scour solid rock (figure 13.5). Stream erosion widens, deepens, and modifies the channel proper and, over time, alters the geometry of the valley surrounding the channel.

Streambed Erosion

A stream eroding downward into its bed is engaged in **downcutting.** How rapidly downcutting occurs is partly a function of the nature of the streambed material: solid rock is generally much more resistant to erosion than is sediment. If the stream itself is transporting sediment, that sediment can abrade the channel bottom and sides and hasten erosion. The higher the velocity of stream flow, the more rapid the erosion, whether or not quantities of sediment are carried by the stream. The water can also attack soluble rocks chemically. A limestone, for example, would be especially susceptible to dissolution. Where a stream is engaged in relatively rapid downcutting, it characteristically carves a narrow, steep-walled, V-shaped valley (figure 13.6).

V-shaped valley of a rapidly downcutting stream: the Grand Canyon of the Yellowstone River.

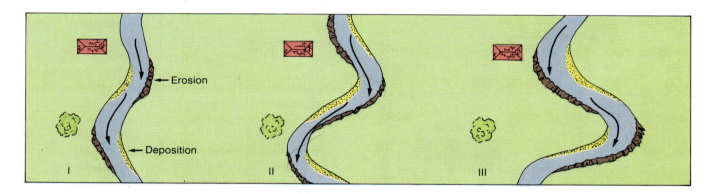

Figure 13.7
Map view of development of meanders. Channel erosion is greatest on the outside of curves, on the downstream side; deposition occurs in the sheltered area inside of curves. Through time, meanders migrate both laterally and downstream.

Meanders

Where downcutting is less rapid and stream velocity slower, the straight channel begins to develop lateral displacements. Small irregularities in the channel cause local variations in flow velocity, which in turn contribute to variations in erosion and sediment deposition from point to point. Bends, or **meanders,** begin to develop along the channel (figure 13.7).

Once meanders begin to form, they tend to move both laterally and along the length of the stream. This can be understood in terms of flow velocity within the channel (figure 13.8). The water strikes with greatest force the outer, downstream bank of the meander, which causes increased erosion in that area. Flow is

Figure 13.8

Here the path of fastest water flow is marked by surface turbulence (bright reflection). Note its shift from side to side in the channel. The stream is flowing toward the camera.

slower on the inside and upstream bank of a meander, so sediment deposition tends to occur there, for reasons described in the next section. Over time, then, meanders can enlarge sideways, and shift downstream, by this combined erosion and sedimentation.

> On an intuitive level, meandering makes sense. It is more difficult to explain why many streams develop regular, sinuous meanders of similar size and shape over a considerable length of the stream. No fully satisfactory theoretical model has been developed to account for this. Nor is it as easy to understand how meanders become established in bedrock as it is to visualize them forming in loose, unconsolidated sediment or soil.

Meanders do not enlarge indefinitely. When large, looping meanders have developed along a stream, they represent sizeable detours for the flowing water. Eventually the stream may make a shortcut and cut off the meanders, abandoning the longer, irregular channel for a shorter, more direct route (figure 13.9). The cutoff meanders are termed **oxbows**. If they remain filled with water, they are *oxbow lakes*.

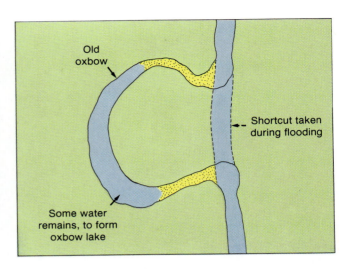

Figure 13.9

Map view of formation of oxbow by meander cutoff.

Meander cutoff most commonly occurs during flood events, when there is a larger volume of water, flowing faster, in the channel. The added momentum of the flood waters may be enough to carve the shortcut, by-passing the meander.

The Floodplain

Over time, the processes of lateral erosion associated with meandering, sediment deposition behind migrating meanders, and additional sedimentation during flood events when the stream has overflowed its banks collectively work to create a **floodplain** (figure 13.10). A floodplain is a flat or gently sloping region around a stream channel into which the stream flows during flood events (figure 13.11). It is quite common for the meandering stream at any given time to occupy only a portion of the breadth of the floodplain. The overall width of the floodplain will be a function of many factors, including the size of the stream, the relative rates of meander migration and of downcutting, and the strength of the valley walls.

While climatic and tectonic conditions remain fairly constant, slow lateral channel migrations over time tend to maintain a single floodplain with a fairly level surface. If these external conditions are suddenly changed significantly, relative rates of meandering and downcutting may also be changed in such a way as to change the consequent floodplain geometry. If there is regional uplift, for instance, downcutting may be rapid and the old floodplain cut up by development of a new, narrower, deeper one within the broader valley of the old. This is one mechanism that has been proposed for the formation of **terraces**, steplike plateaus at some higher elevation above the

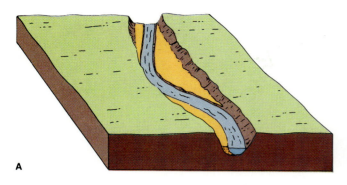

A

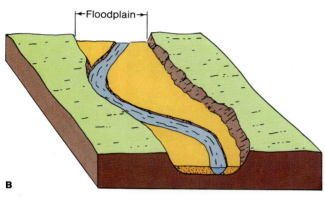

B

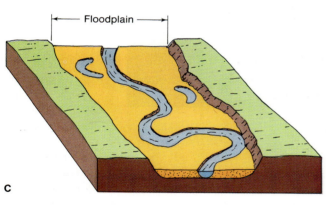

C

Figure 13.10

A floodplain is broadened by meandering.

Figure 13.11

Floodplain around Crooked Creek, Mono County, California.

Figure 13.12

Terraces within a valley, above the present floodplain.

Photo courtesy E. E. Brabb, courtesy U.S. Geological Survey.

Figure 13.13

Incised meanders in rock, with no floodplain surrounding the channel.

present floodplain (figure 13.12). Multiple levels of terraces may be found within a single broad valley. It may also be that none of the old terrace surfaces represents a former floodplain level of the present stream at all. The processes of terrace formation are complex and not fully understood.

Another somewhat problematic erosional process is the formation of **incised meanders** (figure 13.13). As their name suggests, these are meanders deeply cut into rock. The stream is not surrounded by a floodplain at nearly the same elevation but is entrenched within a deep, winding, steep-sided valley

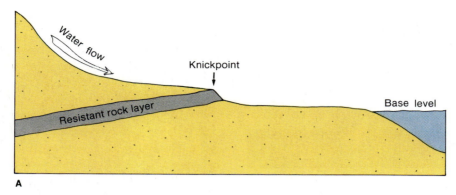

Figure 13.14

Knickpoints in the profile. *(A)* Schematic: slower erosion of a resistant rock layer produces a cliff. *(B)* The Lower Falls of the Yellowstone are there because of a resistant basalt sill within softer volcanics.

closely matching the meandering course of the stream. Several modes of formation have been proposed, and it is not always possible to determine the one responsible in each instance. Some streams, especially slowly flowing ones on gently sloping surfaces, exhibit meanders from their inception. If downcutting is relatively rapid and the valley wallrocks strong enough to maintain steep cliffs, these early meanders may be dug deeper and deeper. Alternatively, meanders may be developed in soft or unconsolidated materials by a once-straight stream, by processes described above. These meanders may be subsequently fixed as the stream cuts down below the softer rock into rocks less prone to lateral erosion. The meander pattern in this instance is imposed on the underlying rock after formation in softer layers above.

Stream Piracy

Along a stream channel there may be steep drops in elevation, known as **knickpoints** (figure 13.14A). They can arise from stream flow across different rock types of different erodability, or by faulting. Above the junction with a tributary the channel may be cut less deeply than below, where an abruptly greater volume of water is present. Major knickpoints may be marked by waterfalls (figure 13.14B). Once established, such scarps tend to erode headward, or upstream, as continued bed erosion wears them back, or as material is carried downstream from the base of the scarp and the undercut scarp slumps and fails. Niagara Falls is a prominent example of such a knickpoint that is migrating upstream through time in this way. Such **headward erosion** at the head of the stream valley may also lengthen the valley from its upper end. In

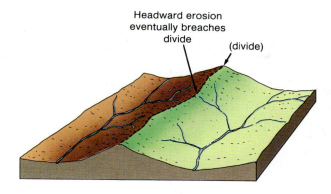

Headward erosion
eventually breaches
divide
(divide)

Some drainage
diverted

Figure 13.15

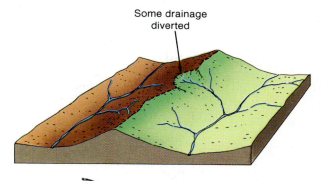

Stream piracy (schematic). Headward erosion of one stream valley penetrates the divide between drainage basins, diverting some of the drainage.

Figure 13.16

Suspended sediment clouds the tributary at right. Note that the two streams do not immediately mix where they join.

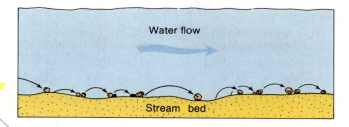

Water flow

Stream bed

Figure 13.17

Saltation (schematic). Heavier particles move in a series of short hops, jostled up into the flowing water, then sinking again to the bottom.

time, headward erosion may cut through a **divide**, a topographic high separating two different drainage basins. When this occurs, some part of one stream system is diverted into the other, in a process known as **stream piracy** (figure 13.15). The redistribution of water accompanying stream piracy can be expected to result in changes in erosion, sediment transport and deposition, and channel geometry in both stream systems.

Sediment Transport and Deposition

Streams can move material in several ways. The heaviest debris may be rolled or pushed along the stream bed. This material is described as the **bed load** of the stream. The stream's **competence** at any point is the largest size of particle it can move in the bed load, which is largely a function of its velocity. Stream **capacity** is the total quantity of material that can potentially be moved, which depends on discharge (that is, on both velocity and quantity of water present). The **suspended load** consists of material light or fine enough that it can be moved along suspended in the stream, supported by the flowing water. The maximum suspended load that can be transported is also dependent on discharge. Suspended sediment clouds a stream and gives the water a muddy appearance (figure 13.16). Material of intermediate size may be carried in short hops along the streambed by a process called **saltation** (figure 13.17). Finally, some substances can be completely dissolved in water (**dissolved load**). The total quantity of material that a stream transports by all these methods is called, simply, its **load.**

How much of a load is actually transported depends on the availability of sediments or soluble material: a stream flowing over solid, insoluble bedrock will not be able to dislodge very much material, while a similar stream flowing through sand may move considerable material. When the stream eventually deposits its load, the resulting deposit is called **alluvium**, a term derived from the French and Latin for "to wash over." The term *alluvium* is a general one for any stream-deposited sediment. Certain alluvial deposits take distinctive forms that are given more specific names.

Velocity and Sediment Sorting

Variations in a stream's velocity along its length are reflected in the sediments deposited at different points. The sediments found motionless in a streambed at any point are those too big or heavy for that stream to move at that point. Where the stream flows most quickly, gravel and even boulders may be swept along with it, together with the finer sediments. If velocity controls the maximum particle size that can be moved, it follows that as the stream slows down it selectively drops the heaviest, largest particles first, continuing to move along the lighter, finer materials. As velocity continues to decrease, successively smaller particles are dropped. In a very slowly flowing stream, only the finest sediments and dissolved materials are still being carried. This link between the velocity of water flow and the size of particles moved accounts for one common characteristic of alluvial sediments in channels: they are often well sorted by size or density, with materials deposited at a given point tending to be similar in size or weight.

Depositional Features in Channels

As a meander migrates laterally, eroding its outer bank, there is generally corresponding deposition along the inner bank where the water flow is slower. The sedimentary feature thus built up is a **point bar** (figure 13.18). Where water slows along the channel bed, through friction, it may blanket the bottom with sediment. If the slowdown is a localized one—caused, for example, by obstacles such as isolated boulders on the bottom—the sediment deposition may be correspondingly localized into an island within the channel. Once such a feature has formed, its very presence provides a greater impediment to stream flow, and it may therefore tend to enlarge as water continues to be slowed and more sediments are deposited.

Where the sediment load carried is large in relation to capacity, channel islands may grow until they reach the water surface, where they may be further stabilized if vegetation begins to grow on them. The stream channel has become divided into two channels. This process is termed *braiding*. A **braided stream** is one with several (perhaps many) channels that divide, then rejoin downstream (figure 13.19).

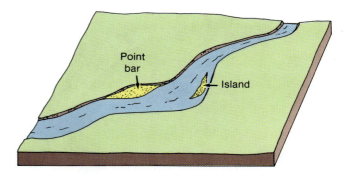

Figure 13.18

Point bars and islands form in a sediment-laden stream where the water flow is slowed.

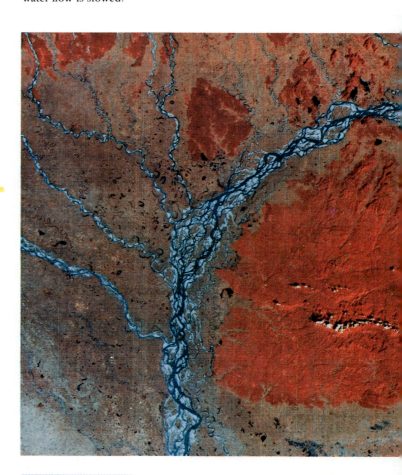

Figure 13.19

A braided stream, with many channels dividing and rejoining: the Brahmaputra River in India. (Satellite photograph.)

© NASA

Patterns of braided channels may become very complex. Where many shallow, braided channels cross a broad area of easily eroded sediment, the channels may shift constantly and the patterns change daily or even hourly.

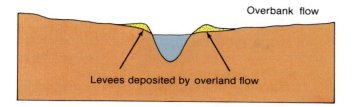

Natural levees form along the margin of a stream channel as a result of sediment deposition during flooding.

Depositional Features in Floodplains

A floodplain is itself a depositional feature, in part. As channels shift, point bars build land where stream waters flowed. Additional deposition in the floodplain can occur outside the channel, during flood events. The whole floodplain can be blanketed in sediment as the channel overflows and the floodwaters are slowed by vegetation or other obstacles. Such sediments are descriptively termed **overbank deposits.** Close to the channel, the overbank deposits may assume a distinctive form, making ridges along the edge of the channel. These are natural **levees** (figure 13.20). Development of levees is generally most pronounced where the suspended sediment load is relatively coarse. Streams carrying primarily fine, silty sediments may not form obvious levees.

Deposits at Stream Mouths

It has already been noted that a decrease in velocity is typically accompanied by deposition, beginning with the coarsest sediments. If the drop in velocity is abrupt, a large deposit may result. This is not uncommon at the **mouth,** or end, of a stream, where it flows into another body of water or terminates in a dry valley or plain.

Sedimentary deposits at the mouths of streams are characteristically fan- or wedge-shaped. Such a deposit formed where a tributary stream flows into another, more slowly moving stream is an **alluvial fan** (figure 13.21). Alluvial fans also form where fast-moving mountain streams flow out onto plains or into flatter valleys, with a consequent sharp drop in water flow velocity.

If a stream flows into a body of standing water, such as a lake or ocean, and the flow velocity drops to zero, all but the dissolved portion of the load will be deposited. The resulting sediment wedge is called a **delta** (figure 13.22). Over time, deposition of delta sediments may actually build the land outward into the water, increasing the land area.

An alluvial fan deposited where a fast-moving, sediment-bearing mountain stream flows into a plain, and slows.
Photo by E. E. Brabb, courtesy U.S. Geological Survey.

The Mississippi River delta (satellite photo). Suspended sediment clouds the water beyond.

© NASA

Stream Channels and Equilibrium

In order for stream waters to flow, there must be a difference in elevation between the **source, where the first perceptible channel indicates the existence of the stream, and the mouth.** The **gradient** of the stream is a measure of the steepness of the slope of the channel. It is the difference in elevation between two points along the stream divided by the length of the channel between them. The higher the gradient, the steeper the channel and, all else being equal, the higher the water velocity.

The water level at the mouth of a stream, where it generally flows into an ocean, lake, or other stream, is the stream's **base level.** Base level represents the lowest level to which a stream can cut down. Once elevation upstream reaches base level, the gradient is zero, velocity should become zero, and no further erosion is possible.

Streams are dynamic systems, their forms changing in response to changes in environmental factors. If over time, regional precipitation and stream discharge increase, more erosion will occur and scour a larger channel to accommodate the larger volume of water. An influx of sediment may result in increased sedimentation. If land in the drainage basin is uplifted, increasing stream gradient, the consequent increases in velocity and rate of downcutting will hasten the stream's re-approach to its base level.

Equilibrium Profile

Over time, provided that environmental factors remain approximately constant, streams and their channels will tend to become adjusted to the prevailing conditions. They will approach a condition of **dynamic equilibrium.** In terms of erosion and deposition, this means that sediment transport is just sufficient to balance the input of sediment from the drainage basin, and erosion of and deposition in the stream channel are equal. A stream in which the channel and its gradient have been thus adjusted, so that input and outflow of sediment are just equal, is sometimes described as a **graded** stream.

It has been found that a graded stream tends to show a characteristic pattern of decreasing gradient from source to mouth. This is illustrated in the longitudinal profile of figure 13.23. Generally, the gradient is steepest near the source, flattening toward the mouth. However, in natural streams velocity does not necessarily decrease toward the mouth, despite the decreasing gradient. The explanation lies in the increase in water volume downstream, associated with

addition of tributaries to the stream and increasing drainage area. The added weight of water being pulled down toward base level by gravity partially compensates for the decrease in gradient. Overall discharge does increase downstream.

The geology of natural systems can cause deviations from this idealized profile. A common deviation is the occurrence of knickpoints, as discussed earlier.

The concept of a stream in perfect equilibrium is an abstraction. Over the long term, gravity will be tending to level the land, and ultimately, in the absence of tectonic uplift, volcanic eruption, or other mountain-building activities, a net downcutting should be expected until the land is flattened to sea level. However, the approach to equilibrium can be so close that on a human time scale the net change is negligible.

Effect of Changes in Base Level; Dams and Reservoirs

Tectonic uplift can raise the upper portion of a drainage basin relative to the base level of the stream system (an adjacent ocean, for example). The net effect is to increase the overall gradient of the system and the rate of downcutting toward the base level. Also, changes in sea level, such as are caused by glaciation (see chapter 16), move base level up or down and decrease or increase average gradients accordingly.

When human activities alter base levels, more complex modifications of the stream system are possible. Consider the construction of a dam with a reservoir behind it along a previously graded stream (figure 13.24). The water level in the reservoir becomes a new local base level for the stream above the dam. Its average gradient is decreased, and any downcutting should likewise decrease. Moreover, when the stream reaches the still reservoir and flow ceases, the stream drops its (nondissolved) sediment load. This begins to fill in the reservoir, reducing the available volume for water storage, whether for irrigation, water supply, flood control, or whatever. Below the dam, the water released is free of suspended sediment or bed load. It is therefore capable of increased erosion through uptake of more sediment. The net result of all of these changes is reflected in a modified longitudinal profile, as shown.

Flooding

In most (moderately humid) climates, the size of a stream channel becomes adjusted to accommodate the average maximum annual discharge. Much of the year, the water level (stream **stage**) may be well below the

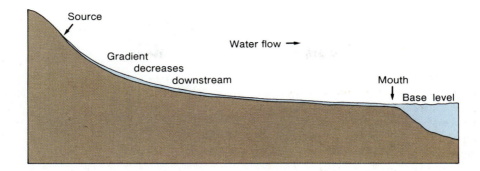

Figure 13.23

Longitudinal profile of a graded stream, concave upward, with steeper gradient near the source.

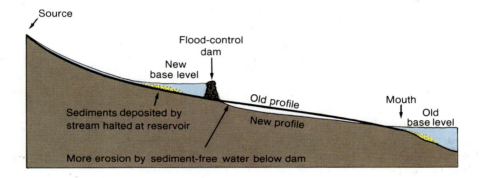

Figure 13.24

Effects of a dam and reservoir on a stream profile. The new base level of the reservoir causes deposition in the reservoir and upstream from it. Erosion increases below the dam.

stream bank height. From time to time the stream will reach **bankfull stage.** Should the water rise above bankfull stage, overflowing the channel, the stream is at **flood stage.** Floods will occur, in other words, when the rate at which water reaches some point in a stream exceeds the bankfull discharge there. It should be realized that floods are not unnatural or particularly unusual events. Flooding is a perfectly normal and, to some extent, predictable phenomenon.

Factors Influencing Flood Severity

Many factors together determine whether a flood will occur. The quantity of water involved and the rate at which it is put into the stream system are among the major factors. Worldwide, the most intense rainfall events occur in Southeast Asia, where storms have dumped up to 200 centimeters (80 inches) of rain in less than three days. (To put such numbers in perspective, that amount of rain is more than double the average *annual* rainfall over the United States!) In the United States, several regions are especially prone to

heavy rainfall events: the southeastern states, vulnerable to storms from the Gulf of Mexico; the western coastal states, subject to prolonged storms from the Pacific Ocean; and the midcontinent states, where hot moist air from the Gulf of Mexico collides with cold air sweeping down from Canada. Streams that drain the Rocky Mountains are particularly likely to flood during snowmelt events, especially when rapid spring thawing follows a winter of unusually heavy snow.

A factor that may moderate the risk of flooding is the rate of infiltration, which in turn is controlled by the soil type and how much soil is exposed. Soils, like rocks, vary in porosity and permeability. A very porous and permeable soil can allow a great deal of water to sink in relatively rapidly. If the soil is less permeable, the proportion of the water that runs off over the surface is greater. No matter how permeable the soil, of course, once it is saturated with water and all the pore space is filled (as, for instance, by previous storms), any additional moisture is necessarily forced to become part of the surface runoff. Topography also influences the extent of surface runoff: the steeper the terrain, the more readily water runs off over the surface, and the less it tends to sink into the soil.

Water that infiltrates into the soil, like surface runoff, also tends to flow downhill, and may in time reach the stream anyway. However, the underground runoff water, flowing through soil or rock, generally moves much more slowly than the surface runoff. The more gradually the water reaches the stream, the better the chances that the stream discharge will be adequate to carry the water away without flooding. Therefore the relative amounts of surface and subsurface runoff, which are strongly influenced by the near-surface geology of the drainage basin, are fundamental factors affecting the severity of stream flooding.

The presence of vegetation may decrease flood hazards somewhat: by providing a physical barrier to surface runoff, slowing it down; through plants' root action, keeping the soil looser and more permeable, increasing infiltration and thus decreasing surface runoff; by soaking up some of the water, later releasing it by evapotranspiration from foliage. Secondly, vegetation can be critical to preventing soil erosion. When the vegetation is removed and erosion increased, much more soil may be washed into a stream. There it can begin to fill in the channel, decreasing its volume and thus reducing the ability of the stream to carry water away quickly without flooding.

Parameters for Describing Flooding

During a flood, not only does stream stage increase; velocity and discharge increase too, as a greater mass of water is pulled downstream by gravity. The magnitude of the flood can be described either by the maximum discharge measured or by the maximum stage. The stream is said to **crest** when the maximum stage is reached. This may occur within minutes of the influx of water, as with flooding just below a failed dam. However, in places far downstream from the water input, or when surface runoff has been delayed, the stream may not crest for several days after the flooding episode begins. In other words, the worst is not necessarily over just because the rain has stopped.

Flooding may affect only a few kilometers along a small stream, or a region the size of the whole Mississippi River drainage basin, depending on how widely distributed the excess water is. Floods that affect only small, localized areas (or streams draining small basins) are sometimes termed **upstream floods.**

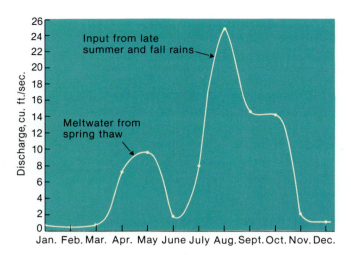

Figure 13.25

Sample hydrograph for Horse Creek near Sugar City, Colorado, spanning a one-year period.

Source: U.S. Geological Survey Open-file Report 79–681.

These are most often caused in nature by sudden, localized, intense rainstorms, and artificially by events like dam failure. Even if the total amount of water involved is moderate, the rapidity with which it is poured into the stream can temporarily exceed the stream channel capacity. The resultant flood is typically brief, though it can also be severe. Because there is no excess of water farther downstream in such a case, there is unfilled channel capacity below to accommodate the extra water. That water can thus be carried rapidly down from the upstream area, quickly reducing the stream's stage. Floods that affect large stream systems and large drainage basins are termed **downstream floods.** These more often result from prolonged heavy rains over a broad area, or extensive regional snowmelt. In such cases the whole stream system is choked with excess water, so these floods are generally of longer duration than upstream floods.

Fluctuations in stream stage or discharge through time can be plotted on a **hydrograph** (figure 13.25). Hydrographs spanning long periods of time are very useful in constructing a picture of the normal behavior of a stream, and of that stream's response to flood-causing events. Logically enough, a flood shows as a peak on the hydrograph. The height and width of that peak, and its position in time relative to the water-input event(s), will depend in part on where the measurements are being taken relative to where the excess water is coming into the system. Upstream, where the drainage basin is smaller and the water need not travel so far to reach the stream, the peak is likely to be sharper—higher crest, more rapid rise and fall of the water level—and to occur sooner after the influx of water. Downstream, in a larger drainage basin where

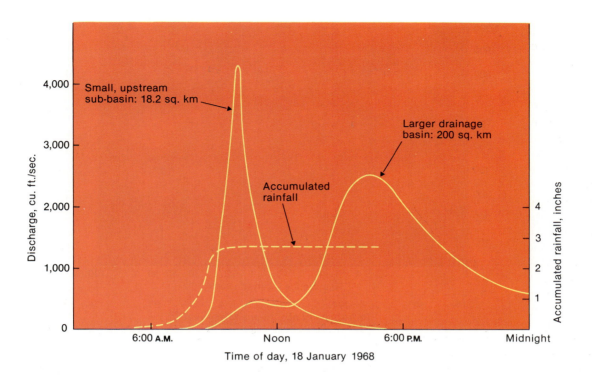

The accumulated rainfall curve and discharge curves for a small upstream sub-basin (18.2 sq. km) and a larger drainage basin (200 sq. km), plotted against time of day on 18 January 1968.

Figure 13.26

Flood hydrographs for two points along Calaveras Creek near Elmendorf, Texas. The flood has been caused by heavy rainfall. Upstream, flooding quickly follows rain. Downstream in the basin, response to water input is more sluggish.

Source: U.S. Geological Survey Water Resources Division.

some of the water must travel a considerable distance to the stream, the arrival of the water spans a longer time. The peak will be spread out, so that the hydrograph will show both a later and a broader, gentler peak (see figure 13.26). Similarly, in the case of an upstream flood, measurements made near the point where the excess water is coming into the system will show an earlier, sharper peak. By the time that water pulse has moved downstream to a lower point in the drainage basin, it will also have dispersed somewhat, so the peak on the hydrograph will again be later, lower, and of longer duration. Of course, a short event like a severe cloudburst will tend to produce a sharper peak than a more prolonged event like several days of steady rain or snowmelt, even if the same amount of water is involved.

Flood-Frequency Curves

Another way of looking at flooding is in terms of the frequency or probability of flood events of differing severity. The availability of long-term records makes it possible to construct a curve of discharge against recurrence interval, or probability, for a particular stream or section of one (figure 13.27). The sample flood-frequency curve shown indicates that, on average, an event producing a discharge of 675 cubic feet per second would occur once every 10 years, an event with discharge of 350 cubic feet/second once every 2 years, and so on. Thus a given flood event can be described by its **recurrence interval** (how frequently a flood of that severity would occur, on average, for that stream). For the stream of figure 13.27, a flood with discharge of 675 cubic feet/second would be called a *10-year flood,* meaning that a flood of that size would occur about once every 10 years; a discharge of 900 cubic feet/second would correspond to a *40-year flood,* and so on. An alternative way of looking at flooding is in terms of probability, which is inversely related to recurrence interval. That is, a 10-year flood has a 1/10 (or 10%) probability of occurrence in any one year; a 2-year flood has a 1/2 (50%) probability in any year, and so on.

Flood-frequency curves can be extremely useful in assessing flood hazards in a region. If one can estimate the severity of the 100-year or 200-year flood in terms of discharge, then even if such a rare and serious event has not occurred within memory, scientists and planners can, with the aid of topographic maps, project how much of the region would be flooded in such events. This in turn is useful in siting new construction projects so as to minimize the risk of flood damage, or in making property owners in threatened areas aware of dangers.

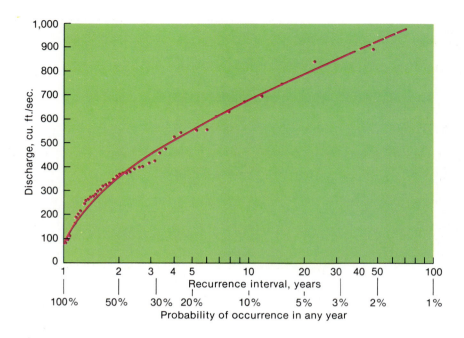

Figure 13.27

Sample flood-frequency curve for the Eagle River at Red Cliff, Colorado. Records span 46 years, so fairly accurate assessments can be made of the likelihood of moderate flood events.

Source: U.S. Geological Survey Open-file Report 79–1060.

Unfortunately, the best constrained part of the curve is by definition that for the lower discharge, more frequent, less serious floods. Often, considerable guesswork is involved in projecting the shape of the curve at the high-discharge end. Much of the United States has been settled for only a century or less. Reliable records of stream stages, discharges, and the extent of past floods typically extend back only a few decades. Many areas, then, may never have recorded a 50- or 100-year flood. Moreover, when the odd severe flood does occur, how does one know whether it was a 60-year flood, a 110-year flood, or whatever? The high-discharge events are rare, and even when they happen their recurrence interval can often only be estimated by educated guess. Considerable uncertainty can exist, then, about just how serious a particular stream's 100-year or 200-year flood might be. This problem is explored further in box 13.1.

Another complication is that streams in areas where people have settled in large numbers are in turn being affected by human activities. The way a stream responded to 10 centimeters of rain from a thunderstorm a hundred years ago may be quite different from the way it responds today. The flood-frequency curves are therefore changing with time. Except for measures specifically designed for flood control, most human activities have tended to aggravate flood hazards. What may have been a 100-year flood two centuries ago might be a 50-year flood now, or even a more frequent one.

Finally, it should be borne in mind that the recurrence intervals assigned to floods are *averages*. Over many centuries a 50-year flood should occur an average of once every 50 years; or, in other words, there is one chance in 50 that that discharge will be exceeded in any one year. However, that does not mean that two 50-year floods could not occur in successive years, or even in the same year. It would therefore be foolish to assume that just because a very severe flood has recently occurred in an area, it is in any sense safe for awhile. Statistically it is most likely that another such severe flood will not happen again for some time, but there is no guarantee.

The probability terminology may be less misleading to the nonspecialist, for it carries no implication of a long time interval between major floods. However, the recurrence-interval terminology is more commonly used by many public agencies dealing with flood hazards, such as the U.S. Army Corps of Engineers.

Box 13.1

How Big is the 100-Year Flood?

The difficulty of knowing the true recurrence intervals of floods of various magnitudes can be illustrated by an example that demonstrates the need for long-term records.

The usual way of estimating the recurrence interval of a flood of given size is as follows: Suppose that records of maximum discharge (or stage) reached by a particular stream each year have been kept for N years. Each of these yearly maxima can be given a rank M, ranging from 1 to N, 1 being the largest, N the smallest. Then the recurrence interval R of a given annual maximum is defined by $R = (N+1)/M$. For example, in table 13.1 are tabulated the maximum one-day mean discharges of the Big Thompson River, as measured near Estes Park, Colorado, for 25 consecutive years, 1951–75. If these values are ranked, 1 to 25, the 1971 maximum of 1030 cubic feet/second is the seventh largest and therefore has an estimated recurrence interval of $(25+1)/7 = 3.71$ years.

Suppose, however, that only 10 years of records were available, for 1966–75. The same 1971 flood happened to be the largest in that period of record. On the basis of the shorter record, its estimated recurrence interval would be $(10+1)/1 = 11$ years. Alternatively, looking only at the first ten years of record, 1951–60, we would estimate the recurrence interval of the 1958 maximum discharge of 1040 cubic feet/second, a flood of nearly the same size, to be 2.2 years. Which is right? Perhaps none of these estimates, but the very fact of their differences illustrates the need for long-term records to smooth out short-term anomalies in streamflow patterns.

Table 13.1

Calculated recurrence intervals for discharges of Big Thompson River at Estes Park, Colorado

Year	Maximum mean 1-day discharge (cu ft/sec)	For 25-year record		For 10-year record	
		Rank	R (years)	Rank	R (years)
1951	1220	4	6.50	3	3.67
1952	1310	3	8.67	2	5.50
1953	1150	5	5.20	4	2.75
1954	346	25	1.04	10	1.10
1955	470	23	1.13	9	1.22
1956	830	13	2.00	6	1.83
1957	1440	2	13.0	1	11.0
1958	1040	6	4.33	5	2.20
1959	816	14	1.86	7	1.57
1960	769	17	1.53	8	1.38
1961	836	12	2.17		
1962	709	19	1.37		
1963	692	21	1.23		
1964	481	22	1.18		
1965	1520	1	26.0		
1966	368	24	1.08	10	1.10
1967	698	20	1.30	9	1.22
1968	764	18	1.44	8	1.38
1969	878	10	2.60	4	2.75
1970	950	9	2.89	3	3.67
1971	1030	7	3.71	1	11.0
1972	857	11	2.36	5	2.20
1973	1020	8	3.25	2	5.50
1974	796	15	1.73	6	1.83
1975	793	16	1.62	7	1.57

Source: Data from U.S. Geological Survey Open-file Report 79–681.

Box 13.1 continued on next page

Box 13.1 *Continued*

The point is further illustrated in figure 1. It is rare to have 100 years or more of records for a given stream, so the magnitude of 50-year, 100-year, or larger floods is commonly estimated from a curve of discharge against recurrence interval, determined as above. Curves *A* and *B* in the figure are based, respectively, on the first and last 10 years of data for the Big Thompson River (last two columns of table 13.1). These two data sets would give estimates for the size of the 100-year flood that would differ by more than 50%. These results would in turn both differ from an estimate based on the full 25 years of data (curve *C*). The figure is a graphic illustration of how important long-term records can be in the projection of recurrence intervals of larger flood events.

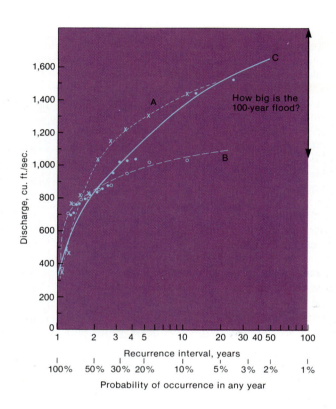

Figure 1
Short-term records (curves A and B) can be misleading about the true recurrence intervals or probabilities of floods of different magnitudes. Compare with curve C, based on a 25-year record.

Flooding and Human Activities

People live in floodplains for many reasons. One may be simple ignorance of the extent of the flood hazard. Someone living in a stream's 100-year or 200-year floodplain may be unaware that the stream could ever rise that high, if historical flooding has all been less severe. In mountainous areas, floodplains may be the only flat or nearly flat land on which to build, and construction is generally far easier and cheaper on nearly level land than on steep slopes. Around a major river like the Mississippi, the 100- or 200-year floodplain may include most of the land for miles around, and it may be impractical to leave that much real estate entirely vacant. Farmers have settled in floodplains since ancient times, because the fine alluvial sediment deposited by flooding streams replenishes nutrients in the soil, making it more fertile. Where rivers are used for transportation, cities may have been built deliberately as close to the water as possible. And, of course, many streams are very scenic features. Obviously, the more people settle and build in floodplains, the more damage flooding will do. What people often fail to realize is that floodplain development can in turn increase the likelihood or severity of flooding.

Figure 13.28

Filling in floodplain land and building structures can result in higher flood stages for the same discharge in later flood events.

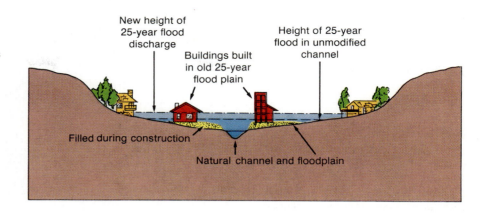

New height of 25-year flood discharge

Buildings built in old 25-year flood plain

Height of 25-year flood in unmodified channel

Filled during construction

Natural channel and floodplain

Effects of Development on Flood Hazards

One factor affecting flood severity is the proportion of surface runoff. The materials extensively used to cover the ground when cities are built, such as asphalt and concrete, are relatively impermeable. Therefore when a considerable area is covered by these materials, surface runoff tends to be much more concentrated and rapid than before, increasing the risk of flooding. The very fact of putting buildings in a floodplain can increase flood heights: the buildings occupy volume that water formerly could fill, and a given discharge will then correspond to a higher stage. Filling in floodplain land for construction similarly decreases the volume available to stream water and aggravates the situation further (figure 13.28). City storm sewers are installed to keep water from building up on streets during heavy rains, and often the storm water is channeled straight into a nearby stream. This works fine if the total flow is moderate enough, but by decreasing the time taken by the water to reach the stream channel, such measures increase the probability of flooding.

Strategies for Reducing Flood Damage

A basic strategy for reducing the risk of flood damage is to identify, as accurately as possible, the area at risk. Careful mapping coupled with accurate stream discharge data should allow identification of those areas threatened by floods of different recurrence intervals. Land that could be inundated often—by the 25-year floods, perhaps—might best be restricted to land uses not involving much building. Such land might be used as grazing pasture for livestock or for parks or other recreational purposes. Where land is at a premium and there is pressure to build even in the floodplain (as in many urban areas), buildings can be raised on stilts so

that the lowest inhabited floor is raised above the expected 200-year flood stage. This at least minimizes both the danger of damage to the building and its contents and the interference by the structure with flood flow. Another possibility would be to provide compensating capacity elsewhere for every loss of water capacity caused by filling in floodplain land. A major limitation of any floodplain zoning plan, whether it involves prescribing design features for buildings or banning them entirely, is that it almost always applies only to new construction, not to scores of older structures already at risk.

If there is open land available, flood hazards along a stream may be greatly reduced by the use of **retention ponds.** These are large basins that catch some of the surface runoff, keeping it from flowing immediately into the stream. They may be elaborate artificial structures, old abandoned quarries, or, in the simplest cases, fields dammed up by dikes of piled-up soil. Often the land can still be used for other purposes, such as farming, except on those rare occasions of heavy runoff when it is needed as a retention pond. Use of retention ponds is frequently a relatively inexpensive option, and has the advantage that it does not attempt to alter the character of the stream itself.

Channelization is a general term for various kinds of modifications of the stream channel, usually intended to increase the velocity of water flow, or the volume of the channel, or both. This in turn increases the discharge of the stream and hence the rate at which surplus water is carried away. The channel might be widened or deepened, especially where soil erosion and subsequent sediment deposition in the stream had partially filled in the channel. Alternatively, a stream channel might be rerouted—for example, by cutting off meanders deliberately to provide a more direct path for water flow. Such measures indeed tend to decrease the flood hazard upstream from where they are carried out. But a meandering stream often tends to keep meandering, or to revert to old meanders. Channelization is not a one-time effort. Constant maintenance is required to limit channel bank erosion and to keep the river in the redirected channel. (In some urban areas, it has been found necessary to line the new channel with concrete or other erosion-resistant materials.) Also, it has sometimes been found that by causing more water to flow downstream faster, channelization in one spot has increased flood hazards downstream.

This last complication is also often true with the building of artificial levees to raise the height of the stream banks so that the water can rise higher without flooding the surrounding country. This is a very ancient technique, one practiced thousands of years ago on the Nile by the Egyptian Pharaohs. Confining the water to the channel rather than allowing it to flow out into the floodplain effectively shunts it downstream faster during high-discharge events, increasing flood risks downstream. It artificially raises the stage for a given discharge, which can increase the risks upstream too. Building levees may make people feel so safe living in the floodplain that development will be far more extensive than if the stream were allowed to flood naturally from time to time. If the levees have not, in fact, been built high enough, and a very severe flood overtops them, or if they simply fail and are

Figure 13.29

When insufficiently high levees are overtopped, water is trapped behind them. Several days after this flood on the Kishwaukee River in DeKalb, Illinois, the stage of the river (right) had dropped considerably. Water level behind the levee (left) remained much higher, prolonging flooding and increasing flood damage.

breached during a high-discharge event, far more lives and property may be lost as a result. Also, if the levees are overtopped, they then trap water *outside* the stream channel, where it may stand for some time after the stream stage subsides, until infiltration returns it to the stream (figure 13.29).

Box 13.2

Geologic Hazards and Public Policy: Flood Insurance

When natural disasters occur, federal disaster-relief funds are often a part of the rescue, recovery, and rebuilding operations. In other words, all taxpayers pay for the damage suffered by those who, through ignorance or by choice, have been living in areas of high geologic risk. The perceived injustice of this was part of the motivation behind legislation establishing national flood insurance in the late 1960s and early 1970s. The measures provide federally subsidized flood insurance for property owners in identified flood-hazard areas, whether stream floodplains or coastal zones. Those most at risk, then, are paying for

insurance against their possible flood losses. The idea was that eventually flood insurance would replace after-the-fact disaster relief.

The law does *not* ban further development in floodplains, and generally does not provide for purchase and abandonment of properties damaged by flooding, unless structural damage is nearly total. An unintended consequence of the availability of flood insurance has been to encourage people to build and rebuild, sometimes several times, in severely flood-prone areas. Maintenance and even expansion of development in areas at risk has resulted.

In practice, the costs to the government to pay claims arising in high-risk areas can far exceed the premiums paid. Thus to an extent the many are still paying for the choices of the few to live at risk, and there is little incentive *not* to build in flood-prone areas. Dissatisfaction with the effects of the program has led Congress to withdraw flood insurance availability from some particularly unstable coastal zones, and to consider making insurance unavailable for new development elsewhere. The dilemma lies in deciding the extent to which to hold individuals responsible for making sound, informed decisions about where to live.

Yet another approach to moderating streamflow in such a way as to prevent or minimize flooding is through the construction of flood-control dams at one or more points along the stream. Excess water is held in the reservoir formed upstream from such a dam, and may then be released at a controlled rate so as not to overwhelm the capacity of the channel beyond. However, we have already noted the effects of such an artificial change in local stream base level. If the stream normally carries a high sediment load, silting up of the reservoir will force repeated dredging, which can be expensive and presents the problem of where to dump the dredged sediment. Sometimes the reduction in sediment carried to the stream mouth is also a problem there: so many dams along the Mississippi

River system have impounded sediment that, overall, its delta is no longer building up, and is eroding in many places. Some large reservoirs, such as Lake Mead behind Hoover Dam, have even been found to cause earthquakes. The water in the reservoir represents an added load on the rocks, increasing the stresses on them, while infiltration of water into the ground under the reservoir increases pore fluid pressures in rocks along old faults. Since Hoover Dam was built, at least 10,000 small earthquakes have occurred in the vicinity. Earthquakes caused in this way are usually low magnitude, but their foci are naturally also close to the dam. This raises concerns about the possibility of catastrophic dam failure caused, in effect, by the dam itself.

Summary

A stream is any body of flowing water usually confined within a channel. The size of a stream, as measured by its discharge, is a function of climate, near-surface geology, and the size of the drainage basin from which the water is drawn. The force of the flowing water may cause both downcutting and lateral erosion of banks, including meandering. Meander formation and migration, coupled with deposition of alluvial sediments in and around the stream, contribute to the creation of floodplains. Material is moved by streams in solution, in suspension, or along the streambed. The faster the water flow, the coarser the particles moved and the greater the capacity of the stream to transport sediment. Alluvium deposited where water velocity decreases may take the form of bars or islands in the channel, overbank deposits (including natural levees) in the floodplain, or alluvial fans and deltas at stream mouths. Over time, most streams tend toward an equilibrium longitudinal profile, with gradient decreasing from source to mouth. When sediment influx and outflow are in balance, the result is a graded stream.

Floods are the stream's response to an input of water too large and/or too rapid for the discharge to be accommodated within the channel. The risk of flooding is a function of climate, soil character, presence or absence of vegetation, and topography. Human activities may unintentionally increase flood hazards by filling in the channel or floodplain during construction, covering soil with impermeable materials, or draining runoff to streams quickly via sewers. Use of flood-frequency records may help to identify the areas most at risk, but the changes wrought recently by human activities limit the usefulness of historic records. Strategies used to reduce flood hazards include restrictive zoning, building of retention ponds, artificial levees, flood-control dams, and various kinds of channelization, but these practices are not without some disadvantages or even risks.

Terms to Remember

alluvial fan	dissolved load	infiltration	recurrence interval
alluvium	divide	knickpoint	retention pond
bankfull stage	downcutting	levees	saltation
base level	downstream floods	load	source
bed load	drainage basin	longitudinal profile	stage
braided stream	dynamic equilibrium	meanders	stream
capacity	floodplain	mouth	stream piracy
channelization	flood stage	order	suspended load
competence	graded stream	overbank deposit	terraces
crest	gradient	oxbows	trellis drainage
delta	headward erosion	point bar	tributary stream
dendritic drainage	hydrograph	radial drainage	upstream floods
discharge	incised meanders	rectangular drainage	

Questions for Review

1. What is a stream? A drainage basin?

2. What determines the order of a stream?

3. Describe the kind of geologic situation in which you might expect each of the following to develop: (a) rectangular drainage; (b) radial drainage.

4. What is a stream's discharge, and how is it related to stream capacity and load?

5. Briefly describe the development and migration of meanders. How are meanders and oxbow lakes related?

6. Under what circumstances does stream piracy occur?

7. What is a braided stream, and how does braiding develop?

8. Why does a stream commonly deposit a delta at its mouth?

9. Sketch the equilibrium longitudinal profile of a graded stream, indicating base level. How is the profile changed by construction of a dam along the stream?

10. What is a flood? Outline briefly how flood hazards are affected by each of the following: (a) intensity of precipitation; (b) soil type; (c) presence of vegetation; (d) construction that adds extensive impermeable cover.

11. Compare and contrast upstream and downstream floods in terms of the causes, intensity, and duration of flooding.

12. What is a flood-frequency curve? Is the term accurate?

13. Cite and briefly explain three strategies for reducing flood hazards, noting the strengths and weaknesses of each.

For Further Thought

1. Make a point of visiting a nearby stream. Walk along the channel, noting meandering, areas where the bank is eroding or point bars are being deposited, and any evidence of human modification of the channel.

2. Investigate the availability of flood-hazard maps for a nearby stream. How current are the maps and the data upon which they are based? Have any efforts been made to reduce the flood hazards? If so, what have they been, what have they cost, and what negative effects, if any, have they had?

Suggestions for Further Reading

Bolt, B. A.; Horn, W. L.; Macdonald, G. A.; and Scott, R. F. *Geological Hazards.* New York: Springer-Verlag, 1975. Chapter 7 provides a good survey of causes of floods and discussion of flood-hazard mitigation.

Chin, E. H.; Skelton, J.; and Guy, H. P. *The 1975 Mississippi River Basin Flood.* U.S. Geological Survey Professional Paper 937, 1975.

Knighton, D. *Fluvial Forms and Processes.* London: Edward Arnold, 1984.

Leopold, L. B.; Wolman, M. G.; and Miller, J. P. *Fluvial Processes in Geomorphology.* San Francisco: W. H. Freeman and Co., 1964.

Morisawa, M. *Streams, Their Dynamics and Morphology.* New York: McGraw-Hill Book Co., 1968.

Richards, K. *Rivers.* New York: Methuen, 1982.

Ritter, D. F. *Process Geomorphology.* 2d ed. Dubuque, Iowa: Wm. C. Brown Publishers, 1985.

Tank, R. W. *Environmental Geology, Text and Readings.* 3d ed. New York: Oxford University Press, 1983. The section on floods, pages 218–77, includes several readings on responses to flood hazards.

Chapter 14

Groundwater and Water Resources

Introduction

Streams and oceans are water in its most visible form. An increasing proportion of the water we actually *use* comes from below the surface, from groundwater supplies. This chapter begins with an examination of the relationship between surface and subsurface waters, and the global water budget. The occurrence of groundwater is then examined in greater detail. The consequences of the presence and the consumption of subsurface water are reviewed, and the problem of groundwater pollution is briefly outlined. The chapter concludes with a survey of the water-supply situation in the United States, including some options for extending freshwater supplies.

The Hydrologic Cycle; The Global Water Budget

The **hydrosphere** includes all the water at and near the surface of the earth. Most of it is believed to have been outgassed from the earth's interior early in its history, when earth's temperature was higher. Now, except for occasional minor contributions from volcanoes bringing additional water up from the mantle, the quantity of water in the hydrosphere remains essentially constant.

Table 14.1 shows how that water is distributed. Several points emerge immediately from these data. One important point is that there is, relatively speaking, very little fresh liquid water on the earth. By far the largest reservoir is the very salty ocean. Most of the fresh water is locked up as ice, mainly in the large polar ice caps. Even the groundwater beneath the surface of the continents is not all fresh, although it is so classified in the table for lack of analytical data on all groundwater. These facts underscore the need for restraint in the use of fresh water.

The Hydrologic Cycle

All of the water in the hydrosphere is caught up in the **hydrologic cycle,** illustrated in figure 14.1. The main processes in the cycle are evaporation into and precipitation out of the atmosphere. Precipitation onto land areas will either be re-evaporated (directly from the ground surface, or indirectly through plants by evapotranspiration), infiltrate into the ground, or run off over the ground surface. Both surface and subsurface runoff act to return water to streams and, usually, ultimately to the oceans. The ocean is the principal

Table 14.1

The water in the hydrosphere

Reservoir	Percent of total water*	Percent of fresh water†	Percent of unfrozen fresh water†
oceans	97.54	—	—
ice	1.81	73.9	—
groundwater	0.63	25.7	98.4
lakes and streams			
salt	0.007		
fresh	0.009	0.36	1.4
atmosphere	0.001	0.04	0.2

*These figures account for over 99.9% of the water. Some water is also held in organisms (the biosphere) and some as soil moisture.
†This assumes that all groundwater is more or less fresh water, as it is not all readily accessible to be tested and classified.

From Mather, J. R., *Water Resources*. Copyright © 1984 by John Wiley and Sons, Inc. Reprinted by permission of John Wiley and Sons, Inc.

source of evaporated water, given its vast area of exposed water surface. The total amount of water moving through the hydrologic cycle is large, more than 100 million billion gallons per year. A portion of this water is diverted for human use, but eventually it makes its way back into circulation in the global water cycle by a variety of routes, including release of municipal sewage, evaporation from irrigated fields, or discharge of industrial wastewater into streams. It should be noted that water in the hydrosphere may spend extended periods of time—even tens of thousands of years—in storage in one or another of the water reservoirs. Still, from the longer perspective of geologic history, such water can be regarded as moving through the hydrologic cycle.

Subsurface Water

If soil on which precipitation falls is sufficiently permeable, infiltration will occur. Gravity will continue to draw the water downward until an impermeable rock or soil layer is reached, and the water begins to accumulate above it. Immediately above the impermeable material there will be a zone of rock or soil that is water saturated (in which water fills all the accessible pore space). This is the **phreatic zone.** Above that is rock or soil in which the pore spaces are filled partly with water, partly with air (**vadose zone**). All of the water occupying pore space below the ground

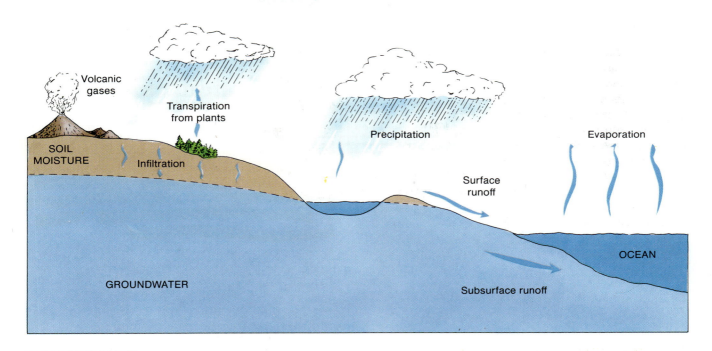

Principal processes in the hydrologic cycle.

surface is, logically, termed **subsurface water.** True **groundwater** is the water in the zone of saturation, or phreatic zone, only. It is distinguished from **soil moisture,** water held in small pores or on grain surfaces in unsaturated soil. The **water table** is defined as the top of the zone of saturation, where the saturated zone is not confined by overlying impermeable rocks (see the section on aquifer geometry later in this chapter). These relationships are illustrated in figure 14.2. Keep in mind that in nature, then, precipitation is the ultimate source of subsurface water.

Groundwater

The water table is not always below the ground surface. Wherever surface water persists, as in a lake or stream, the water table is locally above the ground surface, and the water's surface is the water table. Whether the groundwater feeds the stream or the stream replenishes the groundwater depends on geology and climate. With any perennial stream can be associated a certain level of **baseflow,** supported by groundwater from adjacent rock and soil, and distinguished from the direct surface runoff associated with precipitation or melting events (see figure 14.3). **Springs,** places where water flows out spontaneously onto the earth's surface, are also places where, locally, the water table intersects the surface.

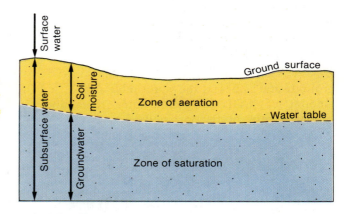

Nomenclature for surface and subsurface waters. *Groundwater* is the water in the zone of saturation, below the water table.

The water table is not necessarily flat like a tabletop, but may undulate with the surface topography and with the distribution of permeable and impermeable rocks underground. The depth to the water table varies, too. The water table is highest when the ratio of input water to water removed is greatest, typically in spring when rain is heavy or snow and ice accumulations melt. In dry seasons, or when human use of

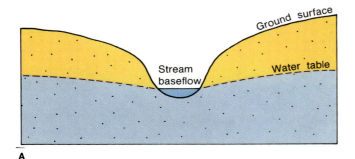

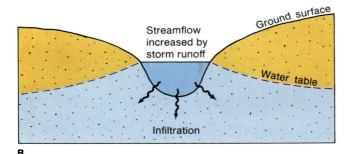

Figure 14.3

The amount of baseflow in a stream is dependent on the height of the regional water table. Baseflow is distinguished from storm runoff. *(A)* In dry conditions, flow is entirely baseflow supported by groundwater. (If the water table is too low, the stream dries up entirely.) *(B)* During storms or melting episodes, surface runoff contributes to streamflow.

groundwater is heavy, the water table drops, and the amount of available groundwater remaining decreases. Groundwater does not move only vertically. It can also flow laterally through permeable soil and rock, from higher elevations to lower, from areas of abundant infiltration to drier ones, or from areas of little groundwater use toward areas of heavy use from which water is being extracted. The processes of infiltration and migration through which groundwater is replaced are collectively termed **recharge.**

Aquifers

As noted in chapter 2, rocks and soils vary greatly in porosity and permeability. If groundwater drawn from a well is to be used as a source of water supply, the porosity and permeability of the surrounding rocks are critical. The porosity controls the total amount of water available. In most places there are not, after all, vast pools of open water underground to be tapped, just the water in the pore spaces in the rocks. This pore water will at most amount to a few percent of the rocks' volume, and usually much less. The permeability will govern both the rate at which water can

be withdrawn, and the rate at which recharge can replenish it as it is used. All the pore water in the world would be of no practical use if it were so tightly locked in the rocks that it could not be pumped out. A rock that holds and transmits enough water to be useful as a source of water is an **aquifer.** Many of the best aquifers are sandstones, but any other type of rock may serve if it is sufficiently porous and permeable. An **aquitard** is a rock in which permeability is low and water flow is very much slower, so that it is not useful as a water source. Its extreme, an **aquiclude,** is a rock that is effectively impermeable on a human time scale, and thus acts as a barrier to water flow. Shales are common aquitards and aquicludes.

Surface Water versus Groundwater as Supply

Surface waters are much more accessible than subsurface waters. Why, then, use groundwater at all? One basic reason is that in many dry areas there is little or no surface water available, while there may be quite a lot of water deep underground. Tapping the latter supply allows people to live and farm in areas that would otherwise be quite uninhabitable. Then too, streamflow varies seasonally. During dry seasons, the water supply may be inadequate. Building dams and reservoirs allows a reserve of water to be accumulated during the wet seasons to be drawn on in dry times. However, we saw some of the negative consequences of reservoir construction in the last chapter. Furthermore, if a region is so dry at some times that a reservoir is necessary, the rate of evaporation of water from the broad, still surface of the reservoir may itself represent a considerable water loss, and aggravate the water-supply problem.

Precipitation (the primary source of abundant surface runoff) varies widely geographically (figure 14.4); so does population density. In many areas, the concentration of people may far exceed what can be supported by available local surface waters, even at the wettest season. Also, streams and large lakes have historically been treated as disposal sites for wastewater and sewage, often unpurified. This makes the surface waters decidedly less appealing as drinking waters. Finally, groundwater is by far the largest reservoir of unfrozen fresh water (recall table 14.1).

For a variety of reasons, then, underground waters may be preferred as a supplementary or even sole water source. When subsurface waters are used, it is groundwater, from the saturated zone, that is tapped. (Trying to pump water out of the unsaturated zone would be a bit like trying to drink root beer by sipping at the foam in the top of the glass: it is not possible to draw up very much liquid that way.) Water

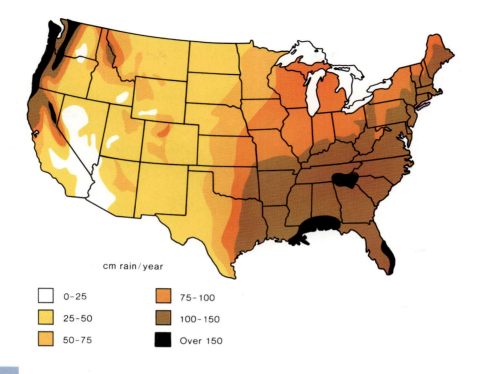

cm rain/year

☐ 0-25	☐ 75-100
☐ 25-50	☐ 100-150
☐ 50-75	■ Over 150

Figure 14.4

Distribution of average annual precipitation over the United States.

Source: U.S. Water Resources Council.

that has passed through the rock of an aquifer has also been naturally filtered to remove some impurities— soil or sediment particles, and even the larger bacteria—although it can still contain many dissolved chemicals. Sandy rocks and sediments can be particularly effective filters.

Aquifer Geometry and Its Consequences

The behavior of groundwater is controlled to some extent by the geology and geometry of the particular aquifer in which it is found.

Confined and Unconfined Aquifers

When the aquifer is overlain only by permeable rocks and soil, it is described as **unconfined** (figure 14.5). In an unconfined aquifer, the water is not under unusual pressure. In a well drilled into an unconfined aquifer, the water will rise to the same height as the water table in the adjacent aquifer rocks. To bring the water up to the ground surface, it must be actively pumped. An unconfined aquifer may be recharged by infiltration over the whole area underlain by that aquifer, as there is nothing to stop the downward flow of water from surface to aquifer.

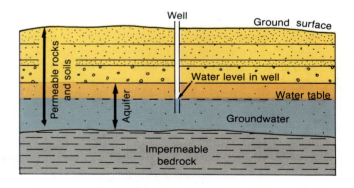

Figure 14.5

An unconfined aquifer system.

When the aquifer is overlain by an aquitard or aquiclude, the aquifer is described as **confined**. Because the vertical movement of water is restricted, water in a confined aquifer may be under considerable pressure from overlying rocks or as a consequence of lateral changes in aquifer elevation. At some places within the aquifer, the water level in the saturated zone may be well below the height to which it would rise if the water could flow freely (figure 14.6). If a well is drilled into a confined aquifer, the water can rise above its level in the aquifer because

Figure 14.6

Water in a confined aquifer, sandwiched between aquitards, builds up extra hydrostatic pressure and may exhibit artesian conditions.

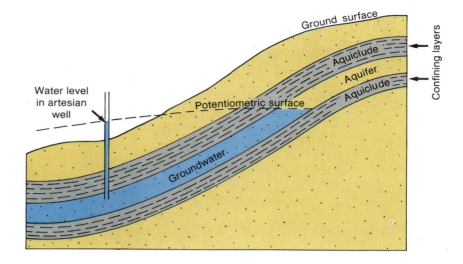

of this extra **hydrostatic** (fluid) pressure. This is an **artesian** system, one in which groundwater can rise above its (confined) aquifer under its own pressure. The water may or may not rise all the way to the ground surface; some pumping may still be necessary to bring it to the surface for use. Advertising claims notwithstanding, this is all there is to "artesian water." It is no different chemically, no purer or better tasting or more wholesome than any other groundwater; it is just under natural pressure.

> Artesian behavior can be demonstrated very simply using a water-filled length of rubber tubing or hose, held up at a steep angle. The water is confined by the hose, and the water in the low end is under pressure from the weight of the water above it. Poke a small hole in the top side of the lower end of the hose and a stream of water will shoot up to the level of the water at the upper end of the hose.

The Potentiometric Surface

In a natural aquifer system, some of the pressure is dissipated through friction between the water and the aquifer rocks through which it flows; the rate of water flow is slower than in the above demonstration, but the principle is similar. In the case of a confined aquifer, rather than describing the height of the water table, it is more meaningful to describe the height of the **potentiometric surface,** the height to which the water's pressure would raise it if it were unconfined. This level will be somewhat higher than the top of the confined aquifer where its rocks are saturated, and may be above the ground surface (see again figure 14.6). In an artesian well, the water will rise to the elevation of the potentiometric surface.

Perched Water Table

Local geologic conditions can be considerably more complex than the simple cases so far considered and may make it difficult to determine the availability of groundwater without thorough study. For example, locally occurring lenses or patches of impermeable rocks within otherwise permeable ones may result in a **perched water table** (figure 14.7). Immediately above the aquiclude will be a local saturated zone far above the true regional water table. Someone drilling a well above this area could be deceived about the apparent depth to the water table, and might find that the perched saturated zone contained very little total water. The quantity of water available there would also be especially sensitive to local precipitation levels and fluctuations. Its quality would be especially subject to degradation by pollution.

Further Effects of the Presence of Subsurface Water

An abundance of surface and near-surface water may indicate an ample water supply, but it also means more water available to dissolve rocks. Most rocks are not very soluble, so this is not a concern in all areas. A few rock types, however, are extremely soluble. The most common of these is limestone.

Caverns and Sinkholes

Over long periods of time, underground water may dissolve large volumes of limestone, creating and slowly enlarging underground caverns (figure 14.8). In the process, the support for the land above is slowly removed. There may be no obvious evidence at the surface of what is taking place, until the ground collapses abruptly into the void, **producing a sinkhole**

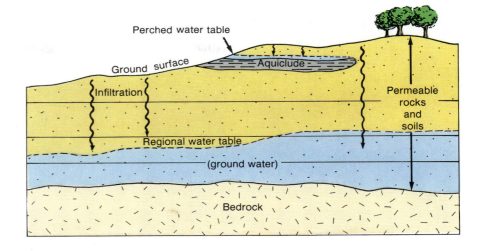

Figure 14.7

A perched water table, formed as infiltrating water is trapped by a patch of impermeable rock.

Figure 14.8

A cavern formed by solution of limestone.

Figure 14.9

Home lost to a sinkhole in Bartow, Florida. This sinkhole was over 150 meters (nearly 500 feet) long, 40 meters wide, and 20 meters deep.

Photo courtesy U.S. Geological Survey.

(figure 14.9). The collapse of a sinkhole may be triggered by a drop in the water table as a result of drought or water use, which leaves rocks and soil that were previously buoyed up by water pressure unsupported. Failure may also be caused by rapid input of large quantities of water (as from heavy rains), washing overlying soil down into the cavern, or by an increase in subsurface water flow rates. Sinkholes come in many sizes. The larger ones are quite capable of swallowing up many houses at a time; they may be over 50 meters deep and cover several tens of acres. A single sinkhole in a developed area can cause millions of dollars in property damage. Sudden sinkhole collapses beneath bridges, roads, and railways have caused accidents and deaths.

Karst Terrain

Sinkholes are rarely isolated phenomena. Limestone most often is formed from chemical sediments deposited in shallow seas, and limestone beds commonly cover broad areas. Therefore, where there is one sinkhole in a terrain underlain by limestone, there are likely to be others. Thousands have formed in Alabama alone since the year 1900. An abundance of sinkholes in an area is a strong indication that more can be expected. Areas with many sinkholes commonly are characterized by the presence of soluble

Telltale karst topography of many round lakes formed in sinkholes: satellite photograph of central Florida.

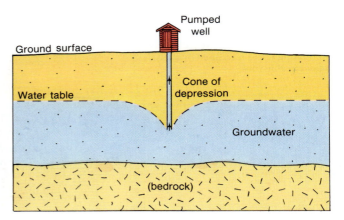

Cone of depression formed around a pumped well in an unconfined aquifer.

rocks close to the surface, and a high water table or abundant precipitation, leading to extensive solution of those rocks.

The situation may sometimes be easily recognized through aerial or satellite photography (figure 14.10), which can show a region to be pockmarked with the circular lakes or holes characteristic of the solution-dominated **karst topography.** Clearly in such an area it makes sense to investigate the subsurface situation before buying or building a home or business. Circular patterns of cracks on the ground or conical depressions in the ground surface may be early signs of trouble developing below.

Some Consequences of Groundwater Withdrawal

When groundwater must be pumped from an aquifer for use, the rate at which water flows in from surrounding rock to replace that which is extracted is generally slower than the rate at which water is taken out. In an unconfined aquifer, the result is a circular lowering of the water table immediately around the well, which is termed a **cone of depression** (figure 14.11). A similar feature is produced on the surface of a liquid in a drinking glass when it is sipped hard through a straw. With a great many closely spaced wells, the cones of depression of adjacent wells may overlap, further lowering the water table between wells. If over a period of time, groundwater withdrawal rates consistently exceed recharge rates, the regional water table may drop and continue to do so. A practical clue that this is happening is that wells throughout a region begin to need to be drilled deeper periodically in order to keep the water flowing (see box 14.1).

A dropping water table should be a warning sign, for the process of deepening wells to reach water cannot go on indefinitely. Groundwater does not exist all the way to the center of the earth, even if we could drill that far, or even through the whole crust. In most places it can be found at most a few kilometers into the crust. (In the deep crust and below, pressures on rocks are so great that compression and plastic flow close up any pores that groundwater might fill.) In many areas, impermeable rocks are reached at much shallower depths. Therefore there is a bottom to the groundwater supply, though we do not always know without drilling just where it is. When the water table is lowered to the base of the aquifer system, groundwater use there is terminated for some time. Groundwater flow rates are commonly very slow— millimeters to meters per year—and recharge of significant amounts of groundwater can thus require decades or centuries.

Cones of depression can also develop in potentiometric surfaces, and those surfaces can be lowered too. Again, this reflects groundwater withdrawal at a

The High Plains Aquifer System

The Ogallala Formation, a limestone aquifer, underlies most of Nebraska and sizeable portions of Colorado, Kansas, and the Texas and Oklahoma panhandles (figure 1). The area is one of the largest and most important agricultural regions in the United States. It accounts for about 25% of U.S. feed grain exports and 40% of exports of wheat, flour, and cotton. More than 14 million acres of land are irrigated with water pumped from the Ogallala. Yields on irrigated land may be triple the yields on similar land cultivated by "dry farming" (no irrigation).

The Ogallala's water was for the most part stored during the retreat of the Pleistocene continental ice sheets, described in chapter 16. Present recharge is negligible. The original groundwater reserve in the Ogallala is estimated to have been approximately 2 billion acre-feet, but each year farmers draw from it more water than the entire flow of the Colorado River. (One acre-foot is the amount of water needed to cover an area of one acre to a depth of one foot; it is over 300 billion gallons.) In 1930, the average thickness of the saturated zone of the Ogallala Formation was nearly 20 meters. Currently it is less than 3 meters, with the water table dropping by anywhere from 15 centimeters to 1 meter per year. Overall, it is believed that the Ogallala will be effectively depleted within four decades. In areas of especially rapid drawdown, it could be locally drained in less than a decade.

What then?

Reversion to dry farming, where possible at all, will greatly diminish yields. Reduced vigor of vegetation may lead to a partial return to pre-irrigation, Dust Bowl-type conditions (see chapter 17). Alternative local sources of municipal water frequently do not exist. Planners in Texas and Oklahoma have advanced ambitious

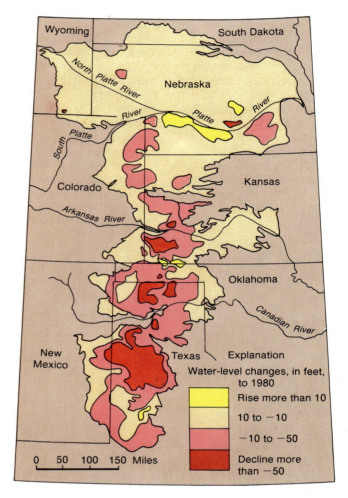

Figure 1

Map extent of the High Plains aquifer system.

After U.S. Geological Survey 1982 Annual Report.

interbasin water-transport schemes as solutions. The more recent of Texas's various proposed alternatives basically all involved transferring Mississippi River water from northeastern Texas across the state to the High Plains of the Panhandle. For various reasons, all these proposals have been abandoned. Oklahoma's Comprehensive Water Plan would draw on the Red River and Arkansas River basins. Each such scheme would cost billions of dollars,

perhaps tens of billions. Water-transport systems of this scale could take a decade or longer to complete, before which time acute water shortages can be expected in the areas of most urgent need. Even if and when the transport networks are finished, the cost of the water may be ten times what farmers in the region can comfortably afford to pay if their products are to remain competitive in the marketplace. Meanwhile, the draining of the Ogallala continues unabated day by day.

Box 14.2

The Great Depression—in Groundwater

We have noted already that in many dry areas, where groundwater use is heavy and withdrawal exceeds recharge, water tables have dropped considerably. What is more surprising is that the effects of excessive pumpage can be seen even in areas widely regarded as receiving adequate moisture. Consider, as an example, northern Illinois.

Sandstones of the Cambro-Ordovician aquifer system have long been used as the principal or sole source of municipal water in many parts of the area. When the height of

the potentiometric surface for the confined aquifer is mapped, as in figure 1, a dramatic drop is seen in the Chicago metropolitan area, from over 750 feet (250 meters) above sea level to, locally, more than 100 feet (30+ meters) below it.

What the contours really represent is a giant cone of depression in the potentiometric surface. It is due to yet another instance of groundwater withdrawal rates far exceeding recharge rates. Just over the last decade, water levels in some wells

have dropped more than 30 meters. Close to Lake Michigan, the situation can be alleviated by using proportionately more lake water. However, nearby communities and rural homesteads lacking ready access to the lake waters are finding it increasingly difficult and expensive to keep the groundwaters flowing. Even if all groundwater withdrawal ceased for a period of time—an unlikely possibility—centuries of recharge at least would be required to restore the water levels.

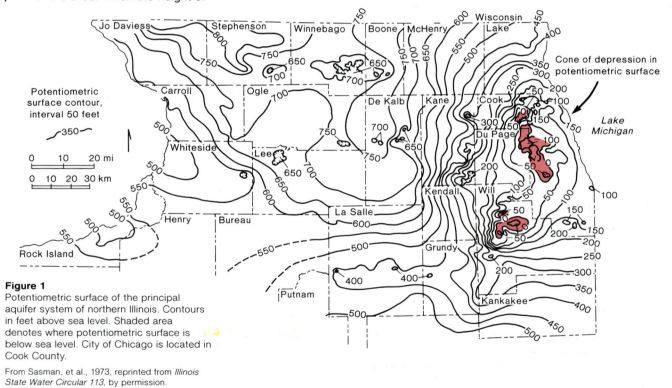

Figure 1
Potentiometric surface of the principal aquifer system of northern Illinois. Contours in feet above sea level. Shaded area denotes where potentiometric surface is below sea level. City of Chicago is located in Cook County.

From Sasman, et al., 1973, reprinted from *Illinois State Water Circular 113*, by permission.

rate exceeding the recharge rate, and gradual depletion of local groundwater. An example can be seen in box 14.2. Where groundwater is being depleted by too much withdrawal too fast, one can speak of "mining" groundwater. The idea is not that the water will never be recharged, but that the rate is so slow on the time scale of human affairs as to be insignificant. From the human point of view we may indeed use up groundwater in some heavy-use areas. Also, human activities in some places have reduced or halted natural recharge, so groundwater consumed may not be replaced, even slowly.

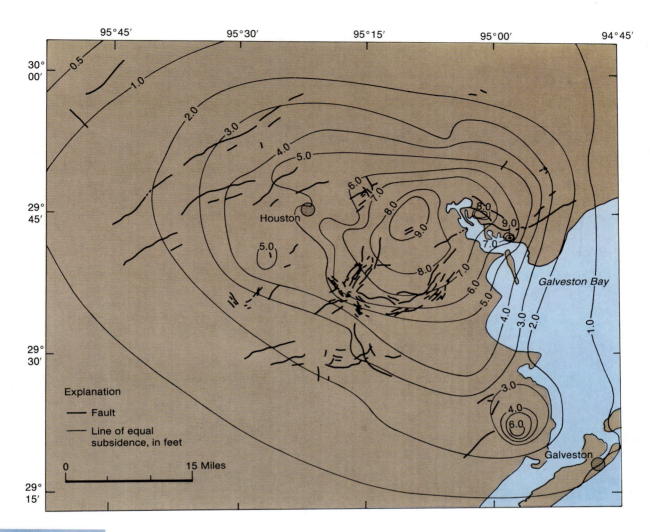

Figure 14.12

Surface subsidence near Galveston Bay since 1906.

Source: U.S. Geological Survey 1982 Annual Report.

Compaction and Surface Subsidence

One secondary effect of lowering the water table may be to contribute to sinkhole formation, as previously noted. Another possible result is that the aquifer rocks, no longer saturated with water, may become compacted from the weight of overlying rocks. This will decrease their porosity, so that their water-holding capacity is permanently reduced, and may also decrease their permeability. At the same time, as the rocks below compact and settle, the ground surface itself may subside. Where water use is heavy the surface subsidence may be several meters, as it has been, for example, in the Houston/Galveston area of Texas (figure 14.12).

At high elevations, or in inland areas, this subsidence causes only structural problems, by disrupting building foundations. In low-elevation coastal regions, the subsidence may lead to extensive flooding, as well as to increased rates of coastal erosion. The city of Venice, Italy, is in one such slowly sinking coastal area. Many of its historic architectural and artistic treasures are threatened by the combined effects of the gradual rise in worldwide sea levels, the tectonic sinking of the Adriatic coast, and surface subsidence from extensive groundwater withdrawal. Drastic and expensive engineering efforts will be needed to save them. In such areas, simple solutions (like pumping water back underground) are unlikely to work. The rocks may have been permanently compacted. Also, what water is to be used? If consumption

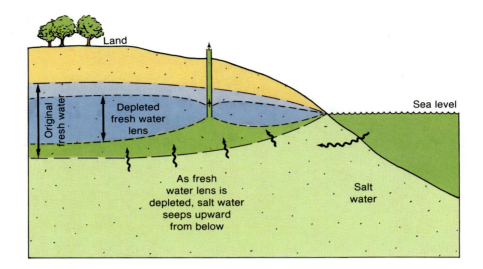

Figure 14.13

Saltwater intrusion and upconing as freshwater lens is depleted in a coastal zone.

of groundwater is heavy, it is probably because there simply is no great supply of fresh surface water. Pumping in salt water, in a coastal area, will in time make the rest of the water in the aquifer salty, too. And presumably a supply of fresh water is still needed for local use, so groundwater withdrawal cannot easily or conveniently be curtailed.

Saltwater Intrusion

Aside from the possibility of surface subsidence, a further problem arises from groundwater use in coastal regions particularly. This is the problem of saltwater intrusion (figure 14.13). When rain falls into the ocean, the fresh water promptly becomes mixed into the salt water. However, fresh water falling on land will not mix so readily with saline groundwater at depth, for water in pore spaces in rock or soil is not vigorously churned by currents or wave action. Fresh water is also less dense than salt water. So the fresh water accumulates in a lens that floats above the denser salt water. If water use approximately equals the rate of recharge, the freshwater lens stays about the same thickness.

However, if consumption of fresh groundwater is heavier, the freshwater lens thins, and the denser saline groundwater moves up to fill in pores emptied by removal of fresh water. Upconing of salt water below cones of depression in the freshwater lens may also occur. Wells that had been tapping the freshwater lens may begin pumping unwanted salt water instead, and the limited freshwater supply gradually de-

creases. Salt water intrusion destroyed useful aquifers beneath Brooklyn, New York, in the 1930s, and is a serious problem in many coastal areas of the south-eastern and Gulf coastal states and some densely populated parts of California.

Urbanization and Groundwater Supply

Obviously an increasing concentration of people means an increasing demand for water. We noted in the last chapter that urbanization may involve extensive modification of surface-water runoff patterns and stream channels. Insofar as it modifies surface runoff and the ratio of runoff to infiltration, urbanization also influences groundwater hydrology.

Loss of Recharge

Placing an expanse of impermeable cover in one part of a broad area underlain by an unconfined aquifer will have relatively little impact on the recharge to that aquifer. Infiltration will continue over most of the region. In the case of a confined aquifer, however, the available recharge area may be very limited. Above most of the aquifer, the overlying confining layer will prevent direct infiltration downward into the aquifer. If impermeable cover is built over the restricted recharge area, recharge can be considerably reduced, thus aggravating the water-supply situation (figure 14.14).

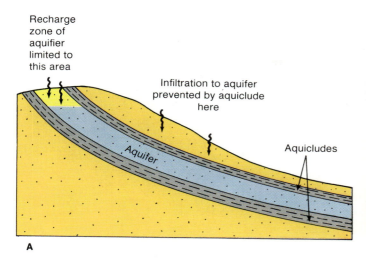

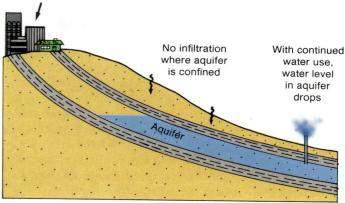

Figure 14.14

Recharge to a confined aquifer. *(A)* The recharge area of this confined aquifer is limited to the area where permeable rocks intersect the surface. *(B)* Construction over that area reduces recharge and, with continued use, leads to depletion of water in the aquifer.

Filling in wetlands is a common way to provide more land for construction. This too can interfere with recharge. Especially if surface runoff is rapid elsewhere in the area, a stagnant swamp, holding water for long periods, can be a major source of infiltration and recharge. Filling it in so water no longer accumulates there and, worse yet, topping the fill with impermeable cover, again may greatly reduce local recharge of groundwater.

Artificial Recharge

In steeply sloping areas or those with low-permeability soils, well-planned construction in the form of artificial recharge basins can aid in increasing groundwater recharge (figure 14.15). The basin acts similarly to a flood-control retention pond, in that it is designed to catch some of the surface runoff during high-runoff events (heavy rain or snowmelt). Trapping the water allows more time for slow infiltration, and thus more recharge in an area from which the fresh water might otherwise be quickly lost to streams and carried away. Recharge basins are a partial solution to the problem in areas where groundwater use exceeds natural recharge rates, but of course they are only effective where there is surface runoff to catch. Where natural infiltration is slow, or where the aquifer to be recharged is confined, artificial recharge may require the use of injection wells to pump water back into the aquifer.

Figure 14.15

This ballfield on Long Island doubles as an artificial recharge basin in wet weather.

Photo courtesy U.S. Geological Survey.

Water Quality

We noted earlier that most of the water in the hydrosphere is in the very salty oceans, while of the remaining few percent the major portion occurs as ice. That leaves relatively little surface or subsurface water as a potential source of freshwater supplies. Moreover, much of the water on and in the continents is not strictly fresh either. Even rainwater, long the standard for pure water, contains dissolved chemicals of various kinds, especially in industrialized areas with substantial air pollution. Once precipitation reaches the ground, it begins to react with soil, rock, and organic debris, dissolving still more chemicals naturally, aside from the possibility of pollution from human activities. Thus water quality must be a consideration in evaluating water supplies.

Measurement of Water Quality

Water quality may be described in a variety of ways. A common approach is to express the amount of a dissolved chemical substance present as a concentration in parts per million (ppm) or, for very dilute substances, parts per billion (ppb). These units are analogous to percentages (which are really "parts per hundred") but are used for lower (more dilute) concentrations. That is, if water contains 1% salt, it contains a gram of salt per hundred grams of water, or one ton of salt per hundred tons of water, or whatever unit one wants to use. Likewise, if the water contains 1 ppm salt, it contains one gram of salt per million grams of water (about 250 gallons), and so on.

One way to express overall water quality is in terms of **total dissolved solids** (TDS), the sum of the concentrations of all dissolved solid chemicals in the water. How low a level of TDS is required or acceptable will vary with the application. Standards might specify a maximum of 500 or 1000 ppm TDS for drinking water; 2000 ppm TDS might be acceptable for watering livestock; in industrial applications where water chemistry is important the water might need to be even purer than normal drinking water. Still, describing water in terms of total content of dissolved solids is not giving the whole story, for at least as important as the quantities of impurities present is *what those impurities are.* If the main dissolved component is calcite from a limestone aquifer, the water may taste fine and be perfectly wholesome with well over 1000 ppm TDS in it. Conversely, if what is dissolved is iron or sulfur, even a few ppm may be enough to make the water taste bad. Many synthetic chemicals that have leaked into water through improper waste disposal are toxic even at concentrations of 1 ppb or less.

Table 14.2

Concentrations of some dissolved constituents in average rain, river water, and seawater

Constituent	Average concentrations (in ppm)		
	Rain- water	River waters	Sea- water
silica (SiO_2)	—	13	6.4
calcium (Ca)	1.41	15	400
sodium (Na)	0.42	6.3	10,500
potassium (K)	—	2.3	380
magnesium (Mg)	—	4.1	1350
chloride (Cl)	0.22	7.8	19,000
fluoride (F)	—	—	1.3
sulfate (SO_4)	2.14	11	2700
bicarbonate (HCO_3)	—	58	142
nitrate (NO_3)	—	1	0.5

Source: Data from J. D. Hem, "Study and Interpretation of the Chemical Characteristics of Natural Water," 2nd ed., U.S. Geological Survey Water-Supply Paper 1473, 1970.

Overall, groundwater quality is highly variable. It may be nearly as pure as rainwater, or saltier than the oceans. Some representative analyses of different waters in the hydrosphere are shown in table 14.2 for reference.

Hard Water

Aside from the issue of health, water quality may be of concern because of the particular ways certain dissolved substances alter water properties. In areas where water supplies have passed through soluble carbonate rocks, like limestone, the water may be described as "hard." Hard water is simply water containing substantial amounts of dissolved calcium and magnesium. There are degrees of hardness; there is no single cutoff figure for calcium and magnesium concentrations that divides "hard" water from "soft," although the level at which hardness becomes objectionable is in the range of 80 to 100 ppm. Perhaps the most troublesome routine problem with hard water is the way it reacts with soap, preventing it from lathering properly, so bathtubs develop rings, laundered clothes retain a grey soap scum, and so on. Hard water or water otherwise high in dissolved minerals may also leave mineral deposits in plumbing and in appliances such as coffeepots and irons. These are principal reasons why many people in hard-water areas use water softeners, which remove calcium, magnesium, and certain other ions from water in exchange for sodium ions. The sodium ions are replenished from the salt (sodium chloride) supply in the water softener. (While softened water containing sodium ions in moderate concentration is unobjectionable in taste or for household use, it may be of concern to those on diets requiring restricted sodium intake.)

Groundwater Pollution

Beyond the natural addition of soluble minerals to water, there are various kinds of pollution arising through human activities. Any soluble material discharged into the air, left exposed on the ground surface, or buried unsealed underground has the potential to pollute groundwater supplies (figure 14.16). Surface-water pollution, too, may lead to groundwater pollution—for example, where a polluted stream contributes to groundwater recharge.

Air pollutants can react with or be dissolved in rainwater. When that rain falls and infiltrates into the soil, the pollutants may be carried along. The acid rain formed when sulfurous exhaust gases (such as from the burning of coal) react with oxygen and water vapor to make sulfuric acid can acidify both surface and groundwater supplies. Volatile metals, such as lead and mercury, are a particular problem in the air near smelters and other metal-processing plants; these metals too may be scavenged out of the air by rain and eventually added to groundwater supplies.

Sewage is a major waste disposal problem. In less densely populated areas, sewage disposal by underground septic tanks on individual homesites is common. A normally functioning septic tank releases sewage slowly into surrounding permeable soil where, ideally, it is decomposed through reaction with oxygen in the soil pore spaces and by the action of soil microorganisms. In an improperly designed septic system, where the water table is too high, too close to the septic system, sewage may mingle with groundwater before it has been fully decomposed, contaminating the groundwater.

Dumps and landfill sites, legal or illegal, are potential sources of groundwater pollution. If they are underlain by permeable rock and soil, liquid wastes can seep out and down toward groundwater supplies below. Rainwater infiltrating from above and dissolving additional soluble chemicals from the wastes compounds the problem. It is not only disposal sites for industrial wastes that are a hazard in this regard, though they may represent especially concentrated sources of toxic chemicals. Consider the variety of dangerous household chemicals in common use likely to be found in municipal landfills: cleansers, paints

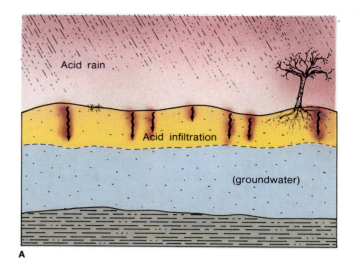

A

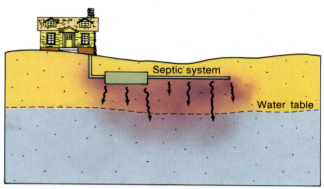

B

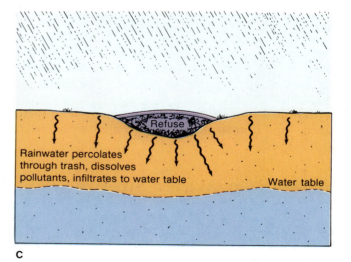

C

Figure 14.16

Some sources of groundwater pollution. *(A)* Acid rain. *(B)* Septic tank too close to the water table. *(C)* Leakage from a landfill/dump site.

and solvents, pesticides, and many others. In principle, toxic wastes in disposal sites can be isolated by impermeable materials (clay layers, plastic liners) above and below. In practice, many carefully designed sites have subsequently leaked, and there is now some question as to whether a landfill site can ever be truly securely sealed.

Road salt applied in winter washes off, dissolves in rain or meltwater, and seeps into the soil. Herbicides and pesticides from farmland dissolve and infiltrate, perhaps down to the water table. The sources of groundwater pollution are many. Furthermore, once it has occurred, it may be difficult to solve or even to detect.

Smoke belching from an exhaust stack is very visible. Oddly colored water discharged into a stream likewise attracts attention. Groundwater pollution is hidden, insidious, often unnoticed until long after the fact. Pollutants escaping from an old dump site may first be detected in well water some distance away, long after the contamination has begun. By the time groundwater pollution is detected, it may be very widespread. The exact extent of the problem may not be readily determined without the drilling of many monitoring wells across the affected area. Even the source can be hard to identify unless the chemistry of the contamination is distinctive.

Treatment at that point in time may simply not be possible, at least while the water remains underground. Natural filtration by passage through an aquifer can remove particles but not dissolved chemicals. Groundwater moves slowly and mixes sluggishly. A polluted lake may be treatable by mixing in chemicals to remove or neutralize the pollutants. An equivalent treatment of groundwater is impractical: how can the added chemicals be dispersed quickly throughout the aquifer as needed? The usual approach to treating groundwater pollution is to halt the pollution at the source—if that source can be identified—and to treat the water only after withdrawal, as it is used.

Water Use, Water Supply

Compared to our actual physical needs for water, we in the United States use a large amount of water. Biologically, we require about a gallon a day per person, or about 250 million gallons per day for the country. We divert some 450 *billion* gallons of water for use each day—for cooking, washing or other household use, for industrial processes, for livestock and irrigation. Even that figure does not include the several trillion gallons of water that flow through U.S.

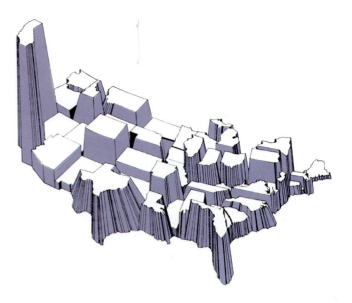

Figure 14.17

Regional variations in water use in the United States.
Source: U.S. Geological Survey Water Supply Circular 1001.

hydroelectric plants every day. Of the total withdrawn, over 100 billion gallons per day are consumed, meaning that the water is not returned as wastewater. Most of the consumed water is lost to evaporation; some is lost in transport.

Regional Variations in Water Use

Water withdrawal varies regionally (figure 14.17), as does water consumption. Aside from hydropower generation, four principal categories of use can be identified: public supplies (home use and some industrial use in urban and suburban areas), rural use (supplying domestic needs for rural homes and watering livestock), irrigation, and self-supplied industrial use (use by industries for which water supplies are separate from municipal sources). The quantities withdrawn for and consumed by each of these categories of users are summarized in figure 14.18.

Regional variations in population distribution do not coincide with variations in water supply. Indeed, some of the larger population centers are very poorly supplied with water. This leads to interstate or interregional political disputes over water rights. A recent example is the effort by officials in states bordering the Great Lakes to declare the water in the lakes "off-limits" to dry regions in the southwestern United States.

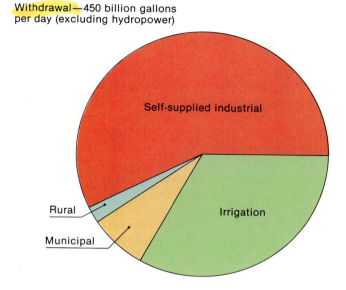

Withdrawal—450 billion gallons per day (excluding hydropower)

Consumption—total 104 billion gallons per day

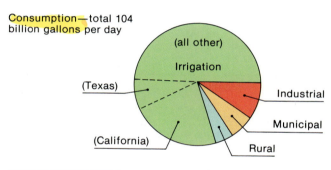

Water withdrawal and consumption by sector.

Industrial versus Agricultural Use

A point that quickly becomes apparent from figure 14.18 is that industry may be called the major water user, but agriculture is the big water *consumer*. Self-supplied industrial users account for more than half the water withdrawn, but nearly all their wastewater is returned, as liquid water, at or near the point in the hydrologic cycle from which it was taken. Most of these users are diverting surface waters and dumping wastewaters back into the same lake or stream. All told, they *consume* only about 10 billion gallons per day, or 10% of total consumption.

Irrigation water—83 billion gallons per day—is nearly *all* consumed, lost to evaporation, through transpiration from plants or as leakage from ditches and pipes. Moreover, 40% of the water used for irrigation is groundwater. Most of the water lost to evaporation drifts with air currents out of the area, to come

down as rain or snow somewhere far removed from the site of the irrigation. It then does not contribute to recharge of the aquifers or runoff to the streams from which the water was drawn. Where irrigation use of water is heavy, water tables have dropped by tens of meters and streams have been drained nearly dry, while we have become increasingly dependent on the crops. The case of the High Plains aquifer system (box 14.1) is just one example.

Extending the Water Supply

The most basic approach to improving the water-supply situation is conservation. In particular, the heavy drain by irrigation must be moderated if the rate at which water supplies are being depleted is to be reduced appreciably. For example, the raising of crops requiring a great deal of water could be shifted, in some cases at least, to areas where natural rainfall is adequate to support those crops. Irrigation methods can be made more efficient too—for example, by replacing open ditches or sprinkler systems with water distribution through perforated pipes at or below ground level, to reduce evaporation losses.

Interbasin Water Transfer

Part of the supply problem is purely local. That is, people persist in settling and farming in areas that may not be especially well supplied with fresh water, while other areas with abundant water go undeveloped. If the people cannot be persuaded to move to the water, perhaps the water can be redirected. This is the idea behind interbasin transfers, moving surface waters from one stream system's drainage basin to another's where demand is higher. California pioneered the idea with the Los Angeles Aqueduct. Completed in 1913, it carried nearly 150 million gallons of water per day from the eastern slopes of the Sierra Nevada to Los Angeles. New York City draws on several reservoirs in upstate New York. Dozens of interbasin transfers of surface water have been proposed, including transfers of water from little-developed areas of Canada to high-demand areas of the United States and Mexico. Such proposals, which could involve transporting water over distances of thousands of kilometers, are not only expensive (one such scheme, the North American Water and Power Alliance, had a projected cost of $100 billion); they may also presume a continued willingness on the part of other nations to share their water with us.

Desalination

Another alternative is to improve the quality of waters not now used, purifying them sufficiently to make them useable. Desalination of seawater, in particular, would allow coastal regions to tap the vast water reservoir of the oceans. There are also groundwaters not presently used for water supplies because they contain excessive concentrations of dissolved materials. There are two basic methods used to purify water of dissolved minerals: filtration and distillation.

In a filtration system, the water is passed through fine filters or membranes that screen out dissolved impurities. An advantage of the method is that it can operate rapidly on a great quantity of water. A disadvantage is that the method works best on water that does not contain very high levels of dissolved minerals. Anything as salty as seawater quickly clogs the filters. This method, then, is most useful for cleaning up only moderately saline groundwaters, or lake or stream waters.

Distillation takes advantage of the fact that when water full of dissolved minerals is heated or boiled, the water vapor driven off is pure water, while the minerals stay behind in what remains of the liquid. This is true regardless of how concentrated the dissolved minerals are, so the method works as well on seawater as on less saline waters. A difficulty with this method is the nature of the necessary heat source. Any conventional fuel will be costly in large quantity. Some solar distillation facilities now exist, but their efficiency is limited by the fact that solar heat is low-intensity heat. If a large quantity of desalinated water is required rapidly, the water to be heated must be spread out shallowly over a large area. A major city might need a solar facility covering thousands of square kilometers to provide adequate water, and construction on such a scale would be prohibitively expensive even if sufficient space were available.

Summary

All the water at and near the earth's surface moves through the hydrologic cycle, the principal component processes of which are evaporation, precipitation, and surface and groundwater runoff. The oceans collectively are the largest single water reservoir by far. Most of the fresh water is stored in ice sheets. Groundwater is the largest reservoir of unfrozen fresh water.

A rock that is sufficiently porous and permeable to be useful as a source of groundwater is an aquifer; relatively impermeable rocks are aquitards or aquicludes. The top of the saturated zone is the water table, which may locally be above the ground surface. In an unconfined aquifer, water in an unpumped well will rise to the height of the water table. In a confined aquifer, additional pressure acting on the water may create artesian conditions, in which the water will rise in an unpumped well to the level of the potentiometric surface, which may be far above the aquifer and even the ground surface. The presence of groundwater in soluble rocks contributes to the formation of caverns and sinkholes. Groundwater withdrawal exceeding recharge can lead to surface subsidence,

compaction of aquifer rocks, and saltwater intrusion, as well as to lowering of the water table and general groundwater depletion. When urbanization modifies runoff patterns or covers recharge areas, it may ultimately decrease groundwater supplies.

Groundwater quality varies widely. Natural water quality is diminished by the solution of soluble rocks and minerals. Hard water is common where limestones are abundant. Additional, sometimes toxic chemicals are added by human activities, especially through improper waste disposal or runoff from highways and farmland. Once established, groundwater pollution can be slow to appear, difficult to trace or monitor, and impossible to treat while the water remains in its aquifer.

Industry accounts for the largest share of water withdrawn for use in the United States, but most of that water is not consumed during use. By contrast, most of the water withdrawn for irrigation *is* consumed, especially through evaporation. About 40% of irrigation water is groundwater, which may not be undergoing recharge at present. Possible strategies for extending water supplies in areas of shortage include conservation, interbasin water transfer, and desalination of either seawater or saline groundwater.

Terms to Remember

aquiclude	groundwater	phreatic zone	subsurface water
aquifer	hard water	potentiometric surface	total dissolved solids
aquitard	hydrologic cycle	recharge	unconfined
artesian	hydrosphere	saltwater intrusion	upconing
baseflow	hydrostatic pressure	sinkhole	vadose zone
cone of depression	karst topography	soil moisture	water table
confined aquifer	perched water table	spring	

Questions for Review

1. Briefly summarize the principal processes and reservoirs of the hydrologic cycle. (You may find it helpful to use a sketch.)

2. How is groundwater distinguished from subsurface water?

3. What is a water table, and how is it related to a potentiometric surface? What is a cone of depression? How does it develop?

4. Why is groundwater frequently preferable to surface water as a source of water supply?

5. What is an aquifer? A confined aquifer?

6. Explain the nature of artesian conditions and how they arise.

7. What is karst topography? Under what conditions does it tend to develop?

8. Groundwater withdrawal that exceeds recharge may lead to surface subsidence. Explain.

9. What is saltwater intrusion, and in what circumstances does it occur?

10. How might urbanization reduce recharge of a confined aquifer? Describe one strategy for increasing recharge to an aquifer (confined or unconfined).

11. What is hard water? Is it harmful?

12. Cite and explain any two ways in which groundwater may become polluted. Is it practical to treat groundwater pollution in place? Why or why not?

13. Industry is the pricipal water *user;* agriculture is the major water *consumer.* Explain.

14. Compare and contrast distillation and filtration as means of desalinating water, noting the advantages and disadvantages of each.

For Further Thought

1. Where does your water come from? What is the quality at the source? Is the water treated before consumption? If the source is groundwater, is there evidence that the supply is being depleted? If so, what plans exist for averting future shortages?

2. If there is a landfill or other waste-disposal site near you, investigate what monitoring is undertaken to detect possible groundwater pollution. What tests are made and how often? If appreciable pollution has been detected in the past, what steps have been taken to reduce or eliminate it?

Suggestions for Further Reading

Berghinz, C. "Venice Is Sinking into the Sea." *Civil Engineering* 41 (1971): 67–71.

Bowen, R. *Ground Water.* London: Applied Science Publishers, 1980.

Dunne, T., and Leopold, L. B. *Water in Environmental Planning.* San Francisco: W. H. Freeman and Co., 1978.

Francko, D. A., and Wetzel, R. G. *To Quench Our Thirst.* Ann Arbor, Mich.: University of Michigan Press, 1983.

Howe, C. W., and Easter, K. W. *Interbasin Transfers of Water, Economic Issues and Impacts.* Baltimore, Md.: Johns Hopkins Press, 1971.

Leopold, L. B. *Water—a Primer.* San Francisco: W. H. Freeman and Co., 1974.

Mather, J. R. *Water Resources.* New York: John Wiley and Sons, 1984.

Sasman, R. T. et al. "Water Level Decline and Pumpage in Deep Wells in the Chicago Region, 1966–1971." Illinois State Water Survey Circular #113, 1973.

Solley, W. B.; Chase, E. B.; and Man, W. B., IV. "Estimated Use of Water in the United States in 1980." U.S. Geological Survey Circular 1001, 1983.

U.S. Water Resources Council. *The Nation's Water Resources 1975–2000.* Washington, D.C.: U.S. Government Printing Office, 1978.

Coastal Zones and Processes

Introduction

Coastal areas vary greatly in character and in the kinds and intensities of geologic processes that occur in them. These areas may be dynamic and rapidly changing under the interaction of land and water, or they may be comparatively stable. In this chapter we will review briefly some of the processes that occur along coasts, examine several different types of coastlines, and consider the impacts of various human activities on them (and vice versa).

Although the terms *coastline* and *shoreline* are sometimes used interchangeably, the **shoreline** is technically the line made by the water's edge on the land, a small, local feature that is constantly shifting and changing shape. The term **coastline** is properly used to describe the overall geometry of the margin of the land, which encompasses a much larger region of the coast and is a somewhat more permanent feature.

Waves and Tides

Waves and associated currents are the principal forces behind change along shorelines. Waves, in turn, are produced as a consequence of the flow of wind across the water surface. Small, local differences in air pressure create undulations in the water surface. The alternating rise and fall of the water surface is a **wave.** If the wind flow continues, the undulations may begin to move laterally across the water surface, and even to enlarge.

Waves and Breakers

The shape and apparent motion of a wave is just a reflection of the form of the water surface. The actual motion of the water molecules is different (figure 15.1). In open, deep water, the water moves in circular orbits, relatively large near the surface and decreasing in diameter with increasing depth. The water surface takes on a rippled form. The peak or top of each ripple is a wave **crest;** the bottom of each intervening low is a **trough.** The **height** of the waves is simply the difference in elevation between crest and trough, and the **wavelength** is the horizontal distance between adjacent wave crests (or troughs). The **period** of a set of waves is the time interval between the passage of successive wave crests (or troughs) past a fixed point.

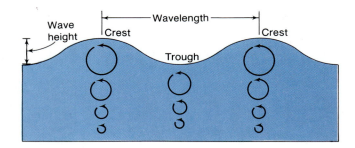

Figure 15.1

Waves and wave terminology. Note the motion of water beneath the surface.

Only when a wave begins to interact with the bottom does it begin to develop into a *breaker.* This first occurs when the depth of water has decreased to about half the wavelength. Friction with the bottom distorts and ultimately breaks up the orbits, and the water arches over in a breaking wave (figure 15.2). Bottom topography thus plays a major role in the nature of waves reaching the shore. If the coast consists of cliffs and the water deepens sharply offshore, breaker development may be minimal. The **surf zone** extends outward from the beach to the outermost limit of the occurrence of breakers.

Wave Refraction

The influence of interaction with the ocean bottom is also seen in the phenomenon of **wave refraction** (figure 15.3). As waves approach a coast, they are slowed first where they first touch bottom, while continuing to move more rapidly elsewhere. As a result, the line of crests is bent, or refracted. Assuming that the slope of the bottom is similar all along the coast, the effect is to deflect the waves toward projecting points of land, concentrating their energy there. The refracted waves tend to approach the shoreline more squarely, with wave crests more nearly parallel to the shore, as they move toward it.

Waves, Tides, and Surges

As noted previously, waves are localized water-level oscillations set up by wind. **Tides** are broader, regional changes in water level caused primarily by the gravitational pull of sun and moon on the envelope of water surrounding the earth (figure 15.4).

If one takes a bucketful of water and swings it quickly in a circle at arm's length, the water stays in the bucket, because forces associated with the circular motion push the water outward from the center

Figure 15.2

Breakers form as undulating waters approaching shore touch bottom. *(A)* (Schematic) Note distortion of water orbits as water shallows. *(B)* Evenly spaced breakers form as successive waves touch bottom along a regularly sloping shore.

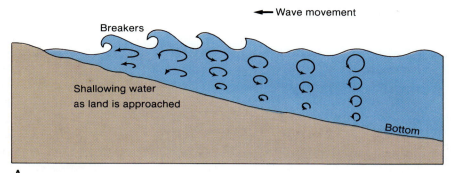

← Wave movement

Breakers

Shallowing water
as land is approached

Bottom

A

B

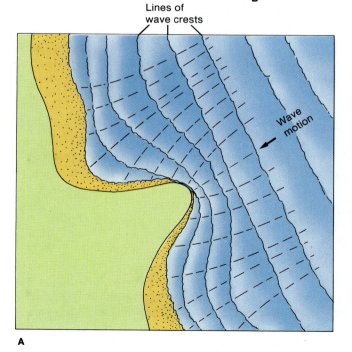

Lines of
wave crests

Wave motion

A

B

Figure 15.3

Wave refraction. *(A)* The energy of waves approaching a jutting point of land is focused on it by refraction. *(B)* Wave refraction along a shoreline.

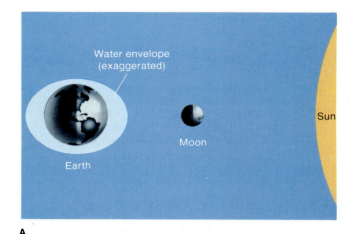

A

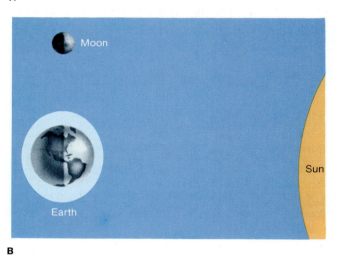

B

Figure 15.4

Tides. *(A)* Spring tides: sun, moon, and earth are all aligned (near times of full and new moons). *(B)* Neap tides: when moon and sun are at right angles, tidal extremes are reduced.

of the circle (and thus into the bucket). Similarly, as the earth rotates, matter would be flung outward from it, were it not for the restraining influence of gravity. Even so, the earth and the watery envelope surrounding it that makes up the oceans show the effects of these rotational forces. The velocity of the moving surface of the earth is greatest near the equator and decreases toward the poles, so earth and water deform to create an equatorial bulge.

Superimposed on this effect of the earth's rotation are the effects of the gravitational pull of the sun and especially the moon. The closer one is to an object, the stronger its gravitational attraction. The moon therefore pulls most strongly on matter on the side of the earth facing it, least strongly on the opposite side.

The combined effects of gravity and rotation cause two bulges in the water envelope: one facing the moon, where its gravitational pull on the water is greatest, and one on the opposite side of the earth where the gravitational pull is weakest and rotational forces dominate (figure 15.4).

As the rotating earth spins through these two bulges of water each day, overall water level at a given point on the surface rises and falls twice daily. This is the phenomenon recognized as tides. The tidal extremes are most significant when sun, moon, and earth are all aligned, and the sun and moon are thus pulling together. The resultant tides are **spring tides.** (They have nothing in particular to do with the spring season of the year.) When the sun and moon are pulling at right angles to each other, the difference between high and low tides is minimized. These are **neap tides.** The magnitude of water level fluctuations in any one spot is also controlled in part by the underwater topography. The oscillations of waves are superimposed on the tidal regime.

Waves rise and fall in seconds, tides over several hours. There is a phenomenon of intermediate duration, usually associated with storms, known as a **surge.** It results from some combination of significant drop in air pressure over an area (which causes a local bulge or rise in water elevation beneath it) and strong onshore winds. Surges from severe storms can easily be several meters high, and are most serious when they coincide with the already high spring-tide water levels. On top of that, of course, strong storms can also cause unusually large waves. The combination of surges and high waves can make coastal storms especially devastating, as illustrated recently by Hurricane Gloria in September of 1985.

Beaches

A **beach** is a gently sloping shore covered by silt, sand, or gravel that is washed by waves and tides. Most beaches are sandy, and the following discussion will focus on this type of beach. Beaches vary in detail, but certain features are commonly observed.

The Beach Profile

A representative profile or cross section of a beach is shown in figure 15.5. The **beach face** is that portion of the beach exposed to direct overwash of the surf. Its slope varies with the grain size of the beach sediment and the energy of the waves. To the seaward side of the beach face, there is typically a more shallowly sloping zone, below the low-tide level. Within this

Figure 15.5

A typical beach profile.

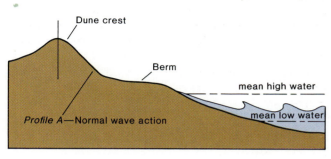

Profile A—Normal wave action

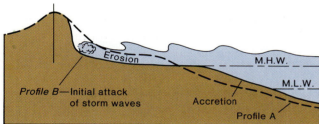

Profile B—Initial attack of storm waves

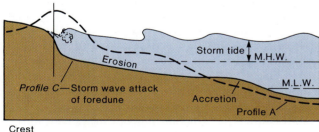

Profile C—Storm wave attack of foredune

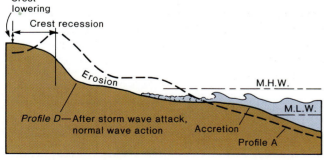

Profile D—After storm wave attack, normal wave action

Figure 15.6

Alteration of beach profile due to accelerated erosion by unusually high storm tides.

Source: "Shore Protection Guidelines," U.S. Army Corps of Engineers.

zone, sediment may be moved by currents, but it is not actually under attack by waves. Landward of the beach face there may be a flat or landward-sloping terrace, or **berm,** backing up the beach. Not all beaches have berms; some have more than one. The landward limit of the beach can be defined in several ways, depending on the particular situation—for example, by the presence of a rocky cliff, or by a zone of permanent vegetation.

It should be emphasized that beach profiles are not static. Even a single beach may not always be characterized by the same slope or other features throughout the year. The surges and increased wave action of storms bring about changes that are both sudden and dramatic (see figure 15.6). As storms vary seasonally in intensity and frequency, so beach geometry may vary seasonally also.

Sediment Transport at Shorelines

As with streams, coastal currents move material more efficiently the faster they move, and sediment-laden waters deposit their load as they are slowed. One consequence of wave refraction along an irregular coastline is that wave energy is dissipated in recessed bays between projecting points of land, and these tend to be sites of sediment deposition (figure 15.7).

On any beach, the rush of water up the beach face after waves have broken (**swash**) tends to push sediment upslope toward the land. Gravity and the retreating **backwash** together carry it down again. A beach profile is described as being in equilibrium when its slope is such that the net upslope and downslope transport of sediment are equal.

If waves approach the beach at an oblique angle, there may be a movement of sediment along the length of the beach, rather than just up and down the beach face perpendicular to the shoreline. Wave refraction will tend to deflect waves so that the wave crests approaching the beach are more nearly parallel to the shoreline, although they may still strike the beach face

Figure 15.7

Deposition occurs in a recessed bay where water energy is lower.

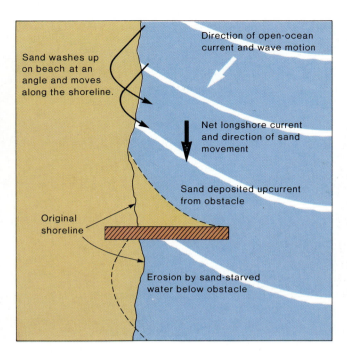

Sand washes up on beach at an angle and moves along the shoreline.

Direction of open-ocean current and wave motion

Net longshore current and direction of sand movement

Sand deposited upcurrent from obstacle

Original shoreline

Erosion by sand-starved water below obstacle

Figure 15.8

Longshore currents and their effect on sand movement. Dashed profile is shoreline after modification.

at a small angle. As the water washes up onto and down off the beach, then, it also moves laterally along the coast. This is a **longshore current** (figure 15.8). Likewise, any sand caught up by and moved along with the flowing water will not be carried straight up the beach, but will be transported at an angle to the shoreline. The net result is **littoral drift,** sand movement along the beach in the same general direction as the motion of the longshore current. Over time, currents tend to move consistently in certain preferred directions on any given beach, which means that through time there will be a continual transport of sand from one end of the beach to the other. On many natural beaches where this occurs, the continued existence of the beach is assured by a fresh supply of sediment produced locally by wave erosion, or delivered by streams from farther inland.

Littoral drift is a common and natural beach process. However, beachfront property owners who become aware of littoral drift often also become concerned about the danger that their beaches (and perhaps their homes or businesses) will wash away. They therefore erect structures to try to "stabilize" the beach, which generally further alter its geometry (figure 15.8). A method commonly used is the construction of one or more groins, jetties, or breakwaters (figure 15.9), which are long, narrow obstacles

set more or less perpendicular to the coastline. By disrupting the usual flow and velocity patterns of the currents, their presence changes the coastline. Currents slowed by encountering such a barrier will tend to drop their load of sand upcurrent from it. Below (downcurrent from) the barrier, the water will be capable of picking up more sediment to replace the lost load, and the beach will be eroded. The common result is that a formerly stable, straight shoreline develops an unnatural scalloped shape. The beach is built out in the area upcurrent of the groins and eroded landward below. Beachfront properties in the eroded zone may be more severely threatened after construction of the so-called stabilization structures than they were before.

Any interference with sediment-laden waters can cause redistribution of sand along beachfronts. A marina built out into such waters may cause some deposition of sand around it, and perhaps within the protected harbor. Further along the beach, the now unburdened waters can more readily take on a new sediment load of beach sand. Even modifications far from the coast can affect the beach. One notable example is the practice, increasingly common over the last century or so, of damming large rivers for flood control, power generation, or other purposes. We have seen that one consequence of the construction of artificial reservoirs is the trapping behind dams of the

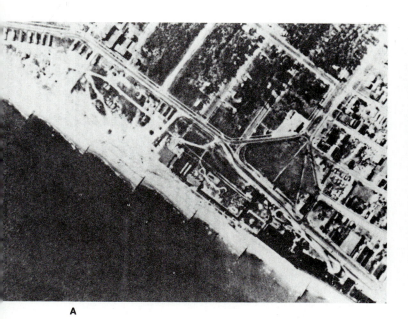

A

B

Figure 15.9

Construction of groins and jetties, obstructing longshore currents, causes deposition of sand. *(A)* Willoughby Spit, Virginia. *(B)* Cold Spring Inlet, New Jersey.

Photos courtesy U.S. Army Corps of Engineers.

sediment load carried by the stream. This further results in the cutoff of sediment supply to coastal beaches near the mouth of the stream. This will lead to erosion of the sand-starved beaches. It may be a difficult problem to solve, too, because there may be no readily available alternate supply of sediment near the problem beach areas.

> Flood-control dams constructed on the Missouri River are believed to be the main reason why the sediment load delivered to the Gulf of Mexico has dropped by more than half over the last 35 years. This may in turn explain the recently observed coastal erosion in parts of the Mississippi delta.

When beach erosion is rapid and development (especially tourism) is widespread, efforts have sometimes been made to import replacement sand to maintain wide beaches. The initial cost of such an effort can be millions of dollars for every kilometer of beach so restored. Moreover, if no steps are taken to reduce the erosion rates, or if in the particular area there are no clearcut steps that *can* be taken to slow or halt the erosion, the replenishment of the sand will clearly just have to be undertaken over and over, at ever-rising cost. In many cases, too, it has not been possible (or been thought necessary) to duplicate the mineralogy

or grain size of the sand originally lost. The result has sometimes been further environmental deterioration. When coarse sands are replaced by finer ones, softer and muddier than the original sand, the finer material more readily stays suspended in the water, clouding it. This increased water **turbidity** is not only unsightly but can be deadly to organisms. Off Waikiki Beach in Hawaii and off Miami Beach, delicate coral reef communities have been damaged or killed by such increased turbidity, resulting from beach "nourishment" programs.

Seacliff Erosion

It is not only through littoral drift along beaches that coasts are eroded. Even rocky cliffs are vulnerable, and they are attacked in a variety of ways. One is direct pounding by waves. Another is the grinding effect of sand, pebbles, and cobbles propelled by waves, which is termed **milling**. Solution and chemical weathering of the rock may bring about still more rapid erosion. As wave action is concentrated at the waterline, so cliff erosion is often most vigorous there (figure 15.10). Also, erosion of projecting points of land is accelerated by the effects of wave refraction, concentrating wave attack there (recall figure 15.3A).

Figure 15.10

Sea arch formed by wave erosion at the waterline on a lava coast, Hawaii.

Sediments are obviously much more readily eroded than are solid rocks, and erosion of sandy cliffs may be especially rapid. Removal of material at and below the waterline undercuts the cliff, leading in turn to slumping and sliding of sandy sediments and the swift landward retreat of the shoreline.

Strategies for Limiting Cliff Erosion

Many who build on coastal cliffs are unpleasantly surprised to discover the very rapid rate at which the coastline can change. Unanticipated cliff erosion can be an especially dramatic threat (figure 15.11), with the *rate* at which it occurs often the largest surprise. Sandy cliffs may in fact be cut back by several meters a year. Various measures, often as unsuccessful as they are expensive, may be tried to halt the erosion.

A common practice is to place some type of barrier at the base of the cliff to break the force of wave impact (figure 15.12). The protection may take the form of a solid wall (*seawall*) of concrete or other material, or a pile of large boulders or other blocky

Figure 15.11

House built on the shore of Lake Michigan about to be lost to cliff erosion.

Photo courtesy EPA/National Archives.

debris (*riprap*). If the obstruction is placed only along a short length of cliff directly below a threatened structure, it is quite possible that the water will merely wash in beneath, around, and behind it, rendering it largely ineffective. Erosion will continue unabated on either side of the barrier; loss of the protected structure will be delayed but probably not prevented. Wave energy also may be reflected off a short length of smooth seawall and attack a nearby unprotected cliff with greater force.

An alternative approach to direct shoreline protection is to erect breakwaters further away from and parallel to the shore, to reduce the energy of the pounding waves reaching shore. This may slow the erosion but again is unlikely to stop it, especially if the cliffs are sandy or made of other weak or unconsolidated materials. Moreover, a breakwater will probably alter the patterns of sediment deposition offshore, which may prove inconvenient, or damaging to marine life.

B

Figure 15.12

Examples of cliff-protection structures. *(A)* Seawall at Galveston, Texas. *(B)* Riprap at base of cliff, El Granada, California. Structure was built in 1973; it may have to be abandoned within decades as erosion continues unabated on either side of riprap.

Photos: *(A)* W. T. Lee, courtesy U.S. Geological Survey. *(B)* Courtesy U.S. Geological Survey.

Emergent and Submergent Coastlines

The elevation of water levels relative to coastal land varies, over the short term, as a result of tides and storm surges. Over longer periods, there may be a steady shift up or down resulting from tectonic processes or from changes related to glaciation. When the coastal land rises or sea level falls, the coastline is described as *emergent*. A *submergent* coastline is found in the opposite case, when the land is sinking relative to sea level or sea level is rising.

Causes of Long-Term Sea Level Change

As plates move, crumple, and shift, continental margins may be uplifted or dropped down (see, for example, figure 15.13). Such movements can shift the land by several meters in a matter of seconds to minutes. The results are abrupt and cause permanent changes in the geometry of the land/water interface and the patterns of erosion and deposition.

In regions that were overlain and weighted down by massive ice sheets in the last ice age, the lithosphere was downwarped by the load, as noted in chapter 11. Tens of thousands of years later, it is still slowly rebounding isostatically to its pre-ice elevation. Where thick ice extended to the sea, a consequence is that the coastline is slowly rising out of the

Figure 15.13

Coastal flooding at Portage, Alaska, due to tectonic subsidence resulting from 1964 earthquake.

Photo courtesy U.S. Geological Survey.

sea. Ice caps themselves, as noted in chapter 14, represent an immense reserve of water. As this ice melts, sea levels rise worldwide. Such simultaneous, global changes in sea level are termed **eustatic** changes. In terms of the relative elevation of land and sea level, post-glacial isostatic rebound counters the eustatic sea-level rise resulting from the melting of the ice.

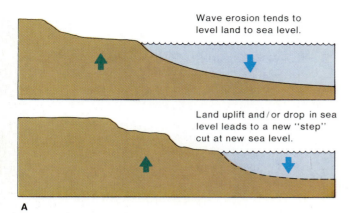

Wave erosion tends to level land to sea level.

Land uplift and/or drop in sea level leads to a new "step" cut at new sea level.

A

B

Figure 15.14

Wave-cut terraces. *(A)* (Schematic) Wave-cut terraces form when land is elevated or sea level falls. *(B)* Wave-cut terraces of the Pacific Palisades area, Los Angeles County, California.

(B) Photo by J. T. McGill, courtesy U.S. Geological Survey.

All these changes in the relative elevation of land and water produce distinctive coastal features, which can be used to identify the kind of vertical movement that is taking place.

Wave-Cut Terraces

Given sufficient time, wave action will tend to erode the land down to the level of the water surface, creating a wave-cut platform at sea level. If the land rises in a series of tectonic shifts, and stays at each new elevation for some time before the next movement, each rise will result in the erosion of a portion of the coastal land down to the new water level. The eventual product will be a series of steplike terraces, called **wave-cut terraces** (figure 15.14). These develop most readily on rocky coasts, rather than along those made of soft, unconsolidated material. The surface of

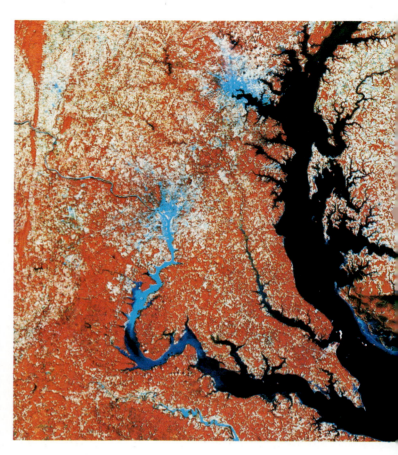

Figure 15.15

A drowned valley: Chesapeake Bay (Landsat satellite photograph). © NASA

each such step, each terrace, represents an old water level marker on the continent's edge. Wave-cut terraces can be formed when the continent is rising relative to the sea, or when sea level is falling with respect to the land.

Drowned Valleys

When, on the other hand, sea level rises or the land drops, one result will be that streams that once flowed out to sea now have the sea rising partway up the stream valley from the mouth. A portion of the floodplain may be filled by encroaching seawater. The feature thus formed is a **drowned valley** (figure 15.15).

A variation on the same feature is produced by the advance and retreat of coastal glaciers. Glaciers flowing off land and into water do not just keep eroding deeper valleys under water until they melt. Ice floats, and glacial ice, made of fresh water, floats especially readily on denser salt water. Therefore the carving of glacial valleys essentially stops at the shore.

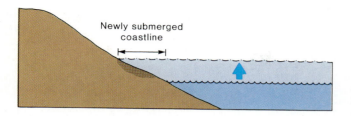

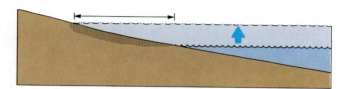

Figure 15.16

Small rises in sea level can have particularly severe effects on a gently sloping coastline: a little increase in depth may mean a large area flooded. Dashed lines indicate new, higher sea level; shading indicates newly inundated land.

During the last ice age, when sea level worldwide was as much as 100 meters lower than it now is, glaciers at the edges of continental landmasses cut valleys into now submerged continental shelf areas. With the retreat of the glaciers and concurrent rise of sea level, these old glacial valleys were emptied of ice and partially filled with seawater. That is the origin of the steep-walled *fjords* so common in Scandinavian countries, British Columbia, and parts of New Zealand.

Present and Future Sea Level Trends

It has recently been realized that much of the coastal erosion presently plaguing the United States is a result of gradual but sustained eustatic sea-level rise, probably the result of the melting of remaining ice in polar regions. The rise is estimated at about 1/3 meter (1 foot) per century. This does not sound particularly threatening, but two additional factors should be considered. First, many coastal areas slope very gently or lie nearly flat, so that a small rise in sea level translates into a far larger inland retreat of the shoreline (figure 15.16). Rates of shoreline retreat (landward movement of the shoreline) due to rising sea level have in fact been measured at several meters per year in some low-lying coastal areas. Second, there is concern that global warming trends will begin to melt remaining ice caps more rapidly, accelerating the sea-level rise (see discussion of the *greenhouse effect* in the next chapter). Consistently rising sea levels are a major reason that efforts to stabilize coastlines repeatedly fail. The problem is not only that the high-energy coastal environment presents difficult engineering

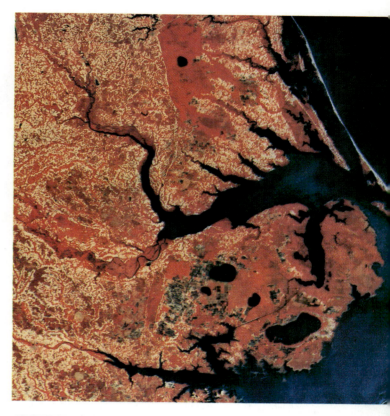

Figure 15.17

Landsat satellite photograph of barrier islands along the Virginia coast.

© NASA

challenges. The problems themselves are intensifying, as shoreline retreat brings water higher and farther inland, pressing ever closer to and more forcefully against more and more structures developed along the coast.

Some Especially Fragile Coastal Environments

Many coastal environments are unstable, but some are particularly vulnerable, either to natural forces or to human interference. This section will consider two such environments, barrier islands and estuaries.

Barrier Islands

Barrier islands are long, low, narrow islands paralleling a coastline (figure 15.17). Exactly how or why they form is not known. Some theories suggest that they have formed through the action of longshore currents on delta sands deposited at the mouths of

Figure 15.18

Development on barrier island, Ocean City, Maryland. Note sand-stabilization structures along the beach and the overall low relief of the island.

Photo courtesy U.S. Geological Survey.

streams. Their formation may also require relative changes in sea level. However they have formed, they now provide important protection for the water and shore inland from them, for they constitute the first line of defense against the fury of high surf from storms at sea. These islands often are so low-lying that water several meters deep may wash completely over them during unusually high storm tides, such as occur during hurricanes, especially along the Gulf and Atlantic coasts of North America.

Because they are usually subject to higher-energy waters on their seaward sides than on their landward sides, most barrier islands retreat landward with time. Typical rates of retreat on the Atlantic coast of the United States are 1/2 to 2 meters per year, but rates in excess of 20 meters per year have been noted, for example, along some barrier islands in Virginia. Clearly such settings represent particularly unstable locations in which to build, yet the aesthetic appeal of long beaches has led to extensive development on privately owned sections of barrier islands (see, for example, figure 15.18). About 1,400,000 acres of barrier islands occur in the United States, and approximately 20% of this total area has been developed. Thousands of structures are at risk, including many in large coastal cities like Miami, Florida, and Galveston, Texas.

In such an environment, shoreline-stabilization efforts—building groins and breakwaters, sand replenishment—tend to be especially expensive and frequently futile. At best the benefits are temporary. Construction of artificial stabilization structures may easily cost tens of millions of dollars. At the same time they may destroy the natural character of the shore, and even the beach that was the principal attraction for developers in the first place. An equally costly alternative is to keep moving buildings or rebuilding roads ever further landward as the beach before them erodes. Expense aside, this is clearly a short-term "solution" only, and in many cases there is no place left to which to retreat. Considering the inexorable eustatic rise of sea level at present, more barrier-island land is being submerged more frequently, so the problems will only intensify.

Federal disaster-relief funds provided in barrier-island and other coastal areas can easily add up to hundreds of millions of dollars after each major storm. Increasingly the question is asked: does it make sense to continue subsidizing and maintaining development in such very risky areas? More and more often the answer is no. From 1978 to 1982, the Federal flood insurance program paid $43 million in damage claims to barrier-island residents, which far exceeded the premiums they had paid. So, in 1982, Congress decided to move toward elimination of federal flood insurance for these areas. It simply did not make good economic sense to encourage development in such naturally unstable areas by providing this kind of protection.

Figure 15.19

The Zuider Zee, a partially filled estuary in the Netherlands (Landsat satellite photo). Patches of reclaimed land (polders) are seen at lower right. Most recent efforts are directed toward reclaiming the west polder, visible as a lighter patch of shallow water at bottom center of photo.

© NASA

Estuaries

An **estuary** is a body of water along a coastline, open to the sea, in which the tide rises and falls, and which contains a mix of fresh and salt water (*brackish water*). San Francisco Bay and Chesapeake Bay are examples. Some estuaries form near the mouths of streams, especially in drowned valleys. Others may be tidal flats in which the water is close to seawater in salinity.

Over time, the complex communities of organisms in estuaries have adjusted to the salinity of that particular water. The salinity itself reflects the balance between freshwater input (usually river flow) and salt water. Any modifications that alter this balance will change the salinity and can have a catastrophic impact on the organisms. Also, water circulation in estuaries is often very limited. This makes them especially vulnerable to pollution: they are not freely flushed out by vigorous water flow, so pollutants can accumulate. In this sense, it may be unfortunate that many of the world's large coastal cities are located beside estuaries.

Where land is at a premium, estuaries are frequently called upon to supply more. They may be isolated from the sea by dikes and pumped dry, or, where the land is lower, wholly or partially filled in. Naturally this further restricts water flow. It generally changes the water's chemistry also, and development may be accompanied by pollution. All of this puts great stress on the organisms in the estuary. Depending on the local geology, additional problems may arise. For example, we have already noted that buildings built on filled land rather than on bedrock suffer much greater damage from ground shaking during earthquakes. This was abundantly demonstrated in San Francisco in 1906.

One of the most ambitious projects involving reclamation of land from the sea in an estuary is that of the Zuider Zee, in the Netherlands (figure 15.19). The estuary itself did not form until a sand bar was breached in a storm in the thirteenth century. Two centuries later, with a growing population wanting more farmland, the reclamation was begun. Initially the access of the North Sea to the estuary was dammed up, and the flow of fresh water into it from streams gradually changed the water in the Zuider Zee from brackish to relatively fresh, which was critical to successful growth of crops. Some portions were filled in to create dry land, while others were isolated by additional dikes, then pumped dry. The latter lands still lie below sea level, and constant pumping is necesary to keep them dry. More than half a million acres of new land have been created in the Zuider Zee. The marine organisms there initially were necessarily eliminated when it was converted to a freshwater system. Continued vigilance over and maintenance of dikes and pumps is required; if the farmlands were to be flooded with salt water they could be unsuitable for agriculture for years.

Trouble on the Texas Coast: A Case Study

The Texas coast, like much of the Atlantic margin of the United States, is rimmed with barrier islands and barrier peninsulas (figure 1). The area is particularly vulnerable to alteration by natural forces because it is frequently subject to severe storms, which are accompanied by storm surges (commonly several meters or more above normal high tides), strong winds, and high waves. Over the last century, an average of one tropical storm a year has made landfall somewhere along the Texas coast, and a hurricane has struck more than once every five years. A particularly fierce storm hit Galveston in 1900, washing away two-thirds of the city's buildings and causing 6000 deaths. Modern meteorological monitoring and improved communications have greatly reduced the number of hurricane-related deaths in recent decades, but the shifting shoreline poses a continuing challenge to structures.

A period of quiet weather in the 1950s was accompanied by a rush of development along the Texas coast. Another period of accelerated building occurred in the 1970s. Building continues today. Meanwhile, the landward retreat of the beaches progresses at average rates of 2 to 7 meters per year. In some especially unstable areas, shoreline changes of over 20 meters per year have occurred.

As in other dynamic environments, stabilization efforts have been undertaken, with mixed results. In 1902, in response to the devastation of the 1900 hurricane, the Galveston seawall was built. The nearly 6-meter-high structure, built at a cost of $12 million, has indeed offered valuable protection to structures on land, especially during subsequent storms. However, it has also demonstrated some of the permanent changes that can result from seawall construction. As noted

Figure 1
Barrier islands along the Texas shore (Landsat satellite photo).
© NASA

in the text, a portion of wave energy is reflected back from smooth-faced seawalls, so the sand in front of them will tend to be more actively eroded. Longshore currents are commonly strengthened along a seawall, and its presence cuts off the supply of sand from any dunes at the back of the beach to the area in front of the seawall. The result, typically, is gradual loss of the sandy beach in front of the seawall, within 1 to 50 years after construction. The beach in front of the seawall at Galveston, once up to several hundred meters wide, is gone now (figure 15.12A). Moreover, the unprotected area at either end of the seawall is being eroded very rapidly. Changes caused by the presence of the seawall have made it likely that that seawall will eventually have to be replaced by another larger, more expensive structure.

Beach replenishment has not often been practiced on the Texas coast, partly for lack of funds, partly for lack of suitable replacement sand. Such an effort was undertaken at Corpus Christi, however, using sand from a nearby river. It has succeeded in the sense that a sandy beach has been preserved there. On the other hand, the river sand is much coarser-grained than was the original beach sand. It has stabilized at a steeper slope angle, both above and below the waterline, than characterized the original beach, and the beach has consequently become less suitable for use by small children. Also, the replenishment efforts must continue if the new beach is not to be eroded away in its turn.

Both of these examples illustrate again the principles that coastal engineering projects permanently change the coastline, and that once such engineering efforts are begun, they must be continued, unless structures or beaches are ultimately to be abandoned altogether.

Figure 15.20

Building about to be lost to cliff erosion. This cliff, at Moss Beach in San Mateo County, California, has retreated over 50 meters in a century. Past positions are known from old maps and photographs; arrows indicate position of cliff at corresponding dates.

Photo by K. R. Lajoie, courtesy U.S. Geological Survey.

Recognition of Coastal Hazards

It is often possible to identify the most unstable or threatened areas along a coast, and avoid building in them. The ability to do so depends both on observations of present conditions and on some knowledge of the area's history.

If one were considering building near a beach or on an island, for instance, the best setting would be at a relatively high elevation (5 meters or more above normal high tide to be above the reach of storm surges), in a spot protected by many high dunes between the proposed building site and the water. Thick vegetation, if present, will help to stabilize the beach sand. It would be very important to know what has happened in major storms in the past. (Was the site flooded? Did the overwash cover the whole island? Were protective dunes destroyed?)

On either beach or cliff sites, one very important factor is the rate of erosion (if any). Information might be gained from neighbors who have lived in the area for some time. A better and more reliable guide would be old ground or aerial photographs, if available from the United States or state geological survey, county planning office, or other sources. Detailed maps made

some time ago, which show how the coastline looked in past times, could also be used. If one knows when the photos were taken or maps made (information that is usually available) one can compare the old coastline with the present one and estimate the rate of erosion (see, for example, figure 15.20). It should be kept in mind that shoreline retreat in the future may be more rapid than it has been in the past, as a consequence of more rapidly rising sea levels. On cliff sites, too, there is landslide potential to consider, and in a seismically active area the dangers are greatly magnified.

It is advisable to find out what shoreline modifications are in place or planned, not only close to the site of interest but elsewhere along the coast. These can have impacts on locations a considerable distance away. Sometimes aerial or even satellite photographs make it possible to examine the patterns of sediment distribution and movement along the coast, which should help in assessing the likely impact of any shoreline modifications. The fact that such structures exist or are contemplated may itself be a warning sign! A history of repairs to or rebuilding of structures suggests not only a very active coastline, but also the possibility that in future, protection efforts might be abandoned for economic reasons.

Summary

Coastal areas are by nature dynamic geologic environments, and many are undergoing rapid change. Erosion is a major factor, with waves and currents the principal agents. Waves are generated by the action of wind on the water surface. They develop into breakers as they approach shore, where they may also be deflected by wave refraction in shallowing water. Storms intensify wave erosion, both by increasing wave heights and by causing storm surges that raise overall water levels. In addition to net erosion or deposition of sediment, many beaches are subject to lateral transport of sand by longshore currents. Interference with this process of littoral drift—for example, through construction of piers or jetties—results in redistribution of sediment and altered patterns of sediment erosion and deposition. The same is often true of other shore protection structures. Even changes far from the shore may affect it, as when trapping of stream sediments by flood-control dams starves beaches of sand and accelerates their disappearance.

An overall rise or fall of the land relative to the sea may result, respectively, in the formation of wave-cut terraces, or drowned valleys and fjords. At present, eustatic sea-level rise, probably caused by the ongoing melting of ice caps, is intensifying many coastal problems, especially in low-lying, vulnerable areas such as barrier islands. Another kind of delicate coastal community is the estuary, characterized by a mix of fresh and salt water and a restricted water flow. Prospective residents of coastal areas can investigate the long-term stability of the coastline through old maps or photographs, in addition to observing present processes active along the coast.

Terms to Remember

backwash	drowned valley	neap tides	tides
barrier island	estuary	period	trough
beach	eustatic	shoreline	turbidity
beach face	height (wave)	spring tides	wave
berm	littoral drift	surf zone	wave-cut terrace
coastline	longshore current	surge	wavelength
crest	milling	swash	wave refraction

Questions for Review

1. Sketch a cross section of several waves and indicate the following: wave crests, wavelength; how water is moving beneath the surface.

2. Do breakers typically form in the open ocean? Explain.

3. Briefly describe the origin of tides.

4. Under what circumstances does littoral drift occur? Why does it not necessarily result in complete removal of a beach's sand over time?

5. Describe how a shoreline is altered when groins or breakwaters are erected to slow littoral drift.

6. Cite two possible problems associated with artificial beach sand replenishment.

7. What is an emergent coastline? Does it necessarily reflect uplift of the coastal land? Cite one erosional feature characteristic of some emergent coastlines.

8. Describe one cause of eustatic sea-level changes.

9. What is a barrier island? Describe how such islands commonly migrate through time, and comment on the impact of rising world sea levels.

10. What is an estuary? Note a situation in which an estuary might develop.

11. Name and describe any three factors that might be considered before undertaking development on a coastal site.

For Further Thought

1. Consider the barrier-island development shown in figure 15.18. Might it have any stabilizing effect on the island? How might you expect the distribution of sand around the island to change over time? What would you expect to be the effect of storm surges?

2. Choose any major coastal city and investigate the extent of flooding that could be expected in the case of a eustatic sea-level rise of (a) 1 meter and (b) 5 meters. For an example of possible problems and responses to them, investigate what is taking place in the city of Venice, Italy, where a combination of tectonic sinking, surface subsidence, and sea-level rise is causing increasing flooding problems.

Suggestions for Further Reading

American Geological Institute. "Old Solutions Fail to Solve Beach Problems." *Geotimes* (December 1981): 18–22.

Barnes, R. S. K., ed. *The Coastline*. New York: John Wiley and Sons, 1977.

Beer, T. *Environmental Oceanography*. New York: Pergamon Press, 1983.

Bird, E. C. F. *Coasts*. Cambridge, Mass.: M.I.T. Press, 1969.

Davies, J. L. *Geographical Variation in Coastal Development*. 2d ed. New York: Longman, 1980.

Fisher, J. S., and Dolan, R., eds. *Beach Processes and Coastal Hydrodynamics*. Stroudsburg, Pa.: Dowden, Hutchinson, and Ross, 1977.

Heikoff, J. M. *Marine Shoreland Resources Management*. Ann Arbor, Mich.: Ann Arbor Science, 1980.

MacLeish, W.H., ed. "The Coast." *Oceanus* 23, no. 4 (1981).

Morton, R. A.; Pilkey, O. R., Jr.; Pilkey; O. R., Sr.; and Neal, W. J. *Living with the Texas Shore*. Durham, N.C.: Duke University Press, 1983.

Pethick, J. *An Introduction to Coastal Geomorphology*. London: Edward Arnold, 1984.

U.S. Army Corps of Engineers. "Shore Protection Guidelines." Washington, D.C.: U.S. Government Printing Office, 1971.

Glaciers

Introduction

The significance of glaciers is not restricted to the fact that they represent the world's largest reservoir of fresh water. Nor are their effects generally sudden or hazardous, or even experienced over most of the earth at present. Ice now covers only about 10% of the continental land area. However, ice has sculptured much of the present landscape. Features attributed to glacial episodes are found over about three-quarters of the continental surface. Only a few tens of thousands of years ago, sheets of ice covered major portions of North America, Europe, and Asia. Melting glaciers recharged aquifers we use today for water; glacial sediments from older ice advances themselves make up some of the aquifers. As they look to the future, many climatologists anticipate further melting of the ice remaining—and not with enthusiasm. In this chapter we will survey the nature of glaciers, the distinctive erosional and depositional features they produce, and other characteristics of glaciated or cold regions. We will also consider how the extent of glaciers has differed in the past (ice ages and their possible causes), and may change in the future (through the alteration of global climate by human activities).

Ice and the Hydrologic Cycle

Approximately three-quarters of the fresh water on earth is found as ice in glaciers. Water enters this reservoir as precipitation (usually snowfall) and leaves it by evaporation or by melting. The size of this water reservoir can be appreciated through the realization that the world's supply of glacial ice is the equivalent of 60 years' precipitation over the whole earth. To put it another way, that quantity of water represents 900 years' flow of all the world's rivers at their present discharge.

In states where glaciers are large or numerous, glacial meltwater may be the principal source of summer streamflow. By far the most extensive glaciers in the United States are those of Alaska, which cover about 3% of the state's land area. Summer streamflow from these glaciers is estimated at nearly fifty trillion gallons of water. Even in less extensively glaciated states like Washington, Montana, California, and Wyoming, streamflow from summer meltwater amounts to tens or hundreds of billions of gallons.

It follows that anything that modifies glacial melting patterns can profoundly affect water supplies in such regions. Dusting the ice surface with a thin layer of dark material, such as coal dust, will increase the heating of the surface and hence the rate of melting and water flow. Conversely, cloud seeding over glaciated areas could be used to increase precipitation and the amount of water stored in the glacier. Increased meltwater flow can be useful not only in increasing water supplies but in achieving higher levels of hydroelectric power production. Techniques for modifying glacial meltwater flow are not now being used in the United States, in part because the majority of glaciers in the United States are in national parks, primitive areas, or other protected lands. Some of these techniques are being practiced in parts of the Soviet Union and China.

The Nature of Glaciers

Fundamentally, a glacier is a mass of ice, on land, that moves under its own weight. A quantity of snow sufficient to form a glacier does not fall in a single winter, so in order for glaciers to develop the climate must be cold enough that some snow and ice persist year round. This in turn requires an appropriate combination of elevation and latitude. There is a tendency to associate glaciers with the extreme cold of polar regions. However, temperatures also generally decrease at high elevations, so glaciers can exist in mountainous areas even in tropical or subtropical regions. There are three mountains in Mexico that have glaciers, all at elevations above 5000 meters (over 16,000 feet). Similarly, there are glaciers on Mount Kilimanjaro in East Africa, but only at elevations above 4400 meters (14,500 feet).

Glacier Formation

Several requirements must be met in order for a glacier to form. There must be sufficient moisture in the air to provide the necessary precipitation. The amount of winter snowfall must exceed summer melting, so that the snow accumulates year by year. Most glaciers start in the mountains, as snowpatches that survive the summer. North-facing slopes, protected from the strongest sunlight, favor this survival. So do gentle slopes, on which snow can pile up thickly instead of plummeting down periodically in avalanches.

As the snow accumulates, it is gradually transformed into ice. The weight of overlying snow packs it down, drives out much of the air, and causes it to recrystallize into coarser, denser, interlocking ice crystals (figure 16.1). Alternate thawing and freezing after deposition hasten the transformation. (Glacial ice could, in fact, be regarded as a very-low-temperature metamorphic rock.) The material intermediate in texture between freshly fallen snow and compact, solid

Figure 16.1

Progression from fluffy, fresh snow to dense glacial ice.

Figure 16.2

An alpine glacier—Sherman Glacier, Alaska. Note the streaks of sediment being carried along by the flowing ice, and tributary glaciers joining the main glacier from the right.

Photo by A. Post, courtesy U.S. Geological Survey.

ice is called firn. It is not unlike the coarse frozen material that forms along roadsides in snowy areas during winter and early spring, from piles of snow plowed off the roads. Complete conversion of snow into glacial ice may take from 5 to 3500 years, depending on such factors as climate and rate of snow accumulation at the top of the pile. Eventually the mass of ice will become large enough that, if there is any slope to the terrain, it will begin to slide or flow downhill.

Types of Glaciers

Glaciers are divided into two types on the basis of size and occurrence. The most numerous today are the alpine glaciers, typically found in mountainous regions, most often at relatively high elevations. Many

occupy valleys in the mountains and are termed, logically, *valley glaciers* (see figure 16.2). Most of the estimated 70,000 to 200,000 glaciers in the world today are alpine glaciers. The larger and rarer continental glaciers (figure 16.3) are also known as *ice caps* (generally less than 50,000 square kilometers in area) or *ice sheets* (larger). They can cover whole continents and reach thicknesses of a kilometer or more. Though they are far fewer in number than the alpine glaciers, the continental glaciers comprise far more ice. At present the two principal continental glaciers are the Greenland and the Antarctic ice sheets.

A continental glacier, or ice sheet—aerial view of a portion of the Antarctic ice sheet. Mount Overlord, in foreground, rises to 3400 meters (10,000 feet) above sea level, but most of it is below the ice surface.

Photo by W. B. Hamilton, courtesy U.S. Geological Survey.

(The Arctic polar ice mass is not a true glacier, as it is not based on land.) The Antarctic ice sheet is so large that it could easily cover the 48 contiguous United States. The geologic record indicates that at several times in the past, still more extensive ice sheets existed on earth.

Movement of Glaciers

The movement of a glacier may be a nearly imperceptible 20 meters (about 65 feet) per year or, for brief periods at least, a glacier may *surge* at up to 10 kilometers per year (about 30 meters, or close to 100 feet, per day). Glacial movements can occur in several ways. The pressure of the ice at its base may be sufficient to melt a little of it (just as ice melts under a skater's blade even in cold weather), and the ice mass as a whole may slide on this meltwater.

Internal deformation within the ice, including plastic flow, is another mechanism by which glaciers move downslope. The existence of this deformation is demonstrated in part by the fact that not all portions of a glacier move equally quickly. Typically, flow is fastest at the top and center (figure 16.4). This is in part because friction at the base of the glacier (and the sides of a valley glacier) slows it down. The effect is analogous to the effect of friction within the channel on stream-flow velocity. Also, the internal deformation is most pronounced in the deeper zones of the glacier, which are under the greatest confining pressure. The uppermost layers of the glacier, riding on more plastic layers below, behave rigidly or brittlely, fracturing to make the crevasses, or cracks, that are the peril of climbers who cross glaciers (figure 16.5).

The Glacial Budget

Matter is continually cycled through a glacier. Where it is cold enough for snow to fall, fresh material accumulates, adding to the weight of ice and snow that pushes the glacier downhill. At some point the advancing edge of the glacier terminates. This may be either because it flows out over water and breaks up, creating icebergs by a process known as calving (figure 16.6), or more commonly, because it has flowed to a place that is warm enough that ice loss (ablation) by melting, evaporation, or calving is at least as rapid as the rate at which new ice flows in to replace it (figure 16.7).

Figure 16.4

Differential rates of movement of different parts of a glacier indicate internal deformation.

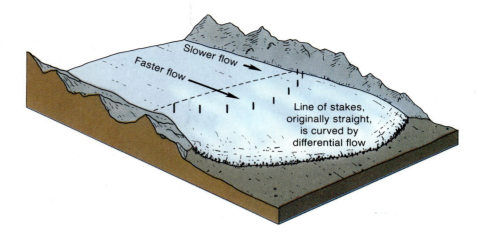

Slower flow

Faster flow

Line of stakes, originally straight, is curved by differential flow

Figure 16.5

Crevasses form in the brittle upper portion of the glacier.

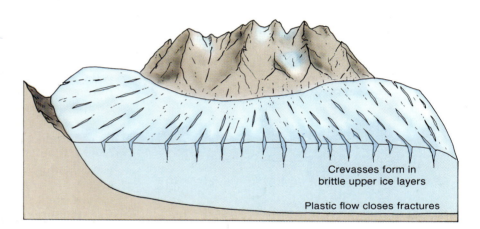

Crevasses form in brittle upper ice layers

Plastic flow closes fractures

Zone of accumulation → ← Zone of ablation →

New snow added

Direction of flow

Material lost

Figure 16.7

Longitudinal cross section of a glacier (schematic). In the zone of accumulation, the addition of new material exceeds loss by melting or evaporation; the reverse is true in the zone of ablation, where there is a net loss of ice.

© NASA

Figure 16.6

Calving of icebergs from an Alaskan glacier.

Photo by M. Dalecheck, courtesy U.S. Geological Survey.

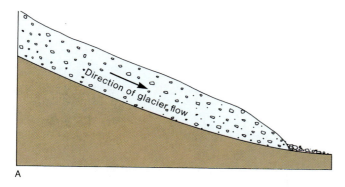

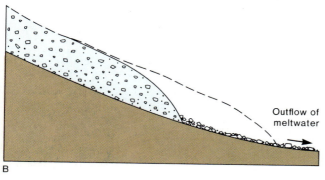

Figure 16.8

Winter advance of glacier carries both ice and sediment downslope *(A)*. In summer, rapid melting causes apparent retreat of glacier, accompanied by deposition of till from melting ice *(B)*.

Over the course of a year, the size of a glacier varies (figure 16.8). In winter the rate of accumulation increases, and melting and evaporation decrease. The glacier becomes thicker and extends farther from its source; it is then said to be *advancing*. In summer, snowfall is reduced or halted and melting accelerates; ablation exceeds accumulation. The glacier is then described as *retreating*, though of course it does not flow backwards, uphill. The leading edge simply melts back faster than the glacier moves forward.

Over many years, if the climate remains stable, the glacier will achieve a sort of dynamic equilibrium, in which the winter advance just balances the summer retreat, and the average seasonal limits of the glacier remain constant from year to year. An unusually cold (or snowy) period spanning several years would be reflected in a net advance of the glacier over that period, and vice versa. Such phenomena have been observed in recent history. From the mid-1800s to the early 1900s, a worldwide retreat of many alpine glaciers far up into their valleys was observed. Beginning about 1940, the trend seemed to have reversed, and glaciers have generally advanced. It should

be noted, however, that all glaciers worldwide do not necessarily advance or retreat simultaneously, for local climatic conditions may deviate from overall world trends.

Glacial Erosion

The mass and solidity of a glacier make it a very effective agent of both erosion and sediment transport, more so than either wind or liquid water.

Sediment Production and Transport

Rocks fall from valley walls onto the ice and are subsequently carried along with it. Note the dark stripes of sediment along the margins of the alpine glacier in figure 16.2. Additional material becomes frozen into the ice at the base and sides of the glacier. Ice itself is too soft to erode the rocks over which it flows, but these rock fragments frozen into the ice cause erosion by abrasion of the surrounding rocks at the base of the glacier or along valley walls. The result is sediment, produced with little alteration of the original chemical and mineralogical character of the parent rock. Continued pulverizing into ever finer fragments eventually produces a powdery, silt-sized sediment termed rock flour.

Because ice is solid, it transports sediments with equal efficiency regardless of particle size. Everything is moved together and, when the ice melts, everything is deposited together, as shown in figure 16.8. Thus sediments deposited by melting glaciers are typically poorly sorted with respect to particle size or density, in contrast to most water- or wind-deposited sediments. This poorly sorted glacial sediment is called till (figure 16.9). Its lithified equivalent is a tillite.

The Glacial Valley

The differences in the character of erosion by water and by ice also result in differences in shape between glacial and stream valleys. Glacial valleys are characteristically U-shaped in cross section (figure 16.10). They are broadened and deepened partly by abrasion by rocks frozen into the ice, and partly by a process known as plucking (figure 16.11). Plucking occurs as water seeps into cracked rocks and freezes, becoming one with the glacier itself. As the glacier moves on, it may tug apart the fractured rock and pluck away chunks of it. The process is accelerated by the fact that water expands as it freezes in the cracks, driving apart rock fragments by ice wedging.

Figure 16.9

Till deposited by Grinnell Glacier, Glacier National Park (note drum and hiker for scale).

Figure 16.10

Typical U-shaped glacial valley in Grand Teton National Park.

Ice freezes into cracks, enlarging them by wedging

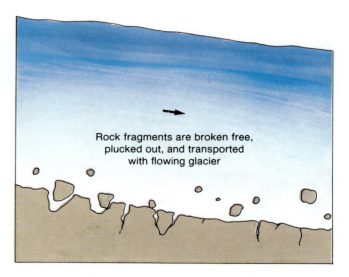

Rock fragments are broken free, plucked out, and transported with flowing glacier

Figure 16.11

Plucking by glacial ice (schematic).

Just as streams may have tributaries, so may alpine glaciers, as valleys join and the ice of several glaciers merges (see figure 16.2). When the main and tributary glaciers melt, the tributary glaciers leave **hanging valleys** (figure 16.12). These smaller, shallower valleys are abruptly truncated by the deeper valley of the main ice mass.

Plucking at the head of an alpine glacier, combined with weathering of the surrounding rocks, produces a rounded or bowl-like depression known as a **cirque.** The cirque becomes a hospitable setting for more snow accumulation to feed the glacier. When the glacier melts away, a cirque bottom may stay filled with water, making a small, rounded lake called a **tarn** (figure 16.13). Other erosional features are formed when more than one glacier erodes a mountain. Where several alpine glaciers flow down in different directions from the same high peak, each chipping away at its head, the result may eventually be a pointed **horn** chiseled from the peak (figure 16.14). Glaciers flowing in parallel, each forming a steep-sided valley, may form an **arête**, a sharp-spined ridge between the valleys (figure 16.15).

Figure 16.12

Hanging valley above the floor of Rock Creek Valley, Montana.

Figure 16.13

Tarns in old glacial cirques, Beartooth Mountains, Montana.

Figure 16.14

A horn formed by the erosion of several glaciers around a single peak, Glacier National Park.

Figure 16.15

An arête formed between two parallel glacial valleys.

Figure 16.16

A glacier was here; striations on a rock surface show direction of flow. Notice that the grooves extend continuously across the boundary between different rock types. They are clearly a surface feature, not a characteristic of either rock.

Other Erosional Features

Gravel and boulders frozen into the base of the ice and dragged along as it moves act as a natural coarse sandpaper. They scrape parallel grooves, or **striations,** in softer rocks over which they move (figure 16.16). Striations are more than evidence of the past presence of a glacier; they also indicate the direction in which it flowed (parallel to the striations). Striations can be produced by both continental and alpine glaciers.

Many alpine glaciers begin by flowing down the valleys of mountain streams. Even a continental glacier may do so locally, widening and deepening the stream valley in the process. The Finger Lakes of upstate New York (figure 16.17) formed in this way, as an ice sheet gouged deeper into old stream valleys. The lakes, too, are elongated in the direction of glacier flow.

The basins now occupied by the Great Lakes were deepened by continental glaciation and filled with glacial meltwater as the glaciers retreated. The basic drainage network of the Mississippi River system was established at the same time, as it carried that meltwater to the sea. Even mountains can be scoured into elongated remnants by the immense mass of ice represented by a continental ice sheet.

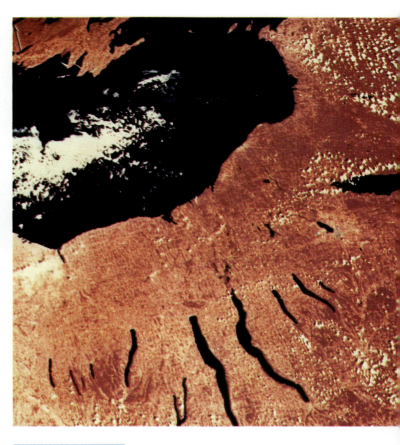

Figure 16.17

Landsat satellite photograph of the Finger Lakes region, New York state.
© NASA

Depositional Features of Glaciers

Sooner or later, the melting ice deposits its load of sediment, either directly or through the action of meltwater. This can produce a variety of features. One of the simplest is the isolated large boulder that has been carried along in the ice far from its parent rock. If it cannot be identified as having been derived from the local bedrock, it is a glacial **erratic.**

Direct Deposition: Moraines

A body of till, carried along in and ultimately deposited directly by melting of the glacier, is a **moraine.** Moraine comes in several forms. The concentrations of rock debris along the sides of a valley glacier, which fall onto it from the valley walls above or are ground out by the ice, are **lateral moraines.** Where tributary glaciers join a valley glacier, the lateral moraines toward the inside join as a ribbon of moraine within the combined ice mass, **a medial moraine** (figure 16.18).

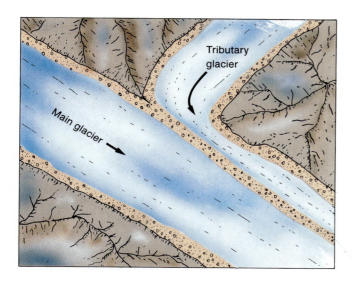

Lateral moraines join to become a medial moraine as a tributary glacier joins the main ice mass.

Formation of a terminal moraine occurs by a combination of "bulldozing" by an advancing glacier and repeated cycles of deposition of till by melting ice.

The merging of many tributaries eventually gives the resulting composite glacier a striped appearance (recall figure 16.2). When the glacier retreats, recognizable ridges of lateral or medial moraine may be left behind. A broad blanket of ice-deposited till with no particular form is simply **ground moraine.**

Glacial sediment transport continually brings a fresh supply of sediment to the glacier's end. If the extent of the glacier remains the same for several years, a ridge of till accumulates there that is known as an **end moraine.** The formation of a ridgelike landform can also be enhanced by an advancing glacier, acting like a bulldozer on previously deposited ground moraine (figure 16.19). A single glacier may leave multiple end moraines (figure 16.20). In such a case, the one marking the farthest advance of the tongue of ice is the **terminal moraine.** End moraines deposited during glacial recession in periods when the ice front was temporarily stationary are **recessional moraines.**

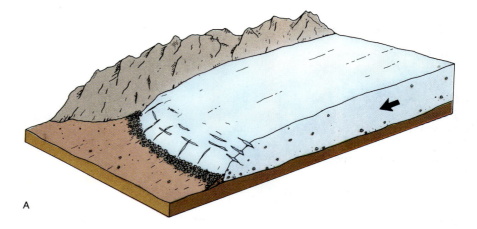

A

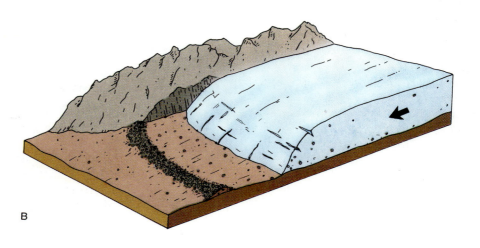

B

Figure 16.20

Maclaren Glacier, Alaska, showing concentric loops of moraine (foreground) left by retreat of the glacier.

Photo by T. L. Péwé, courtesy U.S. Geological Survey.

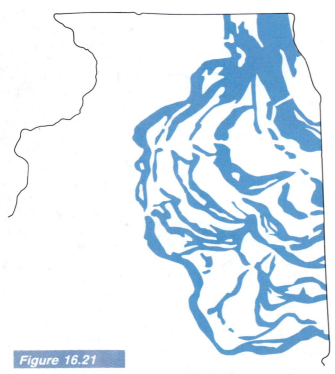

Figure 16.21

Terminal moraines of northeastern Illinois. These numerous features reflect the advances of many separate tongues of ice over hundreds of thousands of years.

Source: H. B. Willman and J. C. Frye, Glacial Map of Illinois, Illinois State Geological Survey, 1970.

Continental glaciers also may leave terminal moraines, by which advances of the great ice sheets can be mapped (figure 16.21). Alternatively, the ice may override previously deposited moraines, streaking them out into oval mounds elongated in the direction of ice flow. These are **drumlins** (figure 16.22).

Meltwater Deposition: Outwash

When glacier ice melts, it not only deposits sediment but releases volumes of water that may in turn redistribute the sediment. The resultant water-deposited sediment is glacial **outwash.** As a water-laid sediment, outwash tends to be better sorted and more obviously layered than till. The very fine rock flour can be transported especially readily, for its fine size tends to keep it in suspension (figure 16.23). It is also easily redistributed later still by the action of wind, to make a distinctive fine sediment called *loess* (see the next chapter).

As in moraines, several distinct landforms can occur in outwash. One is a winding, snakelike outwash ridge deposited by meltwater streams flowing within and beneath a melting glacier. The resultant feature—a stream deposit without a stream valley—

Figure 16.22

Drumlins shaped by the flow of continental glaciers over mounds of sediment (Landsat photo of part of the Canadian Shield). © NASA

Figure 16.23

Lakes in glacial valley clouded and colored by the presence of suspended rock flour. Note variation in color from lake to lake.

is an **esker** (figure 16.24). Glacial retreat can also strand blocks of ice in thick outwash deposits. As the ice melts, it leaves behind a hole in the outwash, called a **kettle** (figure 16.25).

Glaciers and Lakes

End moraines can act as dams within a glacial valley, trapping meltwater to form a lake. Sediment deposition in such a lake typically shows an annual cyclicity. The principal flushing of sediment into the lake occurs during the accelerated melting of summer; in winter, the addition of sediment is minimal. The coarser sediments settle out rather quickly, but the finest rock flour may take months to settle. At the same time, microorganisms such as algae may flourish in the warmer, sunnier summer months, then die and settle to the bottom with the coming of winter. Each year, then, a distinct, two-part layer of sediment is deposited that is coarser at the bottom (spring/summer deposition), finer grained and richer in organic matter

Figure 16.24

An esker formed by a stream flowing within or under a glacier, depositing sediment.

Photo by J. H. Hartshorn, courtesy U.S. Geological Survey.

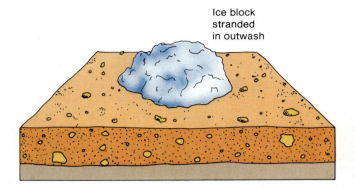

Ice block
stranded
in outwash

Hole left
in outwash
after ice
melted

A

B

Figure 16.25

Kettles. *(A)* Formation of a kettle as a stranded block of ice melts. *(B)* Example of kettle hole in moraine.

(B) W. B. Hamilton, courtesy U.S. Geological Survey.

at the top (fall/winter). Each such annual sediment couplet is a **varve**. Hundreds of these annual depositional layers may be preserved in varved sediments, and these can be counted, somewhat like tree rings, to determine the age of the sediments.

On a grander scale, the difference in climate associated with continental glaciation left visible traces even beyond the area actively glaciated. The glaciated areas, which now have temperate climates, were icy; nearer the equator, where hot, dry conditions now prevail, the weather was more temperate and moister. Abundant rainfall led to the formation of **pluvial lakes** (so named from the Latin for "rain"). The most impressive of these, glacial Lake Bonneville, covered an area about one-third the size of Utah. As the climate dried out, the lake dried up, leaving behind remnant shorelines (by which its extent has been mapped), and ultimately the Bonneville salt flats. (The Great Salt Lake of Utah seems to be a younger lake formed on the old salty bed of Lake Bonneville, deriving its high salt content by solution of that older salt.)

Past Ice Ages and Their Possible Causes

During the late nineteenth century, Professor Louis Agassiz was among the first to use the term *Ice Age*. Drawing on his own field observations and the work of previous investigators, he promoted the concept that continental glaciation on a massive scale must have affected the continents in the not-too-distant geologic past. Evidence of the erosional and depositional features characteristic of glaciers was widespread, even in some equatorial regions, and did not permit any other convincing explanation. With the advent of radiometric dating techniques and detailed studies of tillites and other glacial products in the rock record, it was subsequently recognized that there have been several periods of extensive continental glaciation—around 270, 430, 670, 750, and 960 million years ago, and earlier—in addition to the most recent such episode, for which we have particularly well-preserved evidence. The term **ice age** (uncapitalized) is used for any such period; when capitalized (*Ice Age*), it generally refers only to the last ice age, the one taking place during the Pleistocene, ending with the last glacial retreat only about 10,000 years ago. During that period, up to one-third of the earth's surface was ice-covered.

Box 16.1

Box 16.1

Permafrost

In temperate climates, where temperatures drop below freezing in only a few months of the year, the ground freezes shallowly during this period but thaws during warmer weather. In alpine and arctic climates, at high elevations or high latitudes, the winter freeze is so deep and pervasive that summer's thaw does not penetrate the whole frozen zone, and may barely thaw the surface. Below lies a layer of more or less permanently frozen ground, permafrost. The top of the frozen zone is sometimes referred to as a *permafrost table,* and it behaves

somewhat like a water table, rising and falling with seasonal climatic variations (figure 1).

One consequence of the persistent freezing is the *patterned ground* commonly observed in arctic regions (figure 2). Contraction of the soil during freezing causes it to break up into polygonal chunks somewhat analogous to basalt columns.

As long as permafrost stays frozen, it constitutes a fairly solid base for structures. If disturbed and warmed, whether by natural thawing or by construction, some of the ice

melts. Below is still frozen soil, so, depending on the topography, the water may drain slowly or not at all. The result is a mucky, sodden mass of saturated soil that is difficult to work in or with, and structurally weak (figure 3). This was a major problem during the construction of the Trans-Alaska pipeline. In fact, where the pipeline passes underground, it actually had to be refrigerated in some places to keep the warm oil from melting the surrounding permafrost, which might cause the pipeline to sag and perhaps to break.

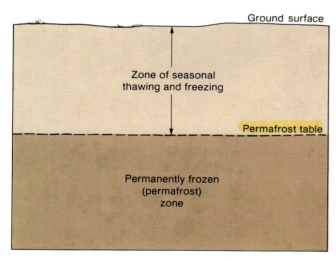

Figure 1
Permafrost and the permafrost table (schematic).

Figure 2
Patterned ground, broken into polygons, in an alpine climate.

A

B

Figure 3
Consequences of the melting of permafrost. *(A)* Tractor mired in mud from melted permafrost. *(B)* These warped railroad tracks, near Copper River, Alaska, had to be abandoned in 1938.

Photos *(A,B)* by O. J. Ferrians, Jr., courtesy U.S. Geological Survey.

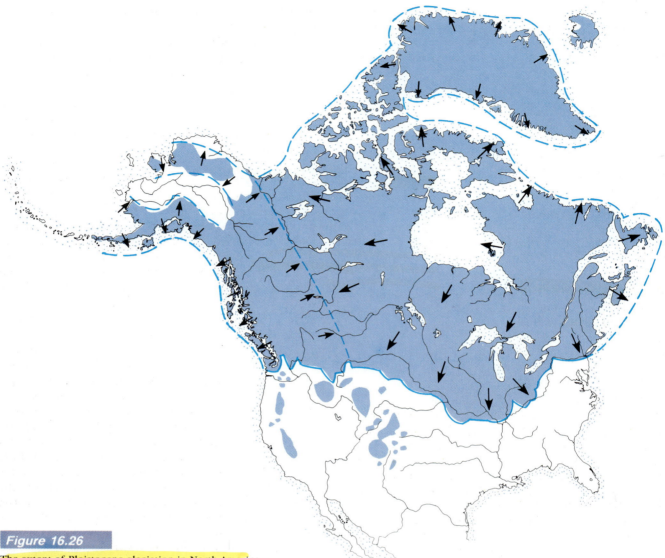

Figure 16.26

The extent of Pleistocene glaciation in North America.

Source: After C. S. Denney, U.S. Geological Survey, National Atlas of the United States.

Nature of Ice Ages

An ice age does not consist of a single cycle of an advance of ponderous continental glaciers, followed, after a time, by a single melting event. The ice sheets advance and retreat many times. At one time, it was thought that the Pleistocene Ice Age consisted (in North America) of four major ice advances, interspersed with warmer interglacial periods. It has recently been recognized that the sequence of events is still more complex. We can say, as a general observation, that any given ice age is likely to have consisted of many ice advances, with warmer interglacial periods between them. Each cycle of advance and retreat could span thousands or even tens of thousands of years; a whole ice age might span millions of years. The maximum extent of Pleistocene ice sheets in

North America is shown in figure 16.26. The ice covered essentially all of Canada and much of the United States, and in many places was well over a mile thick.

A further result of ice-sheet advance on a large scale is a significant drop in sea level. Considering the global water budget, it is evident that the oceans are the only water reservoir of sufficient size to supply the quantity of water represented by extensive ice sheets. During the Pleistocene, eustatic lowering of sea level by over 130 meters (400 feet) must have occurred. We know this in part from finding glacial deposits and stream channels on the continental shelves (chapter 12).

Extreme deviations from present temperatures are not required to cause an ice age. Cooling of 5° to 10° C, or 10° to 20° F, is probably adequate. However, the cold periods would have to last for centuries or millennia, not just a few years, in order for sufficient snow to accumulate. What could cause such sustained cooling is problematic. A variety of possible causes of prolonged climatic fluctuations have been proposed; none has been proven. The proposed causes of ice ages can be classified into two groups: those that involve events external to the earth, and those for which the changes arise entirely on earth.

Possible Causes of Ice Ages—Extraterrestrial

One possible external cause, for example, would be a significant change in the energy output from the sun. We know that there are at present cycles of sunspot activity that cause variations in the intensity of sunlight. Since the earth's surface is warmed mainly by sunlight, these variations should logically result in fluctuations of surface temperature worldwide. However, the variations in solar energy output would have to be about ten times as large as they are in the modern sunspot cycle to account even for the short-term temperature fluctuations observed on earth. In order to cause an ice age of major proportions, lasting thousands of years, any cooling trend would also have to last much longer than the 11 years of the present sunspot cycles. Another problem with linking solar activity fluctuations with past ice ages is simply lack of evidence. Although means exist for estimating temperatures on earth in the past, scientists have yet to conceive of a way to determine the pattern of solar activity in ancient times. There is thus no way to test the theory, to prove or disprove it.

Another external cause for ice ages that has been proposed is related to the observed variation in the tilt of the earth's axis in space. The earth is known to "wobble" as it rotates. Changes in tilt would not change the total amount of sunlight reaching earth, but they would affect its distribution. If for a time the poles were tilted farther away from incident sunlight, the polar region might become enough colder that a large ice sheet could begin to develop.

Possible Earth-Based Causes of Ice Ages

Since the advent of plate-tectonic theory, a novel suggestion for an earth-based cause of ice ages has invoked continental drift. Prior to the breakup of Pangaea, all continental land was together in a single large unit. This concentration of continents would have meant much freer circulation of ocean currents elsewhere on the globe, and in particular, more circulation of warm equatorial waters to the poles. With the breakup of Pangaea, the oceanic circulation patterns would have been disrupted. The poles could have become substantially colder as their waters became more isolated. This in turn could make sustained development of a thick ice sheet possible. No global cooling is involved in this case, only a change in heat distribution resulting in a greater contrast in temperature between polar and equatorial regions.

A limitation of this mechanism is that it can readily be used only to account for the Pleistocene glaciation. There is in fact considerable evidence for extensive glaciation over much of the southern hemisphere—India, Australia, southern Africa and South America—between 200 and 300 million years ago, just *prior* to the breakup of Pangaea, and the distribution of the deposits of this ice age have been of fundamental importance to the pre-drift reassembly of Pangaea. Beyond that, we do not know enough about the positions of all the continents still further back in time to know whether continental drift and the resulting changes in oceanic circulation patterns can be used to explain earlier ice ages.

If incoming solar radiation were partially blocked by something in the atmosphere, the cooling might be adequate to induce the start of an ice age. We know that dust in the atmosphere can cause measurable cooling. Within modern times, as noted in chapter 9, explosive volcanic eruptions of large volcanoes—Krakatoa, El Chichón, and others—have put many tons of dust and volcanic ash and gas into the air, causing redder sunsets and cooling of a few degrees over the earth. In none of these cases was the cooling serious enough or prolonged enough to cause an ice age. Even some of the larger prehistoric eruptions, such as those that occurred in the vicinity of Yellowstone Park, or the explosion that produced the crater in Mount Mazama that is now Crater Lake, would have been inadequate individually to produce the required environmental changes: the dust and debris from these events too would have settled out of the atmosphere within a few years. However, there have been periods of more intensive and frequent volcanic activity in the geologic past. The cumulative effects of multiple explosive eruptions during such an episode might have included a sustained cooling trend and, ultimately, ice-sheet formation. However, there is no definite evidence to show that the cooling would be sufficient to initiate a major ice advance.

The foregoing has been a brief and incomplete sampling of the processes and phenomena that may have caused climatic upheavals in the past. The possibilities are many, and there is as yet insufficient evidence to select among them. Of more immediate concern are the present human activities that may also have begun to alter the climate irrevocably.

Box 16.2

The "Little Ice Age" and Climatic Oscillations

Major ice advances may be separated by hundreds of thousands of years. Short-term climatic cycles of warming and cooling also occur on a scale of years to centuries. Perhaps the best known of the modern fluctuations is the period known as the "Little Ice Age," which lasted from about A.D. 1450 to 1850. Worldwide temperatures were not actually vastly colder than at present. In eastern Europe, for example, average temperatures were just 1° to 1.5° C (2° to 3° F) lower. However, the practical impact of this seemingly small drop in temperature was profound. Alpine glaciers in the Alps and elsewhere advanced dramatically, as paintings from the period attest. At times one could walk between Staten Island and

Manhattan. The river Thames froze more than two dozen times during the Little Ice Age; it has not completely frozen again since the winter of 1813–14. The early American colonists apparently had to endure winters considerably more severe than what we usually experience today. Reduced evaporation from the world's oceans led to devastating droughts in many parts of the world. Then, beginning about 1850, sustained warming began. The warming trend lasted nearly a century until, as noted in the text, global temperatures started to decline again about 1940.

Such evidence as the extent of alpine glaciers and the distribution of fossil plants has been used to reconstruct estimated global

temperature trends over the last 10,000 years. The data suggest variations of up to 6° C (11° F) in average annual temperatures over warming/cooling cycles spanning centuries. Such global temperature oscillations, occurring independently of human activities, complicate the detection of any increased greenhouse-effect heating from atmospheric CO_2, for the two influences are superimposed. Nevertheless, many climatologists believe that by the year 2000, if not sooner, the effects of the CO_2 will be clearly demonstrable over the background effects of natural climatic variations. Thereafter the heating trend might be expected to dominate, ultimately making the earth warmer than it has been for a million years or more.

Ice and the Greenhouse Effect

The evolution of a technological society has meant increasing energy consumption. Most of that energy has been supplied by the fossil fuels—coal, oil, and natural gas—and it is likely that these will continue to be important energy sources for several decades at least. One product that all these fuels have in common is carbon dioxide gas, CO_2. Carbon dioxide is an inert, odorless, tasteless, nontoxic gas. Nevertheless, its production may be a matter of serious concern because of the associated **greenhouse effect**.

The Greenhouse Effect

On a sunny day, it is a good deal warmer inside a greenhouse than outside it. Light enters through the glass, is absorbed by the ground, plants, and pots inside. They in turn radiate heat—infrared radiation, not visible light. Infrared rays cannot readily escape

through the glass panes; they are trapped, and the air inside the greenhouse warms up (figure 16.27). The same effect warms the interior of a car parked out in the sun.

In the atmosphere, carbon dioxide molecules (and water vapor and other gases, to a lesser extent) act similarly to the greenhouse's glass. Light reaches the earth's surface, warming it, and the earth radiates infrared rays back. But the infrared rays are trapped by the carbon dioxide and water molecules, and a portion of the radiated heat is thus trapped in the atmosphere. As a result, the atmosphere stays warmer than it would if the heat were radiated freely back into space. Hence the term *greenhouse effect.*

Excess water in the atmosphere readily falls out as rain or snow. Some of the excess CO_2 is removed by geologic processes, such as solution in the ocean to make carbonate sediments. But in the past century or so, since the start of the so-called Industrial Age, there has been an increase of more than 10% in the concentration of CO_2 in the air. That concentration

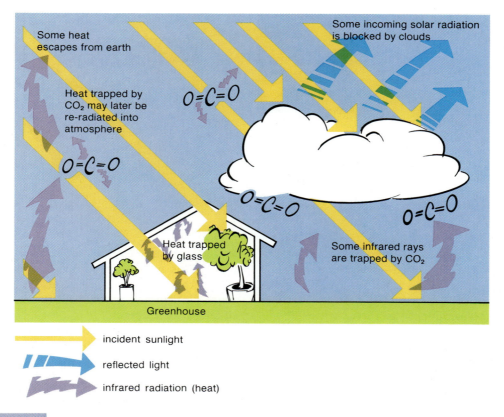

Some heat escapes from earth

Some incoming solar radiation is blocked by clouds

Heat trapped by CO_2 may later be re-radiated into atmosphere

$O=C=O$

$O=C=O$

$O=C=O$

$O=C=O$

Heat trapped by glass

Some infrared rays are trapped by CO_2

Greenhouse

incident sunlight

reflected light

infrared radiation (heat)

Figure 16.27

The greenhouse effect (schematic). Both glass and air are transparent to visible light. Like greenhouse glass, CO_2 molecules in the atmosphere trap infrared rays radiating from the sun-warmed surface of the earth.

continues to climb. If the heat-trapping by CO_2 is proportional to the concentration of CO_2 in the air, it is feared that the increased carbon dioxide will gradually cause increased greenhouse-effect heating of the earth's atmosphere.

So far greenhouse-effect heating has not occurred to any clearly measurable extent, but for an unreassuring reason: particulate air pollution. Human activities have also put soot, dust, and other solid particulates into the air. These, like volcanic ash, reflect away some sunlight before it ever reaches the ground, reducing the heating at the surface and counteracting the effects of rising CO_2 levels in the atmosphere. That may be why the climate has actually cooled since 1940, at least in low latitudes. As efforts continue to reduce the levels of soot and other unwelcome particulates released into the air, greenhouse-effect heating should become correspondingly more significant.

Possible Consequences of Atmospheric Heating

One potential consequence of atmospheric heating is agricultural. In many parts of the world, because of hot climate and little rain, agriculture is already only marginally possible. A temperature rise of only a few degrees could easily make living and farming in some of these areas impossible. Also, some climatologists expect that temperature *extremes* will become more pronounced, resulting in part in more prolonged droughts, which will be still more harmful to agriculture.

A second result comes back to the subject of this chapter: glaciers. Just as more extensive ice sheets mean a lowering of sea level, so more melting of the remaining ice sheets would mean a rise in sea level. Estimates of the resulting sea-level change vary. Upper limits put the projected sea-level rise at over 80 meters (about 250 feet). In such a case, about 20% of the world's land area would be lost. Many millions,

Figure 16.28

An 80-meter rise in sea level would flood many major cities in the United States, and elsewhere.

perhaps billions, of people now living in coastal or low-lying areas would be displaced—a large fraction of major population centers grew up along coastlines and rivers. The Statue of Liberty would be up to her neck in water. The consequences to the continental United States are illustrated in figure 16.28. A more conservative estimate, which takes into account isostatic sinking of fuller oceans as well as the volume of meltwater produced, suggests a net sea-level rise of 45 meters. Even this could be catastrophic for many coastal areas.

Such large-scale melting of ice sheets would certainly take time, perhaps several thousand years. On a shorter time scale, the problem may still become significant. It has been estimated that continued intensive fossil-fuel use could easily double the level of CO_2 in the atmosphere before the middle of the next century. That would produce a projected rise in temperature of 4° to 8° C (7° to 14° F), which would be sufficient to melt at least the West Antarctic ice sheet completely, over a few hundred years. The resulting 3- to 6-meter rise in sea level, although it sounds small, would nevertheless be enough to flood many coastal cities and ports, along with most of the world's present beaches. This would be both inconvenient and extremely expensive. (The only consolation is that the displaced inhabitants would have decades over which to adjust to the changes.)

If we take the much longer view, some believe that another ice age might ultimately be *caused* through this process. The reasoning is as follows: As heating proceeds, there will be more evaporation of water, especially from the oceans. More water in the air means more clouds. More clouds, in turn, would reflect more sunlight back into space before it could reach and heat the surface. The world might then cool off, perhaps enough to initiate another ice advance.

This, of course, is highly speculative, and in any event would not happen for many centuries. The short-term threat posed by ice sheets melting from greenhouse-effect heating, however, is generally accepted by most scientists studying the subject. What is less widely agreed upon is the extent (rate) of greenhouse-effect heating that can be expected in the near term, and whether or not other atmospheric pollutant gases generated by human activities may also have an impact on atmospheric heating.

Summary

Glaciers past and present have sculptured the landscape not only in mountainous regions but over wide areas of the continents. Formed from many seasons' accumulated snow gradually converted to ice, they flow downslope by internal deformation or on a basal meltwater layer, under the influence of gravity. Glaciers leave behind U-shaped valleys, striated rocks, and poorly sorted sediment (till) in a variety of landforms (moraines). Terminal moraines serve to mark glaciers' maximum extent; the orientation of striations and drumlins indicates the direction of flow. Further outwash of sediments by meltwater produces water-laid sediments of glacial origin. Most present glaciers are alpine glaciers. The two major ice sheets remaining are on Greenland and Antarctica.

At several times in the past, sustained periods of colder temperatures have led to major advances of ice sheets over the continents (ice ages). The last of these was in the Pleistocene, within the past 2 million years. Ice advances were accompanied by corresponding eustatic lowering of sea level. Proposed causes of ice ages include plate tectonics, volcanism, wobble of the earth on its axis, and changes in solar activity, but no single cause has been proven conclusively. In modern times, the burning of fossil fuels has been increasing the amount of carbon dioxide in the air. The resultant greenhouse-effect heating may begin to melt the remaining ice sheets, causing a rise in global sea level and ultimately the flooding of many coastal areas.

Terms to Remember

ablation	end moraine	ice age	pluvial lakes
abrasion	erratic	ice wedging	recessional moraine
alpine glacier	esker	kettle	rock flour
arête	firn	lateral moraine	striations
calving	glacier	medial moraine	tarn
cirque	greenhouse effect	moraine	terminal moraine
continental glacier	ground moraine	outwash	till
crevasses	hanging valley	permafrost	tillite
drumlins	horn	plucking	varve

Questions for Review

1. What proportion of the earth's fresh water is currently found as ice? Describe how this water cycles through a glacier.

2. Describe three conditions that encourage glacier formation.

3. What is the distinction between continental and alpine glaciers?

4. What is rock flour, and how is it produced?

5. How does glacial till differ, in terms of sorting, from many water-laid sediments? Why?

6. Describe the origin of (a) hanging valleys and (b) striations.

7. What is moraine? Distinguish among the following: ground moraine, terminal moraine, medial moraine.

8. How do outwash deposits originate? What is an esker?

9. Describe the annual depositional cycle of a glacial lake that gives rise to varves.

10. Cite and briefly summarize any three proposed causes of ice ages.

11. What is the greenhouse effect? Why is it a subject of increasing concern in modern times?

For Further Thought

1. How would you distinguish an esker from sediment deposited by a stream not associated with a glacier? From a natural levee?

2. Assume that during the Pleistocene glaciation, one-third of the continental land area was covered by ice. If the total areal extent of the continents is 149 million square kilometers, that of the oceans is 361 million square kilometers, and sea level was lowered by approximately 130 meters, what was the average thickness of ice over the glaciated areas (ignoring isostatic effects)? You may also find it interesting to examine a bathymetric (depth) chart of the oceans to see how much new land would have been created by that much lowering of sea level.

Suggestions for Further Reading

Barth, M. C., and Titus, J. J., eds. *Greenhouse Effect and Sea Level Rise.* New York: VNR Co., 1984.

Bentley, C. R. "If West Sheet Melts Rapidly—What Then?" *Geotimes* (August 1980): 20–21.

Embleton, C., and King, C. A. M. *Glacial and Periglacial Geomorphology.* 2d ed. London: Edward Arnold, 1975.

Eyles, N., (ed.) *Glacial Geology.* New York: Pergamon Press, 1983.

Flint, R. F. *Glacial and Quaternary Geology.* New York: Wiley-Interscience, 1971.

Goldthwait, R.P., ed. *Glacial Deposits.* New York: Dowden, Hutchinson, and Ross, 1975.

Goudie, A. *Environmental Change.* 2d ed. New York: Oxford University Press, 1983.

Imbrie, J., and Imbrie, K. P. *Ice Ages.* Short Hills, N.J.: Enslow Publishers, 1979.

John, B. S. *The Ice Age, Past and Present.* London: Collins, 1977.

Meier, M., and Post, A. "Glaciers: A Water Resource." U.S. Geological Survey publication, 1980.

Woodwell, G. M. "The Carbon Dioxide Question." *Scientific American* 238 (January 1978): 34–43.

Wind and Deserts

Introduction

Like water and ice, wind acts as an agent shaping the surface of the land. It can erode, transport, and deposit material. However, wind is considerably less efficient than ice or water in modifying the surface. On a worldwide average, for example, winds move only a few percent as much material as do streams. Collectively, mass transport by winds is comparable to mass transport by glaciers, though the latter are areally far more restricted. Wind lacks the ability to attack rocks chemically, by solution, or by wedging during freezing and thawing. Even in many deserts, more sediment is moved during the brief periods of intense surface runoff following occasional rainstorms than by wind during the prolonged dry periods. However, it is in deserts and semiarid regions that the effects of wind action are usually most pronounced and most easily observed—hence the grouping of deserts and wind together in this chapter.

The Origins of Wind

The flow of streams is driven by the downward pull of gravity. Gravity likewise pulls downward on the atmosphere, but this is not the principal driving force behind the horizontal flow of air at the earth's surface. Air moves over the surface primarily in response to differences in pressure, commonly corresponding to differences in temperature.

A basic factor in surface temperature variation is latitude. The sun's rays fall most directly, most intensely, on the earth's surface near the equator. Solar radiation is more dispersed near the poles. On a nonrotating earth with a uniform surface, the result would be as shown in figure 17.1. The surface would be heated more strongly near the equator, less strongly near the poles; air over the equator would, correspondingly, be warmer than polar air. Warmer, less dense (lower pressure) air would rise at the equator, while cooler, denser (higher pressure) air from the poles would move in toward the lower-pressure region. The rising warm air would spread out laterally, cool, and eventually sink. Large circulating air cells would develop, moving somewhat like mantle convection cells, cycling air from equator to poles and back.

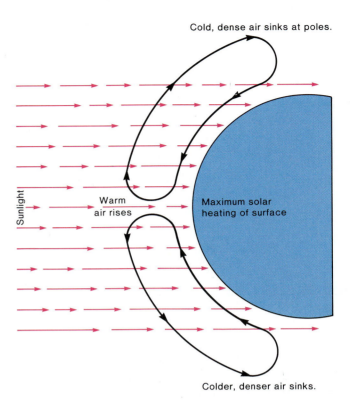

Figure 17.1

Hypothetical wind circulation on a uniform earth with no rotation.

The actual situation is considerably more complicated. For one thing, land and water are heated differentially by sunlight. Moreover, surface temperatures over the continents generally fluctuate much more on a daily basis than temperatures over adjacent oceans. Thus the irregular distribution of land and water will influence the distribution of high- and low-pressure regions, necessarily modifying air flow. Another factor is that the earth rotates on its axis, which adds an east/west component to air movement as viewed from the surface. The earth's surface is not flat; terrain irregularities introduce further irregularities in air circulation. Friction between moving air masses and the surface alters both wind direction and wind speed. For example, the presence of tall, dense vegetation can reduce near-surface wind speeds by 30% to 40% over the vegetated areas.

A generalized view of large-scale global air circulation patterns is shown in figure 17.2. Different latitude belts are characterized by different customary prevailing winds. Most of the United States is in a zone of westerlies, in which winds generally blow from west/southwest to east/northeast. Local weather conditions and geography of course produce local deviations from this pattern on a day-to-day basis.

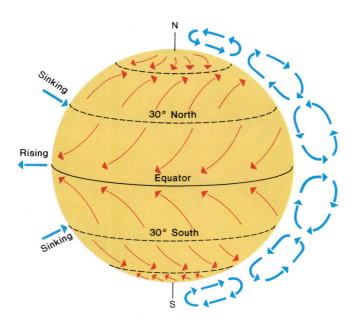

Figure 17.2

Principal present atmospheric circulation patterns.

From Plummer, Charles C., and David McGeary, *Physical Geology,* Third Edition. © 1979, 1982, 1985 Wm. C. Brown Publishers, Dubuque, Iowa. All Rights Reserved. Reprinted by permission.

If wind is predominantly from one direction, rocks will be planed off or flattened on the upwind side.

With a persistent shift in wind direction, additional facets are cut in the rock.

A

B

Wind Erosion and Sediment Transport

Flowing air and flowing water have much in common as agents of sediment transport. Both can move particles by rolling them along the surface, by saltation, or in suspension. Both also transport coarser and denser material the faster they move. Water, being denser and more viscous, is more efficient at transporting quantities of material than is wind, but the processes involved are quite similar.

Like water, wind erodes sediment more readily than solid rock. In fact, wind alone makes little impact on rock. However, wind-transported sediment can wear away at rock by the process of **abrasion.** Wind abrasion is a sort of natural sandblasting, similar to milling by sand-laden waves. Abrasion during dust storms is capable of stripping paint and frosting and pitting the glass of car windshields, as well as eroding rocks. If winds blow consistently from one or a few directions, exposed cobbles and boulders may be planed off where they face the wind (figure 17.3), in time taking on a faceted shape. Grooves may also be cut in the direction of wind flow. If wind speed is too low to lift the transported sediment very high above the ground, tall rocks may show undercutting close to ground level (figure 17.4). Rocks sculptured by abrasion are termed **ventifacts**—literally, "wind-made" rocks (from the Latin).

Figure 17.3

Ventifacts. (*A*) Ventifact formation by abrasion from one or several directions. (*B*) Examples of ventifacts.

(*B*) Photo by W. N. Lockwood, courtesy U.S. Geological Survey.

Wholesale removal of unconsolidated sediment by wind action is **deflation.** It is naturally most active where winds are unobstructed and the sediment exposed, unprotected by vegetative or artificial cover: deserts, beaches, unplanted farmland between crops. In some dry areas, the selective removal of finer sediments by wind, often assisted by seasonal surface

Undercutting of granite boulder by near-surface wind abrasion.

Photo by K. Segerstrom, courtesy U.S. Geological Survey.

runoff, produces a surface sediment of residual coarser material. This **desert pavement** (figure 17.5) effectively protects finer sediments below from further erosion. A desert pavement surface, once established, can be very stable. If the protective layer of coarser gravel and boulders is disturbed, however, the newly exposed fine sediment may be subject to rapid wind erosion.

The key role of vegetation in retarding wind erosion of sediment and soil was demonstrated especially dramatically in the United States in the early twentieth century. After the Civil War, there was a major westward migration of farmers to the Plains states. They found flat or gently rolling land, much of it covered by prairie grasses and wildflowers rather than thick forests, so it was easy to clear and adapt for farming. Over much of the area, native vegetation was removed and the land plowed and planted to seasonal crops. Elsewhere, grazing livestock cropped the prairie. In the 1930s, several years of drought killed the crops, leaving the soil bare and vulnerable to erosion by the unusually strong winds that followed. This was the Dust Bowl episode. Hundreds of millions of tons of soil were removed, transported, and then dumped as the winds lulled, burying homes and farms.

Figure 17.5

Effects of deflation. Example of desert pavement, formed by selective removal of finer sediments.

Photo by J. R. Stacy, courtesy U.S. Geological Survey.

The native prairie vegetation had been adapted to the climate, while many of the crops were not. Once the crops died, there was nothing left to hold down the soil and protect it from the west winds sweeping unobstructed across the plains. The problem of minimizing soil erosion from farmland will be considered further in the next chapter. (See also box 17.1.)

Eolian Deposits

Wind-related processes and products are also termed **eolian** (or, in older literature, *aeolian;* named for Aeolus, Latin god of the winds). There are relatively few kinds of eolian sediment deposits. One feature that they tend to share is that they are commonly well sorted. As with water-laid sediments, this is a consequence of the influence of flow velocity on maximum particle size moved. Winds selectively move finer, lighter materials; as they slow, they drop coarser, denser grains first.

Dunes

The **principal eolian depositional landform is the dune,** a low mound or ridge of sediment. When they hear "dune," most people think "sand dune," but in fact dunes can be formed of sediment of different sizes, and even of snow (figure 17.6). Dunes begin to form when sediment-bearing winds encounter an obstacle to slow them down. With reduced velocity, the wind

A

B

Figure 17.6

Examples of dunes. (A) Sand dune, Great Sand Dune National Monument. Note ripples, like miniature dunes, on dune surface. (B) Snow dunes on a barren Midwestern field in winter. Note ground bared by deflation between dunes.

Figure 17.7

Occasionally dunes can grow very large—Great Sand Dune National Monument, with adjacent mountains for scale. The dunes have accumulated here because winds are slowed by these mountains.

begins to drop the coarsest, heaviest fraction of its load, which in turn creates a larger obstacle that constitutes more of a windbreak, causing more deposition, in a self-reinforcing cycle. Once started, a dune can grow very large. Typical dunes range from 3 to 100 meters in height, but dunes 200 meters or more high are occasionally found (figure 17.7). What ultimately limits the size of dunes is not known. One consideration is that a dune is not a permanent, static object. Once formed, dunes tend to migrate, particularly if winds continue to blow predominantly from a single direction.

Figure 17.8

Erosion has brought out the eolian crossbeds in this outcrop of Navajo Sandstone. One can almost imagine the blowing wind shaping these now-lithified dune sands.

Dune Migration

A dune assumes a characteristic profile in cross section (cut parallel to the direction of wind flow): gently sloping on the side facing the wind, steeper on the downwind side. With continued wind flow, particles are rolled or moved by saltation up the shallower upwind slope and tumble down the steeper face, or **slip face.** The slope of the slip face will be characteristic of the sediment of the dune; for sand, it is commonly 30 to 35 degrees from the horizontal. Layering will often develop on the slip face. If winds and depositional patterns shift, eolian crossbeds result (figure 17.8).

The net effect of the movement of many individual grains under sustained winds from one direction is that the dune itself is moved slowly downwind (figure 17.9). In a desert setting away from civilization, this is largely of academic interest. However, migrating dunes—especially large ones—can be a real menace as they march across roads, through forests,

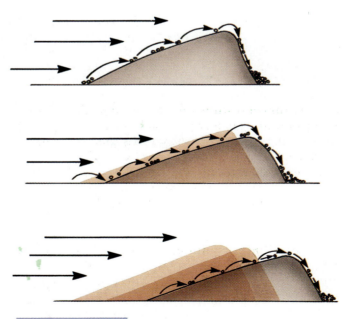

Figure 17.9

Dune migration takes place as a result of many individual grain movements.

A

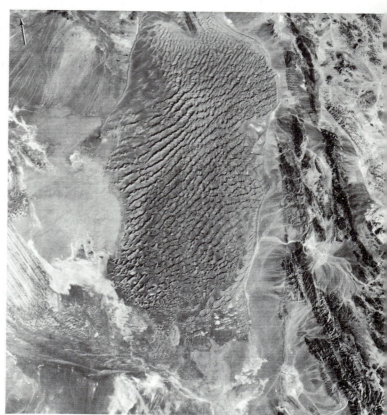

B

Figure 17.10

Marching sand dune encroaching on a forest in Cape Henry, Virginia.

Photo by W. T. Lee, courtesy U.S. Geological Survey.

and even over buildings (figure 17.10). The costs to clear and maintain roads along sandy beaches can be high, for dunes can move several meters or more in a year. The usual approach to trying to stabilize dunes where they threaten construction is to plant vegetation. But many dunes exist where they do in part because the climate is too dry to support vegetation. Quite aside from any water limitations, young plants may be difficult to establish in shifting dune sands because their tiny roots cannot secure a hold.

Dune Forms

There are several types of dune forms. In a given area, one type will predominate, depending on the particular balance among sediment supply, wind characteristics, and abundance of vegetation.

Transverse dunes are elongated perpendicular to the prevailing wind direction (figure 17.11). Many of these have a crescent shape, with arms or "horns" pointing downwind (in the direction of the slip face); these are **barchan dunes**. Very long, narrow barchan dunes grade into continuous *transverse ridges,* with the same cross-sectional profile but not appearing as discrete mounds of sand. Transverse dunes are common where sediment supply is abundant.

Longitudinal dunes (figure 17.12) occur where sediment supply is limited and winds are relatively strong. These dunes are elongated parallel to the direction of wind flow. They may form as the limited quantity of sediment is strung out gradually by the wind.

Figure 17.11

Transverse dunes. *(A)* Crescent-shaped barchan dunes. (See also figure 17.6B.) *(B)* Satellite photo of transverse ridges in a desert.

(A) Photo by G. K. Gilbert, courtesy U.S. Geological Survey. *(B)* © NASA

Figure 17.12

Satellite photo of longitudinal dunes.

© NASA

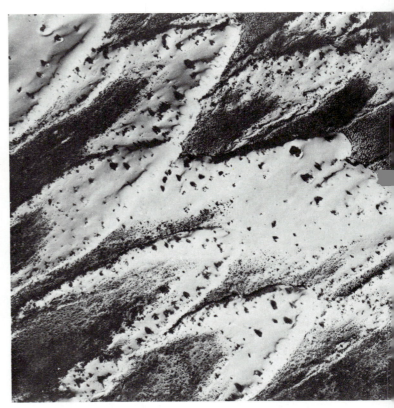

Figure 17.13

Parabolic dunes, the arms of the dunes anchored by vegetation.

Photo by E. D. McKee, courtesy U.S. Geological Survey.

Where vegetation is more abundant, though not so plentiful as to prevent dune formation altogether, **parabolic dunes** tend to form (figure 17.13). At first glance, they appear similar in shape to the barchan dunes, but the arms of parabolic dunes point *upwind*. This appears to be a result of vegetation anchoring the arms of the dune, partially holding sediment in place by slowing wind velocity around the vegetation, while the bulk of the dune marches on.

Loess

Rarely is the wind strong enough to move sand-sized or larger particles very far or very rapidly. Fine dust, on the other hand, is more easily suspended in the wind and can be carried many kilometers before it is dropped. A deposit of windblown silt is known as **loess** (figure 17.14). The rock and mineral fragments in loess are in the range of 0.01 to 0.06 millimeters (0.0004 to 0.0024 inches) in diameter.

The principal loess deposits in the United States are in the central part of the country, and their spatial distribution provides a clue to their source (figure 17.15). They are concentrated around, and particularly to the east sides of, major rivers of the Mississippi basin. Apparently they represent the finest size fraction of sediments originally deposited along the rivers' banks and in their floodplains, later moved eastward by the prevailing west winds.

Figure 17.14

Example of loess.

Photo by L. B. Buck, courtesy U.S. Geological Survey.

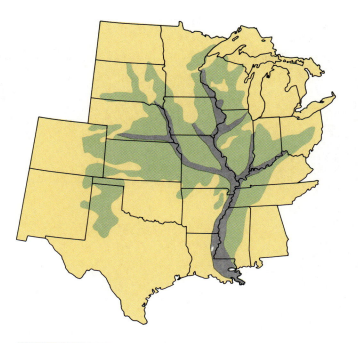

Figure 17.15

Distribution of loess deposits in the United States, and location of major glacier-supplied stream valleys.

Source: After J. Thorp and H. T. U. Smith, "Pleistocene eolian deposits of the United States, Alaska, and parts of Canada," Geological Society of America map, 1952.

Those same rivers drained away much of the meltwater from retreating ice sheets in the last ice age. Glacially produced sediment was washed down and deposited along the river valleys. The lightest material, much of it rock flour, was further redistributed by the wind. Dry glacial erosion does not involve as much chemical breakdown as does stream erosion, as noted in the last chapter. Therefore many soluble minerals are preserved in glacial rock flour. These minerals provide some valuable plant nutrients to the farmland soils now developed on the loess. Newly deposited loess is also quite porous and open in structure, so it has good moisture-holding capacity. These two characteristics together make the farmlands developed on Midwestern loess particularly productive.

Not all loess deposits are ultimately of glacial origin. Some in the western United States may have formed from dust blown off the southwestern deserts. Loess derived from the Gobi Desert covers large areas of China, and additional loess deposits are found downwind of other major deserts. Loess may also be derived from the finest fractions of volcanic ash deposits.

Loess does have drawbacks with respect to applications other than farming. While its light, open structure is reasonably strong when dry and not heavily loaded, it may not make a suitable foundation material. Loess is subject to *hydrocompaction,* meaning

Figure 17.16

The Painted Desert of the southwestern United States.

that when wetted it tends to settle, crack, and become denser and more consolidated, to the detriment of any structures built upon it. The very weight of a large building can also cause settling and collapse of loess.

Deserts

A **desert** is a region having so little vegetation that no significant population can be supported on that land (figure 17.16). It need not be hot or even, technically, dry. Polar ice caps are a kind of desert. In more temperate climates, deserts are indeed characterized by very little precipitation, but they may be consistently hot, cold, or quite variable in temperature with the season or time of day. The distribution of the arid regions of the world (exclusive of polar deserts) is shown in figure 17.17.

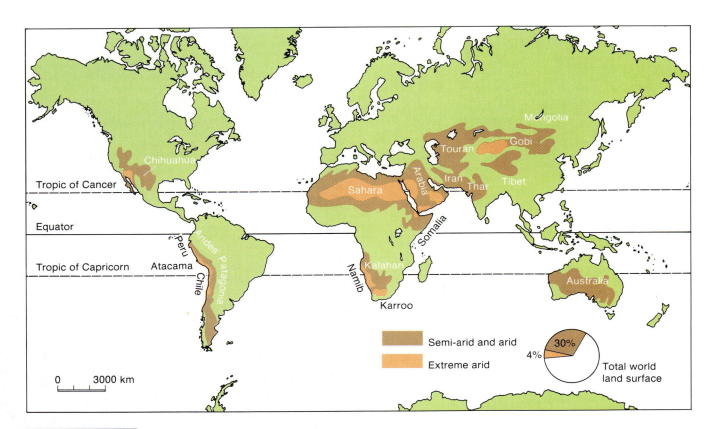

Figure 17.17

Distribution of the world's arid lands.

From Gouldie, A., and J. Wilkinson, *The Warm Desert Environment.*
Copyright © 1977 Cambridge University Press. Reprinted by permission.

Causes of Natural Deserts

A variety of factors contribute to formation of a desert. One is indeed moderately high surface temperatures. Most vegetation, under such conditions, requires abundant rainfall and/or slow evaporation of what precipitation does fall. The availability of precipitation is in turn governed in part by the global air circulation patterns shown in figure 17.2.

Warm air holds more moisture than cold. When the pressure on a mass of air is increased, again it can hold more moisture. Air spreading outward from the equator at high altitudes is chilled and at low pressure, as air pressure and temperature decrease with increasing altitude. When that air circulates downward, at about 30° north and south latitudes, it is warmed as it approaches the surface and also subjected to increasing pressure from the deepening column of air above it. It can then hold considerably more water, so when it reaches the surface it causes quite rapid evaporation. Note in figure 17.17 that many of the world's major deserts fall in belts close to these zones of sinking air at 30° north and south of the equator.

Topography plays a role in controlling the distribution of precipitation. A high mountain range along the path of principal air currents between the ocean and a desert area may be the cause of the latter's dryness. As moisture-laden air from over the ocean moves inland across the mountains, it is forced to higher altitudes where the temperatures are colder and the air thinner (lower pressure). Under these conditions, much of the moisture originally in the air mass is forced out as precipitation, and the air is much drier when it moves inland and down out of the mountains. In effect, the mountains cast a **rain shadow** on the land beyond (figure 17.18). Rain shadows cast by the Sierra Nevada of California and, to a lesser extent, by the southern Rockies contribute to the dryness of the western United States.

Because the oceans are the major source of moisture in the air, simple distance from the ocean (in the direction of air movement) can be a factor. The longer an air mass is in transit over dry land, the greater the opportunity for it to lose some of its moisture through precipitation. On the other hand, even coastal areas can have deserts, under special circumstances. If the land is hot and the adjacent ocean cooled by cold currents, the moist air coming off the ocean will also be

Figure 17.18

Rain shadow cast by mountains lying in the path of the winds from the ocean.

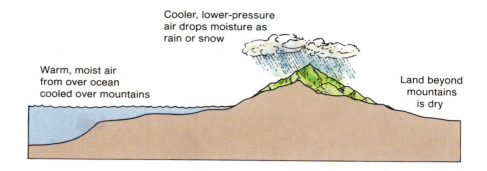

Cooler, lower-pressure air drops moisture as rain or snow

Warm, moist air from over ocean cooled over mountains

Land beyond mountains is dry

Figure 17.19

An alluvial fan formed as mountain streams flow into a dry plain.

Photo courtesy U.S. Geological Survey.

cool and carry less moisture than warmer air over an ocean. As that cooler air warms over the land and becomes capable of holding still more moisture, it causes rapid evaporation from the land rather than precipitation. This phenomenon is observed along portions of the western coasts of Africa and South America.

Desert Landforms and Phenomena

We have already looked at the nature of the dunes that constitute the principal landform of many deserts. Several other features result from the climatic conditions of arid deserts, especially the limited precipitation and sporadic streamflow.

Streams in deserts, which may flow only during a short rainy season or even just for a matter of hours after a cloudburst, are termed **ephemeral streams.** Their channels lie so far above the water table that there is no baseflow from groundwater to keep them flowing between precipitation events. These storm-fed streams may be characterized by **flash flooding,** a very

sharp rise and fall of water level during and immediately after the cloudburst. As the stream rushes down its channel, there is minimal interaction between the rapidly flowing water and the channel sides, little meandering or lateral channel movement. There is usually rapid infiltration into the parched ground, which may ultimately consume all the water. While streamflow persists, the rapid flow can move a large bed load.

Should the stream flow out of the mountains into a plain, especially if it is constantly losing water to infiltration, its sediment load may be deposited in an **alluvial fan** at the edge of the plain (figure 17.19). Multiple fans may overlap as several streams flow out of the same mountain range.

Slow erosion of bedrock at the mountain front may also lead to formation of a bedrock platform, sloping gently away from the foot of the mountains down into the desert. This is a **pediment,** its name derived from the Latin for "foot." From a distance, a

The Dust Bowl

The Dust Bowl area proper, although never exactly defined, comprises close to 100 million acres of southeastern Colorado, northeastern New Mexico, western Kansas, and the Texas and Oklahoma panhandles. The farming crisis there during the 1930s resulted from a combination of factors, some of which were noted in this chapter: clearing or close grazing of natural vegetation, drought (rainfall less than 50 centimeters per year for several years), sustained winds (averaging over 15 kilometers/hour, with velocities ranging much higher during storms), and poor farming practices, including widespread disregard of wind erosion as a potential problem. There had been droughts in the area before, but they had had a much less severe impact on agriculture. It was only in the late 1800s and the early decades of this century that mechanization of farming made possible the rapid expansion of cultivated acreage, from about 12 million acres in 1879 to over 100 million acres by 1929. This greatly increased the area threatened by adverse conditions.

Figure 1

Example of a 1930s dust storm in the Dust Bowl.

Photo courtesy U.S.D.A. Soil Conservation Service.

The action of the wind was most dramatic during the fierce dust storms, which began in 1932. They were described as "black blizzards" that blotted out the sun (figure 1). As windblown dust spread east, black rain fell in New York state, black snow in Vermont. In May of 1934, one 36-hour storm whipped up a dust cloud more than 2000 kilometers long. People choked on the dust, some dying of suffocation or of a

pediment may appear very similar to a series of alluvial fans, but pediments are erosional features, made of bedrock with little or no sediment cover, while alluvial fans are deposits of sediment.

Many deserts are characterized by **internal drainage,** in which streams terminate in enclosed basins rather than flow out of the region. The base level of these streams is a plain, not a sea. Into the basin the streams carry fine sediments, which are deposited as the waters evaporate or infiltrate. To these clastic sediments are added salts crystallized out of solution from the stream waters as they evaporate and the dissolved salts become more concentrated. The resultant "dry lakes" are termed **playas** in North America (nomenclature varies worldwide). The fine-grained sediments of playas are especially subject to shrinkage as they dry out, so mud cracks are common surface features.

Desertification

Climatic zones, like many other surface features, shift through time. Deserts have not always been in the same places. Topography changes, global temperatures change, plate motions move landmasses to different latitudes. The changes have included the develop-

Box 17.1 *Continued*

A

B

C

Figure 2

Some results of wind transport in the Dust Bowl era. *(A)* "Bruce and Thad on farm abandoned to dust", Morton County, Kansas, 1940. *(B)* Farm buildings and implements buried by windblown dust. *(C)* Windblown dust on field smothers crops.

Photos: *(A)* S. W. Lohman, courtesy U.S. Geological Survey. *(B)* and *(C)* Courtesy U.S.D.A. Soil Conservation Service.

"dust pneumonia" similar to the silicosis miners develop from breathing rock dust.

By the late 1930s, concerted efforts by individuals and state and federal governmental agencies to improve farming practices so as to reduce wind erosion—together with a return to more normal rainfall—had considerably reduced the problems. However, drought struck again in the 1950s, and tens of millions of acres of cropland were damaged by wind erosion in 1954. In a dry period in the winter of 1965, high winds raised

dust 10,000 meters into the air and carried some of it east as far as Pennsylvania. Millions of acres more were damaged in the mid-1970s. In recent years, the extent of wind damage has been limited somewhat by the widespread use of irrigation to maintain crops in dry areas and dry times. However, as noted in chapter 14, some of the important sources of that irrigation water are rapidly becoming depleted. Future spells of combined drought and wind may produce more scenes like those of figure 2.

---

ment of new deserts in areas that had previously had more extensive vegetative cover. But the term **desertification** is generally restricted to apply only to the relatively rapid development of deserts caused by the impact of human activities.

The exact definition of the lands at risk is difficult. The *arid* and *semi-arid* lands are commonly taken to be those with annual rainfall less than 60 centimeters (24 inches), though the extent to which vegetation will thrive in low-precipitation areas also

depends on additional factors such as temperature and local evaporation and infiltration rates. Many of the arid lands border true desert regions. It should be emphasized that desertification does not involve the advance or expansion of desert regions as a result of forces originating within the desert. Rather, desertification is a patchy conversion of dry-but-habitable land to uninhabitable desert as a consequence of land-use practices (perhaps accelerated by natural factors, such as drought).

Some Causes of Desertification

Vegetation in dry lands is by nature limited. At the same time it is a precious resource, which may in various cases provide food for people or for livestock, wood for shelter or energy, and protection for the soil to reduce erosion. Desertification typically involves severe disturbance of that vegetation. The environment is not a resilient one, and its deterioration, once begun, may be irreversible and even self-accelerating.

On land used for farming, native vegetation is routinely cleared to make way for crops. While the crops thrive, all may be well. If they fail, or if the land is left unplanted for a time, several consequences follow. One, as in the Dust Bowl, is erosion. A second, linked to the first, is loss of soil fertility. It is the topmost soil layer, rich in organic matter, that is most nutrient rich, and this is lost first to erosion. A third result may be loss of soil structural quality. Under the baking sun typical of many dry lands, with no plant roots to break it up, the soil may crust over, becoming less permeable. This increases surface runoff, correspondingly decreasing infiltration by what precipitation does fall and thus decreasing reserves of soil moisture and groundwater on which future crops may depend. All of these changes together make it that much harder for future crops to succeed, and the problems intensify.

Similar results follow from the raising of numerous livestock on the dry lands. In drier periods, vegetation may be reduced or stunted. Yet it is precisely during those periods that livestock, needing the vegetation not only for food but also for the moisture it contains, will put the greatest grazing pressure on the land. The soil may again be stripped bare, with resultant deterioration and reduced future growth of vegetation, as described for cropland.

Natural drought cycles thus play a role in desertification. However, in the absence of intensive human land use, the degradation of the land during drought is typically less severe, and the natural systems in the arid lands can recover when the drought ends. On a human time scale, desertification—permanent conversion of marginal dry lands to deserts—is generally observed only where human activities are also significant.

Impact of Desertification

The principal reason for concern about desertification is that it is effectively reducing the amount of arable (cultivatable) land on which the world depends for food. An estimated 600 million people worldwide now live on the arid lands. All of those lands, in some measure, are potentially vulnerable to desertification. More than 10% of those 600 million people live in areas identified as actively undergoing desertification now. Some projections suggest that by the end of this century, one-third of the world's once arable land will be rendered useless for the culture of food crops as a consequence of desertification and attendant soil deterioration. The recent famine in Ethiopia may have been precipitated by drought, but it will be prolonged by desertification brought on by overuse of land incapable of supporting concentrated human or animal populations.

Summary

Wind arises from differential heating of the atmosphere, its movement complicated by topography and by the earth's rotation. Globally, it is a far less efficient agent of surface change than is water. Its effects are particularly prominent in regions of limited vegetation and precipitation. Wind erodes material through abrasion by wind-carried sediment, producing ventifacts, or through deflation, the removal of unconsolidated material. The principal depositional eolian landform is the dune. Dune shapes vary with the relative importance of sediment supply, wind action, and presence of vegetation. If winds blow consistently in one direction, dunes may also migrate. Where a source of abundant fine sediment exists, selective transport and redeposition of this material by wind produces loess, which often contributes to excellent farmland. Deserts, defined as regions incapable of supporting enough vegetation to sustain significant human or animal populations, form in a variety of climates. The majority are in warm or temperate regions and are dry. Many deserts are characterized by internal drainage of ephemeral streams, which deposit alluvial fans of sediment and form playas within the basins into which they drain. The amount of land area classified as desert is presently increasing as a result of desertification accelerated by human activities in arid lands.

Terms to Remember

abrasion	desert pavement	internal drainage	playa
alluvial fan	dune	loess	rain shadow
barchan dune	eolian	longitudinal dune	slip face
deflation	ephemeral stream	parabolic dune	transverse dune
desert	flash flooding	pediment	ventifact
desertification			

Questions for Review

1. What is the principal cause of wind? Briefly describe three factors that influence the direction of wind flow.

2. Compare and contrast the processes of sediment transport by streams and by wind, and the sizes of the materials moved.

3. Describe the nature of wind abrasion and the formation of ventifacts.

4. What is deflation, and how is it involved in the formation of desert pavement?

5. Summarize the process of dune formation and migration.

6. How do barchan and parabolic dunes differ? Why?

7. What is loess? Name three kinds of material from which loess deposits may be derived.

8. What is a desert? Are all deserts found in hot climates? Explain.

9. Describe the production of a rain shadow by a mountain range.

10. Explain the phenomenon of flash flooding and the formation of alluvial fans and playas in terms of common drainage characteristics in and near deserts.

11. What is desertification? Explain two ways in which it is accelerated by human activities.

For Further Thought

1. Obvious graded bedding is less common in wind-laid than in water-laid sediments; sorting of wind-deposited sediments is commonly very good. Suggest explanations for these observations.

2. If you live in a snowy area, make an inspection tour around your home or school buildings after a snowstorm with high winds. Note the distribution of snow and try to relate it to the distribution of obstacles to wind flow. (Readers in sparsely vegetated areas might do the same after a windstorm in dry weather, although the sediment distribution patterns may be more subtle and require closer scrutiny.)

Suggestions for Further Reading

Battan, L. J. *Fundamentals of Meteorology.* Englewood Cliffs, N.J.: Prentice-Hall, 1979.

Brookfield, M. E., and Ahlbrandt, T. S., eds. *Eolian Sediments and Processes.* New York: Elsevier, 1983.

Cooke, R. U., and Warren, A. *Geomorphology in Deserts.* Berkeley: University of California Press, 1973.

Goudie, A., and Wilkinson, J. *The Warm Desert Environment.* New York: Cambridge University Press, 1977.

Greeley, R., and Iversen, J. D. *Wind as a Geological Process.* Cambridge: Cambridge University Press, 1985.

Hurt, R. D. *The Dust Bowl.* Chicago: Nelson-Hall, 1981.

Secretariat of the U.N. Conference on Desertification, Nairobi. *Desertification: Its Causes and Consequences.* New York: Pergamon Press, 1977.

Wells, S. G., and Haragan, D. R., eds. *Origin and Evolution of Deserts.* Albuquerque, N. Mex.: University of New Mexico Press, 1983.

Weathering, Erosion, and Soil

Introduction

Broadly defined, **soil is the surface accumulation of rock and mineral fragments** and, **usually, some organic matter, that covers the underlying solid rock of most areas.** Soil scientists make a **distinction between true soil,** which they further define as capable of supporting plant growth, and **regolith, which is a more general term for all surface sediment accumulations in place,** regardless of their agricultural promise. By and large, the same weathering processes are involved in the formation of the material in either case, and most geologists and many engineers use the terms soil and regolith interchangeably. This chapter surveys principal weathering processes, controls on the composition of the resulting soil, and aspects of soil erosion and its minimization.

Note that the conventional definition of soil implies little or no transportation away from the site of formation. **Once the solid products of weathering are transported and redeposited, they fall into the broader category of** *sediment.*

Weathering

The term **weathering encompasses a variety of chemical, physical, and biological processes that act to break down rocks in place.** The relative importance of different kinds of weathering processes is largely determined by climate. Climate, topography, the composition of the bedrock on which the soil is formed, and time determine a soil's final composition.

Mechanical Weathering

Mechanical weathering is the physical breakup of rocks, without changes in their composition. It requires that some physical force or stress be applied. Often it is caused by water, as when water in cracks repeatedly freezes and expands, forcing rocks apart, then thaws and contracts or flows away. This process is termed **frost wedging** (figure 18.1). Crystallizing salts in cracks may have the same wedging effect, as may plant roots forcing their way into crevices. In extreme climates like those of deserts, where the contrast between day and night temperatures is very large, many cycles of daily thermal expansion of rocks baked by sunlight and contraction as they cool at night might cause enough stress to break them up, although the phenomenon has not been reproduced in the laboratory.

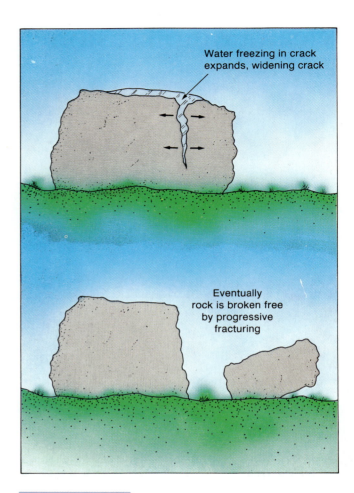

Water freezing in crack expands, widening crack

Eventually rock is broken free by progressive fracturing

Figure 18.1

Frost wedging (schematic).

The **removal of stress, too, can break up rocks** (figure 18.2). When a rock once deeply buried and under pressure is uplifted and the overlying rock eroded away, the rock is *unloaded.* It will tend to expand and may fracture in concentric sheets in the process. As **these sheetlike layers of rock flake off, the original rock mass is broken up by exfoliation.** Granites, which would necessarily have crystallized at some depth initially, **often weather by exfoliation** when exposed at the surface.

Whatever the cause, th**e principal effect of mechanical weathering** is to break large chunks of rock into smaller ones, and in so **doing to increase the total exposed surface area of the particles** (figure 18.3). Further breakup of fragments may also occur as sediment undergoes transport from the site of the source bedrock.

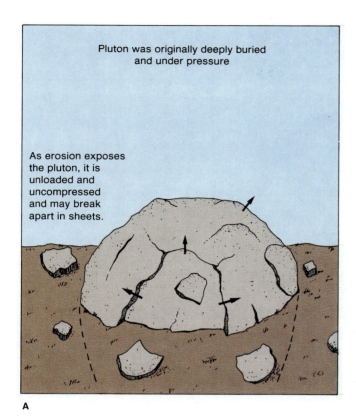

Pluton was originally deeply buried and under pressure

As erosion exposes the pluton, it is unloaded and uncompressed and may break apart in sheets.

A

B

Figure 18.2

Unweighting, and the associated pressure release, causes exfoliation. *(A)* Exfoliation (schematic). *(B)* Example of granite boulder weathering by exfoliation.

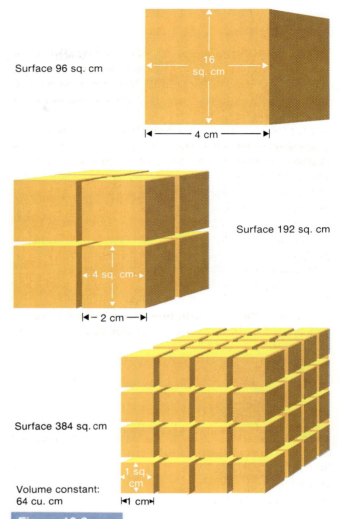

Surface 96 sq. cm

16 sq. cm

4 cm

Surface 192 sq. cm

4 sq. cm

2 cm

Surface 384 sq. cm

1 sq. cm

Volume constant: 64 cu. cm   1 cm

Figure 18.3

Mechanical breakup increases surface area and surface-to-volume ratio.

Chemical Weathering

Chemical weathering means the breakdown of minerals by chemical reaction with water, with other chemicals dissolved in water, or with gases in the air. Minerals differ in the kinds of chemical reactions they undergo, as well as in how readily they weather chemically.

Calcite (calcium carbonate) dissolves completely, leaving no other minerals behind in its place. This is a contributing factor to the formation of underground limestone caverns. Solution of calcite takes place rather slowly in plain water, more rapidly in acidic water, such as is produced by acid rainfall. This is becoming a serious problem where limestone and its metamorphic equivalent, marble, are widely used

Table 18.1

Some chemical weathering reactions

Solution of calcite (no solid residue)
  $CaCO_3 + 2 H^+ = Ca^{2+} + H_2O + CO_2$ (gas)
Breakdown of ferromagnesians (possible mineral residues include iron compounds, quartz, clays)
  $FeMgSiO_4$ (olivine) $+ 2 H^+ = Mg^{2+} + Fe(OH)_2 + SiO_2$
  $2 KMg_2FeAlSi_3O_{10}(OH)_2$ (biotite) $+ 10 H^+ + 1/2 O_2 = 2$
    $Fe(OH)_3 + Al_2Si_2O_5(OH)_4$ (kaolinite, a clay) $+ 4 SiO_2 + 2$
    $K^+ + 4 Mg^{2+} + 2 H_2O$
Breakdown of feldspar (clay and quartz are the common residues)
  $2 NaAlSi_3O_8$ (sodium feldspar) $+ 2 H^+ + H_2O =$
    $Al_2Si_2O_5(OH)_4 + 4 SiO_2 + 2 Na^+$
Solution of pyrite (making dissolved sulfuric acid, H_2SO_4)
  $2 FeS_2 + 5 H_2O + 15/2 O_2$ (gas) $= 4 H_2SO_4 + Fe_2O_3 \cdot H_2O$
  Note that hundreds of possible reactions could be written; the above are only examples of the kinds of processes involved.
  All ions (charged species) are dissolved in solution; all other substances, except water, are solid unless specified otherwise.

Figure 18.4

Sculptural details on this gargoyle are being eaten away by acidic rainfall.

for outdoor sculptures and building stones. Cement and concrete also contain calcite, and are therefore also susceptible to damage from acid rain. Calcite dissolution is gradually destroying delicate sculptural features and eating away at the very fabric of many buildings in urban areas where acid rain is particularly common (figure 18.4).

Actually, all rain is naturally somewhat acidic, though there is much talk now of the "acid rain" caused by human activities (see box 18.1). The acidity of a solution is proportional to the concentration of hydrogen ions (H^+) in it. In air, some carbon dioxide (CO_2) reacts with water vapor (H_2O) to make carbonic acid, a weak acid (H_2CO_3). The carbonic acid dissociates in solution, breaking up into H^+ ions and HCO_3^- (bicarbonate) ions, thereby increasing rain acidity.

Silicates tend to be somewhat less susceptible to chemical weathering and to leave other minerals behind when they are attacked. Feldspars principally weather into clay minerals. Ferromagnesian silicates leave behind insoluble iron oxides and hydroxides and sometimes clays, depending on the specific ferromagnesian mineral in question, while other chemical components are dissolved away. Those residual iron compounds are responsible for the reddish or yellowish colors of many soils. In most climates, quartz is extremely resistant to chemical weathering, dissolving only slightly. A summary of some of the common chemical weathering reactions is given in table 18.1.

The susceptibility of many of the silicates to chemical weathering can be inferred from the conditions under which they originally formed. Given several silicates that have crystallized from the same magma, those that formed at the highest temperatures tend to be the least stable, or most easily weathered, at the low temperatures of soil formation at the earth's surface, and vice versa. The relative resistance to chemical weathering of many common silicates can be estimated by inverting Bowen's Reaction Series (see figure 18.5). A rock's tendency to weather chemically is determined in turn by the minerals that make it up. For example, a gabbro, formed at high temperatures and rich in ferromagnesian minerals and calcic plagioclase, generally weathers more readily than a granite rich in quartz and low-temperature feldspars. In outcrop, differential susceptibility to weathering is reflected in differential relief between adjacent rock units (figure 18.6).

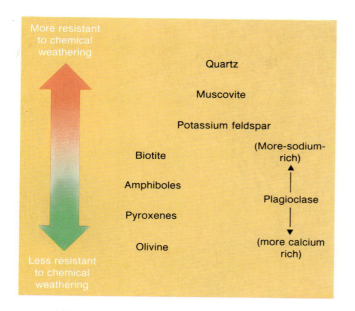

A

Figure 18.5

Bowen's Reaction Series and susceptibility to chemical weathering.

B

Figure 18.6

Differential susceptibility of rock units to weathering results in differential relief. *(A)* These granite veins are more resistant than the sedimentary rock they crosscut. *(B)* In Bryce Canyon, the more resistant sedimentary layers are prominent as shelves in the weathered landscape.

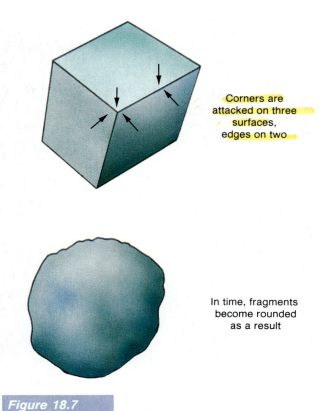

Corners are attacked on three surfaces, edges on two

In time, fragments become rounded as a result

Figure 18.7

Angular fragments are rounded by weathering.

Climate plays a major role in the intensity of chemical weathering. Most of the relevant chemical reactions involve water, or at least dissolved chemicals. All else being equal, the more water, the more chemical weathering. Also, most chemical reactions proceed more rapidly at high temperatures than at low. Therefore, warmer climates are conducive to more rapid chemical weathering. Plants, animals, and microorganisms develop more abundantly and in greater variety in warm, wet climates too. Many organisms produce compounds that react with and dissolve or break down minerals. Such chemical weathering involving biological activity is thus more intense in warm, moist climates, just as inorganic chemical weathering generally is.

The rates of chemical and mechanical weathering are interrelated. Chemical weathering may accelerate the mechanical breakup of rocks if the minerals being dissolved are holding the rock together by cementing the mineral grains, as in some sedimentary rocks. Increased mechanical weathering may in turn accelerate chemical weathering by increasing exposed surface area, since it is only at grain surfaces that minerals, air, and water interact. The higher the ratio of surface area to volume—that is, the smaller the particles—the more rapid the chemical weathering.

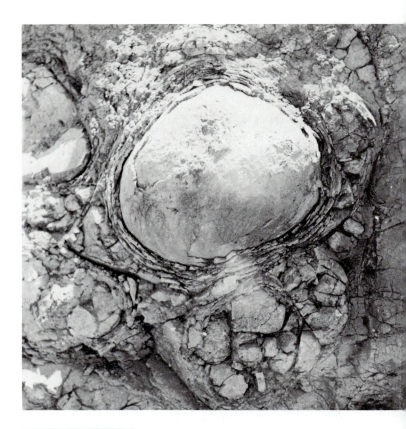

Figure 18.8

Spheroidal weathering.
Photo by K. Segerstrom, courtesy U.S. Geological Survey.

The relationship of surface area to weathering rate can also be seen in the tendency of angular chunks of rock to become more rounded and spheroidal with progressive weathering (figure 18.7). The corners of a block are under attack on several surfaces; edges are being attacked from two sides; the faces of the block are being weathered from only one direction. Therefore the corners are rounded most rapidly. Once the rock mass has assumed a rather spheroidal shape, further weathering may proceed in curved layers inward from the surface, in a process known as **spheroidal weathering** (figure 18.8).

Soil

The result of all the foregoing processes, together with the accumulation of decaying remains from organisms living on the land, is the formation of a layer between bedrock and atmosphere that we call soil. Digging down through the soil we encounter in most areas a series of layers of different colors, compositions, and physical properties. The number of layers and the thickness of each vary.

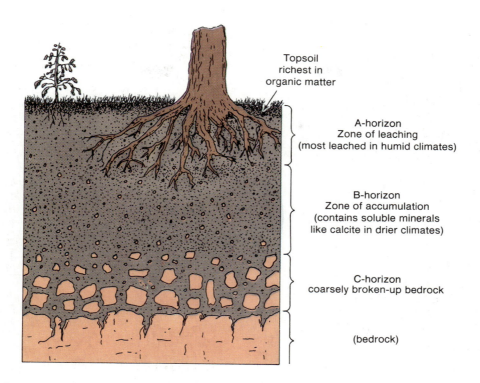

Topsoil
richest in
organic matter

A-horizon
Zone of leaching
(most leached in humid climates)

B-horizon
Zone of accumulation
(contains soluble minerals
like calcite in drier climates)

C-horizon
coarsely broken-up bedrock

(bedrock)

Figure 18.9

A representative soil profile, showing the various soil horizons.

The Soil Profile

A basic, generalized soil profile is shown in figure 18.9. At the top is the **A horizon.** It consists of the most intensively weathered rock material, being most exposed to surface processes. It also contains most of the organic remains. Unless the local water table is exceptionally high, precipitation infiltrates down through the A horizon. In so doing the water may dissolve soluble minerals and carry them away with it. This process is known as **leaching,** and the A horizon is known as the *zone of leaching.* Fine-grained minerals, like clays, may also be carried physically downward with percolating water.

Many of the minerals leached from the A horizon accumulate in the layer below, the **B horizon,** also known as the *zone of accumulation* or *zone of deposition.* Soil in the B horizon is often coarser-grained than the A-horizon soil because it has been somewhat protected from surface processes. Organic matter from the surface has also been less well mixed into the B horizon.

Below the B horizon is a zone consisting principally of very coarsely broken up bedrock and little else. This is the **C horizon,** which does not resemble our usual idea of soil at all. Sometimes the bedrock or parent material itself is designated as a fourth horizon, the D or (in more current usage) R horizon.

The boundaries between soil horizons may be sharp, or indistinct. One horizon may be divided into several recognizable subhorizons—for example, the top of the A horizon might consist of a layer of topsoil particularly rich in organic matter. There may even exist such an organic-rich layer at the top of the profile that a distinct "O horizon" can be designated. Subhorizons gradational between A and B or B and C horizons may be recognized. One or more horizons may be absent from the soil profile.

Variations in soil characteristics arise from the different mix of soil-forming processs and starting materials found from place to place. The overall total thickness of soil above the bedrock is partly a function of the local rate of soil formation, and partly a function of the rate of soil erosion. The latter reflects the work of wind and water, the topography, and the extent and kinds of human activities.

Soil Composition

Mechanical weathering merely breaks up rock without changing its composition. In the absence of pollution, wind and rainwater rarely add many chemicals, and runoff water may carry away some leached chemicals in solution. Thus chemical weathering tends to

Box 18.1

Acid Rain and Weathering

It seems clear that interaction of CO_2 and H_2O in the atmosphere, producing carbonic acid, will cause rain naturally to be somewhat acid. As the amount of CO_2 in the air is increased by the burning of fossil fuels, some increase in the acidity of rainfall may be expected. Nitrogen gases, produced in internal-combustion engines, also react to form some nitric acid (HNO_3). What concerns those who focus on so-called **acid rain,** however, is the acidity resulting not from these but from sulfur in the air.

Sulfur released into the air reacts with oxygen and water vapor to produce sulfuric acid (H_2SO_4), a strong acid that is both corrosive and highly irritating to eyes and lungs. Human activities contribute sulfur to the air principally through the burning of coal (see chapter 21). Downwind of coal-burning furnaces particularly, rainfall is significantly more acidic than normal; this is acid rain. It causes more rapid chemical weathering of rocks and buildings, acidifies lakes and streams (sometimes to the detriment of wildlife), and may more readily leach toxic metals such as lead and cadmium out of soils and into water supplies. (In places, measured rainfall

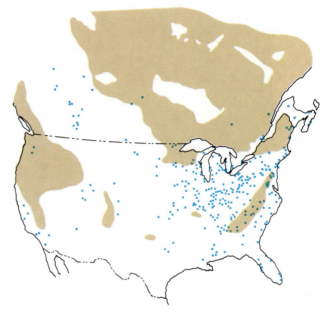

Figure 1
Regions downwind of major sources of atmospheric sulfur (dots) are, in many cases, regions underlain by granitic bedrock (shaded in brown).

Courtesy of Walter W. Roberts. Originally appeared in *Science 80,* Vol. 1, No. 5, 74–79.

acidity has been comparable to that of the stomach acid that digests our food!)

Local geology may either moderate or aggravate the problems. Limestone tends to neutralize acid, while waters in granitic rocks and soils tend already to be somewhat acidic. Unfortunately, many areas of

North America that lie downwind (northeast) of major sulfur sources have granitic bedrock (figure 1). This means that the effects of acid rain are not moderated by interaction with the rock or soil after the rain falls, and problems caused by acid rain may be especially severe in such areas.

involve a net subtraction of elements from rock or soil. A primary control on a soil's composition, then, is the composition of the bedrock (or sediment) on which it is formed. If the bedrock is low in certain critical plant nutrients, the soil produced from it will be so also, and chemical fertilizers may be needed in order to grow particular crops on that soil even if it has never been farmed before.

The extent and balance of the weathering processes that are involved in soil formation determine the extent of further depletion of the soil in various elements. Mechanical weathering usually dominates only in cooler, drier climates; under other conditions, chemical weathering, often accelerated by biological activity, will be the dominant process. The weathering processes also influence the mineralogy of the soil, the compounds in which its elements occur. The physical properties of the soil are affected by its mineralogy, the texture of the mineral grains (coarse or fine, well or poorly sorted, rounded or angular, and so on), and any organic matter present.

Early attempts at soil classification emphasized compositional differences among soils and thus principally reflected the effects of chemical weathering. Two broad categories of soil were recognized. The pedalfer soils were seen as characteristic of more humid regions. Where the climate is wetter, there is naturally more extensive leaching of the soil, and what remains is enriched in the less soluble oxides and hydroxides of aluminum and iron, and, especially in the B horizon, clay. The term *pedalfer* comes from the Latin prefix *pedo-*, meaning "soil," and the Latin words for aluminum (*alumium*) and iron (*ferrum*). In North America, pedalfer-type soils would be found in higher-rainfall areas like the eastern United States and most of Canada. Pedalfer soils are typically acidic.

Where the climate is drier, such as in the western and especially the southwestern United States, leaching is much less extensive. Even quite soluble compounds like calcite remain in the soil, especially in the B horizon. From this observation came the term pedocal for the soil of a dry climate. The presence of calcium carbonate makes pedocal soils more alkaline.

The pedocals are thus not highly leached by infiltrating water from above. Moreover, in many of these dry areas, net water flow is *upward*. Where evaporation near the surface is rapid, soil moisture may be drawn upward along the grain surfaces by capillary action. The effect is similar to the way in which kerosene is drawn up the wick in a kerosene lamp as fuel is burned at the top of the wick. In the case of pedocal soil, the water rises, carrying with it any dissolved salts, and then as it evaporates, it deposits those salts. Some pedocal soils become so salty that they are unsuitable for agriculture.

Toward More Precise Classification

One problem with the simple pedalfer/pedocal classification scheme is that in order for it to be strictly applied, the soils it describes must have formed over suitable bedrock. For example, if the bedrocks and early soils in a region contained little calcium, then the later soils would not be likely to contain a great deal of calcite either, regardless of how little leaching had occurred. Conversely, a rock poor in iron and aluminum, such as a limestone, will not leave an iron- and aluminum-rich residue, no matter how extensively it is leached. Still, the terms *pedalfer* and *ped-*

Table 18.2

Soil orders of the Seventh Approximation, with general descriptions

Order	Description
1. Entisols	Soils without layering, except perhaps a plowed layer.
2. Vertisols	Soils with upper layers mixed or inverted because they contain expandable clays (clays that expand when wet, contract and crack when dry).
3. Inceptisols	Very young soils with weakly developed soil layers, and not much leaching or mineral alteration.
4. Aridosols	Soils of deserts and semiarid regions, and related saline or alkaline soils.
5. Mollisols	Grassland soils, mostly rich in calcium; also forest soils developed on calcium-rich materials. Characterized by a thick surface layer rich in organic material.
6. Spodosols	Soils with a light, ashy grey A horizon, and a B horizon containing organic matter and clay leached from the A horizon.
7. Alfisols	Includes most other acid soils with clay-enriched subsoil.
8. Ultisols	Similar to Alfisols but with weathering more advanced; includes some lateritic soils.
9. Oxisols	Still more weathered than the Ultisols; includes most laterites.
10. Histosols	Bog-type soils.

From *Geology of Soils*, by C. B. Hunt. W. H. Freeman and Company. Copyright © 1972. Reprinted by permission.

ocal can be used generally to indicate, respectively, more and less extensively leached soils.

Modern soil classification has become considerably more sophisticated and complex. Various schemes may take into account characteristics of present soil composition and texture, the type of bedrock on which it was formed, the present climate of the region, the degree of chemical maturity of the soil, or the extent of development of the different soil horizons. Different countries have adopted different approaches. The new UNESCO world map uses 110 different soil map units. This is in fact a small number by comparison with the number of distinctions made under certain classification schemes.

The United States comprehensive soil classification, known as the Seventh Approximation, has ten major categories (orders) that are subdivided through five more levels of classification into a total of some 12,000 soil series. A short summary of the ten orders is presented for reference in table 18.2. Certain of the

Figure 18.10

Lateritic soil in Hawaii, showing the bright red color of iron oxide minerals.

orders are characterized by a particular environment of formation and the distinctive soil properties resulting from it—for example, the Histosols, which are bog soils. Others are characterized principally by physical properties—for example, the Entisols, which lack horizon zonation, and the Vertisols, in which the upper layers are mixed because these soils contain expansive clays. Most of the Oxisols and some Ultisols, on the other hand, are a type of soil that has serious implications for agriculture, especially in much of the Third World: lateritic soils.

Lateritic Soil

One widely recognized soil type is sufficiently distinctive and important, especially to many less developed nations, to justify special consideration. This is

lateritic soil. A *laterite* may be regarded as an extreme kind of pedalfer. Lateritic soils develop in tropical climates with high temperatures and heavy rainfall, so they are extensively leached. Even quartz may have been dissolved out of the soil under these conditions. The lateritic soil often contains very little besides the insoluble aluminum and iron oxide compounds (figure 18.10).

Soils of lush tropical rain forests are commonly lateritic. This might suggest that lateritic soils have great potential as farmland to feed the world's growing population. Surprisingly, the opposite is true, for two reasons.

The highly leached character of lateritic soils is one reason. Even where the vegetation is dense, the soil itself has few soluble nutrients left in it. The forest holds a huge reserve of nutrients, but there is no corresponding reserve in the soil. Further growth is supported by the decay of earlier vegetation. As one plant dies and decomposes, the nutrients it contained are quickly either taken up by other plants, or leached away. If the rain forest is cleared for the planting of crops, most of the nutrients are cleared away with it, leaving little in the soil to nourish the crops. Many natives of tropical climates practice a "slash-and-burn" agriculture, cutting and burning the jungle to clear the land. Some of the nutrients in the burned vegetation do settle into the topsoil temporarily, but the relentless leaching by the warm rains soon depletes them. Lateritic soil may be nutrient-poor and infertile within a few growing seasons.

A second problem with lateritic soil is suggested by the term *laterite* itself, which is derived from the Latin for "brick." A lush rain forest shields the soil from the drying and baking effects of the sun, while vigorous root action helps to keep the soil well broken up. When its vegetative cover is cleared and it is exposed to the baking tropical sun, lateritic soil can quickly harden into a solid, bricklike consistency that resists infiltration by water or penetration by crops' roots. What crops can be grown provide little protection for the soil and do not slow the hardening process very much. Within five years or less a freshly cleared field may become completely uncultivable. The tendency of laterite to bake into brick has been used to advantage in some tropical regions, where blocks of hardened laterite have served as building materials. Still, the problem of how to maintain crops remains. Often the only recourse is to abandon each field after a few years and clear a replacement. This results over time in the destruction of vast tracts of rain forest where only a moderate expanse of farmland is needed. Once the soil has hardened and efforts to farm it have been given up, it may revegetate only slowly, if at all. This compounds the loss of arable land resulting from desertification, described in the previous chapter.

In Indochina, in what is now Kampuchean jungle, are the remains of the Khmer civilization that flourished from the 9th to the 16th century. There is no clear evidence of why that civilization vanished. It has been suggested that a major reason was the difficulty of agriculture in the region's lateritic soil. Perhaps the Mayas, too, moved north into Mexico to escape the problems of lateritic soils. Today, some countries with lateritic soils achieve agricultural success only because frequent flooding deposits fresh, nutrient-rich soil over the depleted laterite. The floods, however, cause enough problems of their own that flood-control efforts are in progress or under serious consideration in many of these areas of Africa and Asia. Unfortunately, successful flood control could create an agricultural disaster for such regions.

Figure 18.11

The impact of a single raindrop loosens many soil particles. Photo courtesy U.S.D.A. Soil Conservation Service.

Soil Erosion

A distinction can be made between *weathering* (the breakdown of rock or mineral material in place) and *erosion* (physical removal of material), though weathering may accelerate erosion. We have already touched on aspects of erosion by the action of water, wind, and ice. This section focuses specifically on erosion as it affects soil.

Rain striking the ground helps to break soil particles loose (figure 18.11). Surface runoff and wind together carry loosened soil away. Surface runoff washing down over a slope surface is **sheet wash.** Where the water begins to erode small preferred channels, **rill erosion** occurs. Should the rills enlarge to produce channels so deep that normal cultivation will not erase them, the result is **gullying** (figure 18.12). We have noted that the faster wind and water travel, the larger the particles and greater the load they can move. Therefore high winds cause more erosion than calmer ones, and fast-flowing surface runoff moves more soil than slow. This in turn suggests that, all else being equal, steep and unobstructed slopes are more susceptible to erosion by water, for surface runoff flows more rapidly over them.

Figure 18.12

Gullying on unprotected farmland.

Photo courtesy U.S.D.A. Soil Conservation Service.

Figure 18.12

Flat land lacking vegetation is correspondingly more vulnerable to wind erosion. The physical properties of the soil—for example, how cohesive it is—will also influence its vulnerability to erosion.

Erosion Rates

Rates of soil erosion can be estimated in a variety of ways. Over a large area, erosion resulting from surface runoff may be judged by estimating the sediment load of streams draining the area. On small plots, runoff can be collected and its sediment load measured. Wind is harder to trap or monitor comprehensively, especially over a range of altitudes, so the extent of wind erosion is more difficult to estimate. Generally, it is much less significant than erosion through surface-water runoff, except under drought conditions. Controlled laboratory experiments can also be used to simulate wind and water erosion, and measure their impact.

Estimates of the total amount of soil erosion in the United States vary widely, but U.S. Soil Conservation Service estimates put the figure at over four billion tons per year. The present rates of erosion have been accelerated by human activites, construction and especially farming (figure 18.13; table 18.3). Some 412 million acres of U.S. cropland suffer soil erosion losses from water alone averaging an estimated 4.8 tons per acre per year. In very round numbers, that is about a 0.04-centimeter thickness of soil removed each year, on average.

Figure 18.13

Erosion on a bare construction site.

Photo courtesy U.S.D.A. Soil Conservation Service.

Table 18.3

Erosion by water from undeveloped land in the United States

Land type	Erosion rate (tons/acre/year)		Acres affected (millions)	Total erosion loss (millions of tons)
	Range	Average		
cropland	.83–15	4.8	412	1978
rangeland	.22–6.4	3.4	408	1387
forest land	.02–2.5	1.2	367	440
pastureland	.13–12.6	2.6	133	346
			Total water erosion loss	4151

Source: U.S. Department of Agriculture, Soil Conservation Service.

How does this compare with the rate of soil *formation*? It is difficult to generalize, because the rate of soil formation is so sensitive to climate, the nature of the parent rock material, and other factors. We can put some upper limits on soil-formation rates by looking at soils in areas of the northern United States last scraped clean by the glaciers tens of thousands of years ago. Where the parent material was glacial till, which should weather more easily than solid rock, about one meter of soil has formed in 15,000 years in the temperate, fairly humid climate of the upper Midwest. The corresponding average rate of 0.006 centimeters of soil formed per year is less than one-sixth the average erosion rate from cropland, considering erosion by water alone. Furthermore, soil formation in many areas and over more resistant bedrocks is slower still. In some places in the Midwest, virtually no soil has formed on glaciated bedrocks, and in the drier Southwest, soil formation is likely to be much slower also. Soil erosion, too, can locally be much more rapid than the average. Rates of erosion from cultivated hillsides may be over 90 tons per acre per year where no soil-conservation measures are practiced.

It can be seen from figure 18.14 that erosion during active urbanization (highway construction, building of structures, and the like) is considerably more severe than erosion on any sort of undeveloped land. However, the total area undergoing active development is only about 1 1/2 million acres each year. Also, this disturbance is of relatively short duration, and once the land is covered by buildings, pavement, or established landscaping, erosion rates typically drop below even natural background levels. The problems of erosion during construction are thus very intensive but typically localized and short-lived. The remaining discussion in this section will therefore focus on the volumetrically larger problem of erosion from agricultural land.

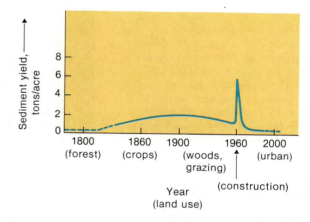

Figure 18.14

Rates of erosion vary dramatically as a function of land use.

From Wolman, M. G., "A Cycle of Sedimentation and Erosion in Urban River Channels," in *Geografiska Annaler,* Vol. 49, Series A. © Swedish Society of Anthropology and Geography. Reprinted by permission.

Consequences of Soil Erosion

If all the soil in an area is lost, farming clearly becomes impossible. Even while some remains, the present rapid erosion rates are cause for concern. It is the topsoil, with its higher content of organic matter and nutrients, that is especially fertile and suitable for agriculture, and naturally this is lost first, as the soil erodes. The organic-matter-rich topsoil usually has the best structure for agriculture, too—it is more permeable, more readily infiltrated by water, and retains moisture better than deeper soil layers. Soil erosion from cropland, then, leads to reduced productivity and crop quality, and reduced income for farmers.

Also, the soil eroded from one place is deposited, sooner or later, somewhere else. If a large quantity is moved and dumped on other farmland, the crops

Box 18.2

Off-Road Vehicles and Erosion

The use of vehicles elsewhere than on planned, prepared paths may cause accelerated erosion. The problems are not unique to the twentieth century; the westward parade of many wagon trains along the Oregon Trail succeeded in carving a path out of stone (figure 1).

However, the increased use of off-road recreational vehicles (ORVs) in recent years is becoming an increasingly significant cause of rapid soil erosion. ORVs are often used on lightly vegetated slopes. In dry climates, the vegetation thereby destroyed may not easily re-establish itself (figure 2). The result is typically a slope prone to more rapid soil erosion, and perhaps also to gullying during heavy rains. This is a major reason for restricting ORV use to prescribed areas designated for the purpose.

Figure 1
Oregon trail ruts.

Figure 2
Modern off-road vehicle tracks in dry country in the western United States.

Windbreaks retard wind erosion by decreasing wind velocity.

Photo courtesy U.S.D.A. Soil Conservation Service.

on which it is deposited may be stunted or destroyed (although *small* additions of fresh topsoil may, as noted earlier, enrich cropland and make it more productive). A subtle consequence of soil erosion in some places has been increased persistence of toxic residues of herbicides and pesticides in the soil: loss of nutrients and organic matter through topsoil erosion may decrease the activity of soil microorganisms that normally speed the breakdown of toxic agricultural chemicals.

Another major soil-erosion problem is sediment pollution. In the United States, about 750 million tons of the eroded sediment ends up in lakes and streams. This decreases the water quality and may harm wildlife. The problem is still more acute when those sediments contain toxic chemical residues. A secondary consequence of this sediment load is the filling-in of stream channels and reservoirs, restricting navigation, and decreasing the volume of reservoirs and thus their usefulness, whether for water supply, hydropower, or flood control.

Strategies for Reducing Erosion

A wide variety of approaches exist for reducing erosion on farmland. Basically, they all involve either reducing the velocity of an eroding agent, or protecting the soil from its effects. Under the latter heading are such practices as leaving stubble in the fields after a crop has been harvested, and planting cover crops in the season between cash crops. In either case, the plants' roots help to hold the soil in place while the main crop is not growing, and the plants to some extent shield the soil from wind and rain.

The lower the wind or water runoff velocity, the less material carried. Wind can be slowed down by the use of *windbreaks* along the borders of fields. These can be trees or hedges planted in rows perpendicular to the dominant wind direction (figure 18.15), or low fences similarly arrayed. This does not altogether stop the soil movement, as can be seen by the ridges of soil that pile up along windbreaks. However, it does reduce the distance over which soil is transported, and some of the soil caught at field edges can be redistributed over the fields.

Figure 18.16

Contour plowing to slow sheet wash and channelized water runoff.

Photo courtesy U.S.D.A. Soil Conservation Service.

Surface runoff may be slowed on moderate slopes by *contour plowing* (figure 18.16). Plowing rows parallel to the contours of the hill, perpendicular to the direction of water flow, creates a ridged land surface down which water does not rush as readily. Other slopes may require *terracing* (figure 18.17). A single slope is broken up into a series of shallower slopes or even steps that slant backward into the hillside. Again, surface runoff making its way down the slope does so more slowly, if at all, and therefore carries far less sediment with it. Terracing has been practiced since ancient times.

Other strategies can be applied to minimize erosion in nonfarm areas. In the case of urban construction projects, one reason for the severity of the resulting erosion is that it has been common practice to clear a whole area, such as an entire housing-development site, at the beginning of the work, even though only a small portion of the site will be worked actively at any given time. Doing the clearing in stages as needed, and thus minimizing the length of time the soil is exposed, is a way to reduce urban erosion. Stricter mining regulations requiring reclamation of strip-mined land (see chapter 20) are already significantly reducing soil erosion and related problems in mined areas.

Figure 18.17

Terracing also reduces the velocity of surface runoff. (A) Terraces in Pakistan, where the practice dates back many centuries. (B) Modern terraces in Midwestern farmland.

Photos: (A) G. C. Taylor, Jr., courtesy U.S. Geological Survey. (B) Courtesy U.S.D.A. Soil Conservation Service.

A

B

Summary

Soil forms from the breakdown of rock materials through chemical and mechanical weathering. Weathering rates are closely related to climate. The character of the resultant soil is a function of the nature of the parent material and the kinds and intensities of weathering processes through which the soil is formed. The wetter the climate, the more leached the soil; the hotter the climate, the more vigorous the chemical weathering, provided that water is present. The pedalfer soils are the somewhat leached soils of temperate climates with moderate rainfall. Pedocal soils, formed in drier climates, retain many more soluble minerals. The lateritic soils of tropical climates are particularly unsuitable for agriculture: not only are they highly leached of nutrients, but when exposed by clearing of the forest cover, they harden to brick-like solidity.

Soil erosion by wind and water is a natural part of the rock cycle. Where accelerated by human activity, it can also be a serious problem, especially on farmland, or, locally, in areas subject to construction or strip-mining. Erosion rates far exceed inferred rates of soil formation in many places. Loss of soil fertility or productivity is a particular concern on farmland. A secondary problem associated with soil erosion is the resultant sediment pollution of lakes and streams. Strategies to reduce soil erosion focus on reducing wind and water velocity—through windbreaks, terracing, or contour plowing—or providing vegetative cover to protect and anchor soil in place.

Terms to Remember

acid rain	C horizon	leaching	rill erosion
A horizon	exfoliation	mechanical weathering	sheet wash
B horizon	frost wedging	pedalfer	soil
capillary action	gullying	pedocal	spheroidal weathering
chemical weathering	lateritic soil	regolith	weathering

Questions for Review

1. What is the distinction between soil and regolith? Between soil and sediment?

2. Define *mechanical weathering* and give two examples.

3. What is exfoliation, and why is it thought to occur?

4. Chemical weathering is more rapid in warmer and wetter climates than in cooler and drier ones. Why?

5. Rank the following minerals in terms of their expected resistance to chemical weathering (most resistant first): biotite, calcite, quartz, calcium-rich plagioclase, amphibole.

6. Sketch a generalized soil profile. Indicate the A, B, and C horizons, zone of accumulation, and zone of leaching. Explain the latter two terms.

7. What is the distinction between a pedalfer and pedocal soil, and in what kind of climate does each occur?

8. What is a laterite? Where are laterites common? Are they fertile soils for agriculture? Explain.

9. Cite and briefly explain any three strategies for reducing soil erosion.

10. What is customarily meant by the phrase "acid rain"? Is the impact of acid rainfall the same everywhere in the United States? Explain.

For Further Thought

1. Considering the reactions in table 18.1, suggest why underground caverns form in limestone but not in granite, even if the granite is weathered.

2. Rates of weathering, and the relative resistance of different rock types to weathering, can often be evaluated with more accuracy using stone buildings, monuments, or even gravestones in cemeteries than by examining natural rocks in the field. Why? Suggest some possible limitations on the use of such artificial structures to estimate weathering rates.

3. Inspect a local construction site or farmland and look for evidence of soil erosion. Consider what might be done to reduce any problems that you identify.

Suggestions for Further Reading

Beasley, R. P. *Erosion and Sediment Pollution Control.* Ames, Iowa: Iowa State University Press, 1971.

Bridges, E. M. *World Soils.* 2d ed. Cambridge, England: Cambridge University Press, 1978.

Buol, S. W.; Hole, F. D.; and McCracken, R. J. *Soil Genesis and Classification.* 2d ed. Ames, Iowa: Iowa State University Press, 1980.

Hunt, C. B. *Geology of Soils.* San Francisco: W. H. Freeman and Co., 1972.

Lindsay, W. L. *Chemical Equilibria in Soils.* New York: John Wiley and Sons, 1979.

Morgan, R. P. C. *Soil Erosion.* London: Longman Group, 1979.

Thompson, L. M., and Troeh, F. R. *Soils and Soil Fertility.* 4th ed. New York: McGraw-Hill, 1978.

Troeh, F. R.; Hobbs, J. A.; and Donahue, R. L. *Soil and Water Conservation.* Englewood Cliffs, N.J.: Prentice-Hall, 1980.

Mass Movement, Mass Wasting

Introduction

The internal heat of the earth drives mountain-building processes. Just as inevitably, the force of gravity acts to tear the mountains down again. Gravity is the great leveler. It pulls constantly downward on every mass of material everywhere on earth, causing a variety of phenomena collectively termed **mass wasting,** in which geological materials are moved downslope from one place to another. Erosion is one form of mass wasting. The movement associated with mass wasting can be slow, subtle, almost undetectable on a day-to-day basis; but if continued over days or years, the cumulative effects can be very large. Alternatively, that movement can be sudden, as in a rockslide or avalanche, and the resultant changes swift, devastating, and obvious.

Landslide is a general term for rapid mass movement. In the United States alone, over one billion dollars in property damage is done each year by landslides and related mass movements, though losses of human life are relatively small. Many landslides occur quite independently of human activities. In some areas, active steps have been taken to control downslope movement or limit the resultant damage. On the other hand, certain human activities only serve to aggravate local landslide dangers, usually as a result of ignorance and failure to take that hazard into account. Large areas of this country are potentially at risk, and not all of these are particularly mountainous regions (figure 19.1).

Types of Mass Movements

Mass movements may be subdivided on the basis of the type of material moved and the nature of the movement, including the rate of movement. The material moved can be either unconsolidated, fairly fine material (soil, snow) or large, solid blocks or sheets of rock. It may be quite uniform in size, or a mix of fine sediment and boulders. *Landslide* is a nonspecific term for rapid mass movements, in rock or in soil. A few examples may clarify the different types of mass movements.

Falls

A **fall** is a free-falling action in which the moving material is not always in contact with the ground below. Falls are most often *rock falls* (figure 19.2). They frequently happen on very steep slopes or cliffs, when rocks high on the slope, weakened and broken up by weathering, lose support as materials under them are eroded away. Repeated rock falls in one place over a period of time result in the accumulation of piles of blocky debris, termed **talus** (figure 19.3).

Slumps and Slides

In a **slide,** a fairly coherent unit of material slips downward along a clearly defined surface or plane. Rockslides most often occur by movement along a bedding plane between successive layers of sedimentary rocks, or slippage where differences in the original composition of the sedimentary layers result in a

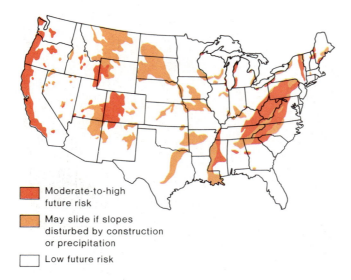

Moderate-to-high future risk

May slide if slopes disturbed by construction or precipitation

Low future risk

Figure 19.1

Landslide hazards in the United States.

Source: U.S. Geological Survey Professional Paper 950.

Figure 19.2

Rockfall (schematic).

weakened layer or a surface with little cohesion (figure 19.4). If the movement of the rock or soil mass is over a short distance only, the slide may be termed a **slump.**

Slumps in soil, rather than rock, are often characterized by a rotational movement of the soil mass accompanying the downslope movement. The surface at the top of the slide may be relatively undisturbed. The clifflike scar at the head of the slump is a type of **scarp.** The lower part of the slump block may end in a flow (figure 19.5).

Flows and Avalanches

Landslides involving unconsolidated, noncohesive material are extremely common. These are **flows,** in which material moves in a chaotic, disorganized fashion like fluid, with mixing of particles within the

Figure 19.3

Accumulation of talus from multiple rockfalls.

A

B

Figure 19.4

Rockslides. *(A)*(Schematic) A coherent mass of rock moves as a unit when inadequately supported. *(B)* ''Fallen City,'' a complex rockslide in the Bighorn Mountains. Here, sliding has occurred along bedding planes and joints in the limestone.

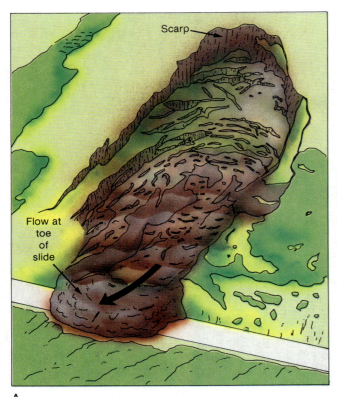

A

B

Figure 19.5

Slumps in soil. *(A)* (Schematic) Most of the soil mass moves coherently, and the surface is disturbed relatively little. *(B)* An example of a soil slump, ending in a flow across the road below. Note scarp at head of slump.

(B) Courtesy U.S. Geological Survey.

flowing mass. Flows need not involve only soils. Snow avalanches (figure 19.6) are one kind of flow; nuées ardentes are another, associated only with volcanic activity (see chapter 9). Where soil is the flowing material, flow phenomena may be described as *earth-flows* (fairly dry soil) or *mudflows* (if saturated with water) (see figure 19.7). A kind of wet-soil flow especially characteristic of alpine regions is a result of near-surface melting of frozen soil, while layers below remain frozen and impermeable. The resultant movement of sodden material over solid, impermeable ground is **solifluction.**

Often a wide variety of materials—soil, rocks, trees, and so on—will be incorporated together in a single flow. The result is a **debris avalanche** (figure 19.8). Regardless of the nature of the materials moved, all flows have in common the chaotic movement of the particles or objects in the flow.

Figure 19.6

Beginning of a snow avalanche, set off by the slight disturbance of a skier passing over an unstable patch of snow.

Photo by Ludwig, courtesy U.S. Forest Service/U.S. Geological Survey.

A

Figure 19.7

Recent earthflows in grassland, Washington County, Pennsylvania.

Photo by J. S. Pomeroy, courtesy U.S. Geological Survey.

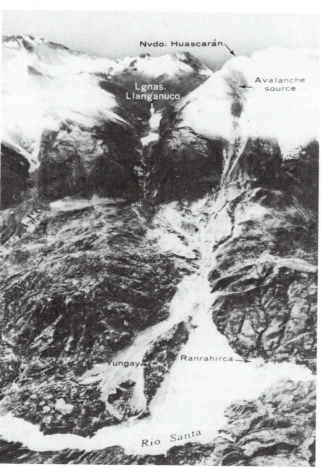

B

C

Figure 19.8

Debris avalanche. *(A)* (Schematic) The moving mass is a chaotic jumble of varied materials. *(B)* The Nevados Huascarán debris avalanche, 1970, aerial view. *(C)* A 700-ton block of granite that was part of the Nevados Huascarán debris.

Photos: *(B)* T. H. Nilsen, courtesy U.S. Geological Survey. *(C)* Courtesy U.S. Geological Survey.

Scales and Rates of Movement

Mass movements can occur on a variety of scales, as well as at a variety of rates. They may involve a few cubic meters of material, or more than a billion cubic meters. Total displacement may be only a few centimeters or even millimeters; or the falling, sliding, or flowing material may travel several kilometers.

In the most rapid mass movements, which include most rock falls, avalanches, and mudflows, the moving materials can travel at speeds up to several tens of meters per second. Such events thus happen quickly, leaving little time for anyone to react once they start. The greatest proportion of mass-movement casualties are caused by such events. Slides generally move at more moderate rates, in the range of a few meters per week to meters per day. Slow slides, slumps, and earthflows may move as slowly as a millimeter per year, or less. Extremely slow movement is described as **creep.** Soil creep is more common than rock creep. The principal impact of the slower kinds of mass wasting is property damage.

Some Causes and Consequences of Mass Movements

Basically, mass movements occur whenever the downward pull of gravity overcomes the forces resisting sliding or flow. Sudden movements may be set off by a triggering mechanism. Mass movements, in turn, may cause secondary problems, such as flooding (see later in this section).

Effects of Slope

Steepness of slope is one major factor contributing to the probability of a landslide: all else being equal, the steeper the slope, the greater the downslope pull, or **shearing stress** (see figure 19.9). Counteracting the shearing stress is *friction,* in the case of unconsolidated material, or **shear strength** in the case of solid rock. When shearing stress overcomes friction or shear strength, sliding occurs.

For unconsolidated material, one can define an **angle of repose,** which is the maximum slope angle at which the material is stable (figure 19.10). This angle will vary with the material. Smooth, rounded particles will tend to support only very low-angle slopes (imagine making a heap of marbles or ball bearings), while rough, sticky, or irregular particles can be piled more steeply without becoming unstable.

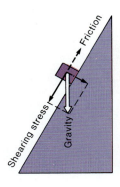
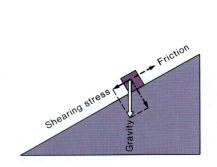

Figure 19.9

Effects of slope geometry on slide potential. The mass of the block, and thus the total downward pull of gravity, is the same in both cases, but the steeper the slope, the greater the shearing stress component.

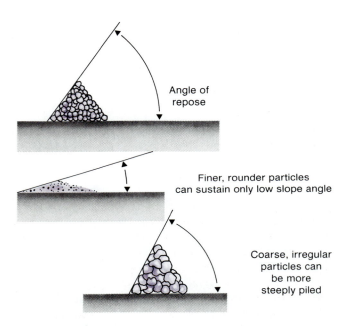

Figure 19.10

Angle of repose, an indicator of an unconsolidated material's resistance to sliding. Coarser and rougher particles can maintain steeper stable slope angles.

Solid rock can be perfectly stable even at a vertical slope, but may lose its strength if it is broken up by weathering or fracturing. Also, in layered sedimentary rocks, there may be weakness along bedding planes, where different rock units are imperfectly held together; or some units may themselves be weak or even slippery (clay-rich layers, for example). Such planes of weakness are potential slide planes.

Landslides following winter storms in California. Increased erosion at the base of this sea cliff and the extra mass of and lubrication by rainwater were all contributing factors in triggering the sliding.

Photo courtesy U.S. Geological Survey.

Slumping in expansive clay seriously damaged this road near Boulder, Colorado.

Photo courtesy U.S. Geological Survey.

Slopes may be **steepened to unstable angles** by **natural erosion by water** or ice. Erosion can also undercut rock or soil, removing the support beneath a mass of material and thus leaving it susceptible to falling or sliding. This is a common contributing factor to landslides in coastal areas. Over long periods of time, slow tectonic deformation can also alter the angles of slopes and bedding planes, making them steeper or shallower. This is most likely to be a significant factor in young, active mountain ranges, such as the Alps, or the Coast Ranges of California. Steepening of slopes by tectonic movements was suggested as the cause of a large rockslide in Switzerland in 1806, which buried the township of Goldan under a block of rock nearly 2 kilometers long, 300 meters wide, and 60 to 100 meters thick.

Effects of Fluid

Aside from its role in erosion, water can greatly increase the likelihood of mass movements in other ways. It can seep along bedding planes in layered rock, reducing friction and making sliding more likely. As noted in connection with earthquakes and faulting, an increase in pore water pressure in rocks has been found to decrease their resistance to shear stress, which can also facilitate sliding.

In unconsolidated materials, the role of water is variable. A little moisture may add some cohesion—consider that it takes damp sand to build sand castles. However, substantial increases in water content both increase pore pressure and reduce the friction between particles that provides strength. The very mass of water in saturated soil may add enough extra weight, enough additional downward pull, to set off a landslide on a slope that was stable when dry (see figure 19.11).

Frost wedging and the associated fracturing can weaken rocks. Some soils rich in clays absorb water readily; one type of clay, montmorillonite, may absorb 20 times its weight in water and form a weak gel. Such material fails easily under stress. Other clays expand when wet, contract when dry, and can destabilize a slope in the process (figure 19.12). In terms of property damage, expansive clays are, in fact, the most costly geologic hazard in the United States (see figure 19.13).

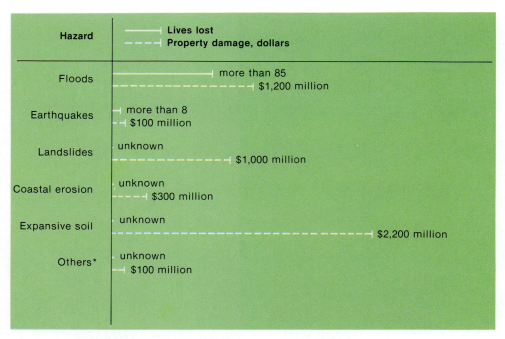

Hazard	——— Lives lost / ‐ ‐ ‐ Property damage, dollars		
Floods	more than 85		
	$1,200 million		
Earthquakes	more than 8		
	$100 million		
Landslides	unknown		
	$1,000 million		
Coastal erosion	unknown		
	$300 million		
Expansive soil	unknown		
	$2,200 million		
Others*	unknown		
	$100 million		

*Includes volcanic eruptions, tsunamis, subsidence, creep, and other phenomena.

Figure 19.13

Relative magnitude of annual loss of life and property damage in the United States from various geologic hazards.

Source: U.S. Geological Survey Professional Paper 950.

Earthquakes

Landslides are a common consequence of earthquakes in hilly terrain, as noted in chapter 8. Seismic waves passing through rock stress and fracture it. The added stress may be as much as half that already present due to gravity. Ground shaking also jars apart soil particles and rock masses, reducing the friction holding them in place. One of the most lethal earthquake-induced landslides happened in Peru in 1970. One slide had already occurred below the steep, snowy slopes of Nevados Huascarán, the highest peak in the Peruvian Andes, in 1962, unprompted by an earthquake; it had killed approximately 3500 people. In 1970, a magnitude 7.7 earthquake centered 130 kilometers to the west shook loose a much larger debris avalanche that buried most of the towns of Yungay and Ranrahirca, and more than 18,000 people with them (recall figures 19.8B,C). Some of the debris was estimated to have moved at 1000 kilometers per hour. In the 1964 Alaskan earthquake, the Turnagain Heights section of Anchorage was heavily damaged by landslides. California contains not only many fault zones but many sea cliffs and hillsides prone to landslides when earthquakes occur.

Quick Clays

A geologic factor that contributed to the 1964 Anchorage landslides, and which continues to add to the landslide hazards in Alaska, California, parts of northern Europe, and elsewhere, is a material known as **quick clay** or **sensitive clay**. True *quick clays* are most common in northern latitudes. They are formed from glacial rock flour deposited in a marine environment. When this extremely fine sediment is later uplifted above sea level by tectonic movements, it contains salty pore water. The sodium chloride in the pore water acts as a "glue" holding the clay particles together. Fresh water subsequently infiltrating the clay washes out the salts, leaving a delicate honeycomb-like structure of particles not firmly held together. Vibration from seismic waves breaks the structure apart, reducing the quick clay to a fraction of its original strength, creating a finer-grained equivalent of quicksand that is highly prone to sliding. So-called sensitive clays are similar in behavior but differently formed. Weathering of volcanic ash, for example, can produce a sensitive clay sediment, *bentonite*. Such deposits are locally common in the western United States, where there has been much relatively recent volcanism.

Effects of Vegetation

The presence of vegetation tends to stabilize slopes. Plant roots, especially those of larger shrubs and trees, can provide a strong interlocking network to hold unconsolidated materials together and prevent flow. (This benefit may continue for several years after trees are cut down, before the roots decompose.) Actively growing vegetation also takes up moisture from the upper layers of soil, and can thus reduce the overall moisture content of the mass.

Secondary Consequences of Mass Movements

Just as landslides can be a result of earthquakes, other events, notably floods, can be produced by landslides. A stream may create unstable slopes in the process of cutting a valley. Subsequent landslides into the valley can dam up the stream, creating a natural reservoir. The filling of the reservoir will make the area behind the earth dam uninhabitable, though this will usually happen slowly enough that lives are not lost. This was one effect of the largest landslide of recent history. It occurred in 1911, when an earthquake of magnitude 7.0 struck a remote area in the Pamir Mountains of the Soviet Union. The resulting landslide, involving some 2.5 billion cubic meters of material, overwhelmed the village of Usoy, killing 54 people. It also dammed the river Murgab. The lake that formed behind the slide eventually grew to be nearly 300 meters deep and 53 kilometers (over 30 miles) long, and drowned the site of another town, Sarez. (See also figure 19.14.)

A further danger is that the unplanned dam formed by the landslide will later fail. In 1925, an enormous rockslide a kilometer wide and over 3 kilometers long in the valley of the Gros Ventre in Wyoming blocked that valley, and the resulting lake stretched to 9 kilometers long. After spring rains and snowmelt, the "dam" failed. Flood waters swept down the valley below. Six people died, and the toll would have been far worse in a more populous area.

Landslides in and near planned dams and reservoirs can also cause problems. In the reservoir disaster at Vaiont, Italy, the presence of the reservoir aggravated the local landslide hazards, and the subsequent slides and floods destroyed several towns (see box 19.1).

An Anthropogenic Landslide: The Vaiont Dam Disaster

The Vaiont River flows through an old glacial valley in the Italian Alps. The valley is underlain by a thick sequence of sedimentary rocks, principally limestones with some clay-rich layers, which were folded and fractured during the building of the Alps. The beds on either side of the valley dip down toward the valley floor (figure 1). The rocks themselves are relatively weak, and are particularly prone to sliding along the clay-rich layers. Evidence of old rockslides can readily be seen in the valley and was noted in drill cores taken in the early 1960s, shortly after construction of the dam was completed. Extensive solution of the carbonates by groundwater has further weakened the rocks, producing sinkholes and underground caverns and channels.

The Vaiont Dam, built for power generation, is the highest "thin-arch" dam in the world, over 265 meters (875 feet) high at its highest point. Modern engineering methods were used to stabilize the rocks in which it is based. The capacity of the reservoir behind the dam was 150 million cubic meters—initially.

As the water level in the reservoir rose following completion of the dam, the pore pressures of groundwater in the rocks of the reservoir walls rose also. This tended to buoy up the rocks and swell the clays of the clay-rich layers, further decreasing their strength and making sliding easier. In 1960, a block of 700,000 cubic meters of rock slid from the slopes of Mount Toc, on the south wall, into the reservoir. Creep was noted over a still greater area. A set of monitoring stations was established on the slopes of Mount Toc to track any further movement. In 1960-62,

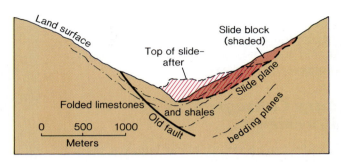

Figure 1
Cross section of the Vaiont River valley.

From G. A. Kiersch, "The Vaiont Reservoir Disaster," *Mineral Information Service 18, #7*, pp. 129–138, 1965. Reprinted by permission.

measured creep rates occasionally reached 25 to 30 centimeters (nearly 1 foot) per week.

Late summer and fall of 1963 were times of heavy rainfall in the Vaiont valley. The saturated rocks represented more mass pushing downward on zones of weakness. Groundwater flow increased, while the water table and reservoir level rose by over 20 meters. Measured creep rates increased. What was still not yet realized was that the rocks were slipping not as many small blocks, but as a single coherent unit. On October 1, animals that had been grazing on those slopes of Mount Toc moved off the hillside, and would not return.

The rains continued, and so did the creep, rates of which increased to 20 to 30 centimeters per *day*. Finally, on October 8, the engineers realized that all the creep-monitoring stations were moving together. They also discovered that a far larger area of hillside was moving than they had thought. They began to try to reduce the water level in the reservoir by opening the gates of two outlet tunnels. Yet the water level continued to rise: the silently creeping mass had begun to encroach on the

reservoir. On October 9, creep rates as high as 80 centimeters per day were measured. At about 10:40 that night, in the midst of another downpour, disaster struck.

A resident of Casso later reported that at first there was a sound of rolling rocks, ever louder. Then a gust of wind hit his house, smashing windows and raising the roof so that the rains poured in. Just as abruptly, the wind ceased and the roof collapsed.

A 240-million-cubic-meter chunk of hillside had slid into the reservoir (figure 2). The shock of the slide was detected on seismometers in Rome, Brussels, and elsewhere in Europe. The sudden movement created the shockwave of wind that rattled Casso and drew the water 240 meters upslope out of the reservoir after it, on the far bank. The displaced water then crashed over the dam, in a wall 100 meters above the dam crest, and rushed down the valley below. The water wave was still over 70 meters high some 1 1/2 kilometers downstream where the Vaiont flows into the Piave River. The energy of the rushing water was such that some of it flowed *upstream* in the Piave for more than 2 kilometers.

Box 19.1 *Continued*

Within a period of about five minutes, nearly 3000 people were drowned and entire towns obliterated.

It is a great tribute to the designer of the Vaiont Dam that the dam itself held through all of this, resisting forces far beyond its design specifications. However, those who chose the site had every reason to realize the landslide risks. There was ample evidence of persistent slope instability. In fact, the dam builders in this instance were later found guilty of gross negligence and imprisoned.

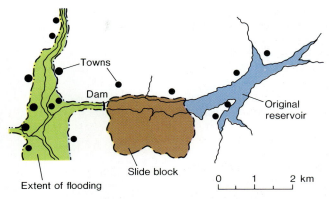

Figure 2
The Vaiont Reservoir slide block. Note areas and towns inundated by the resulting flooding.

From G. A. Kiersch, "The Vaiont Reservoir Disaster," *Mineral Information Service 18, #7*, pp. 129–138, 1965. Reprinted by permission.

Landslides and Human Activities

Just reviewing the factors that influence landslide hazards, it is possible to see many ways in which human activities can increase the risk of landslide. One is by the clearing of stabilizing vegetation. In some instances, where clearcutting logging operations have been undertaken and sloping soil exposed, earthflows and mudflows have become far more frequent, or been more severe than before.

Construction

Many types of construction lead to oversteepening of slopes. Highway roadcuts, quarrying or open-pit mining operations, and construction of stepped home-building sites on hillsides are among the activities that can cause problems (figure 19.15). Where dipping layers of rock are present, removal of material at the lower end of the layers may leave large masses of rock unsupported, held in place only by friction between the layers. Slopes cut in unconsolidated material at angles higher than the angle of repose of that material are by nature unstable, especially if no attempt at planting stabilizing vegetation is made (figure 19.16).

The very act of putting a house above a naturally unstable or artificially steepened slope adds weight to the slope, thereby increasing the shear stress acting on the slope.

Other activities connected with the presence of housing developments on hillsides can increase the risk of landslides in more subtle ways. Watering the lawn, using a septic tank for sewage disposal, and even putting in an in-ground swimming pool from which water can seep slowly are all activities that increase the moisture content of the soil and can render the slope more susceptible to sliding. Irrigation and the use of septic systems increase the flushing of fresh water through soils and sediments. In areas underlain by quick or sensitive clays, this may hasten the washing-out of salty pore waters and the destabilization of the clays. On the other hand, the planting associated with deliberate landscaping can reduce the risk of sliding.

Recognizing Past Mass Movements

Rockfalls tend to be quite obvious, especially in a generally vegetated area. Talus slopes are inhospitable to most plants, so rockfalls tend to remain barren

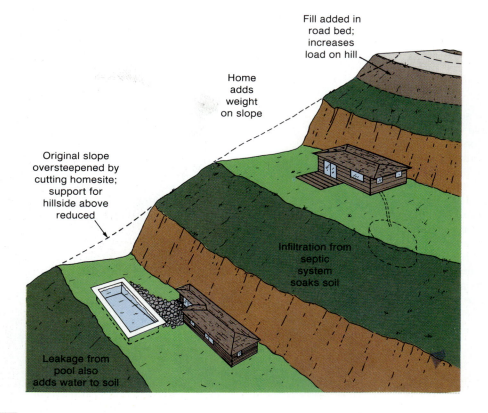

Fill added in
road bed;
increases
load on hill

Home
adds
weight
on slope

Original slope
oversteepened by
cutting homesite;
support for
hillside above
reduced

Infiltration from
septic
system
soaks soil

Leakage from
pool also
adds water to soil

Figure 19.15

Effects of construction and human habitation on slope stability.

Figure 19.16

Failure of steep, unplanted roadcut now threatens the cabin above.

of trees and plants, as in figure 19.3. Lack of vegetation may also mark the paths of past debris avalanches or other soil flows or slides (figure 19.17). These scars on the landscape point plainly to slope instability.

Landslides are not the only kinds of mass movements that recur in the same places. Snow avalanches disappear when the snow melts, but historical records of past avalanche occurrence can still be used to pinpoint particularly dangerous areas. Similarly, records of the character of past volcanic activity, and an examination of the typical products of a particular volcano, can be used to assess that volcano's tendency to produce nuées ardentes.

Very large slumps and slides may be less obvious, especially when viewed from ground level. The coherent nature of rock and soil movement in most slides means that vegetation growing atop the slide may not be greatly disturbed by the movement. Aerial photography or high-quality topographic maps can be helpful in such a case. In a regional overview, the mass movement often shows very clearly, revealed by a

A

B

Figure 19.17

Areas prone to landslides may be recognized by the failure of vegetation to establish itself on unstable slopes.

scarp at the head of the slump, or an area of hummocky, disrupted topography relative to surrounding, more stable areas.

With creep, individual movements are short distance and the whole process is slow, so vegetation may continue to grow in spite of the slippage. More detailed observation can still reveal the movement. For example, trees are biochemically programmed to grow vertically upward. If the soil in which they are growing begins to creep downslope, the tree trunks may be tilted, and this indicates the soil movement (figure 19.18A). Further growth will continue to be vertical. If slow creep is prolonged over a considerable period of time, during which growth proceeds, curved tree trunks may result (figure 19.18B). Inanimate objects can reflect soil creep also. Slanted utility poles and fences and the tilting of once-vertical gravestones or other monuments likewise indicate that the soil is moving (figure 19.19). The ground surface itself may show cracks parallel to (across) the slope.

Figure 19.18

Tilted tree trunks are a consequence of creep. *(A)* "Drunken forest" of trees tilted by soil creep, Cerro de Canijuata, Mexico. *(B)* This curved tree trunk has developed on an imperfectly stabilized slope above a retaining wall.

(A) K. Segerstrom, courtesy U.S. Geological Survey.

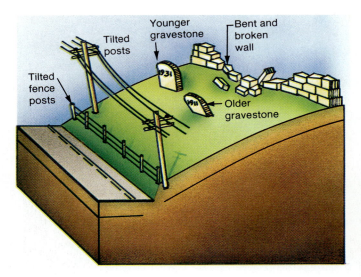

Figure 19.19

Other signs of creep: tilted monuments, fences, utility poles.

Figure 19.20

Chain-link fencing draped over roadcut, protecting road from rockfall.

There are additional signs that a prospective home buyer can look for as indications of unstable land underneath. Ground slippage may have caused cracks in driveways, garage floors, or brick or concrete walls; doors and windows that jam may do so because the frame has warped due to differential movement in the soil and foundations. If there has already been enough movement to cause obvious structural damage, it may not be possible to stabilize the slope adequately, except perhaps at great expense. It may be especially important to investigate site stability if the site itself has a slope of more than 15%, or if much steeper slopes lie above or below it. Note, however, that steep slopes are not required in order for sliding to occur (see box 19.2).

Some Possible Preventive Measures

Sometimes local officials resign themselves to the existence of a problem, and perhaps take some steps to limit the resulting damage (figure 19.20). In places where the structures to be protected are few or small, and the slide zone is narrow, it may be economically feasible to bridge structures and simply let the slides flow over them. This might be done to protect a railway line or road running along a valley from avalanches—snow or debris—below a particularly steep slope (figure 19.21). This solution would be far too expensive on a large scale, however, and no use at all if the

Figure 19.21

Avalanche protection structure built over railroad track or road in snow avalanche area.

Box 19.2

Portuguese Bend, Los Angeles County, California

The Portuguese Bend section of the Palos Verdes Peninsula has a history of landslides extending back for decades, perhaps centuries. Early aerial photographs of the region show clearly that much of the south side of the peninsula is slide material. The appearance of the area has been described as ''that of a rumpled carpet thrown on the hills'' (Bolt et al. 1975, 194).

Housing developments were begun there in the 1950s. In 1956, a 1-kilometer-square area began to slip, although the slope in that vicinity was less than 7°. Cracks developed within the slide; houses on or near it were damaged or destroyed; a road built across its base had to be rebuilt repeatedly (see figure 1). Once begun, the movement has continued over the ensuing decades, at a slow but inexorable average of 3 meters per year.

The responsibility for activating this slide is unclear. Most of the homes used cesspools for sewage disposal, which added fluid to the ground. The county highway department, in building a road across what became the top of this slide, had added a lot of fill, and thus weight. A court decided that the County of Los Angeles was at least partly to blame, and awarded a homeowners' association more than $5 million in damages. Perhaps the most fundamental conclusion is that development in such a demonstrably unstable area was unwise from the outset.

Figure 1
Damage from the Portuguese Bend landslide.
Photos by E. F. Patterson, courtesy U.S. Geological Survey.

A

B

Figure 19.22

Slope stabilization, Lake Tahoe. *(A)* More natural-looking retaining structures are achieved by using irregular rocks formed into building blocks by wire mesh. *(B)* Side view of resultant retaining wall.

base on which the structure was built were sliding also. Avoiding the most landslide-prone areas altogether would greatly limit damage, but as is true with fault zones, floodplains, and other hazardous settings, developments may already be in place in areas at risk, and economic pressure for more development can be strong. Certain steps to reduce the potential for mass movements may then be undertaken.

Slope Modification

If a slope is too steep to be stable under the load that it carries, then one can either (1) reduce the slope angle, (2) place additional supporting material at the foot of the slope to prevent a slide or flow at the base of the slope, or (3) reduce the load (weight) on the slope by removing some of the rock or soil (or artificial structures) high on the slope. See, for example, figures 19.22 and 19.23. These measures may be used in combination. Depending on the instability of the slope, they may need to be executed cautiously. If earthmoving equipment is being used to remove soil at the top of a slope, for instance, there is some chance that the added weight of the equipment itself, or the vibrations from it, may trigger a landslide. Where retaining walls are placed below a slope, the greatest success has generally been achieved with low, thick walls placed at the toe of a relatively coherent slope; high, thin walls have been less successful, given the distribution of stresses on an unstable slope.

To stabilize exposed near-surface soil, ground covers or other vegetation may be planted. The preferred varieties are fast-growing plants with sturdy and extensive root systems. Where plants are an impractical solution, surface application of thin sheets of concrete or other material may be tried as an alternative. This is not always effective (see box 19.3).

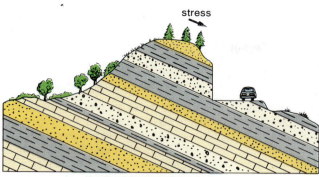

Before: Roadcut leaves steep,
unsupported slope

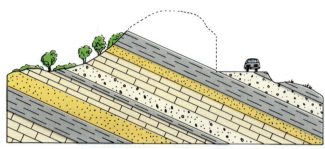

After: Material removed to
reduce slope angle and load

Figure 19.23

Removal of some of the material on a hillside reduces load and slope angle, increasing stability.

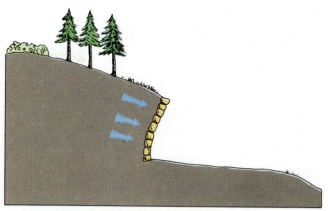

A   Water trapped in soil causes movement, pushing down retaining wall.

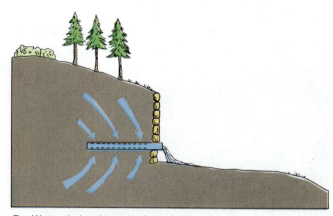

B   Water drains through pipe, allowing wall to keep slope from moving.

Figure 19.24

Improved drainage also enhances slope stability, by reducing both load and pore pressure.

Removing Fluid

Considering the major role that water can play in mass movements, the other principal strategy in reducing landslide hazards is to reduce the water content, or pore pressure, of the rock or soil, thereby increasing frictional resistance to sliding. This might be done by covering the surface completely with an impermeable material and diverting surface runoff above the slope. Alternatively, subsurface drainage might be undertaken (figure 19.24). As a preliminary step, old water wells in the area can be vigorously pumped. Additional networks of underground boreholes can then be drilled to increase drainage, and pipelines installed to carry the water out of the slide area. All such moisture-reducing techniques naturally have the greatest impact where rocks or soil are relatively permeable. Where the rock or soil drains only slowly, hot air may be blown through the boreholes to dry out the ground more rapidly.

Other Stabilization Methods

Other techniques that have been tried include driving vertical piles into the foot of a shallow slide to hold the sliding block in place. This procedure can be expected to work only where the slide is comparatively solid (loose soils simply flow between piles), with thin slides (so that piles can be driven deep into stable material below), and on low-angle slopes (otherwise the shearing stresses may just snap the piles). It has been tried on several European slides, generally not very effectively. Greater success has been achieved with the

Box 19.3

Landslide Control in the Rockies: An Example

If the terrain is basically unstable, and one builds there anyway, the result may be a nonstop battle against natural forces. Such is the case with the road on the east side of Beartooth Pass that winds sharply down to the town of Red Lodge, Montana. Roads in this area are few, and it is easy to see why (a sample of the topography is shown in figure 19.17). The slopes are steep, and the guard rails along the highway are dented from falling boulders. Spring is an especially dangerous time, as the soil wet from the spring thaw slides particularly readily, often taking sections of road with it. The road, however, is critical to local transportation and has therefore been regularly rebuilt after each slide.

Several years ago, an attempt was made to stabilize some of the most slide-prone slopes by applying concrete to the slope surfaces (figure 1A). Several large expanses of slope were so treated. Realizing the potential problems associated with possible water accumulation behind these concrete sheets, the engineers provided for drainage holes (B).

A return visit three summers later suggested that this solution was not an unqualified success. The slopes had continued to move, and the thin concrete sheets had buckled and cracked in many places (C, D). Furthermore, some very new asphalt below one patch (C) showed that the spring washout had claimed the soil underneath the roadway there as well.

A

B

C

D

Figure 1
Slope stabilization effort east of Beartooth Pass, Montana.

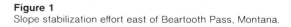

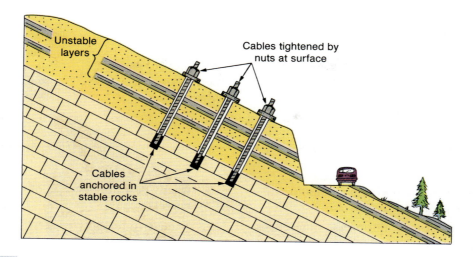

Figure 19.25

Use of rock bolts to stabilize a rock slide.

use of *rock bolts* to stabilize rocky slopes and, occasionally, rockslides. Rock bolts have long been used in tunneling and mining to stabilize rock walls. It is sometimes also possible to anchor a rockslide with giant steel bolts driven into stable rocks below the slip plane (figure 19.25). Again this works best on thin slide blocks of very coherent rocks on low-angle slopes. Additional procedures occasionally used on unconsolidated materials include hardening unstable soil by drying and baking it with heat (this can work well with clay-rich soils), or treating with portland cement. By far the most common strategies, however, are modification of slope geometry and load, or dewatering, or a combination of these two. The more

ambitious engineering efforts are correspondingly more expensive and usually reserved for large construction projects rather than individual homesites.

It should be kept in mind that determining the absolute limits of slope stability is an imprecise science. This was illustrated in Japan, in 1969, when a test hillside being deliberately saturated with water unexpectedly slid prematurely, killing several of the researchers. The same point is made less dramatically every time a slope-stabilization effort fails.

Summary

Mass movements occur when the shear stress acting on rocks or soil exceeds the strength of the material to resist it. Gravity provides the main component of the shear stress. The basic susceptibility of a slope to failure is determined by a combination of factors including geology (nature and strength of materials), geometry (slope angle), and moisture content. Sudden failure usually also involves a triggering mechanism, such as vibration from an earthquake, addition of moisture, steepening of slopes (naturally or through human activities), or removal of vegetation. One possible secondary consequence of mass movements is flooding.

The slowest mass wasting is *creep*. More rapid movements, collectively termed *landslides* when they involve rock or soil material, are subdivided on the basis of the nature of the movement and the materials moved. Slumps and slides involve movement of coherent rock or soil masses. Flows, which include snow avalanches and nuées ardentes as well as debris avalanches and soil flows, occur only in unconsolidated materials, and are characterized by more chaotic movement of particles within the flowing mass. Recognized landslide hazards may be reduced by modifying the slope geometry, reducing the weight acting on the slope, planting vegetation, or dewatering the rocks or soil.

Terms to Remember

angle of repose	flow	scarp	slide
creep	landslide	sensitive clay	slump
debris avalanche	mass wasting	shearing stress	solifluction
fall	quick clay	shear strength	talus

Questions for Review

1. What is the distinction between a fall and a slide? Between a slide and a flow?

2. Name four kinds of flow, noting the differences among them.

3. Rank the following in terms of velocity, from slowest to fastest: soil slump, creep, debris avalanche.

4. What is the driving force behind mass movements? What properties of material resist the shearing stress component?

5. Explain the concept of angle of repose and its relationship to slope stability. To what kinds of material does this concept apply?

6. Briefly describe the role of fluid in mass movements.

7. Name two possible triggers for sudden slope failure.

8. What is a *quick clay*?

9. Describe any four ways in which human activities can destabilize slopes.

10. Suggest one possible approach to stabilizing (a) a steep roadcut in severely weathered granite; (b) a sandy soil slope; (c) a shallow but unstable slope in thinly bedded shale, where the bedding dips parallel to the slope.

For Further Thought

1. What kind of geologic features might you look at in order to obtain an estimate of the angle of repose of (a) sand and (b) volcanic ash? What factors might affect the results of your observations, or limit their applicability?

2. Spring is landslide season in mountainous areas of many temperate climates. Suggest at least three reasons for this.

Suggestions for Further Reading

Bolt, B. A.; Horn, W. L.; MacDonald, G. A.; and Scott, R. F. "Hazards from Landslides." In *Geological Hazards,* chapter 4. New York: Springer-Verlag, 1975.

Fleming, R. W., and Taylor, F. A. *Estimating the Cost of Landslide Damage in the United States.* U.S. Geological Survey Circular 832, 1980.

Kiersch, G. A. "The Vaiont Reservoir Disaster." Mineral Information Service *18* (1965): 129–38.

Radbrich-Hall, D. H.; Colton, R. B.; Davies, W. E.; Skipp, B. A.; Lucchitta, I.; and Varnes, D. J. "Preliminary Landslide Overview Map of the Conterminous United States." U.S. Geological Survey Miscellaneous Field Studies Map MF–771, 1976.

Schuster, R. L., and Krizek, R. J., eds. *Landslides, Analysis and Control.* National Research Council, Transportation Research Board Special Report 176, 1978.

Schuster, R. L.; Varnes, D. J.; and Fleming, R. W. "Landslides." In *Facing Geologic and Hydrologic Hazards,* edited by W. W. Hays, 55–65. U.S. Geological Survey Professional Paper 1240–B, 1981.

Utgard, R. O.; McKenzie, G. D.; and Foley, D. *Geology in the Urban Environment.* Minneapolis: Burgess Publishing Co., 1978.

Voight, B., ed. *Rockslides and Avalanches.* New York: Elsevier, 1978.

Zaruba, Q., and Mencl, V. *Landslides and Their Control.* New York: Elsevier, 1969.

Part Four

Part opener photo © Steve McCutcheon

The title of this part reflects the fact that the subjects treated here are receiving more attention in physical geology courses than they have historically. In part, the shift reflects practical concerns, for example about availability of resources. It also reflects significant increases in knowledge that have come about within the last decade or two.

All of our mineral resources, and most of our energy sources, are geologic in nature—they consist of geologic materials, are formed and modified by geologic processes, and may be concentrated by geologic structures. A survey of these various resources and their occurrence, then, is a natural extension of physical geology. Some of the recent advances in basic geologic understanding, notably the theory of plate tectonics, have profoundly influenced our thinking about how and where some of these resources form; this is especially true of the minerals. In addition, chapters 20 (Minerals) and 21 (Energy) include a look at the supply-and-demand picture for a number of key geologic resources, which is a matter that demands increasing attention from both geologists and nongeologists.

Study of the rest of the solar system was at one time considered to be strictly in the province of astronomy, not geology. However, as our knowledge of other solar-system objects has increased, the boundary between disciplines has become less distinct. Images returned from spacecraft, and samples brought back from the moon and recovered from meteorite falls, have shown both intriguing similarities between terrestrial and extraterrestrial features and materials, and definite differences. By studying the other planets, moons, and smaller objects, we improve our understanding of the earth. Such studies are now well within the scope of modern geology, and therefore chapter 22 has been included in this text.

Mineral Resources

Introduction

It has already been noted that the bulk of the earth's crust is composed of fewer than a dozen elements. Eight chemical elements make up over 98% of the crust. Many of those missing from the list are vitally important to our present society, including industrial and precious metals, essential components of chemical fertilizers, and elements like uranium that serve as energy sources. Some of these are found in very minute amounts in the average rock of the continental crust: copper, 0.006%; tin, 2 parts per million; gold, 4 parts per *billion*. Clearly, then, many useful elements are mined from very atypical rocks. This chapter will begin with a look at the occurrences of a variety of rock and mineral resources, then briefly consider the supply-and-demand picture for the United States and the world. Aspects of that picture suggest a need to develop additional sources of mineral materials for the future; several possibilities are examined. Some environmental impacts of mining activities are also noted.

Ore Deposits

An **ore** is a rock in which a valuable or useful metal occurs at a concentration sufficiently high to make it economically worth mining. A given ore deposit may be described in terms of the **enrichment** or **concentration factor** of a particular metal:

$$\text{Concentration factor (of a given metal in an ore)} = \frac{\text{Concentration of the metal in that ore}}{\text{Concentration of that metal in average continental crust}}$$

The higher the concentration factor, the richer the ore (by definition), and the less of it must be mined in order to extract a given amount of metal. In general, the minimum concentration factor required for profitable mining is inversely proportional to the average crustal concentration. Metals like iron or aluminum, which make up about 6% and 8% of average continental crust respectively, need be concentrated only by a factor of four or five times in order for mining to be profitable. Copper must be enriched about 100 times relative to average rock, while mercury (average concentration 80 ppb) must be enriched to about 25,000 times its average concentration before the ore is rich enough to mine profitably. The exceptions to this rule are those few extremely valuable elements, such as gold, which are so high-priced that a small quantity will justify a considerable amount of mining.

The unit value of the mineral or metal extracted, and its concentration in a particular deposit, are major factors is determining the profitability of mining a specific deposit. Naturally the economics are sensitive to world demand. If demand climbs and prices rise in response, additional, less enriched ore deposits may be opened up; a fall in price will cause economically marginal mines to close. The practicality of mining a specific ore body may also depend on the mineral(s) in which a metal of interest is found, for this will affect the cost of extracting the pure metal. Three iron deposits containing equal concentrations of iron will not be equally economic if the iron occurs mainly in oxides in one deposit, silicates in another, and sulfides in the third.

By definition, ores are somewhat unusual rocks. It is therefore not surprising that the known economic mineral deposits are very unevenly distributed around the world. (See, for examples, figure 20.1.) The United States happens to control about 60% of the world's known molybdenum deposits, and 40% of the lead. But although this country is the major world consumer of aluminum, using about 40% of the total produced, we have virtually no domestic aluminum ore deposits. Australia contains fully one-third of the world's aluminum ore. Zaire and Zambia together account for approximately half the known economic cobalt deposits, while Thailand and Malaysia contain the same proportion of tin. South Africa controls nearly half the world's known gold and platinum ore, and 75% of the chromium. Almost three-fourths of the tungsten deposits are located in the People's Republic of China. The great disparities in mineral wealth among nations should be borne in mind during later discussions of world mineral supply/demand projections.

Types of Mineral Deposits

Deposits of economically valuable rocks and minerals form in a variety of ways. This section will describe some of the more important processes involved.

Magmatic Deposits

Magmatic activity gives rise to several kinds of deposits. Certain igneous rocks, just by virtue of their compositions, contain high concentrations of useful silicate or other minerals. These deposits may be especially valuable if the rocks are coarse-grained *pegmatites* (figure 20.2), from which individual minerals are easily recovered. In some pegmatites, crystals may be over ten meters long. Many pegmatites are also enriched in uncommon elements. Pegmatites commonly crystallize from the residual fluids left after most of a body of magma has solidified; many rare or trace elements not readily incorporated into the crystal

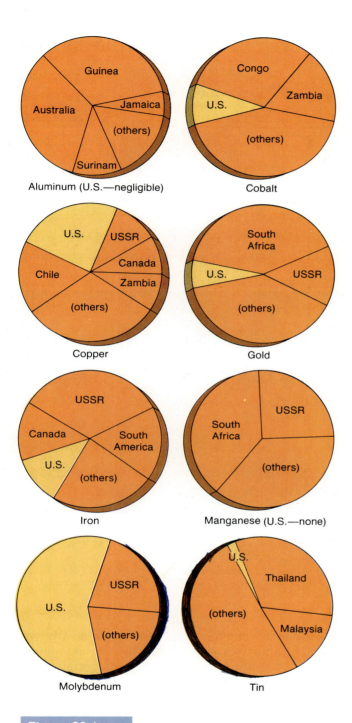

Proportions of world reserves of some nonfuel minerals controlled by various major producers. (Data from Ferguson (1982) and tables 20.2 and 20.3.) Although the United States is a major consumer of most metals, it is a major producer of very few.

Pegmatite, a very coarse-grained plutonic rock, may yield large, valuable crystals.

structures of the common silicates are likewise left over, concentrated in the residual fluids at this stage. Rarer minerals mined from pegmatites include tourmaline and beryl. (Tourmaline is used as a gemstone and for crystals in radio equipment, and also mined for the element boron that it contains. Beryl is mined for the metal beryllium when the crystals are of poor quality, or for the gemstones aquamarine and emerald when the crystals are large, clear, and well colored.)

Other useful minerals may be concentrated within a cooling, crystallizing magma chamber by gravity. If they are more or less dense than the magma, they may sink or float as they crystallize, instead of remaining suspended in the solidifying silicate mush, and accumulate in thick layers that are easily mined (figure 20.3). Chromite (chromium oxide, Cr_2O_3) and magnetite (Fe_3O_4) are both quite dense. In a magma of suitable bulk composition, rich concentrations of these minerals may form in the lower part of the magma chamber by gravitational settling during crystallization (figure 20.4). The dense precious metals, such as gold and platinum, can sometimes also be concentrated during magmatic crystallization. However, these metals are valuable enough that even where they are disseminated throughout an igneous rock body, they may be worth mining if their concentrations are sufficiently high.

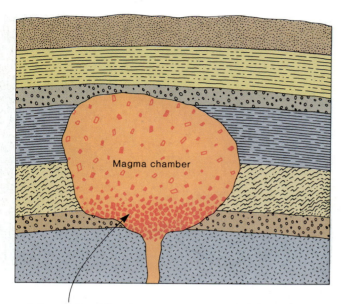

A dense mineral like chromite or magnetite may settle out of a crystallizing magma to be concentrated at the bottom of the chamber.

Figure 20.3

Formation of a magmatic ore deposit by gravitational settling of a dense mineral during crystallization.

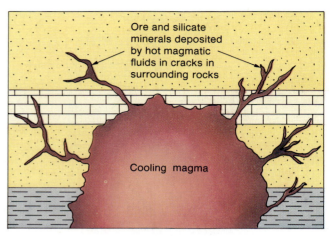

Ore and silicate minerals deposited by hot magmatic fluids in cracks in surrounding rocks

Cooling magma

A

B

Figure 20.5

Hydrothermal ore deposits form in various ways. (A) Deposition in veins around a magma chamber. (B) Veins of hydrothermal sulfide ore permeate this rock.

Diamonds are mined primarily from pipelike plutons called **kimberlites** (named after the diamond-mining locality of Kimberley, South Africa) that originate in the mantle. Even where only a few gem-quality diamonds are scattered within many tons of kimberlite, their high value makes the mining profitable.

Hydrothermal Ores

Not all mineral deposits related to igneous activity form within igneous rock bodies. Magmas have water and other fluids dissolved in or associated with them. Particularly during the later stages of crystallization, the fluids may escape from the cooling magma, seeping through cracks and pores in the surrounding rocks and carrying with them dissolved salts, gases, and metals. These warm fluids can leach additional metals from the rocks through which they pass. In time the fluids

Figure 20.4

Sample of magnetite ore formed by crystal settling in a pluton.

This is also true of diamonds. One reason for the rarity of diamonds is that they must be formed at extremely high pressures, such as are found in the mantle, and then brought up rapidly into the crust before they can break down into the graphite that is the stable polymorph of carbon at the earth's surface.

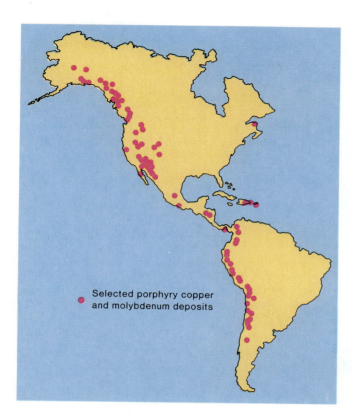

A

B

Distribution of copper and molybdenum deposits. Note close correlation with subduction zones.

From Jensen, M. L. and A. M. Bateman, *Economic Mineral Deposits,* 3d ed. © 1981 by John Wiley and Sons, Inc. Reprinted by permission.

cool and deposit their dissolved minerals, creating a **hydrothermal** (literally, "hot water") ore deposit (figure 20.5). The particular minerals deposited will vary with the composition of the hydrothermal fluids, but a great variety of metals are known to occur in hydrothermal deposits: copper, lead, zinc, gold, silver, platinum, uranium, and others. Because sulfur is a common constituent of magmatic gases and fluids, the ore minerals are frequently sulfides. The fluids need not all originate within the magma. Sometimes circulating subsurface waters will be heated sufficiently by a nearby cooling magma to dissolve, concentrate, and redeposit valuable metals in a hydrothermal ore deposit. The fluid involved may also be a mix of magmatic and nonmagmatic fluids.

Link to Plate Margins

The relationship between magmatic activity and many ore deposits suggests that hydrothermal and igneous-rock deposits should be especially common in regions of great magmatic activity, especially plate boundaries. This is generally true. A striking example of this link can be seen in figure 20.6, which indicates

Figure 20.7

Effects of hydrothermal activity along the East Pacific Rise. *(A)* An active "black smoker chimney" vent. The black is mainly fine-grained iron sulfide. *(B)* Sample of minerals deposited around a black smoker, including metallic sulfides.

Photos by W. R. Normark, courtesy U.S. Geological Survey.

the locations of a particular type of copper and molybdenum deposit of igneous origin in North and South America. Comparison with locations of modern plate boundaries shows that these deposits fall close to present or recent subduction zones.

As noted in chapter 12, recent investigations of underwater spreading ridges have revealed hydrothermal fluids gushing from the sea floor at vents along these ridges, depositing sulfides and other minerals as they cool (figure 20.7). Similar activity accounts for

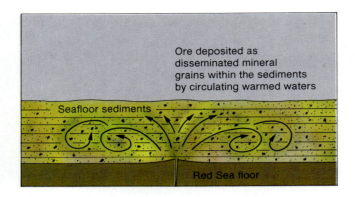

Ore deposited as disseminated mineral grains within the sediments by circulating warmed waters

Seafloor sediments

Red Sea floor

Figure 20.8

Ore deposition occurs in muds over a young spreading ridge in the Red Sea.

Figure 20.9

Sample of layered sedimentary iron ore (folded).

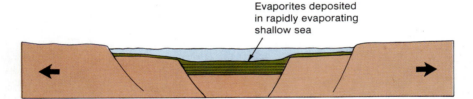

Evaporites deposited in rapidly evaporating shallow sea

Figure 20.10

Evaporites may be deposited in the shallow sea first formed when a continent is thinned during rifting.

the metal-rich muds at the bottom of the Red Sea, which is also a spreading rift zone (figure 20.8). Seawater seeps into the sediments and fractured sea floor at ridges, is warmed, reacts with the seafloor rocks, and dissolves metals. The warmed waters rise and cool, encounter and react with unaltered seawater, and deposit a variety of metallic minerals.

Sedimentary Deposits

Sedimentary processes can also produce economic mineral deposits. Some such ores have been deposited directly as chemical sedimentary rocks. Layered sedimentary iron ores (figure 20.9) are an example. These large deposits, which may extend for tens of kilometers, are for the most part very ancient. Their formation is believed to be related to the development of the earth's atmosphere. Under the oxygen-free conditions believed to have prevailed initially, iron from the weathering of continental rocks would have been very soluble in the oceans. As photosynthetic organisms began producing oxygen, that oxygen would have reacted with the dissolved iron and caused it to precipitate. If the majority of large iron ore deposits

formed in this way, during a time long past when the earth's surface chemistry was very different from what it now is, it follows that additional similar deposits are unlikely to form now or in the future.

Seawater contains a variety of dissolved salts and other chemicals. When a body of seawater trapped in a shallow sea dries up, it deposits these minerals in **evaporite** deposits. Some evaporites may be hundreds of meters thick. Ordinary table salt (mineralogically, halite, NaCl) is one mineral commonly mined from evaporite deposits. Others include gypsum and salts of the metals potassium and magnesium. Evaporite formation requires a marine setting with restricted water flow, and evaporation that is relatively rapid in comparison with the influx of additional water. Under these conditions, water in the restricted basin becomes progressively more concentrated in its dissolved salts, until they begin to precipitate out of the oversaturated solution. One situation in which conditions may be appropriate to evaporite deposition is in the early stages of continental rifting, when the sea first begins to invade the thinned rift zone (figure 20.10). Shallow inland seas that become isolated from the oceans may also be sites of evaporite deposition.

Although evaporites are not now being deposited in the Mediterranean Sea, its floor is thick with salt beds. These probably formed when water inflow through the Straits of Gibraltar was more severely restricted: as evaporation proceeded, the evaporites were precipitated. Appropriate conditions could have resulted from lower eustatic sea level (such as would have occurred during past ice ages), or as a result of small differences in the relative positions of Europe and Africa, related to plate movements.

Other Low-Temperature Ore-Forming Processes

Streams play a role in the formation of mineral deposits. They are rarely the sites of primary formation of ore minerals. However, as noted in chapter 13, streams often deposit sediments that are well sorted by size and density. The sorting action can effectively concentrate certain weathering-resistant, dense minerals in places along the stream channel. Such deposits are termed **placers.** The minerals of interest are typically weathered out of the rocks of the stream's drainage basin, then transported, sorted, and concentrated while other minerals are dissolved or swept away (figure 20.11). Gold, diamonds, and tin oxide (cassiterite) are examples of minerals that have been mined from the sands and gravels of placer deposits.

Even weathering alone can produce useful ores, by leaching away unwanted minerals and leaving a residue enriched in a valuable metal. For example, the extreme leaching of tropical climates gives rise to lateritic soils from which nearly everything has been dissolved except for the aluminum and iron oxide and hydroxide compounds. Many of the aluminum deposits presently being mined were enriched in aluminum by lateritic weathering to the point of being economic. (The iron content of the laterites is generally still not as high as that of the richest sedimentary iron ores described above. However, as the richest of those iron deposits are mined out, it may become profitable to mine more laterites for iron as well as for aluminum.)

Metamorphic Deposits

Finally, metamorphic processes can create economic mineral deposits. The mineralogical changes caused by the heat or pressure of metamorphism can produce secondary minerals of value. Graphite, used in "lead" pencils, in batteries, as a lubricant, and for many applications where its very high melting point is critical, is usually mined from metamorphic deposits. One

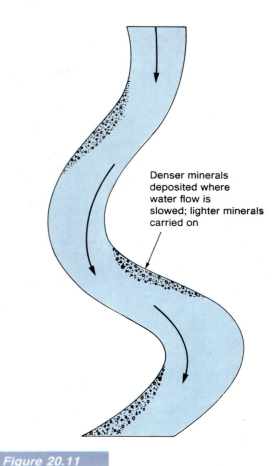

Figure 20.11

Formation of a placer deposit.

Denser minerals deposited where water flow is slowed; lighter minerals carried on

way in which graphite is formed is by the metamorphism of coal, which is also carbon-rich. Asbestos, used less in insulation now but still prized for its heat- and fire-resistant properties, is a hydrous ferromagnesian silicate formed by the metamorphism of mafic igneous rocks, with the addition of water. Garnets, which are used as abrasives as well as semiprecious gems, are common metamorphic minerals also, especially in rocks of moderate to high metamorphic grade.

Mineral and Rock Resources—Examples

Dozens of minerals and rocks have some economic value. This section will briefly survey a sampling of these resources, noting their principal occurrences and applications. Nonfuel mineral resources are commonly divided into the *metallic* minerals (those mined to extract one or more metals) and the *nonmetallic* minerals and rocks (for example, feldspar or granite).

Box 20.1

Dividing the Pie: Antarctica and Mineral Resources

Antarctica has presented a jurisdictional problem for many years. In the decades prior to 1960, seven countries—Argentina, Australia, Chile, France, New Zealand, Norway, and the United Kingdom—had laid claim to portions of the continent, either on the grounds of their exploration efforts there, or on the basis of geographic proximity (figure 1, p. 409). Several of these claims overlapped. None were recognized by most other nations, including five others (Belgium, Japan, South Africa, the United States, and the Soviet Union) that had carried out considerable scientific research in Antarctica.

After a history of disagreement punctuated by occasional violence, these twelve nations signed a treaty in 1961 that set aside all previous territorial claims to Antarctica. It was agreed that the whole continent should remain open; that military activities, weapons testing, and nuclear-waste disposal be banned there; and that every effort be made to preserve the distinctive Antarctic flora and fauna.

The question of mineral resources was not addressed, however, in part because it was not yet realized that significant mineral or petroleum resources might occur in Antarctica, and in part because their exploitation there was then economically impractical in any case. Not until 1972 was the question of minerals even raised at the conventions of the 12 original and 11 later treaty signatory nations. At present, the costs of recovering minerals from Antarctica would still be prohibitively high in comparison with other sources, but this may not be the case indefinitely. Eventually, the question of who has the right to exploit Antarctic mineral resources will have to be resolved.

Metals

The term *mineral resources* usually brings metals first to mind. Overwhelmingly, the metal used in largest quantity is iron. It is also one of the most common metals. Nearly all iron ore is used for the manufacture of iron and steel products. Iron is mined principally from the ancient sedimentary deposits described in the last section, but also from a few laterites and from concentrations of magnetite in plutons.

Aluminum is the second most widely used metal. It is light in weight for its strength, which makes it particularly useful in the transportation and construction industries. Aluminum is also widely used in packaging, especially for beverage cans. Aluminum is the third most common element in the crust, but there it is most often found in silicates, from which it is technologically difficult and economically costly to extract. Most commercial aluminum deposits are *bauxite,* an aluminum-rich laterite in which the aluminum is found predominantly as a hydroxide. Even in this form, the extraction of the aluminum is energy-intensive: 3% to 4% of the electricity used in the United States is consumed just in the production of aluminum metal from aluminum ore.

Many less common but important metals are found in sulfide ore deposits. These include copper (which can also be found as a native metal), lead, zinc, nickel, cobalt, and others. Sulfides, in turn, occur frequently in hydrothermal deposits, and may also be concentrated in igneous rocks. Copper, lead, and zinc may also be found in sedimentary ores; some laterites are moderately enriched in nickel and cobalt. Clearly, these metals may be concentrated into economically valuable deposits in a variety of ways. Copper, an excellent conductor of electricity, is primarily used for electrical applications, and also in the construction and transportation industries. An important use of lead is in batteries; among its many other applications, it is a component of many solders and is used in paints and ceramics. The zinc coating on steel cans (misnamed "tin cans") keeps them from rusting, and zinc is also used in the manufacture of brass and other alloys.

The so-called precious metals—gold, silver, and platinum—have such high unit value that they are used in commerce. But they also have some unique industrial uses. Gold is used for jewelry and in the arts (for gilding and for coloring glass), but also in electronics and in dentistry. It is particularly valued for

Figure 1
Antarctic land claims prior to 1961 treaty. Note overlapping claims.

Adapted by permission from Craddock, C., et al., 1970. Geologic Maps of Antarctica. *Antarctic Map Folio Series*, Folio 12. American Geographical Society.

its resistance to tarnishing. Silver's principal use, accounting for nearly half the silver consumed in the United States, is for photographic materials (the light-sensitive emulsions in film); the next broadest applications are in electronics. Platinum is an excellent *catalyst*, a substance that promotes chemical reactions. Today half the platinum used in the United States goes into automobile emissions-control systems, with the rest finding applications in the petroleum and chemical industries, in electronics, and in medicine. All the precious metals share a resistance to the formation of compounds, so they can be found as native metals. Silver also forms sulfide minerals. Most frequently, the precious metals are mined from igneous or hydrothermal ore deposits. Most gold and silver production in this country is in fact a byproduct of the mining of ores of more abundant metals like copper, lead, and zinc, with which small amounts of the precious metals may be associated.

Nonmetallic Minerals and Rocks

Another byproduct of the mining of sulfides is the nonmetal sulfur. It may also be recovered from petroleum during refining, from volcanic deposits (sulfur is sometimes precipitated as pure native sulfur at fumaroles; see figure 20.12), and from evaporites rich in sulfate minerals such as gypsum. The primary use of sulfur is for the manufacture of sulfuric acid for industrial purposes.

Several important minerals are recovered from evaporite deposits. The most abundant is halite, or rock salt (figure 20.13), used principally as a source of the sodium and chlorine of which it is composed, and secondarily for road salt, either directly or through the production of other salts, like calcium chloride. Halite has many additional lower-volume uses, including, of course, use as table salt. Gypsum, essential to the manufacture of plaster and wallboard for construction, is another evaporite mineral. Others include phosphate rock and potassium-rich potash, key ingredients of synthetic fertilizers.

As noted in chapter 2, "clay" is not a single mineral but comprises a large group of hydrous silicates with a layered or sheet structure. They are formed at low temperature, commonly by weathering, and are abundant in sedimentary deposits. The diversity of clay minerals leads to a variety of applications, from fine ceramics to the making of clay piping and other construction materials, the processing of iron ore, and drilling for oil.

In terms of overall mass, the quantity of various metals and minerals used is insignificant compared to the amount of rock, sand, and gravel we use (figure 20.14). In 1984, in the United States, more than 700 million tons of sand and gravel were used in construction, especially in making cement and concrete. Another 950 million tons of crushed rock was consumed, for fill in foundations and roadbeds and in other applications (limestone in cement-making, for instance). That amounts to nearly seven tons of rock products per person in the United States. In addition, more than a million tons of *dimension stone* were used in construction—slate for flagstones, other attractive and/or durable rocks like marble and granite for monuments and building facings. And tens of millions of tons of relatively pure quartz sandstone were used for glassmaking—silica is the main ingredient in glass.

Figure 20.12

Sulfur deposition around a fumarole.

Photo by J. C. Ratté, courtesy U.S. Geological Survey.

Figure 20.13

Rock salt from an evaporite deposit.

Crushed stone
7900 lb

Sand and gravel
5900 lb

Salt
358 lb

Gypsum
202 lb

Phosphate
365 lb

Potash
51 lb

Iron ore
534 lb

Aluminum
49 lb

Zinc
9 lb

Lead
9 lb

Copper
18 lb

Figure 20.14

U.S. per capita consumption of mineral and rock resources, 1984.

Mineral Supply and Demand

The very uneven global distribution of mineral resources has from time to time meant that political disruption in one or a few countries has seriously disrupted the supply of a particular commodity. From this viewpoint, then, it is of some interest to examine the mineral resource supplies of the United States alone, and relate these supplies to the nation's consumption of mineral materials. From a broader perspective, leaving aside the unanswerable question of the extent to which each country's mineral wealth may be considered available to all, we can also look at the world mineral supply/demand picture.

To this point, we have been using the term *resources* informally, to designate some natural material that is useful and/or economically valuable to our present society. In the context of minerals and fuels, the term has a more precise definition and is distinguished from the related term *reserves*. The amount

represented by the **reserves** is that quantity of a given material that has been found and could be exploited economically with existing technology. The term is usually further restricted to apply only to material not yet consumed. (If one wishes to refer to the quantity of that material already used as well as remaining reserves, the term *cumulative reserves* is used.) The category of **resources** is a much larger one: it includes reserves, plus deposits of the material that are known but that cannot be exploited economically given present prices and technology, plus deposits that have not yet been found but are believed to exist, based on extrapolation from the abundance and geologic occurrence of known deposits. Clearly the reserves are the most conservative estimate of the amount of a given mineral or fuel on which we can count, especially in the near future.

The other side of the supply/demand picture is demand, which includes prediction of what that demand is likely to be in the future. Projection from historical trends may or may not be appropriate. Between 1947 and 1974, the production rates of most minerals grew at between 2% and 10% per year. The effects of such exponential growth in demand can be seen in figure 20.15. If demand increases at 2% per year, it will double not in 50 years but in 35. An increase of 5% per year leads to a doubling of demand in 14 years and a tenfold increase in 47 years! On the other hand, depressed-world economic conditions since the mid-1970s caused demand for many resource materials to level off or even to decline over the short term. Data for the early to mid-1980s suggest that demand is on the upswing again, at least in the industrialized countries such as the United States. In looking at predictions of the longevity of mineral reserves, one should bear in mind the underlying assumptions about future demand.

U.S. Mineral Consumption and Production

All told, the United States accounts for a very large proportion of world consumption of mineral and rock materials, relative both to its population and to its production of most minerals. To put the consumption in a global perspective, table 20.1 presents 1984 U.S. production and consumption data and the percentage of world totals that these represent. Keep in mind in looking at the consumption figures that fewer than 6% of the people in the world live in the United States.

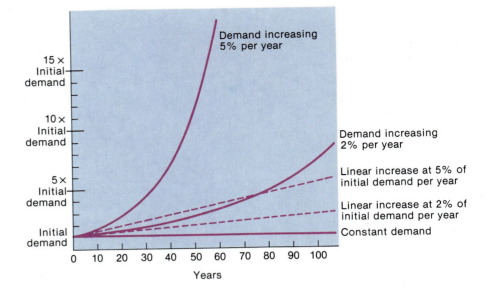

Figure 20.15

Effects of exponential growth in demand on consumption of mineral resources.

Table 20.1

U.S. production and consumption of rock and mineral resources, 1984†

Material	U.S. primary production	U.S. production as % of world	U.S. consumption	U.S. consumption as % of world‡
aluminum	4433	22.6	5833	38.4
chromium	0	0	466	5.0
cobalt	0	0	9.25	31.9
copper	1150	12.9	2100	23.5
iron ore	51,000	6.5	64,000	8.2
lead	340	10.5	1030	31.7
manganese	0	0	740	3.0
nickel	7.5	1.0	225	29.3
tin	negligible	0	57	24.8
zinc	292	4.2	1133	16.2
gold	2300	5.1	4800	10.7
silver	44,000	11.0	170,000	42.5
platinum group	6	0.1	4100	61.2
clay	44,000	*	41,330	*
gypsum	14,300	16.0	24,300	46.3
phosphate	53,900	34.2	43,791	27.8
potash	1600	5.7	6100	21.9
salt	38,750	*	43,000	*
sulfur	11,330	21.6	13,640	26.0
sand and gravel	709,000	*	708,000	*
stone, crushed	950,000	*	950,000	*
stone, dimension	1227	*	2834	*

*World data not available.
†All production and consumption figures in thousands of tons, except for gold, silver, and platinum group metals, for which figures are in thousands of troy ounces.
‡Assumes that overall, world production approximates world consumption.
Source: *Mineral Commodity Summaries 1985*, U.S. Bureau of Mines.

Table 20.2

World production and reserves statistics, 1984*

Material	Primary production	Reserves	Projected lifetime (years)
bauxite	86,680	24,500,000	282
chromium	9210	7,540,000	82
cobalt	29	9200	317
copper	8932	561,000	63
iron ore	785,000	206,300,000	263
lead	3250	135,000	41.5
manganese	24,900	12,000,000	482
nickel	768	111,000	145
tin	230	3300	14
zinc	6985	319,000	45.6
gold	45,000	1,450,000	32
silver	400,000	10,800,000	27
platinum group	6700	1,200,000	179
gypsum	89,300	(large)	
phosphate	157,300	38,300,000	243
potash	27,900	17,000,000	609
sulfur	52,500	2,700,000	51

*All production and reserve figures in thousands of tons, except for gold, silver, and platinum group, given in thousands of troy ounces.
Source: *Mineral Commodity Summaries 1985*, U.S. Bureau of Mines.

World Mineral Supply and Demand

As noted above, world demand for minerals (and energy) has slackened since the mid-1970s. It is unlikely that demand will remain depressed over the longer term. If that were to be the case, it would imply, among other things, no significant economic growth or improvement in standard of living for the billions of people living in underdeveloped countries.

Table 20.2 presents projections for the lifetimes of selected world mineral and rock reserves. It assumes constant demand at 1984 levels, which are generally depressed somewhat from those of prior years by the residual effects of world recession. Note that for many of the metals the present reserves are projected to last only a few decades, even at constant 1984 consumption levels. Consider the implications if consumption were to resume the rapid growth rates of the middle of the twentieth century.

Such projections also presume unrestricted distribution of minerals, so that they can be used as needed where needed, regardless of political boundaries or economic factors such as a nation's purchasing power. It is instructive to compare global projections with corresponding data for the United States only (table 20.3). Even at 1984 consumption rates, the United States has less than half a century's worth of reserves of many of the materials listed. Of some metals, notably aluminum, chromium, manganese, platinum, and tin, we have virtually no reserves at all.

Some relief can be anticipated from the economic component of the definition of reserves. That is, as currently identified reserves are depleted, the law of supply and demand will drive up minerals' prices. This, in turn, will make some of what are presently subeconomic resources profitable to mine, and they will effectively be reclassified as reserves.

However, there will still be technological obstacles to the development of some of the subeconomic resources. It is unclear whether the necessary advances in technology will be achieved rapidly enough to help substantially. Also, if we move toward developing lower- and lower-grade ore, we will by definition have to process more and more rock and disturb more and more land to extract a given quantity of mineral or metal. Mining an ore that is 1% copper means processing five times as much ore for a given yield of copper as mining that copper from a 5%-copper ore.

Table 20.3

Projected lifetimes of U.S. mineral reserves
(assuming complete reliance on domestic reserves)*

Material	Reserves	Projected lifetime (years)
bauxite†	44,000	11.4
chromium	0	0
cobalt	950	102.7
copper	90,000	42.9
iron ore	24,800,000	388
lead	27,000	26.2
manganese	0	0
nickel	2800	12.4
tin	44	0.8
zinc	53,000	46.8
gold	100,000	20.8
silver	1,800,000	10.6
platinum group	16,000	3.9
gypsum	500,000	20.6
phosphate	5,940,000	136
potash	360,000	59
sulfur	192,500	14.1

*Reserves in thousands of tons, except for gold, silver, and platinum group metals, given in thousands of troy ounces.

†Note that bauxite consumption is only a partial measure of total aluminum consumption; additional aluminum is consumed as refined aluminum metal, of which there are no reserves.

Source: *Mineral Commodity Summaries 1985*, U.S. Bureau of Mines.

Minerals for the Future: Some Options Considered

If demand for minerals cannot realistically be reduced, supplies must be increased or extended. We must develop new sources of minerals in either traditional or nontraditional areas, or we must conserve minerals better.

New Methods in Mineral Exploration

Most of the "easy ores," near-surface mineral deposits in readily accessible places, have probably already been discovered. In order to find the less easily located ore deposits, a variety of new methods are being applied. Geophysics, for example, provides some assistance. Rocks and minerals vary in density and in their electrical and magnetic properties, so changes in rock types or distribution below the earth's surface cause gravity and magnetic anomalies that can be measured at the surface, or small variations in the electrical conductivity of rocks. Some ore deposits may be detected in this manner. As a simple example, many iron deposits are strongly magnetic; large magnetic anomalies may indicate the presence, and show the extent of, a subsurface body of iron ore. Radioactivity is another readily detected property, and uranium deposits can be located with the aid of a Geiger counter.

Geochemical prospecting is another method that is increasingly widely used. It may take a variety of forms. Some studies are based on the recognition that soils will to a degree reflect the chemistry of the materials on which they form. The soil over a copper-rich ore body, for instance, may itself be enriched in copper relative to surrounding soils. Surveying soil chemistry over a wide area can help to pinpoint likely spots to drill in search of ores. Occasionally plants can be sampled instead of soil: certain plants tend to concentrate particular metals, making them very sensitive indicators of locally high concentrations of those metals. Water or stream sediments may likewise be sampled and analyzed for high concentrations of the metals being sought. Even soil gases can supply clues to ore deposits. Mercury is a very volatile metal, and high concentrations of mercury vapor have been found in gases filling the pore spaces in soils over mercury ore bodies.

Remote sensing, which includes the use of aerial photographs and satellite imagery, is another useful tool (see also appendix A). The Landsat satellites provide images covering the world every 18 days. This gives at least a preliminary look at many areas that might otherwise be too remote or difficult to reach on the ground. Moreover, the satellite images can be processed through computers to sharpen them or to bring out unusual features in the geology or vegetation that might bear further investigation (figure 20.16). Remote sensing must still be followed by ground-based geologic studies eventually. Perhaps its greatest usefulness at present is in limiting the scope of these slower and more costly detailed studies, by helping to pinpoint areas particularly likely to yield new mineral deposits, and thus allowing further exploration efforts to be concentrated in the most promising areas.

Advances in geologic understanding play a role, too. The development of plate-tectonic theory has helped geologists to recognize the association between particular types of plate boundaries and the occurrence of certain kinds of mineral deposits. This association may be useful in directing the search for

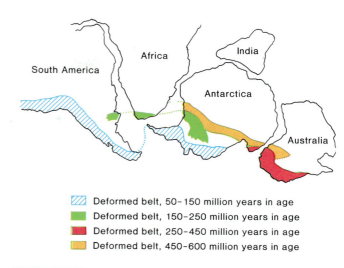

☒ Deformed belt, 50–150 million years in age
■ Deformed belt, 150–250 million years in age
■ Deformed belt, 250–450 million years in age
■ Deformed belt, 450–600 million years in age

Figure 20.17

Pre-continental-drift reassembly of landmasses suggests locations of possible ore deposits in unexplored regions by extrapolation from known deposits on other continents.

Adapted by permission from Craddock, C., et al., 1970. *Geologic Maps of Antarctica. Antarctic Map Folio Series,* Folio 12. American Geographical Society.

A

B

Figure 20.16

Landsat satellite photographs may reveal details of the geology that will aid in mineral exploration. Vegetation, also sensitive to geology, may enhance the image. *(A)* View of South Africa, dry season. *(B)* Same view, rainy season. Recognizable geologic features include a granite pluton (round feature at top center) and folded layers of sedimentary rock (below).

Photos © NASA

new ores. If we know that molybdenum deposits are often found over present subduction zones, for instance, exploration for more molybdenum deposits might logically be concentrated in other present or past subduction zones. Also, the realization that many of the continents were once united suggests likely areas to look for more ores (figure 20.17). If mineral deposits of a particular kind are known to occur near the margin of one continent, similar deposits might be found at the corresponding edge of another continent once connected to the first. If a mountain belt rich in some kind of ore seems to have a counterpart range on another continent once linked to it, the same kind of ore may occur in the matching mountain belt. Such reasoning is part of the basis for supposing that economically valuable ore deposits may exist on Antarctica.

The most successful exploration efforts may integrate several of the methods described above, making use of the particular advantages of each.

Marine Mineral Resources

Increased consideration is also being given to seeking mineral wealth in unconventional places. In particular, the sea may provide partial solutions to some mineral shortages.

Seawater itself contains virtually every chemical element dissolved in it. However, most of the dissolved material is halite. Vast volumes of seawater would have to be processed to extract the small quantities of dissolved copper, gold, and so on. Moreover, we do not have adequate technology to extract, selectively and efficiently, a few specific metals of interest from all that salt water on a routine basis, although the Japanese have been experimenting with extraction of uranium from seawater. Other types of underwater mineral deposits have greater potential.

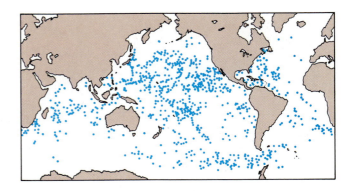

Manganese nodule distribution on the sea floor.
Courtesy *Oceanus* Magazine, Vol. 25, No. 3.

As noted in the chapter on glaciers, during the last ice age, when a great deal of water was locked in ice sheets, eustatic sea levels were lower than at present. Much of the continental shelf area was dry land, and streams flowing off the continents flowed out across the exposed shelves to reach the sea. Where the streams' drainage basins included appropriate source rocks, those streams might have concentrated valuable placer deposits. With the melting of the ice and rising sea levels, any placers on the continental shelves were again submerged. Finding and mining them may be more costly and difficult than land-based mining, but as deposits on land are exhausted, placer deposits on the continental shelves have begun to prove worth seeking.

The hydrothermal ore deposits forming along some spreading ridges are another possible source of needed metals. In many places, the quantity of material being deposited is too small, and the depth of water above would make recovery prohibitively expensive, at least over the near term. However, the metal-rich muds of the Red Sea contain high enough concentrations of metals like copper, lead, and zinc that some exploratory dredging is underway, and several companies are interested in the possibility of mining those sediments. Along a section of the Juan de Fuca ridge, off the west coast of Oregon and Washington, hundreds of thousands of tons of zinc- and silver-rich sulfides have already been deposited, and the hydrothermal activity continues.

Perhaps the most widespread undersea mineral resource, and the most frequently and seriously discussed as a near-term resource, are the manganese nodules. Ranging in size up to about 10 centimeters in diameter, these are lumps composed mostly of manganese oxides and hydroxides. They also contain lesser amounts of iron, copper, nickel, zinc, cobalt, and other metals. Indeed the value of the minor metals may be a greater financial motive for mining manganese nodules than the manganese itself. The nodules are found over much of the deep-sea floor (figure 20.18), in regions where sedimentation rates are slow enough not to bury them. At present, the costs of recovering these nodules are high compared with costs of mining the same metals on land, and the technical problems associated with recovering the nodules from beneath several kilometers of water remain to be worked out, as does the practical question of who owns seabed resources in international waters and who has the right to exploit them (see box 20.2).

Conservation of Mineral Resources

Demand for particularly scarce individual minerals might be moderated by making substitutions. For example, for certain applications one might replace a rare metal with a more abundant one. The extent to which this is likely to succeed is limited by the fact that reserves of *most* metals are relatively restricted, as is clear from table 20.3. Substituting one metal for another reduces the demand for the second while increasing demand for the first—the focus of the shortage problem shifts, but the problem persists. Nonmetals are replacing metals in other applications. Unfortunately, many of these nonmetals are plastics or other materials derived from petroleum, of which none too much remains either.

Box 20.2

The Exclusive Economic Zone

Traditionally, nations bordering the sea claimed territorial limits of 3 miles (5 kilometers) outward from their coastlines. After realizing later that valuable minerals might be found on the continental shelves, some nations individually asserted their rights to the whole width of the adjacent continental shelf. As noted in chapter 12, that is a highly variable extent, even for single countries. Nations bordered by narrow shelves found this action unfair. Some consideration was then given to equalizing claims by extending territorial limits to 200 miles offshore, which would encompass not only all of the continental shelves but, in most places, a considerable expanse of sea floor as well.

After eight years of intermittent negotiations collectively described as the "Third U.N. Conference on the Law of the Sea," a Law of the Sea Treaty emerged in the United Nations that attempted to bring some order out of the chaos of oceanic territorial claims. Among other provisions, the Treaty established **Exclusive Economic Zones** (EEZs) extending up to 200 miles from the shorelines of coastal nations, within which those nations have exclusive rights to mineral-resource exploitation. In addition to possible resources of the continental shelf, some deep-sea manganese nodules would fall within Mexico's EEZ; Saudi Arabia and the Sudan would divide the rights to the metal-rich muds of the floor of the Red Sea. Included in the EEZ off the west coast of the United States would be a part of the East Pacific Rise spreading-ridge system with its

active hydrothermal vents and sulfide deposits. Areas of the ocean basins outside any nation's EEZ would be under the jurisdiction of an International Seabed Authority, and the Treaty includes some provision for financial and technological assistance to developing countries wishing to share in resources there.

The United States did not sign the Treaty, but in March of 1983 President Reagan unilaterally proclaimed an EEZ for the United States that extends to the 200-mile limit offshore. This move expands the undersea territory under U.S. jurisdiction to some 3.9 billion acres, more than one and a half times the land area of the United States and its territories (figure 1). So far, however, exploration of this new domain has been minimal.

United States Exclusive Economic Zone

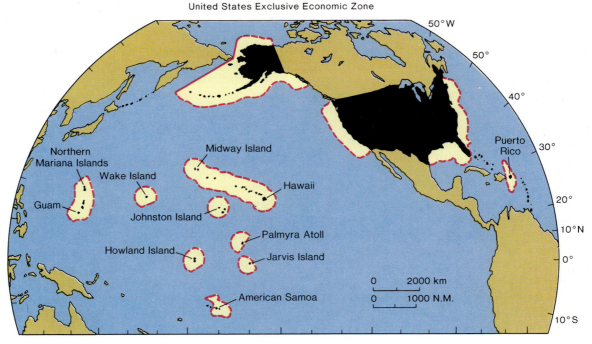

Figure 1

The Exclusive Economic Zone of the United States.

Source: U.S. Geological Survey 1983 Annual Report.

Table 20.4

Recycling of metals in the United States

Metal	Recycled scrap as % of consumption
aluminum	16
chromium	18
cobalt	5
copper	21
iron	negligible
lead	48
manganese	negligible
nickel	11
tin	21
zinc	8
gold	53
silver	19
platinum group	33

Source: Mineral Commodity Summaries 1985, U.S. Bureau of Mines.

Recycling

The most effective way to extend some mineral reserves may be through recycling, which reduces the need for additional primary production from reserves. Some metals are already extensively recycled, at least in the United States (see table 20.4 for examples). Worldwide, recycling is less widely practiced in general, in part because less industrialized countries have accumulated fewer manufactured products from which materials can be retrieved. Two additional benefits of recycling are a reduction in the volume of waste-disposal problems and a decrease in the extent to which more land must be disturbed by new mining activities.

Not all materials lend themselves equally well to recycling. Among those that can most readily be recycled are metals that are used in pure form in large chunks—copper in pipes and wiring, lead in batteries, aluminum in beverage cans. The individual metals are relatively easy to recover and require minimal effort to purify for re-use. Recycling aluminum is also appealing from an energy (and cost) standpoint: it takes only one-twentieth as much energy to produce aluminum by recycling old scrap as it does to extract aluminum metal from bauxite. Sometimes

it is even economical to recover and recycle some of these discrete metallic items from old dump or landfill sites.

Where different materials are intermingled in complex manufactured objects, it is more difficult and costly to extract individual metals. Consider trying to separate the various metals from a refrigerator, a lawnmower, or a TV set. Only in a few rare cases are metals valuable enough that the recycling effort in such a case may be worthwhile: for instance, there is interest in recovering the platinum from catalytic mufflers. Alloys, mixtures of different metals in application-specific proportions, present additional problems.

Also, some materials are not used in discrete objects at all. The potash and phosphorous in fertilizers are strewn across the land and can hardly be recovered. Road salt washes off streets and into soil and storm sewers; lead in gasoline is emitted with exhaust fumes into the atmosphere. Clearly, these things cannot be recovered and re-used. For the foregoing and other reasons, it is therefore unrealistic to expect that all mineral materials can ever be completely recycled.

Re-Mining Old Mines

As supplies tighten relative to demand, old mines with lower-grade ores may become profitable again, and be reopened. It may even happen that the wastes from past ore processing will prove to be resources for the future. In many ore deposits, ore of several grades is present. The country rock surrounding a rich vein of hydrothermal sulfide minerals may itself contain sulfides disseminated through the rock in lower concentrations. When an ore deposit was first exploited, economics may have dictated that only the richest veins be extracted. However, much of the country rock may have been excavated and crushed or ground during mineral processing, along with the highest-grade ore. The waste crushed rock, termed **tailings,** may still contain some of the ore of interest.

If the price of the metal(s) mined has increased sufficiently, it may be worthwhile to go back and reprocess the tailings. One way to do this without even disturbing the tailings piles is by a method known as *heap leaching.* An acid or other chemical capable of dissolving the ore of interest is flushed through the tailings, then recovered for extraction of metal(s). Old tailings from gold mining, for example, can be heap-leached by a cyanide solution to recover more gold. Heap leaching is a relatively inexpensive approach to maximizing the metal extracted from ore bodies already exploited.

Collapse of land surface over an old abandoned copper mine in Arizona. Note roads, trees, and houses for scale.

Photo by F. L. Ransome, courtesy U.S. Geological Survey.

Some Impacts of Mining Activities

Mining and mineral processing activities can modify the environment in a variety of ways. Most obvious is the presence of the mine itself.

Underground Mines

Underground mines are relatively inconspicuous. They disturb a small land area close to the principal shaft(s). There may be some accumulation of waste rock dug out of the mine piled close to its entrance, but in most underground mines the tunnels follow the ore body as closely as possible so as to minimize the amount of non-ore rock to be removed. When mining activities are complete, the shafts can be sealed and the area often returns very nearly to its pre-mining condition. However, occasionally near-surface underground mines have collapsed years after abandonment, when supporting timbers have rotted away or groundwater

has enlarged the underground cavities through solution (figure 20.19). In some cases, the collapse occurred so long after the mining had ended that the existence of the old mines had been entirely forgotten.

Surface Mines

Surface mining activities can be further broken down into open-pit mining (including quarrying) and strip-mining. Open-pit mining, such as is shown in figure 20.20, is practical when a large three-dimensional ore body is located near the surface. Most of the material in the pit is the valuable commodity and is extracted for processing. Thus this procedure permanently changes the topography, leaving a large hole in its wake. The exposed rock may begin to weather and, depending on the nature of the ore body, may release pollutants into surface runoff water.

The world's largest open-pit mine, Bingham Canyon, Utah. Over $6 billion worth of minerals has been extracted from this one mine over the decades of its operation.

Photo courtesy Kennecott.

Strip-mining, more often used to extract coal than mineral resources, is practiced most commonly when the material of interest occurs in a layer near and approximately parallel to the surface. Overlying vegetation, soil, and rock are stripped off, the coal or other material is removed, and the waste rock and soil cover is dumped back as a series of **spoil banks** (figure 20.21). The finely broken-up material of spoil banks, with its high surface area, is very susceptible both to erosion and to chemical weathering. Chemical and sediment pollution of runoff from spoil banks used to be common, and vegetation reappeared on the steep, unstable slopes only gradually, if at all. Now much stricter laws govern strip-mining activities, and reclamation of new strip mines is provided for. Successful reclamation can thoroughly erase the evidence of the past mining activities (figure 20.22).

Mineral Processing

The tailings that are left over from mineral processing may end up heaped around the processing plant to weather and wash away, much like spoil banks. As noted above, traces of the ore are left behind in tailings. Depending on the nature of the ore, rapid weathering of the tailings may leach out such harmful elements as mercury, arsenic, cadmium, and uranium, contaminating surface and ground waters. The chemicals used in processing—for example, the cyanide used on gold ore—are often toxic also. There is concern, for instance, about possible hazards associated with large heap-leaching operations. Smelting to extract metals from ores may release lead, arsenic, mercury, and other potentially toxic, volatile elements along with exhaust gases and ash, unless emissions are tightly controlled. Mineral processing, then, represents a very large potential source of pollutants, if not done with care.

Figure 20.21

Coal strip mine operation in progress. Spoil banks are at right. Note highway and car for scale.

Photo by P. F. Narten, courtesy U.S. Geological Survey.

Figure 20.22

Lush grasses on reclaimed strip-mined land, Mercer County, North Dakota.

Photo courtesy U.S. Geological Survey.

Summary

Economically valuable mineral deposits occur in a variety of geologic settings—igneous, metamorphic, or sedimentary. Valuable minerals may be found disseminated in igneous rocks, concentrated in plutons by gravity, crystallized as coarse-grained pegmatites, or deposited in hydrothermal veins associated with igneous activity. Sedimentary deposits include evaporites, placers, laterites and other ores concentrated by weathering, and sediments and sedimentary rocks themselves (e.g., limestone, sand and gravel, or sedimentary iron ores). Metamorphism plays a role in the formation of deposits of certain minerals, like graphite.

Both the occurrence of and demand for mineral and rock resources are very unevenly distributed worldwide. Even assuming, conservatively, constant resource consumption levels, projections suggest that present reserves of most metals and other minerals could be exhausted within decades, both within the United States and globally. Price increases would lead to the development of some deposits that are presently uneconomic, but these are also limited. Strategies for averting further mineral shortages include applying new exploration methods to find more ores, looking to undersea mineral deposits not exploited in the past, recycling metals to reduce the demand for newly mined material, and reworking old mines and tailings. Decreasing new mining activity would minimize the potential negative impacts of mining, such as disturbance of the land and release of harmful chemicals through accelerated weathering or in connection with mineral processing.

Terms to Remember

concentration	hydrothermal	remote sensing	spoil banks
(enrichment) factor	kimberlite	reserves	strip-mining
evaporite	ore	resources	tailings
Exclusive Economic Zone	placer		

Questions for Review

1. What is an ore? How is its concentration factor defined?

2. Describe two kinds of magmatic ore deposits, and name one mineral mined from each.

3. What are hydrothermal ore deposits? Why are they especially associated with plate boundaries?

4. Under what conditions do evaporites form?

5. What is a placer deposit, and what kinds of minerals are concentrated in these?

6. Of which of the following does the United States use the largest quantity: iron, halite, platinum, or sand and gravel? Name one important application of each of these materials.

7. What is the distinction between *reserves* and *resources?* Under what conditions might some resources be reclassified as reserves?

8. Name and describe two kinds of exploration techniques used in the search for mineral deposits.

9. How has the development of plate-tectonic theory aided in finding more ore deposits?

10. Is seawater a potential source of essential metals? Explain.

11. Describe one marine mineral resource, not presently exploited significantly, that might become important in the future.

12. Why are some old mines and tailings piles being considered as future metal sources?

13. Describe one potential environmental problem or hazard associated with (a) underground mining, (b) surface mining, and (c) mineral processing after mining.

14. What is an Exclusive Economic Zone? List three kinds of mineral deposits that might be encompassed by an EEZ.

For Further Thought

1. Select one of the metals—aluminum, manganese, or tin—of which the United States controls very little or none of the identified world reserves. Investigate its uses in more detail; consider the implications of a cessation of imports. Has the United States any subeconomic deposits of this metal? Of what kind(s) are they, and how much metal do they contain relative to domestic demand?

2. Choose any one metallic mineral resource and investigate its geologic occurrence, its distribution (worldwide and, if applicable, within the United States), present consumption rates and past trends, reserve and resource estimates. Evaluate the impact of the mining and processing methods customarily used to extract it.

Suggestions for Further Reading

Agnew, A. F., ed. *International Minerals: A National Perspective*. Boulder, Colo.: Westview Press, 1983.

Brookins, D. G. *Earth Resources, Energy, and the Environment*. Columbus, Ohio: Charles E. Merrill Publishing Co., 1981.

"Deep Ocean Mining." *Oceanus* 25, no. 3 (1982).

Fergusson, J. E. *Inorganic Chemistry and the Earth*. New York: Pergamon Press, 1982.

Jensen, M. L., and Bateman, A. M. *Economic Mineral Deposits*. 3d ed. New York: John Wiley and Sons, 1981.

Shusterich, K. M. *Resource Management and the Oceans*. Boulder, Colo.: Westview Press, 1982.

U.S. Bureau of Mines. *Mineral Commodity Summaries 1985*.

———. *Minerals Yearbook*. Published annually.

Whitmore, F. C., Jr., and Williams, M. E., eds. *Resources for the Twenty-First Century*. U.S. Geological Survey Professional Paper 1193, 1982.

Zumberge, J. H. "Mineral Resources and Geopolitics in Antarctica." *American Scientist* 67 (January 1979): 68–76.

Energy Resources

Introduction

The earliest energy source used by primitive people was the food they ate. Then fires fueled by wood were employed to produce energy for cooking, heat, light, and protection from predators. As hunting societies turned into agricultural ones, the energy of animals was put to use. The world's total energy demands were quite small and could readily be met by these sources.

Gradually the labor of animals was replaced by that of machines, new and more sophisticated technologies were developed, and demand for manufactured goods rose. These factors all increased the need for energy and spurred the search for new energy sources. It was primarily the fossil fuels that eventually met those needs (figure 21.1). In recent years, however, it has become apparent that our future energy demands may be satisfied by a different mix of energy sources. Fossil-fuel supplies are limited. They are among the **nonrenewable** energy sources. Nonrenewable resources are those that are not being produced at present, or are being produced at rates much slower than current consumption rates. On a human time scale, the supplies of nonrenewable fuels are finite.

Knowing how soon and to what extent alternative energy sources will be needed depends heavily on projections of future energy use. Such projections are extremely sensitive to such variables as the state of world economies, the rate of industrialization of less developed nations, and the extent of energy-conservation efforts in more developed nations. Rather than attempt to make projections, we will instead survey various alternative energy sources and note the potential of each to supply energy for the future.

Fossil Fuels—Oil and Natural Gas

The term *fossil* refers to any remains or evidence of ancient life. The **fossil fuels,** then, are those energy sources that have formed from the remains of once-living organisms. These include oil, natural gas, coal, oil shale, and tar sand. The differences in the physical properties of the various fossil fuels as we find them today arise from differences in the starting materials from which they formed and from differences in what happened to those materials after the organisms died and were buried in the earth.

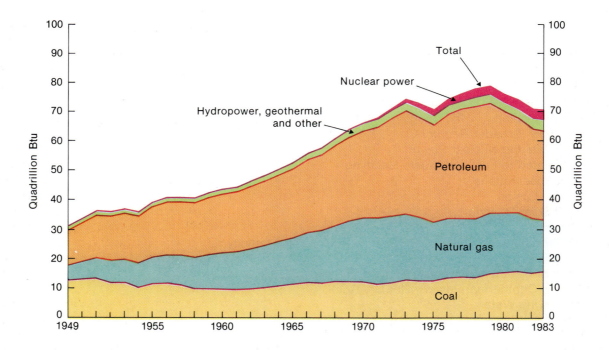

Figure 21.1

U.S. energy consumption over the past several decades. Note the consistent rise up to the time of the OPEC oil embargo of the early 1970s.

Source: Energy Information Administration.

Formation of Oil and Gas

Oil, or petroleum, is not a single chemical compound. Petroleum comprises a variety of liquid hydrocarbon compounds (compounds made up of the elements carbon and hydrogen in various proportions). There are also gaseous hydrocarbons (so-called natural gas), of which the compound methane, CH_4, is the most common. The exact processes by which organic matter is transformed into liquid and gaseous hydrocarbons are not fully understood, but the transformation is believed to occur somewhat as described below.

A basic requirement for the production of a large deposit of any fossil fuel is a large initial accumulation of organic matter, which is rich in carbon and hydrogen. Another requirement is that the organic debris should be buried quickly, so that decay by biological means or by reaction with oxygen in the air will not destroy it.

Microscopic life is abundant over much of the ocean. When these organisms die, their remains settle to the sea floor. Near shore—for example, on many continental shelves—rapid accumulation of terrigenous sediments also occurs. In such a setting, the starting requirements for formation of oil are satisfied: there is an abundance of organic matter, rapidly buried by sediment. Oil and most natural gas deposits are believed to form from such accumulations of marine microorganisms. (Additional natural gas deposits may be derived from the remains of terrestrial plants.)

As burial continues, the organic matter begins to change. Pressures increase as the weight of the overlying pile of sediment or rock increases; temperatures increase according to the geothermal gradient; and slowly, over long periods of time, chemical reactions take place. These reactions serve to break down the large, complex organic molecules into simpler, smaller hydrocarbon molecules. The nature of the hydrocarbons will change with time and continued application of heat and pressure. In the early stages of petroleum formation, the deposit may consist mainly of larger hydrocarbon molecules (so-called heavy hydrocarbons), which have the thick, nearly solid consistency of asphalt. As the petroleum matures through time, and the breakdown of large molecules continues, successively lighter hydrocarbons are produced. Thick liquids give way to thinner ones, from which we derive lubricating oils, heating oils, and gasoline. In the final stages, most or all of the petroleum is further broken down into very simple, light, gaseous molecules—natural gas. Most of the maturation process occurs over the temperature range 50° to 100° C (approximately 120° to 210° F). Above

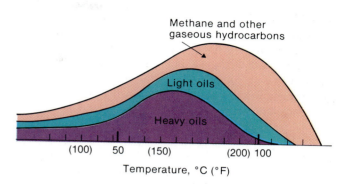

these temperatures, the remaining hydrocarbon is almost wholly methane, and with further increases in temperature it too can be broken down and destroyed (see figure 21.2).

The amount of time required for oil and gas to form is not known precisely. From the fact that virtually no petroleum is found in rocks younger than 1 to 2 million years old, we infer that the process is comparatively slow. Even if it were to take only a few tens of thousands of years (a geologically short period), we are using up the world's oil and gas far faster than significant new supplies could be produced.

Oil and Gas Migration and Concentration

Once the solid organic matter is converted to liquids and gases, these hydrocarbons can migrate out of the rocks in which they formed. In order for them to do so, the surrounding rocks must be permeable. The pores and cracks in rocks are commonly saturated with water at depth. Most oils and all natural gases are less dense than water, so they tend to rise as well as to migrate laterally through the water-filled pores of permeable rocks.

Unless stopped by impermeable rocks, the oil and gas may keep rising right up to the earth's surface. There are many known oil and gas *seeps* where these substances escape into the air or the oceans, or flow out onto the ground. These are not very efficient sources of hydrocarbons for fuel, although asphalt

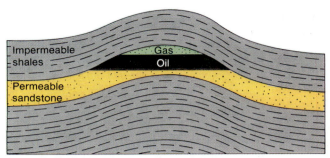

(a) A simple fold trap

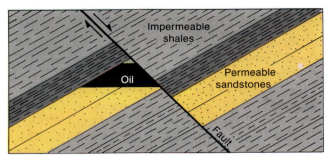

(c) A fault trap

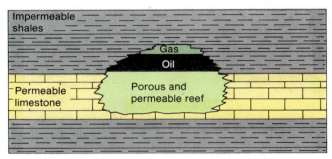

(b) Petroleum accumulated in a fossilized ancient coral reef

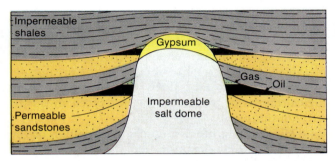

(d) Petroleum trapped against an impermeable salt dome which has risen up from a buried evaporite deposit

Figure 21.3

Types of petroleum traps. *(A)* A simple fold trap. *(B)* Petroleum accumulated in a fossilized ancient coral reef. *(C)* A fault trap. *(D)* Petroleum trapped against an impermeable salt dome that has risen up from a buried evaporite deposit.

from seeps in the Middle East was used in construction as much as 5000 years ago. Commercially, the most valuable deposits are those in which a large quantity of oil and/or gas has been concentrated and confined by impermeable rocks in **traps** (figure 21.3). The **reservoir rocks** in which the oil or gas has accumulated should be relatively porous if a large quantity of petroleum is to be found in a small volume of rock. The reservoir rock should also be relatively permeable, so that the oil or gas will flow out readily once a well is drilled into the reservoir. If the reservoir rocks are not naturally very permeable, it may be possible to fracture them artificially, through the use of explosives or high-pressure fluids, to increase the rate at which oil and gas will flow through them.

Supply and Demand

As with minerals, the most conservative estimate of how much of a fuel we have available is the amount of the *reserves*—known, or "proven," accumulations that can be produced economically with existing technology.

Oil is commonly quantified in terms of units of *barrels* (1 barrel = 42 gallons). Worldwide, over 300 billion barrels of oil have been consumed, with the estimated remaining reserves being close to 700 billion barrels (figure 21.4). Those numbers do not immediately suggest impending shortages, until it is realized that half the consumption has taken place in the last decade or so. Also, global demand continues to increase as more countries advance technologically.

The United States alone consumes nearly 30% of the oil used worldwide. Initially, U.S. oil resources probably amounted to about 10% of the world's total. Cumulative U.S. oil resources are estimated to have been not much above 200 billion barrels. Already close to half of that has been produced and consumed; remaining U.S. resources are estimated at about 120 billion barrels. Out of that, our present proven *reserves* are under 30 billion barrels. For more than a decade, the United States has been using up somewhat more domestic oil each year than the amount of new domestic reserves discovered, or proven, so that net U.S. oil reserves have been decreasing year by year (see table 21.1). Domestic production has likewise been

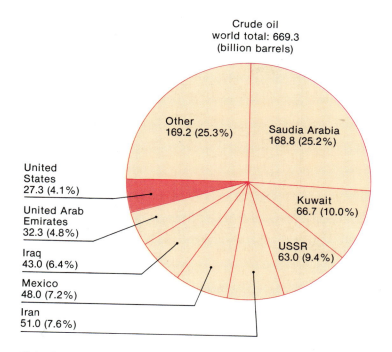

Crude oil
world total: 669.3
(billion barrels)

Other
169.2 (25.3%)

Saudia Arabia
168.8 (25.2%)

United
States
27.3 (4.1%)

United Arab
Emirates
32.3 (4.8%)

Iraq
43.0 (6.4%)

Mexico
48.0 (7.2%)

Iran
51.0 (7.6%)

Kuwait
66.7 (10.0%)

USSR
63.0 (9.4%)

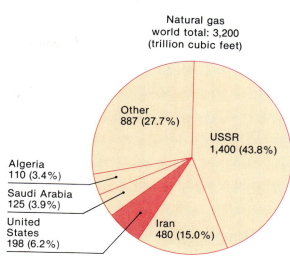

Natural gas
world total: 3,200
(trillion cubic feet)

Other
887 (27.7%)

USSR
1,400 (43.8%)

Algeria
110 (3.4%)

Saudi Arabia
125 (3.9%)

United
States
198 (6.2%)

Iran
480 (15.0%)

Note: Quantities are scaled in proportion to
area according to the Btu content of the
reserves. One billion barrels of crude
oil equals approximately 5.3 trillion
cubic feet of wet natural gas.

Figure 21.4

Estimated world crude oil and natural gas proven reserves,
December, 1983.

Source: *Annual Energy Review* 1983, U.S. Energy Information
Administration.

Table 21.1

U.S. proven reserves of crude oil and natural gas,
1976–82

Year	Crude oil (billions of barrels)	Natural gas (trillions of cu f)
1976	33.5	213.3
1977	31.8	207.4
1978	31.4	208.0
1979	29.8	201.0
1980	29.8	199.0
1981	29.4	201.7
1982	27.8	201.5

Source: Energy Information Administration.

declining. Furthermore, without oil imports, our reserves would dwindle still more quickly. The United
States consumes over 5 billion barrels of oil per year,
to supply approximately half of all the energy we use.
At times, more than 40% of the oil we have consumed
has been imported from other countries; the proportion is still close to 30%. Simple arithmetic demonstrates that the rate of oil consumption in the United
States is very high compared even to the estimated
total U.S. resources available. Some of those resources
have yet to be found, too, and will require time to discover and to develop.

Both U.S. and world oil supplies are likely to be
nearly exhausted within a few decades (figure 21.5),
especially when accelerating world energy demands
are considered. On occasion, explorationists do find
the rare, very large concentration of petroleum—for
example, the deposits on the North Slope of Alaska

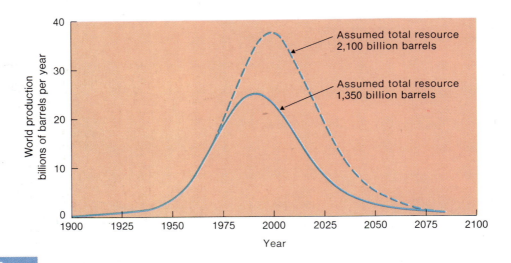

Figure 21.5

When this classic projection by M. King Hubbert was first widely
published in the early 1970s, many people were startled by it,
but in fact the total resource estimates had long existed in the
geologic literature. The supply problem was made acute by the
rapid rise in consumption starting in the 1950s.

From "The Energy Resources of the Earth," M. K. Hubbert. Copyright ©
1971 by Scientific American, Inc. All Rights Reserved.

and the oil beneath Europe's North Sea. Yet even these
make only a modest difference in the long term. The
Alaskan oil, for instance, represents reserves of only
about 10 billion barrels, a great deal for a single region but only about 2 years' worth of U.S. consumption.

The production-consumption picture for conventional natural gas is similar to that for oil (figures
21.1, 21.6; table 21.1). Natural gas presently supplies about 25% of the energy used in the United States.
This country has proven natural gas reserves of about
200 trillion cubic feet; we consume roughly 20 trillion cubic feet per year, and each year find less in new
domestic reserves than the quantity consumed. As with
oil, it is expected that the U.S. supply of conventional
natural gas will be exhausted within decades.

Future Prospects

There is a tendency to assume that, as oil and gas supplies dwindle and prices rise, there will be increased
exploration and discovery of new reserves, so that we
will not really run out of these fuels. Most experts in
the petroleum industry disagree. Price increases, too,
may temporarily moderate consumption, especially
in industrialized countries (recall figure 21.1). However, long-term *world* demand is unlikely to diminish
or even to remain at current levels.

It is true that decreasing reserves have prompted
exploration of more areas. However, most regions not
yet explored have been neglected precisely because

they are unlikely to yield appreciable petroleum. The
high temperatures involved in the formation of igneous and most metamorphic rocks would destroy organic matter, so oil would not have formed or been
preserved in these rocks. Nor do these rock types tend
to be very porous or permeable, so they generally
make poor reservoir rocks as well, unless fractured.
The large cratonic regions underlain predominantly
by igneous and metamorphic rocks are simply very
unpromising places to look for oil.

It should be pointed out as well that despite a
quadrupling in oil prices between 1970 and 1980
(after adjustment for inflation), U.S. proven reserves
have continued to decline. This is further evidence
that higher prices will not automatically lead to pro-
portionate, or even significant, increases in fuel supplies. And each time a temporary excess of production
over demand causes petroleum prices to plummet, as
in early 1986, exploration (as well as the development of new energy sources) comes to a virtual stand-
still.

Enhanced Oil Recovery

A few techniques are being developed to increase petroleum production from known deposits. An oil well
initially gushes on its own because the oil and any associated gas are under pressure from overlying rocks
(*primary recovery*). When flow slackens, water may
be pumped into the reservoir, filling empty pores and
buoying up more oil to the well (*secondary recovery*). Primary and secondary recovery together

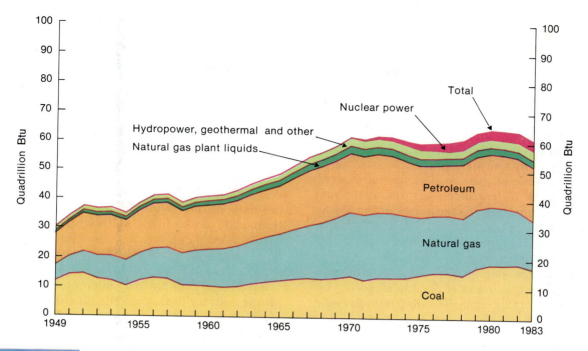

U.S. energy production since the mid-twentieth century, by energy source. Note that overall domestic production has begun to decline, despite sharp rises in fossil-fuel prices.

Source: Energy Information Administration.

extract an average of one-third of the oil in a given trap, although the figure varies greatly with the physical properties of the oil and reservoir rocks in a given oil field. On average, then, two-thirds of the oil in each deposit has historically been left in the ground.

This has created interest in **enhanced-recovery** methods. The permeability of rocks may be increased by deliberate fracturing, using explosives, or water under very high pressure. Carbon dioxide gas under pressure can be used to force out more oil. Hot water or steam may be pumped underground to warm thick, viscous oils so that they flow more easily and can be extracted more completely. Other methods are also being tested. In some cases, the fuel can even be burned in place and the heat energy released used at the site.

All enhanced-recovery methods are expensive compared to primary and secondary recovery. That they are now being used at all is largely a consequence of the dramatic increase in oil prices in the 1970s. Researchers in the petroleum industry believe that from a technological standpoint, up to an additional 40% of the oil initially present in a reservoir might be extractable by enhanced-recovery methods. This would substantially increase oil reserves. An additional attractive feature of enhanced-recovery methods is that they could be applied to old oil fields

that have already been discovered, developed by conventional methods, and abandoned, as well as to new oil fields. In other words, numerous areas in which these techniques could be used (if the economics were right) are already known, not waiting to be found.

Geopressurized Natural Gas and Other Alternate Gas Sources

There is some evidence that additional natural gas reserves may be found at very great depths in the earth. Thousands of meters below the surface, conditions are sufficiently hot that any oil would have been broken down to natural gas. This natural gas, under the tremendous pressure exerted by the overlying rock, may be dissolved in the pore water, much as carbon dioxide is dissolved under pressure in soda in a bottle. Pumping this water to the surface is like taking off the bottle cap: the pressure is released and the gas bubbles out. There may be enormous quantities of natural gas in such **geopressurized zones.** Recent estimates of the amount of potentially recoverable gas in this form range from 150 to 2000 trillion cubic feet.

There are special technological considerations in developing these gas resources. It is difficult and very expensive to drill in such deep, high-pressure environments. Also, many of the pore fluids in which the gas is dissolved are very saline and cannot be disposed of casually without risk of pollution. Geopressurized natural gas may become an important supplementary energy source for the future, but its total potential and the economics of its development are currently poorly known.

In the meantime, some studies are being made of the feasibility of enhanced gas recovery from "tight" (low-permeability) sandstones in the Rocky Mountains and gas-bearing shales in the Appalachians, using fracturing techniques to release the gas. These projects are still in the experimental stages. As with geopressurized gas, the quantity of potentially recoverable gas in these rocks is uncertain. Recent estimates, however, range from 60 to 840 trillion cubic feet.

Oil Spills

Accidental (non-natural) oil spills occur in two principal ways: from accidents during drilling of offshore oil wells, and from wrecks of oil tankers at sea. Oil spills represent by far the largest negative impact from the extraction and transportation of petroleum, although as a source of water pollution they are less significant volumetrically than petroleum pollution from careless disposal of used oil.

Drilling accidents have been a growing concern as more areas of the continental shelves are opened to drilling. Normally the drill hole is lined with a steel casing to prevent lateral leakage of the oil, but on occasion the oil finds an escape route before casing is complete. This happened in Santa Barbara in 1979, when a much-publicized spill produced a 200-square-kilometer oil slick. Alternatively, drillers may unexpectedly hit a high-pressure pocket that causes a blowout; this was the cause of a spill in the Gulf of Mexico in 1979 that released millions of gallons of oil.

Tanker disasters are potentially becoming larger all the time: the largest supertankers now being built are as long as the Empire State Building is high (300 meters)! The largest single marine oil spill ever resulted from the wreck of the *Amoco Cadiz* off the coast of France. The bill for cleaning up that spill came to over $50 million.

When an oil spill occurs, the oil, being less dense than water, floats. The lightest, most volatile hydrocarbons start to evaporate immediately, decreasing the volume of the spill somewhat (and polluting the air). Then a slow decomposition process sets in, due to sunlight and bacterial action. After several months, the spill may be reduced to about 15% of its original volume, and what is left is usually thick asphalt lumps. These can persist for many months more.

In calm seas, if a spill is small, it may be contained by floating barriers and picked up by specially designed ships that can skim up to 50 barrels of oil per hour off the water surface. Some attempts have been made to soak up oil spills with peat moss, wood shavings, even chicken feathers. Large spills or spills in rough seas are more of a problem. Perhaps the best prospect for dealing with future oil spills is the development of "oil-hungry" microorganisms that will eat the spill for food, and thus get rid of it. Scientists are currently developing suitable bacterial strains. (Even this may not be a perfect solution: what happens if these creatures make their way into tanks and oil storage facilities?) For the time being, there really is no good solution to the problems posed by a major oil spill.

Fossil Fuels—Coal

Prior to the discovery and widespread exploitation of oil and natural gas, wood was the most commonly used fuel, followed later by coal as the industrial age began in earnest last century. However, coal was bulky, cumbersome, and dirty to handle and to burn, so it fell somewhat out of favor, particularly for home use, when the liquid and gaseous fossil fuels were developed. Now, facing the prospect of running out of oil and gas, many energy users are looking harder at the potential of coal again.

Formation of Coal Deposits

Coal is formed, not from marine organisms, but from the remains of land plants. A swampy setting, in which plant growth is abundant, where fallen trees, dead leaves, and other debris are protected from decay (either by standing water or by rapid burial under later layers of plant debris), is especially favorable to the initial stages of coal formation. The Carboniferous period of geologic time, named for its abundant coal deposits, was apparently characterized over much of the world by just the sort of tropical climate conducive to the growth of lush, swampy forests in lowland areas.

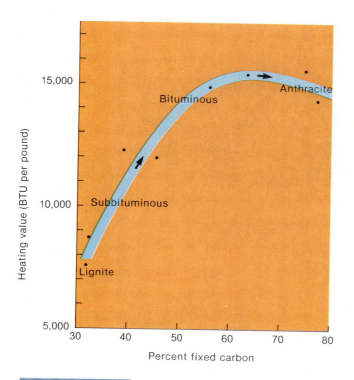

Figure 21.7

Change in character of coal with increasing application of heat and pressure. There is a general trend toward higher carbon content and higher heat value with increasing grade, though heat value declines somewhat as coal tends toward graphite.

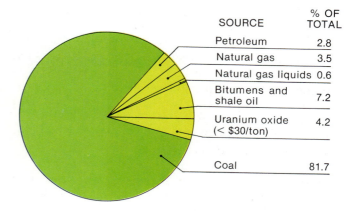

SOURCE	% OF TOTAL
Petroleum	2.8
Natural gas	3.5
Natural gas liquids	0.6
Bitumens and shale oil	7.2
Uranium oxide (< $30/ton)	4.2
Coal	81.7

Figure 21.8

Comparison of the total recoverable energy from various U.S. fossil-fuel reserves. The key word is *recoverable,* which reflects the limitations of present technology.

Source: *Coal Data Book* (1980). The President's Commission on Coal.

Given a suitable setting, the first combustible product formed is *peat*. Peat can form at the earth's surface, and there are places, like the Dismal Swamp of North Carolina or the Malaysian peninsula, where it can be found forming today. It can serve as an energy source—it is so used in Ireland—but it is not a very efficient one. Further burial, with more heat, pressure, and time, gradually dehydrates the organic matter and transforms the spongy peat into soft brown coal (*lignite*), then to the harder coals (*bituminous, anthracite;* see figure 21.7). As the coals become harder their carbon content increases, and so does the amount of heat that can be released by burning a given weight of coal. As with oil, however, there is a limit to the heat to which coals can be subjected; excessively high temperatures lead to metamorphism of coal into graphite.

Like oil, the higher-grade coals seem to require periods of time to form that are long compared to the rate at which we are using coal, so it too can be regarded as a nonrenewable resource. On the other hand, the supply of coal in the world, and in the United States also, represents a total energy resource far larger than that of petroleum (figure 21.8).

Coal Reserves and Resources

Coal resource estimates are less uncertain than are estimates of oil and gas. Coal is a solid, so it does not migrate. It will therefore be found in the sedimentary rocks in which it formed; one need not seek it in igneous or metamorphic rocks. It occurs in well-defined beds, which are easier to map than underground oil and gas concentrations. And it is formed from land plants, which did not become widespread until several hundred million years ago, so one need not look for coal in ancient rocks.

The estimated world reserve of coal is about 650 billion tons; total resources, over 10 trillion tons. The United States controls over 30% of the world reserves, some 200 billion tons (figure 21.9). Total U.S. coal resources may approach ten times the reserves. Furthermore, most of that coal is yet unmined. At present coal provides about 20% of the energy used in the United States. While we have consumed close to 50% of our petroleum resources, we have used up only a few percent of our coal. Counting only the *reserves,* the U.S. coal supply could satisfy our energy needs for several centuries, at current levels of energy consumption, if coal could be used for all purposes. It would last correspondingly longer if used as a supplement to other energy sources (figure 21.10).

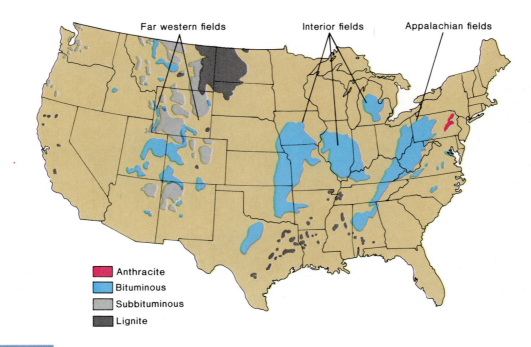

Figure 21.9

Distribution of U.S. coal fields.

Source: U.S. Geological Survey.

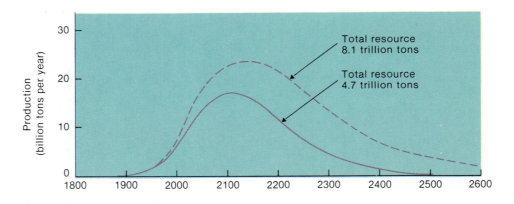

Figure 21.10

Projected world coal production, assuming coal as the principal substitute for petroleum. Note the small fraction of coal actually consumed to date (compare with figure 21.5).

From "The Energy Resources of the Earth," by M. K. Hubbert. Copyright © 1971 by Scientific American, Inc. All Rights Reserved.

Limitations on Coal Use; Gasification and Liquefaction

Solid coal is simply not as versatile a substance as petroleum or natural gas. It cannot be used directly in most forms of transportation. It is a relatively dirty and inconvenient fuel for home heating, which is why people abandoned it for that purpose in the first place.

Therefore we cannot expect to adopt coal as a substitute for oil in all applications, given present technology. Currently, it is used primarily in power plants generating electricity.

It is possible to convert coal to liquid or gaseous hydrocarbon fuels by causing the coal to react with steam or with hydrogen gas at high temperatures. The conversion processes are termed **liquefaction** (when the product is a liquid fuel) and **gasification** (when the product is gaseous). Both types of processes are intended to transform the coal into a cleaner-burning, more versatile fuel, to expand the range of its possible applications.

Commercial coal gasification has existed at some scale for 150 years. At present, only experimental coal-gasification plants are operating in the United States. Current gasification processes yield a gas that is a mixture of carbon monoxide and hydrogen with a little methane. The heat that can be derived from burning this mix is only 15% to 30% of what can be obtained from burning an equal volume of natural gas. The low heat value makes it uneconomic to transport the gas over long distances, so it is typically burned right at the production site. Technology exists to produce high-quality gas from coal, equivalent in heat content to natural gas, but it is presently uneconomic when compared to natural gas.

Pilot projects are also underway to study *in situ* (on site), underground coal gasification, in which coal would be gasified in place and the gas then extracted. The expected environmental benefits of eliminating the step of mining the coal include reduced land disturbance, water use, and air pollution, and less solid waste produced at the surface. Potential drawbacks include possible groundwater pollution and surface subsidence over gasified coal beds. Underground gasification may provide a means of using coals that are too thin to mine economically, or those that would require excessive land disruption to extract.

The practice of generating liquid fuel from coal has a longer history than many people realize. Germany produced gasoline from its abundant coal during World War II; South Africa has a liquefaction plant in commercial operation now, producing gasoline and fuel oil. A variety of technologies exist, and several noncommercial pilot plants are in operation in the United States. However, their products are far from economically competitive with conventional petroleum in this country, and will probably not become so before the late 1990s.

Environmental Impacts of Coal Use

Extensive mining and burning of coal creates a set of major pollution problems. Like all fossil fuels, coal produces CO_2 when burned. The additional pollutant that is of special concern with coal is sulfur. The sulfur content of coal can be over 3%, some in the form of pyrite (FeS_2), some bound in the organic matter of the coal itself. When the sulfur is burned along with the coal, sulfur gases, notably sulfur dioxide (SO_2), are produced. These gases are toxic and can be severely irritating to lungs and eyes. They also react with water in the atmosphere to produce sulfuric acid, which is removed from the air as acid rain.

> Oil also contains sulfur, derived from the original organic matter. But oil is refined to separate the various compounds, and much of the sulfur is removed during the refining process. By the time the oil is burned, it will typically produce only about a tenth as much of the various sulfur gases as the burning of coal.

Some of the sulfur in coal can be removed prior to burning, but the process is expensive and is only partially successful, especially with the organic sulfur. Alternatively, the sulfur gases produced can be trapped by special devices ("scrubbers") in exhaust stacks, but again the process is expensive (in terms of both money and energy) and incomplete. From the standpoint of reducing sulfur emissions, low-sulfur coal (coal containing 1% sulfur or less) is a more desirable fuel than high-sulfur coal. On the other hand, much of the low-sulfur coal in the United States is also lower-grade coal, so more of it must be burned to generate the same amount of energy.

Coal use also produces a great deal of solid waste. The ash residue left after coal is burned typically amounts to 5% to 20% of the original volume. The ash, which consists mostly of noncombustible silicate minerals in the coal, also contains toxic metals. If released with waste gases, the ash fouls the air. If captured by scrubbers or otherwise confined before emission, this ash still must be disposed of. If exposed at the surface, the fine ash, with its high surface-to-volume ratio, can weather very rapidly, and toxic metals can be leached from it, posing a water-pollution problem. Uncontrolled erosion of the ash could cause sediment pollution. The magnitude of this problem should not be underestimated. One coal-fired electric power plant can produce over a million tons of solid waste per year.

Environmental Impacts of Coal Mining

Underground mining of coal is notoriously dangerous, as well as expensive. The rising costs of providing a safer working environment for underground coal miners are largely responsible for a steady shift in mining methods, from about 20% surface mining around 1950, to over 60% surface mining by 1980. No reversal of this trend is in sight. Most of the surface mining is strip-mining. A particular problem with the strip-mining of coal is, again, the associated sulfur. Some of this sulfur is left in the spoil banks, either in the small amount of coal left behind or in the pyritic shales often interbedded with the coal. This sulfur can

Box 21.1

Caution: Fossil Fuels Below!

The potential problems associated with strip mining have received wide publicity. Less well known, but often equally serious, are the possible consequences of underground extraction of fossil fuels.

Withdrawal of oil, like that of water, can cause surface subsidence, resulting in building failure and, in low-lying coastal areas, flooding. In fact, the pumping of water associated with secondary oil recovery often serves the additional purpose of propping up the land surface somewhat as the petroleum is withdrawn.

Surface subsidence over old shallow underground coal mines is not unknown (figure 1). This subsidence typically takes the form of pits or trenches. Surface failure may lag years or decades behind mining, developing as supporting structures decay and subsurface waters weaken rocks through weathering. Areas so affected are particularly common in the northwestern and north-central United States.

As water and oxygen seep into abandoned coal mines, spontaneous ignition may start the remaining coal burning (figure 2). The U.S. Bureau of Mines estimates that some 250 uncontrolled mine fires are presently burning in 17 states. The burning can lead to more settling, more subsidence pits, and increased air supply to the fire. Carbon monoxide and noxious sulfur gases rise to the surface. The flames can in principle be smothered by blocking all pits, shafts, and other openings through which air is reaching the flames, but the geometry of mine workings may be so complex as to make it impossible to locate and seal every airway. As a result, such fires can and do burn for years.

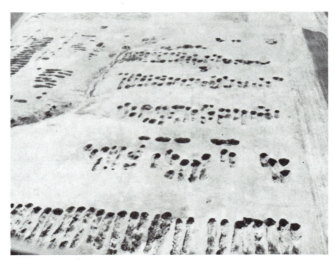

Figure 1
Subsidence pits over old coal mines, Mercer County, North Dakota.
Photo by C. R. Dunrud, courtesy U.S. Geological Survey.

Figure 2
Out-of-control fire in abandoned underground coal mine, Sheridan County, Wyoming.
Photo by C. R. Dunrud, courtesy U.S. Geological Survey.

Figure 21.11

Abandoned coal strip mine, Fulton County, Illinois.

Photo by Arthur Greenberg, courtesy EPA/National Archives.

react with water and air to produce acid runoff water, which slows revegetation of the area (figure 21.11) and may cause a water-pollution problem also. (Even underground coal mines may have acid drainage, but the associated water circulation is generally more restricted.) Mine reclamation is complicated by the need to restore a relatively low-sulfur surface layer in which the vegetation can become established. As with any reclamation, water availability is also a consideration. This is particularly true in the western United States, where half of our coal reserve (and much of the low-sulfur coal) is found: the dryness of the climate makes reclamation a major concern.

New Fossil Fuels—Oil Shale and Tar Sand

Oil Shale

Oil shale is very poorly named. The rock, while always sedimentary, need not be a shale, and the hydrocarbon in it is not oil! The potential fuel in oil shale is a waxy solid, *kerogen,* which is formed from the remains of plants, algae, and bacteria. The rock must be crushed and heated to distill out the "shale oil," which is then refined, somewhat as crude oil is, to produce various liquid petroleum products.

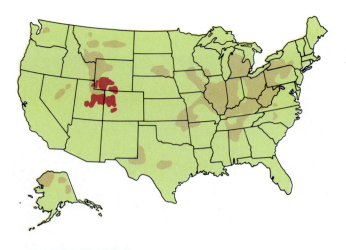

Figure 21.12

Distribution of U.S. oil shale deposits. The richest of these are the Green River formation (red).

From J. W. Smith, "Synfuels: Oil Shale and Tar Sands," in Ruedisili and Firebaugh, *Perspectives on Energy*, Oxford University Press, 1982.

About two-thirds of the world's known oil shale is in the United States (see locations in figure 21.12). The total estimated resource could yield 2 to 5 trillion barrels of shale oil. There are a number of reasons why we are not yet using this apparently vast resource to any significant extent, and may not be able to do so in the near future:

1. Much of the kerogen is widely dispersed through the oil shale, so huge volumes of rock must be processed to obtain moderate amounts of shale oil. Even the richest oil shale yields only about 3 barrels of shale oil per ton of rock processed. The cost of extraction is not presently competitive with that of conventional petroleum.

2. Much of the oil shale is located so near the surface that it would logically be exploited by strip-mining. Most of the richest oil shale, however, is located in dry areas of the western United States (Colorado, Wyoming, Utah), so land reclamation after mining is correspondingly difficult.

3. The water shortage presents a further problem. Current processing technologies require large amounts of water—about 3 barrels of water for every barrel of shale oil extracted. In the western states, there is no obvious abundant source of water to process the oil shale.

4. The volume of the rock increases as a consequence of processing. It is possible for the volume of waste rock to be 20% to 30% larger than the original volume of rock mined. Aside from problems caused by accelerated weathering of the crushed material, there is the basic question of where to put it. Clearly it will not all fit back into the space mined out.

In short, oil shale represents a large energy resource, but there are serious problems of water supply, waste disposal, and land reclamation to be solved before it will make a major contribution to our energy use.

Tar Sand

Tar sands are sedimentary rocks containing a very thick, semi-solid, tarlike petroleum. The heavy petroleum in tar sands is believed to have been formed in the same way and from the same materials as lighter oils. Tar sand deposits may represent very immature petroleum deposits, in which the breakdown of large molecules has not progressed to the production of the lighter liquid and gaseous hydrocarbons. Alternatively, the lighter compounds may have migrated away, leaving this dense, viscous material behind. Either way, the tarry petroleum is too thick to flow out of the rock. Like oil shale, tar sand presently must be mined, crushed, and heated to extract the petroleum, which can then be refined into various fuels.

Many of the environmental problems associated with oil shale likewise apply to tar sand. The tar is disseminated through the rock, so large volumes of rock must be mined and processed to extract appreciable petroleum. Many tar sands are near-surface deposits, so the mining method is commonly stripmining. The processing requires a great deal of water, and the waste rock after processing may take up a larger volume than the original tar sand.

The Athabasca tar sands of Canada may contain several hundred billion barrels of petroleum. Some mining activities are already underway in the province of Alberta, and Canada hopes to be meeting one-third of its oil needs from the tar sands by 1990. The United States, however, has very limited tar sand resources (figure 21.13). We cannot look to tar sand to solve our domestic energy problems, even if the environmental and developmental difficulties could be overcome.

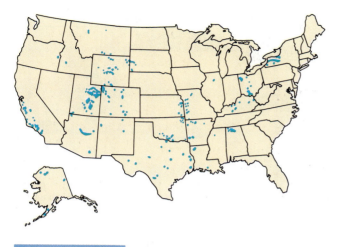

Figure 21.13

Distribution of U.S. tar sands.

From J. W. Smith, "Synfuels: Oil Shale and Tar Sands," in Ruedisili and Firebaugh, *Perspectives on Energy,* Oxford University Press, 1982.

Nuclear Power—Fission

Nuclear power actually comprises two different types of processes, with different advantages and limitations. Only nuclear **fission** is presently commercially feasible. Fission is the splitting apart of atomic nuclei into smaller ones, with the release of energy (figure 21.14). A few isotopes (some 20 out of more than 250 naturally occurring isotopes) can undergo fission spontaneously, and do so in nature. Some additional nuclei can be induced to split apart, and the naturally fissionable nuclei can be made to split more rapidly, increasing the rate of energy release. The fissionable nucleus of most interest in connection with modern nuclear power reactors is the isotope uranium-235.

A uranium-235 nucleus can be induced to undergo fission by the firing of another neutron into the nucleus. It splits into two lighter nuclei (not always the same two), and releases additional neutrons as well as energy. Some of the newly released neutrons can induce fission in other nearby uranium-235 nuclei, which release more neutrons and more energy in breaking up, the process continuing in a **chain reaction.** A controlled chain reaction with continuous moderate release of energy is the basis for fission-powered reactors. The energy released heats cooling water that circulates through the reactor's core. The heat removed from the core is transferred, through a heat exchanger, to a second water loop in which steam is produced. The steam in turn is used to run turbines to produce electricity (figure 21.15).

Only 0.7% of natural uranium is uranium-235. Natural uranium must be processed (enriched) to increase the concentration of this isotope to several

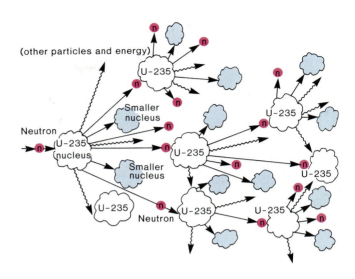

Figure 21.14

Schematic diagram of nuclear fission and chain reaction involving U-235.

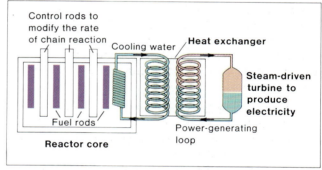

Containment building

Figure 21.15

Schematic diagram of conventional nuclear fission electricity-generating plant.

percent of the total uranium in order to produce reactor-grade uranium capable of sustaining a chain reaction. As the reactor operates, the U-235 atoms are split and destroyed, so in time the fuel becomes so depleted in this isotope, or "spent," that it must be replaced with a fresh supply of enriched uranium.

Uranium Supply; Breeder Reactors

Uranium occurs in a variety of sedimentary and hydrothermal ore deposits. World estimates of available uranium are difficult to obtain, in part because the strategic importance of uranium leads to some secrecy. For the United States, the estimates are also very

Table 21.2

U.S. uranium reserve and resource estimates, 1983

Recoverable costs (per pound U_3O_8)	Reserves (thousands of tons)	Total resources (thousands of tons)
$30	180	1307
$50	576	2642
$100	889	4270

Note: The geology of known uranium occurrences is such that relatively little would be added to the $100/lb estimates by including deposits exploitable in the cost range $100 to $200/lb. Actual uranium prices (in constant 1983 dollars) rose from under $20/lb U_3O_8 in the early 1970s to over $65/lb in 1976, then declined sharply to $20 to $25/lb in 1983.
Source: Data from Annual Energy Review, 1983.

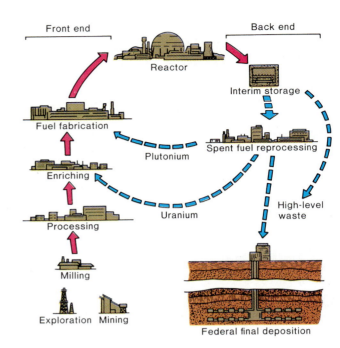

➡ Fuel cycle as it operates currently.

➡ Fuel cycle as it would operate with spent fuel reprocessing and federal waste storage.

Figure 21.16

The nuclear fuel cycle, as it presently operates and as it would function with fuel reprocessing.

Source: Energy Information Administration.

sensitive to the price obtainable for the processed "yellowcake," uranium oxide (U_3O_8), which has oscillated considerably. Table 21.2 summarizes present reserve and resource estimates at different price levels.

Using present technology, nuclear electricity-generating capacity in the United States probably could not be increased to much more than four times present levels (to supply less than 15% of our total energy needs) by the year 2010 without serious fuel shortages, because the rarer isotope U-235 is in such short supply. However, uranium-235 is not the only possible fission-reactor fuel. When an atom of the nonfissionable U-238 absorbs a neutron, it is converted to plutonium-239, which *is* fissionable. During the chain reaction inside the reactor, as freed neutrons move about, some are captured by the abundant U-238 atoms, making plutonium. Spent fuel could be *reprocessed* to extract this plutonium, as well as to re-enrich the uranium in U-235. The way in which reprocessing would alter the nuclear fuel cycle is shown in figure 21.16. Fuel reprocessing with recovery of both plutonium and uranium could reduce the demand for new uranium by an estimated 15%.

To maximize production of fuel, a **breeder reactor** could be used. Breeder reactors produce energy during operation just as conventional reactors do, by fission in a sustained chain reaction within the reactor core. In addition, they are designed so that surplus neutrons not required to sustain the chain reaction are used to produce more fissionable fuels from suitable materials, like plutonium-239 from uranium-238, or thorium-233 from the common isotope

thorium-232. Breeder reactors can synthesize more nuclear fuel for future use than they are actively consuming while generating power. It seems clear that if the nuclear-fission option is to be pursued vigorously into the next century, reprocessing of spent fuels and use of breeder reactors are essential. Yet at present, even reprocessing of spent fuel is minimal in the United States.

Breeder reactor technology is more complex than that of conventional reactors. The core coolant in a breeder is liquid metallic sodium, not water; the reactor operates at much higher temperatures. The costs are substantially higher than for present commercial reactors and might be close to $10 billion per reactor by the time any breeders could be completed in this country. The breeding process is slow, too; the break-even point after which fuel produced would exceed fuel consumed might be several decades after initial operation. Congress recently canceled funding for the experimental Clinch River breeder reactor, in part due to high estimated costs. No commercial breeders currently exist in the United States, and only a handful of breeder reactors are operating in the world.

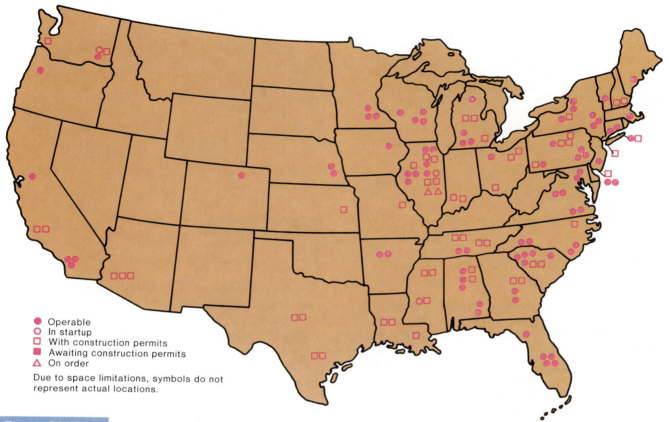

● Operable
○ In startup
□ With construction permits
▣ Awaiting construction permits
△ On order
Due to space limitations, symbols do not
represent actual locations.

Figure 21.17

Distribution of U.S. commercial nuclear power plants, December
1983. There were 79 operable plants, 3 in startup phases, 53 for
which construction permits had been granted, and 2 on order.

Source: Energy Information Administration.

Even conventional reactors are falling out of favor in
the United States. At the end of 1983, there were 79
nuclear power plants operating in the country (out of
294 worldwide), with 65 more under construction
(figure 21.17). However, over the last decade, nuclear
plant cancellations have far exceeded new orders, and
several utilities have revised their plans, to make new
power plants coal-fired, not nuclear. The 79 nuclear
plants represent less than 8% of the 10,602 electricity-generating plants in commercial operation in the
United States.

Concerns Related to Nuclear Power

There are a number of major concerns regarding the
use of fission power that collectively account for its
limited acceptance. One is reactor safety. In normal
operation, nuclear power plants release very minor
amounts of radiation, which are believed to be harmless. We are, after all, constantly exposed to radiation—from cosmic rays, from uranium and other
naturally radioactive elements in rocks, soil, brick,

masonry, and even wood, from the radium in our
watch dials, from medical X rays. (We are, in fact, radioactive ourselves, from the carbon-14 in our
bodies.) The normal radiation release from a nuclear
power plant is negligible by comparison.

There is a small but finite risk of damage to a reactor from accident or sabotage. One of the more serious possibilities is the so-called loss-of-coolant
event, in which the flow of cooling water to the reactor core is interrupted. Resultant overheating of the
core might lead to *core meltdown,* in which the fuel
and core materials would deteriorate into a molten
mass, which might or might not melt its way out of
the containment building (scientific opinion is divided on this). A partial loss of coolant occurred at
Three Mile Island. No full-scale loss-of-coolant accident with core meltdown has occurred in the United
States. A comprehensive study commissioned by the
Nuclear Regulatory Commission in 1975 projected the
risk of a core meltdown at about one in a million per
reactor per year. This is certainly a small risk factor,
particularly considering other risks of daily life; it is
also only an educated guess.

In mid-1986, an explosion and core meltdown, with fatalities and release of significant radiation, did occur at the Chernobyl nuclear power station in the Soviet Union. The extent of release of radioactive material was due in part to the fact that Soviet reactors typically are not placed in containment buildings of the kind used in the United States. Nevertheless, this incident revived many of the concerns about U.S. reactor safety that had been voiced at the time of the Three Mile Island accident. The Chernobyl reactor design is different from that of most commercial reactors in the United States, but some reactors in this country are of similar type.

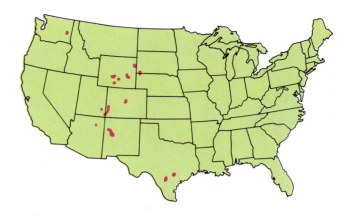

Figure 21.18

Distribution of U.S. uranium deposits.
Source: Energy Information Administration.

Concerns Related to Fuel and Waste Handling

The mining and processing of uranium ore are operations that affect relatively few places in the United States (figure 21.18). Nevertheless, they pose hazards because of the natural radioactivity of uranium. Miners exposed to the higher radiation levels in uranium mines have been shown to have higher rates of occurrence of some types of cancer. Carelessly handled tailings from processing plants may expose others to radiation hazards. (There have been cases in which radioactive tailings were unknowingly mixed into concrete and used in construction; homes built with the radioactive concrete were determined to be unsafe to live in.)

The use of reprocessing and/or breeder reactors involves concerns about plutonium handling. Plutonium itself is both radioactive and chemically highly toxic. As a readily fissionable material, it can also be used to make nuclear weapons. This is a growing concern as terrorist activities increase worldwide. Extensive handling, transport, and use of plutonium would create a significant security problem, requiring very tight inventory control.

The radioactive wastes from the production of fission power are a further consideration. Many waste-disposal schemes have been proposed, ranging from waste disposal in (under) polar ice caps, to rocketing wastes into the sun, to placement of solid or liquid wastes in old abandoned mines. Studies of the feasibility of placing wastes in caverns in bedrock are underway. A prime rock type under consideration is basalt, which (unless fractured) has low permeability. One of the most promising solutions may be disposal in old salt mines. Salt has a high melting point, the better to contain heat-producing wastes; it flows plastically under sustained stress, so it can self-seal if fractured; and salt is relatively impermeable.

Two aspects of the radioactive-waste problem may be highlighted. First, radioactive materials cannot be treated by chemical reaction, heating, or other physical processes in order to make them nonradioactive. In this respect they differ from many toxic chemical wastes that can be broken down by appropriate treatment. Radioactive materials are so because their nuclei are inherently unstable, and chemical reactions or changes in physical state do not affect the atomic nucleus. Second, there is enough indecision about the best method of radioactive-waste disposal and the appropriate site(s) for it that none of the radioactive wastes generated anywhere in the world have been consigned to permanent disposal sites to date. The wastes are in temporary storage while various disposal options are being explored. Acceptable waste-disposal methods will have to be identified and adopted, if only to dispose of wastes already accumulated.

Concluding Observations

Different individuals assess the pros and cons of nuclear fission power in quite different ways. For many the uncertainties and potential risks outweigh the benefits. For others, the problems associated with using coal (the most abundant, readily useable alternative) appear at least as great. Use of nuclear fission power is in any case restricted to the generation of electricity, and thus there is a limit to the extent to which it can supply our energy needs. In the United States, production of electricity accounts for about 30% of total energy consumed, though its proportion of the total is growing with time. Thus, even with a wholehearted commitment to nuclear fission power, it could not realistically become our principal energy source for many decades.

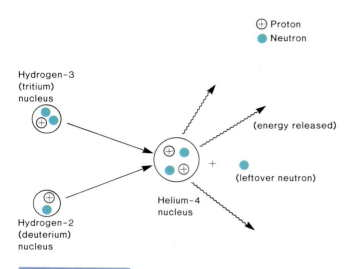

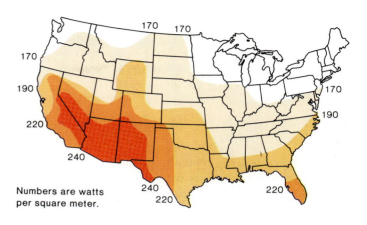

Figure 21.19

Schematic diagram of the nuclear fusion process.

Figure 21.20

Distribution of solar energy over the United States. Maximum insolation occurs over the southwestern part of the country.

Source: Union of Concerned Scientists, Kendall, S. W., and S. J. Nadis (eds.), *Energy Strategies: Toward a Solar Future.* © Ballinger Publishing Co., Cambridge, MA.

Nuclear Power—Fusion

Nuclear **fusion** is the opposite of fission. Fusion is the process by which two or more smaller atomic nuclei combine to form a large one, with an accompanying release of energy (figure 21.19). It is the process by which the sun generates the vast amounts of energy that it radiates. In the sun, simple hydrogen nuclei containing only a single proton are fused to produce helium. For technical reasons, fusion of the heavier hydrogen isotopes deuterium (nucleus containing one proton and one neutron) and tritium (one proton and two neutrons) would be easier to achieve on earth.

Hydrogen is plentiful. The oceans represent a vast reserve of hydrogen, an essentially inexhaustible supply of the necessary fuel. This is true even considering that deuterium is only 0.015% of natural hydrogen. The product of the projected fusion reactions, helium, is a nontoxic, chemically inert, harmless gas. There could be some mildly radioactive light-isotope by-products of fusion reactors, but they are much less hazardous than many of the products of fission reactors.

The principal reason for failure to use fusion power is technology, or lack of it. In order to bring about a fusion reaction, the reacting nuclei must be brought very close together at extremely high temperatures (millions of degrees at least). The necessary conditions can be achieved in the interior of the massive sun, but no known physical material on earth can withstand such temperatures. The techniques being tested in laboratory fusion experiments are elaborate and complex, involving containment of the fusing materials with strong magnetic fields, or use of lasers to heat frozen pellets of hydrogen very rapidly.

At best the experimenters have been able to achieve the necessary conditions for fractions of a second, and the energy required to bring about the fusion reactions has exceeded the energy released thereby. Decades of intensive, and expensive, research will be needed before fusion can become a commercial reality; even then, it will be costly. The ultimate potential contribution of fusion power is also limited, in that it is another means of generating electricity in stationary power plants only. The abundance of the fuel supply and the relative cleanness of the process (as compared both to fission and to fossil-fuel power generation) does make it a potentially desirable option for the next century.

Solar Energy

The sun can be expected to continue shining for approximately 5 billion years, so this resource is effectively inexhaustible, in sharp contrast to uranium or the fossil fuels. Sunlight falls on the earth without any mining, drilling, or other disruption of the land. Sunshine is free, under the control of no company or cartel, and subject to no political disruption. The use of solar energy is essentially pollution-free, producing no hazardous wastes, air or water pollution, or noise. All these features make solar energy an attractive option for the future.

The total solar energy reaching the earth is more than ample to provide for all human energy needs, both now and for the foreseeable future, but it is also spread over a broad surface area. Where large quantities of energy are used, collectors must likewise cover a great deal of area. Sunlight is also variable in intensity, both from region to region (figure 21.20) and from day to day as weather conditions change. For various reasons, the two areas in which solar energy can make the greatest immediate contribution are in space heating and in the generation of electricity, uses that together account for about two-thirds of U.S. energy consumption.

Solar Heating

Solar space heating typically combines direct use of sunlight for warmth with some provision for collecting and storing additional heat for use when the sun is not shining. The structure itself is designed to allow the maximum amount of sunlight to stream in through south and west windows during the colder months. This heats not only the indoor air but other materials inside. Media used for storing heat include water—in barrels, in tanks, even in indoor swimming pools—and the rock, brick, or other dense solids used in the building's construction (figure 21.21). These create a "thermal mass" that radiates heat back to heat the building when needed later. This is passive-solar heating.

Active-solar systems, in addition, use mechanical devices to circulate solar-heated water (figure 21.22). Flat solar collectors on the roof are water-filled boxes with a glass surface to admit sunlight, and a dark lining to absorb it and help heat the water. The warmed water is circulated either directly into a storage tank, or into a heat exchanger through which a tank of water is heated. The solar-heated water can serve to supply both hot-water heat and the building's hot-water supply. If a building already has conventional hot-water heat, incorporation of solar collectors into the scheme may also be relatively inexpensive.

While solar heating may be adequate by itself in mild and sunny climates, a conventional heating system is generally needed as backup in areas subject to prolonged spells of cloudiness or extreme cold. In such areas, use of solar energy can reduce, but not wholly eliminate, the consumption of conventional fuels. It has been estimated that in the United States, 40% to 90% of most homes' heating requirements could be supplied by passive solar heating systems, depending on location.

Solar Electricity

Direct production of electricity using sunlight is accomplished through **photovoltaic cells** (also known simply as *solar cells*; see figure 21.23). They have been

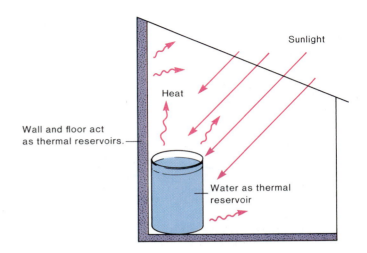

Figure 21.21

Basics of passive-solar heating, with water or structural materials as thermal reservoir.

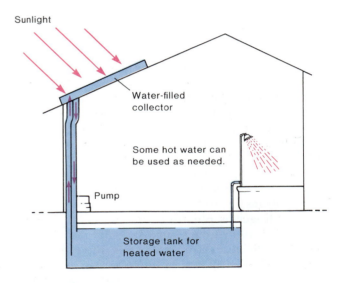

Figure 21.22

A common type of active-solar system, with a pump to circulate water between collector and heat exchanger/storage tank.

used for years as the principal power source for satellites, and in a few remote areas on earth difficult to reach with power lines. Historically, a major limitation on solar-cell use has been cost, which is several times higher per unit of power-generating capacity than with either fossil-fuel or nuclear-powered generating plants.

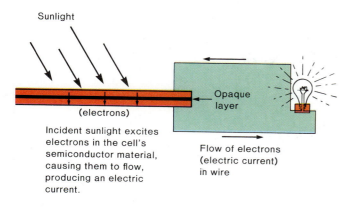

Figure 21.23

Schematic diagram of photovoltaic (solar) cell for the generation of electricity.

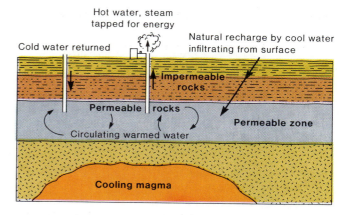

Figure 21.24

Extraction of geothermal energy involves the use of warmed circulating groundwater.

> The low efficiency of present solar cells, together with the diffuse character of sunlight, makes photovoltaic conversion an inadequate option for energy-intensive applications, such as factories. Even in the areas of strongest sunlight, incident radiation is of the order of 250 watts per square meter. The best solar cells are only about 20% efficient, meaning power generation of only about 50 watts per square meter. In other words, to keep one 100-watt light bulb burning would require at least 2 square meters of collector—with the sun always shining! A 100-megawatt power plant would require nearly a square mile of collectors. This represents a large commitment both of land and of mineral resources.

Storing solar electricity is also a more complex matter than storing heat. For individual home users, batteries may suffice. No wholly practical scheme for large-scale storage has been devised. Major improvements in efficiency of generation and in storage technology will be needed before the sun can satisfy more than 10% to 15% of U.S. electricity needs.

Further Environmental Impacts of Commitment to Solar Electricity

The *use* of solar energy is environmentally benign. Unfortunately, the construction of solar facilities is not. The most efficient photovoltaic cells use potentially toxic materials such as gallium and arsenic. These are already health hazards in both the mining and manufacturing phases of the semiconductor industry. The risks will be greatly magnified if that industry is expanded to the power-plant scale.

Siting such a sizeable collector array requires a substantial commitment and disturbance of land. Its presence may alter patterns of evaporation and surface runoff. These considerations may be especially critical in the desert areas that are the most favorable sites for such facilities from the standpoint of intensity and constancy of incident sunlight.

Geothermal Power

Magma rising into the crust from the mantle brings nearer the surface material that is hot compared to average crustal temperatures at the same depth. Heat from the cooling magma will heat any groundwaters circulating nearby (figure 21.24). This is the basis for **geothermal** power. The magma-warmed waters may escape at the surface in geysers and hot springs, signaling the existence of the shallow heat source below. More subtle evidence of the presence of hot rocks at depth comes from measurements of heat flow at the surface. High heat flow and recent magmatic activity go together, and in turn are most often associated with plate boundaries. Therefore most areas in which geothermal energy is being tapped extensively are along or near plate boundaries (figure 21.25).

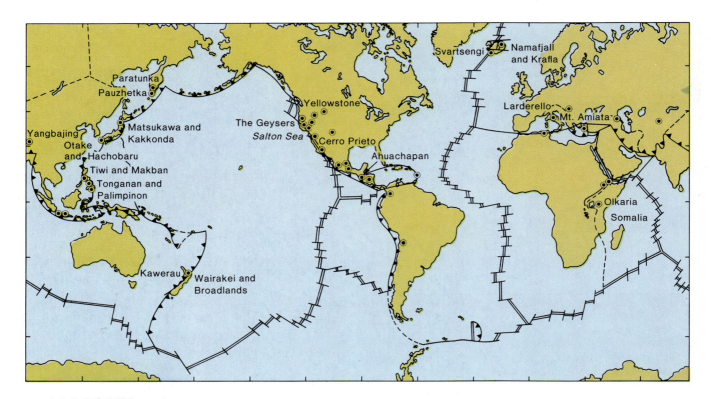

Figure 21.25

Distribution of geothermal power plants worldwide.

From L. J. P. Muffler, "Geothermal Energy," Ch. 26 in Ruedisili and Firebaugh, *Perspectives on Energy,* Oxford University Press, 1982.

Applications of Geothermal Energy

Exactly how the geothermal energy is used depends largely on how hot the system is. In some places the groundwater is warmed, but not enough to turn to steam. Still, the water temperatures may be comparable to what a home heating unit produces, and the water can successfully be circulated directly through homes to heat them. This is being done in Iceland and in parts of the Soviet Union.

Other geothermal areas may be so hot that the water is turned to steam. The steam can be used, like any boiler-generated steam produced using more conventional fuel, to run electric generators. The largest geothermal-electricity operation in the United States is The Geysers, in California, which has operated since 1960. More than a million watts of generating capacity are on line, with more power plants planned. Other steam systems are being used in Larderello, Italy, in Japan, Mexico, the Philippines, and elsewhere. Altogether, there are about 40 sites worldwide where geothermal electricity is actively being developed. Some 18 of the 10,602 electric power plants in the United States in 1983 were geothermal plants.

In a few areas—such as the Texas Gulf Coast—geothermal waters contain a bonus: dissolved methane. The methane can be extracted to provide additional energy beyond the heat of the water itself.

Environmental Considerations

In favorable cases, geothermal power generation is quite competitive economically with conventional methods. The use of geothermal steam is also largely pollution-free. There may be some sulfur gases, derived from the magmatic heat source, mixed with the steam. These certainly pose no more serious a pollution problem than sulfur from coal burning, although the observation that trees may be killed by sulfurous fumes emitted from fumaroles suggests a need for control. There are no ash, radioactive-waste, or carbon-dioxide problems with geothermal power as there are with other fuels. The warm geothermal waters may be a problem. They frequently contain large quantities of dissolved chemicals that can clog or corrode pipes, or pollute local ground or surface waters if allowed

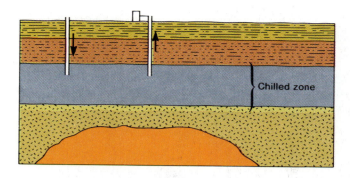

Decrease in productivity of a geothermal field due to chilling of the aquifer and adjacent rocks.

to run off freely. However, these difficulties are relatively small and controllable. Three other limitations restrict the potential contribution of geothermal power.

Limitations on Geothermal Power

First, each geothermal field can be used only for a limited period of time—a few decades, on average—before the rate of heat extraction is seriously reduced. This is a consequence of the fact that rocks conduct heat very poorly. That characteristic of rock can easily be demonstrated by turning over a flat, sun-baked rock on a bright but cool day. The surface exposed to the sun may be hot to the touch, but the heat will not have traveled far into the rock, and the underside will be cool. Conversely, as hot water or steam is withdrawn from a geothermal field, it is replaced by cooler water that must be heated before use. Initially the heating may be rapid, but in time the permeable rocks will become chilled to such an extent that water circulating through them heats too slowly or too little to be useful (figure 21.26). The heat of the magma will not have been exhausted, but it is only slowly transmitted into the chilled permeable rocks. Some time must then elapse before those permeable rocks will have been reheated sufficiently that normal operations can be resumed at the power plant.

A second limitation is that, like geothermal power plants themselves, the resource is stationary. Oil or coal can be moved to power-hungry population centers. Geothermal power plants must be put where the hot rocks are, and long-distance transmission of the power they generate is not technically practical. Most large cities are far removed from major geothermal resources.

Third is a limitation on the total number of suitable sites. Clearly the plate boundaries cover only a small part of earth's surface, and many of them are inaccessible (seafloor-spreading ridges, for instance). Not all have abundant circulating subsurface water, either.

Hot Dry Rock

On the other hand, many areas away from modern plate boundaries have heat flow somewhat above the normal level, and geothermal gradients above the average of 30° C per kilometer (85° F per mile) typical of continental crust. Where thermal gradients are at least 40° C per kilometer, even in the absence of much subsurface water, the region is described as a **hot dry rock** potential geothermal resource. Deep drilling to reach usefully high temperatures must be combined with induced circulation of water pumped in from the surface to make use of these hot dry rocks. The amount of heat potentially extractable from hot dry rocks is estimated at more than ten times that of natural hot-water and steam geothermal fields, just because the former are much more extensive.

Most of the regions identified as possible hot-dry-rock geothermal fields in the United States are in thinly populated western states with restricted water supplies. How much of an energy contribution they may ultimately make is therefore highly uncertain. Certainly hot-dry-rock geothermal energy will be less economical than that of the hot-water or steam fields where circulating water is already present and temperatures are higher. Currently, some experimentation with hot-dry-rock fields is in progress. Commercial development in such areas is not expected before the end of this century. For the near term, and well into the next century, it is most likely that geothermal energy will be a major energy source only in the immediate vicinity of the few areas best suited to its production, supplying them with heat and/or electricity.

Energy from Moving Water

The energy of falling or flowing water has been used for centuries. It is now used primarily to generate electricity. Hydroelectric power has consistently supplied several percent of U.S. energy needs for decades; it currently provides about 10% to 15% of our electricity.

Figure 21.27

Glen Canyon dam hydroelectric project.

Conventional Hydropower

The principal requirements for the generation of substantial amounts of hydroelectric power are a large volume of water and rapid movement of that water. Modern commercial generation of hydropower typically involves damming up a high-discharge stream, impounding a large reserve of water, and releasing it through a restricted outflow as desired, rather than operating subject to great seasonal variations in discharge (figure 21.27).

This is a very clean energy source in operation. The water is not polluted as it flows through the generating equipment. No chemicals are added to it, nor are any dissolved or airborne pollutants produced. The water itself is not consumed during power generation; it merely passes through the generating equipment. Hydropower is renewable as long as the streams continue to flow. That it is economically competitive with other sources is demonstrated by the fact that nearly one-third of the electricity-generating plants in the United States in 1983 were hydropower plants.

The Federal Power Commission has estimated that the potential energy to be derived from hydropower in the United States, if tapped in every possible location, is about triple current hydropower use. In principle, then, hydropower might be made to supply one-third to one-half of our electricity, if consumption were held near present levels. However, it is not clear that development on that scale will ever occur in practice. We have already noted in other chapters some of the problems posed by dam construction, including silting-up of reservoirs, habitat destruction, water loss by evaporation, and even earthquakes. Some sites with considerable power potential may not be available or appropriate for development. Other sites in which the water flows along fault zones may simply not be safe for dam construction. Many sites—in Alaska, for instance—are too remote from population centers to be practical unless power transmission efficiency is improved.

An alternative to the development of many new hydropower sites would be to add hydroelectric generating facilities to dams already in place for flood control, recreational uses, or other purposes. Although release of impounded water for power generation will alter streamflow patterns, it is likely to be less disruptive than either the original dam construction or the creation of new dam/reservoir systems.

Like geothermal power, conventional hydropower is limited by the stationary nature of the resource. There is the additional consideration that hydropower is somewhat more susceptible to natural interruptions than other sources so far considered. Just as 100-year floods are rare, so are prolonged droughts, but they do happen.

For various reasons, then, it is unlikely that many new hydropower plants will be developed. This clean, economical, renewable energy source can continue indefinitely to make a modest contribution to our energy use, but it cannot be expected to supply very much more energy in the future than it now does.

Tidal Power

All large bodies of standing water on the earth, including the oceans and large lakes like the Great Lakes, have tides. Unfortunately, the energy represented by tides is too dispersed, in most places, to be useful. The difference in mean high-tide and low-tide water levels on the average beach is about one meter. A commercial tidal-power electric generating plant requires at least 8 meters' difference between high and low tides for efficient generation of electricity, and a bay or inlet with a narrow opening that could be dammed to regulate water flow in and out. There are very few places in the world where the proper conditions are met. Tidal power is being used at a small plant at Passamaquoddy, Maine, and also at locations in France and Russia. Worldwide, the total potential of tidal power is estimated at only 2% of the energy potential of conventional hydropower, which is itself limited.

Wind Energy

The winds are ultimately powered by the sun, so in a sense, wind energy can be regarded as a derivative of solar energy. It is clean and, like sunshine, renewable indefinitely. Wind power has been utilized to some extent for more than 2000 years—the windmills of the Netherlands are probably the best known example. Today there is considerable interest in making more use of wind power. Presently, windmills are used for pumping groundwater and for the generation of electricity, usually on individual homesites. The cost of commercial wind power on a large scale is not now competitive with conventionally generated electricity, but that may simply be because windmills are not now mass produced and are therefore relatively expensive.

Wind energy shares certain limitations with solar energy. It is dispersed not in two but in three dimensions. Wind is erratic and highly variable in speed, both regionally and locally. The regional variations in potential power supply are even more significant than they might appear from comparing average wind velocities, because windmill power generation increases as the cube of wind speed. So, if wind velocity doubles, power output increases by a factor of 8 (2 x 2 x 2).

It is clear from figure 21.28 that most of the consistently windy places in the United States are rather far removed physically from most of the high-population and heavily-industrialized areas. Even where average wind velocities are great, strong winds do not

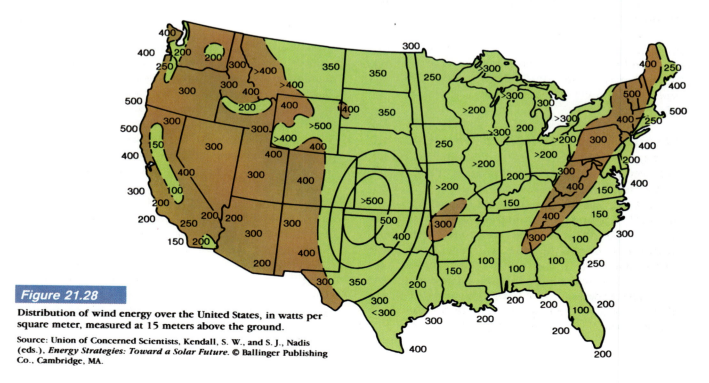

Figure 21.28

Distribution of wind energy over the United States, in watts per square meter, measured at 15 meters above the ground.

Source: Union of Concerned Scientists, Kendall, S. W., and S. J., Nadis (eds.), *Energy Strategies: Toward a Solar Future.* © Ballinger Publishing Co., Cambridge, MA.

that could be committed to windmill arrays, and the distance the electricity could be transmitted without excessive loss in the power grid. It would require about 1000 1-megawatt windmills to generate as much power as a moderately large conventional electric power plant. The windmills would have to be spread out so as not to block each other's wind flow. However, the land would not have to be devoted exclusively to windmills. Farming and grazing of livestock, activities that now go on in the Great Plains, could continue on the same land concurrently with power generation. The storage problem would remain.

Industrial and consulting firms have projected that the United States might produce 25% to 50% of its electricity from wind power by the year 2000, but only with a major national program for wind-power development. No governmental program of such scale is presently active or contemplated. In 1983, the 12 wind-powered electric generating facilities in the United States accounted for only 0.0014% of generating capacity.

Biomass

The term **biomass** technically refers to the total mass of all the organisms living on earth. In an energy context, the term comprises various ways of deriving energy from organisms or from their remains. Possible biomass fuels are many. Wood is a biomass fuel. There were 8 wood-fueled electricity-generating plants in the United States in 1983, with a collective capacity ten times that of the wind-powered plants; at least 25% of U.S. households burn some wood for heat. The production of alcohol from plant materials for use as fuel—either by itself or as an additive to gasoline—is another present use of biomass fuel. Sometimes using biomass energy sources means burning waste plant materials after a crop is harvested (see box 21.2). Certain plants manufacture flammable, hydrocarbon-rich fluids. The possiblity of raising fuel on farms is thus being evaluated. Some microorganisms produce oil-like substances also; they could be cultivated on ponds and the oil periodically skimmed off. Such possibilities represent potential future options.

A biomass fuel increasing in acceptance could be called "gas from garbage." Organic wastes, when decomposed in the absence of oxygen, yield a variety of gaseous products, among them methane (CH_4). The decay of organic wastes in sanitary landfill operations makes these suitable sites for production of methane. Straight landfill gas is too full of impurities to use alone, but landfill-derived gas can be blended with purer, pipelined natural gas to extend the gas supply.

Modern windmill array for power generation, Altamont Hills, California.

Photo courtesy AP/Wide World Photos.

blow constantly. This presents a storage problem that has yet to be solved satisfactorily. In the near future, wind-generated electricity might most effectively be used to supplement conventionally generated power when wind conditions are favorable.

Most schemes for commercial wind-power generation of electricity involve *wind farms*, concentrations of many windmills in a few especially favorable sites (figure 21.29). The Great Plains are the most promising area for inital efforts on a large scale. The limits to wind-power use would then include the area

The city of Los Angeles is among several now making some use of landfill gas. Methane can also be produced by decay of manures, and on some livestock feedlots this is proving to be a partial solution to energy-supply and waste-disposal problems simultaneously.

All biomass fuels are burned to release their energy. They are carbon-rich and thus share the carbon-dioxide pollution problem with the fossil fuels. However, they do have the very attractive feature of being renewable on a human time scale, unlike the fossil fuels.

Box 21.2

New Energy Diet for a Small User: Hawaii

It may be some time before adequate alternatives to the fossil fuels are developed for areas where energy use is intensive, such as cities of the northern United States dominated by heavy industry. Elsewhere, however, a judicious mix of renewable energy sources can serve as the major source of energy. The Hawaiian islands have made great strides in that direction.

The efforts have been motivated in part by the realities of geology. The islands have no fossil-fuel resources; energy costs are high and imported supplies subject to disruption. Geology has also contributed to the development of alternatives. The hot spot beneath the island of Hawaii, and the associated volcanism, make the area a promising one for the generation of geothermal power. A pilot geothermal-electricity plant now taps some of the hottest geothermal steam in the world (360° C, or 675° F). The total potential of the site is estimated at more than five times that island's current electricity needs. (A drawback is the ever-present threat of damage from earthquakes or volcanic eruptions, but the plant has been engineered to minimize potential damage.)

On Oahu, where most of Hawaii's people live, the potential for power generated from the strong, consistent trade winds is substantial. Wind power may supply 10% of the state's energy within a decade. And about one-third of the state's electricity is already being provided by biomass—in this case, through the burning of plant wastes from sugar-cane processing, wastes that were previously dumped into the sea or into landfills. Expanded use of alternative energy sources could make Hawaii energy self-sufficient by the year 2000.

Summary

At present, the world's major energy sources are the fossil fuels, formed from the remains of ancient plants and animals, modified by heat and pressure through time in the earth. All of the fossil fuels are nonrenewable energy sources. The supplies of petroleum and conventional natural gas deposits may be exhausted within decades. Coal resources could potentially last for centuries, but there are significant negative environmental impacts associated with extensive coal use, including acid rain and acid mine drainage, land disturbance by mining activities, and the pollution and disposal problems posed by coal ash. Oil shale deposits, containing the waxy hydrocarbon kerogen, are plentiful in the United States. However, the kerogen is thinly dispersed through the rocks, which would have to be strip-mined; oil shale also requires a great deal of water to process, and most U.S. deposits occur in dry areas. Domestic tar sand resources are negligible.

Conventional nuclear reactors consume the rare isotope U-235, the supply of which is severely limited. Significant expansion of the use of nuclear fission power would require the use of breeder reactors, and of fuel reprocessing to recover plutonium, which poses security problems. All fission reactors present waste-disposal problems and raise concerns about radiation safety. Fusion power, which would be far cleaner, is technologically unfeasible for the present.

The inexhaustible or renewable energy sources include solar energy, geothermal power, hydropower, tidal and wind energy, and various biomass fuels. The principal practical applications of solar energy are for space heating and the generation of electricity; there is a storage problem in both contexts, and it is not yet possible to generate solar electricity efficiently for energy-intensive applications. Present use of geothermal energy, whether for hot-water heat or steam-generated electricity, is limited to a few sites along modern plate boundaries with recent volcanism. In the future, it may be possible to develop hot-dry-rock geothermal areas, in intraplate settings with above-average heat flow. The thermal conductivity of rocks limits the lifetime of each individual geothermal field, typically to several decades. The potential for additional development of either conventional hydropower or tidal power is restricted by the number of suitable sites. Optimum sites for wind-energy development are mostly far removed from major population centers, and wind shares with sunlight the problem of variable power generation with changing weather. There are significant technological, supply, or environmental limitations associated with each of the various alternative energy sources. It seems probable that the energy-use picture in the future will not be dominated by a single energy source; rather, a blend of sources with different strengths and weaknesses is likely to be used.

Terms to Remember

biomass	fission	geopressurized zones	oil shale
breeder reactor	fossil fuels	geothermal	photovoltaic cells
chain reaction	fusion	hot dry rock	reservoir rocks
enhanced recovery	gasification	liquefaction	tar sand
		nonrenewable	traps

Questions for Review

1. What are the two basic initial requirements for forming a fossil-fuel deposit?

2. Briefly outline the process of petroleum formation and maturation.

3. Sketch or describe any two types of petroleum traps.

4. Approximately what proportion of estimated U.S. oil resources have been produced and consumed?

5. Describe any enhanced-recovery method for oil. What are the attractions of such methods?

6. What is geopressurized natural gas, and where is it found?

7. From what materials is coal formed? How does its quality change with progressive heating?

8. Why is there interest in coal gasification and liquefaction processes?

9. Name and describe one environmental problem associated with (a) coal mining; (b) coal use/burning.

10. What is oil shale? How is fuel produced from it?

11. Cite three environmental or developmental problems associated with both oil shale and tar sand.

12. Compare and contrast nuclear fission and fusion processes.

13. If fission power is to be used extensively in the future, breeder reactors will be needed. Explain.

14. The use of breeder reactors will, in turn, require extensive fuel reprocessing to recover plutonium. Name two concerns related to that reprocessing.

15. Cite two potential advantages of fusion power over fission. Why are fusion reactors not now in use?

16. For what two kinds of applications can solar energy make the most immediate contribution? What two principal limitations restrict the usefulness of solar energy for applications requiring a great deal of power, such as factories?

17. Briefly describe the nature of a natural geothermal system. Why are geothermal areas restricted geographically?

18. What are the principal applications of geothermal energy? What are two important limitations on its usefulness?

19. What is a "hot-dry-rock" geothermal resource, and how could it be used?

20. For what is conventional hydropower now used? Discuss the extent to which additional hydropower-generation sites might be developed.

21. What area(s) has (have) the greatest potential for wind-power development? Name at least two factors that now limit the potential contribution of wind power.

22. Describe any two biomass fuel sources. What drawback do they share with the fossil fuels? What advantage do they have?

For Further Thought

1. Increased concern about energy conservation has led many homeowners to insulate their homes more and more snugly, to minimize heat loss by the escape of warmed air. Investigate the problems of "indoor air pollution" that have developed as a result.

2. Compare the cost estimates for shale oil, petroleum from tar sand, and gas from coal gasification with contemporary prices for conventional oil and natural gas. Do this for several different times—for example, 1972, 1978, 1983, and 1986. Determine the extent of oil and gas exploration and of development of the alternative fuels at the same times, and see what patterns, if any, emerge.

3. What energy sources are used by your local electric utility? Has the company changed its fuels, power-plant types, or plans for additional power plants over the last two or three decades? If so, to what extent have (a) economic factors and (b) environmental considerations entered into the decisions?

Suggestions for Further Reading

Brookins, D. G. *Earth Resources, Energy, and the Environment*. Columbus, Ohio: Charles E. Merrill Publishing Co., 1981.

Burk, C. A., and Drake, C. L. *The Impact of the Geosciences on Critical Energy Resources*. Boulder, Colo.: Westview Press, 1978.

Dick, R. A., and Wimpfen, S. P. "Oil Mining." *Scientific American* 243 (April 1980): 182–88.

Energy. San Francisco: W.H. Freeman and Co. A collection of offprints from *Scientific American,* 1970–79.

Hoyle, F., and Hoyle, G. *Commonsense in Nuclear Energy*. San Francisco: W.H. Freeman and Co., 1980.

Hunt, J. M. *Petroleum Geochemistry and Geology*. San Francisco: W. H. Freeman and Co., 1979.

Hunt, V. D. *Handbook of Energy Technology*. New York: Van Nostrand Reinhold Co., 1982.

International Atomic Energy Agency, Vienna, Austria. This agency publishes a quarterly bulletin on many aspects of nuclear power and other applications of radioactive materials.

Kendall, H. W., and Nadis, S. J., eds. *Energy Strategies: Toward a Solar Future*. Cambridge, Mass.: Ballinger Publishing Co., 1980.

Menard, H. W. "Interdependence of Nations and the Influence of Resource Estimates on Government Policy." U.S. Geological Survey Professional Paper 1193, 1979.

National Academy of Sciences. *Energy in Transition, 1985–2010*. San Francisco: W.H. Freeman and Co., 1979.

National Geographic Society. *Special Report on Energy*. Washington, D.C., 1981.

Perlman, E. "Kilowatts in Paradise." *Science 82* 2 (January/February 1982): 78–81.

Ruedisili, L. C., and Firebaugh, M. W., eds. *Perspectives on Energy* 3d ed. New York: Oxford University Press, 1982.

Sivard, R. L. *World Energy Survey*. 2d ed. Leesburg, Va.: World Priorities, 1981.

U.S. Department of Energy, Energy Information Administration. Publishes a wide variety of energy-related data. Sources used in this chapter include the following: "Annual Energy Outlook 1983" (published in 1984), "Annual Energy Review 1983" (1984), "Coal Data Book" (1980), "Commercial Nuclear Power: Prospects for the U.S. and the World" (1983), "Inventory of Power Plants in the U.S. 1983" (1984), "Nuclear Plant Cancellations: Causes, Costs, and Consequences" (1983), "United States Uranium Mining and Milling Industry" (1984), and "U.S. Crude Oil, Natural Gas, and Natural Gas Liquids Reserves" (1983).

Watson, J. *Geology and Man.* Boston: George Allen and Unwin, 1983.

Whitmore, E. C., Jr., and Williams, M. E., eds. *Resources for the Twenty-First Century.* U.S. Geological Survey Professional Paper 1193, 1982.

Chapter 22

Planets and Other Strangers: The Solar System

Introduction

The study of the solar system beyond the Earth was once the sole province of astronomers. As information about other planets and moons was gathered, it became increasingly apparent that the features of many other solar-system objects might have terrestrial analogues. Many of the same internal and surface processes recognized on Earth might act on other bodies. Indeed, the resulting surface features might be better exposed and preserved on bodies lacking the water, vegetation, and geologic activity that obscures the face of the Earth. Also, the Earth and other planets share a common origin. By developing an understanding of the similarities and differences between the Earth and other bodies in the solar system, and the reasons for them, we improve our understanding of the Earth. So planetology has moved within the scope of geology.

The nebular model for the origin of the solar system was reviewed briefly in chapter 1. You may recall that a consequence of that model is that the planets that were formed should vary in composition as a function of their distance from the Sun at the time of formation. The differences in composition have been confirmed by modern data. The planets' subsequent development has also varied, partly as a result of those compositional differences and partly as a result of size differences. In this chapter, we will first briefly survey the basic characteristics of each planet, in order outward from the Sun. Later sections will examine Earth's Moon and several other kinds of solar-system objects, including the planetlike asteroids, comets, and meteorites.

Overview of the Planets

The study of the planets requires some different tools from the study of terrestrial features, but although the other planets are remote, we do have a variety of sources of information about them. Information about mass and density is derived, using principles of physics, from orbital characteristics and the calculated gravitational interactions between planets. Passing spacecraft can confirm mass data for planets and moons, measure planetary magnetic fields, and photograph the surfaces of objects. The light spectra reflected from the surfaces of planets, whether viewed through telescopes or via spacecraft, can be used to determine surface compositions, much as starlight is used to deduce stellar compositions. Those planets having solid surfaces and not-too-severe surface conditions can be probed and sampled by spacecraft that land on their surfaces. These landers can also collect seismic data and information on planetary atmospheres.

According to the nebular model, the planets closest to the Sun should consist predominantly of materials that can condense at relatively high temperatures. Those farther out should contain higher proportions of lower-temperature materials. The accretion temperatures of nebular materials are correlated somewhat with density. That is, the high-temperature iron-nickel alloy that would have been expected to condense early from the solar nebula is also very dense. Silicates, especially hydrous silicates, are less so, and for the most part also would have condensed at lower temperatures. Ices of water, ammonia, and methane, which require very low

Table 22.1

Basic data on the planets

Planet	Mean distance from Sun*	Equatorial radius (km)	Mass relative to Earth†	Density (g/cm³)	Number of moons	Mean surface temperature (°C)
Mercury	58	2440	.0558	5.42	0	350/−170‡
Venus	108	6050	.8150	5.25	0	−33
Earth	150	6380	1.000	5.52	1	22
Mars	228	3400	.107	3.94	2	−23
Jupiter	778	71,400	317.9	1.31	16	−150
Saturn	1425	60,330	95.15	0.69	17	−180
Uranus	2870	25,400	14.54	(1.19)§	14	−210
Neptune	4490	24,300	17.23	1.66	2	−220
Pluto	5900	(1,500)§	.0022	(0.9)§	1	−230

*In millions of kilometers
†(mass of planet)/(mass of Earth)
‡Two surface temperatures are given for Mercury because its rotation is so slow that the same side faces the sun for prolonged periods, and temperature
 differences between the day and night side are extreme.
§Numbers are uncertain.

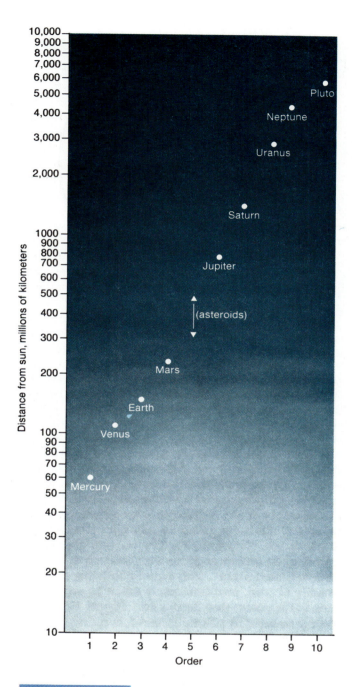

As this graph shows, the spacing of the planets' orbits exhibits a geometric regularity.

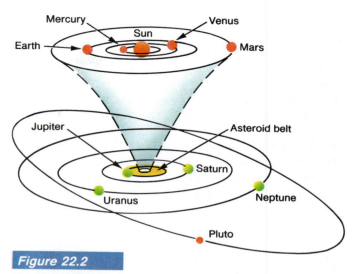

The orbits of the planets, approximately to scale. Note that most lie in the same plane.

proportions of silicate material, with some iron-nickel, and are rather similar in size to Earth. This group includes Mercury, Venus, and Mars, along with Earth. Farther out lie the **Jovian planets,** the Jupiter-like ones, much larger, and less dense because they contain large quantities of gas. The Jovian group includes Jupiter, Saturn, Uranus, and Neptune. Little is known of small, remote Pluto. What little is known indicates that it does not fit well with either group.

With the exception of Pluto, most of the planets' orbits lie in or near the same plane, the **ecliptic** plane, presumably the plane of the rotating disk of material from which the planets condensed. There is a striking regularity in their spacing: if plotted on a logarithmic scale, they fall in an evenly spaced sequence, especially if the asteroid belt is included collectively as a "planet" at its average position between Mars and Jupiter (figure 22.1). Each planet's orbital radius is about 75% of the orbital radius of the next planet outward from it (figure 22.2). It was once believed that this was a primary feature dating from the formation of the solar system. However, analysis of the mechanics of planetary orbits suggests that over long periods of time, gravitational interactions among the planets will cause mutual orbital modifications tending toward this kind of regular spacing.

Mercury

Prior to the Mariner 10 flyby, very little was known about Mercury. Named for the winged-footed messenger of the gods, the planet Mercury travels briskly around the sun, circling it once every 88 days. As the

temperatures to condense, are also quite low in density. Therefore one might expect a decrease in planet densities outward from the Sun, and indeed this seems generally to be true (table 22.1).

It can also be seen from the table that, in terms of density and size, most of the planets fall into two distinct groups. The **terrestrial planets** are the Earthlike ones, in the sense that they contain large

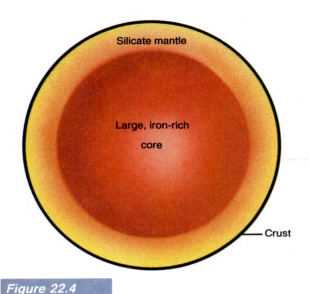

Figure 22.4

Cross section of Mercury consistent with density constraints.

Figure 22.3

The cratered surface of Mercury.
© NSSDC/Goddard Space Flight Center

planet closest to the sun, Mercury is the most difficult to observe astronomically: it is never very far from the sun in the sky, and therefore is most often visible in the day rather than the night sky, when the reflected light from the planet's surface is far outshone by the sun.

Composition

One known quantity is Mercury's density. Its average density is very close to that of the Earth. But Mercury is a far smaller planet, with a volume about one-twentieth that of Earth, meaning that its interior is far

less compressed than that of Earth. That in turn implies that if the densities are corrected for compressional effects, Mercury's density is really much greater than Earth's, indicating a much higher proportion of iron.

Before Mariner 10, it was not known whether Mercury was a uniform agglomeration of metal and silicate grains preserved unchanged since its accretion, or a differentiated body like the Earth. Mariner photographs of the surface showed a striking resemblance to the lunar surface, a rocky one pockmarked with craters (figure 22.3). The relative absence of metal at the surface suggests that Mercury is indeed a differentiated planet, with a core consisting predominantly of iron, proportionately much larger in relation to the planet than Earth's core (figure 22.4). The core is mantled by silicate material, broadly similar in composition to Earth's mantle.

Surface Processes and History

Surface processes on Mercury are sharply different from those of the other terrestrial planets. There is no surface water, and no evidence of features shaped by surface water in the past, such as stream channels or obviously water-laid sediments. There is virtually no atmosphere, except for a few stray atoms of hydrogen and helium caught by Mercury's weak gravitational field as they have streamed outward from the Sun; there are no eolian landforms. The only significant modification of Mercury's surface now is the result of occasional meteorite impacts.

The history of Mercury cannot be known in any detail without direct sampling of the rocks. Still, flyby observations provide some information. There is evidence that some of the flatter plains on the surface originated through volcanism, as flood basalts. This must have been quite early in the planet's history, for two reasons. One, even the volcanic plains show some cratering, and lunar studies indicate that meteorite impacts in the solar system have been rare over the last several billion years. Two, such a small planet has quite a high surface-to-volume ratio, and can thus cool rapidly after differentiation. Once it has cooled to the point that a thick lithosphere has formed, it is impossible for magma to force its way up to the surface. Some residual evidence of the early cooling exists, in the form of long, high, arcuate scarps with as much as 3 kilometers (2 miles) of vertical relief, some of which show evidence of thrust-faulting, or compression. These scarps are believed to have formed as a result of differential contraction during the cooling of the planet.

Magnetic Field

One surprising feature of Mercury is that it has been found to have a (weak) magnetic field, about 1% the strength of Earth's field. How it is produced is something of a mystery. It is not clear that Mercury's core could still be extensively molten, which would be necessary if Mercury's magnetic field is produced in the same way that the Earth's is believed to be. Furthermore, Mercury's rotation on its axis is extremely slow—it rotates only once every 59 Earth days—and therefore it is difficult to imagine vigorous motion in that core, even if it is fluid. An alternative possibility is that Mercury simply preserves a *remanent magnetization,* or paleomagnetism, imposed on it very early in solar-system history, when the Sun's magnetic field would have been much stronger. But during later differentiation to form the iron core, Mercury's interior would surely have been heated above the Curie point of iron, destroying that magnetism.

Venus

Mercury was long concealed from us by its proximity to the Sun; Venus is shrouded in clouds. Consequently, even flyby missions provide limited information about this nearest-neighbor planet to the Earth.

Composition

In some ways, Venus can be regarded as Earth's twin. It is nearly as large, and very similar in density. Given these facts, and the closeness of Venus to the Earth (which implies accretion at comparable temperatures, which ought in turn to result in a similar bulk composition), Venus's interior is believed to resemble that of Earth, with a proportionately sized iron-rich core, mantle, and thin crust. This assumes that the planet's interior has been differentiated. This is likely because of the broad geologic/chemical similarities to the (differentiated) Earth. Also, limited measurements by U.S. and Soviet spacecraft suggest the presence of granitic, granodioritic, and basaltic rocks at the surface, implying a comparable crustal diversity to that of Earth.

But if Venus has a core like Earth's, we are faced with the problem of explaining why it has no significant magnetic field. Two obvious possibilities are (1) that the core is solid, or (2) that it is circulating too slowly. It is difficult, in the former case, to explain why Venus's core should be solid when the Earth's is not. The latter explanation may be the more plausible, in that Venus rotates extremely slowly (one rotation every 243 terrestrial days).

An unusual feature of the rotation of Venus is that its motion is *retrograde,* opposite in direction from the direction of its revolution around the sun. All other planets except Uranus rotate and revolve in the same direction. No satisfactory explanation for the retrograde motion has been advanced.

Given the dense cloud cover, most of the information we have about the nature of the Venusian surface is obtained remotely, through radar mapping (figure 22.5). Most of the surface has now been mapped topographically. Two-thirds of the surface consists of rolling plains; areas of extreme relief are rare. The highest of the highlands, or mountains, rise about 11 kilometers (7 miles) above the plains. The surface is rubbly, again a probable result of meteorite impact.

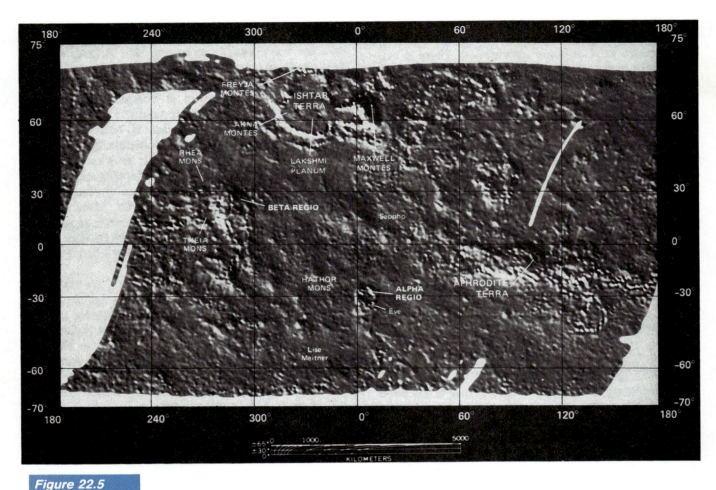

Figure 22.5

Radar relief map of the surface of Venus.

© NSSDC/Goddard Space Flight Center

Atmosphere

The way in which Venus differs most markedly from the Earth is in its atmosphere. That atmosphere is 90 times as dense as the Earth's. Venus's atmosphere is also 95% carbon dioxide, in contrast to Earth's atmosphere, which is about 80% nitrogen, 20% oxygen. That much CO_2 leads to intense greenhouse-effect heating: the surface temperature averages about 475° C, or about 900° F (hot enough to melt lead). This is so even though the atmosphere is thick with clouds, which reflect much of the incident sunlight back into space.

Interestingly, the amount of CO_2 in Venus's atmosphere is comparable to the amount of CO_2 represented by the carbonate rocks of the Earth's crust. The surface temperatures of Venus are hot enough that the carbonate minerals are unstable, so the CO_2 remains in the atmosphere. However, this is a bit of a chicken-and-egg problem: the surface is so hot only because the CO_2 is in the atmosphere rather than in the rocks, and why is the same not true on Earth?

Another significant difference in the atmospheres of Earth and Venus is that Venus's atmosphere contains high concentrations of sulfuric acid, as fine droplets. This naturally makes the atmosphere highly corrosive, and one would expect the weathering of surface rocks to be correspondingly intense. There is no liquid water at the surface of Venus; again, the high surface temperatures keep what little water is present gaseous, in the atmosphere. The atmosphere does rotate very rapidly with respect to the planet, and this coupled with the high atmospheric density gives rise to very strong winds. Wind erosion and transport of surficial materials should be much more significant on the surface of Venus than on any of the other terrestrial planets, including the Earth.

A

B

Mars

Mars, the "red planet," derives its nickname from the red color of its regolith, attributed to iron oxides and rusty clays (figure 22.6). The planet has been observed in greater detail than Mercury and Venus, and the observations have revealed a greater diversity of features. The nature of the Martian interior is still not well known. The density of Mars is somewhat less than that of the Earth and Venus, so it must contain proportionately less iron than these planets. The existence of volcanoes and the chemistry of the surface rocks suggests that some differentiation has occurred, and other physical data indicate a concentration of mass toward the planet's center, consistent with the existence of a distinct core. How large the core may be depends on whether it consists mainly of iron-nickel alloy, or of less-dense iron sulfide. Mars lacks a significant magnetic field, implying little modern circulation in the core.

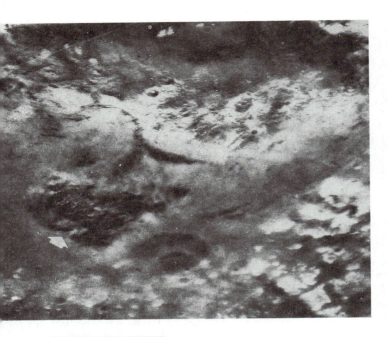

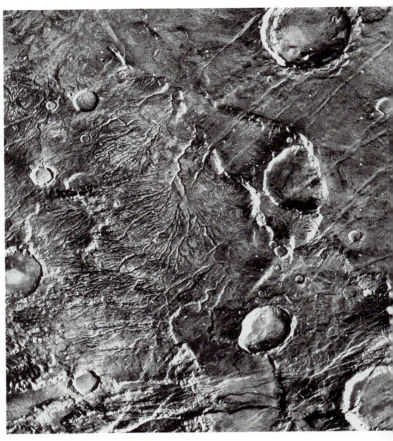

A

Figure 22.9

Possible water-sculpted landforms on Mars. *(A)* Stream channels in dendritic pattern. *(B)* Teardrop shapes might be formed by water sweeping around obstacles.

Martian Atmosphere and Surface Features

The atmosphere of Mars is compositionally similar to that of Venus, dominated by CO_2, but the Martian atmosphere is far less dense. Atmospheric pressure on Mars is only about 1% of the pressure of the terrestrial atmosphere. Nevertheless, vigorous winds can sweep across the planet's surface, raising dust in storms (figure 22.7). Recognizable eolian depositional features include massive dune fields (figure 22.8).

There is currently no surface water on Mars. However, there are surface features that appear to have been formed by running water, such as streamlike channels (figure 22.9). This suggests that water did exist on the Martian surface at one time; the puzzling question is, how? A denser atmosphere would have been required to keep the water from vaporizing too readily, and higher temperatures would also have been

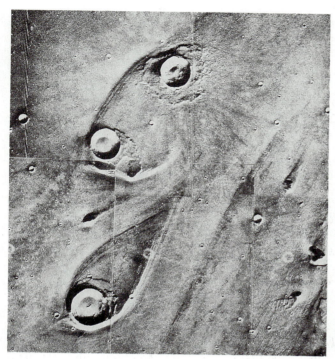

B

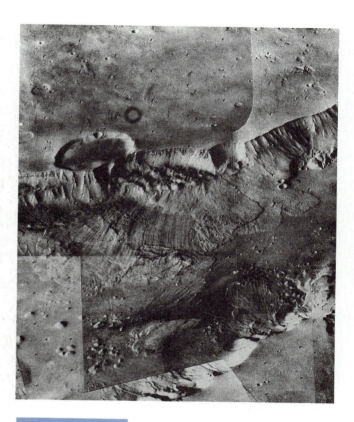

needed, so that any condensed H_2O would have been water rather than ice. It is clear, in short, that at one time, climatic conditions on Mars were very different from what they now are. It is difficult to estimate the age of the stream channels, especially in the absence of returned samples, but the extent of cratering and other modifications of at least some of the channels suggests that they are quite ancient. Some of the eolian features may also have formed under the influence of an earlier, denser atmosphere.

Mass wasting can be observed on the Martian surface (figure 22.10). The slumps and landslides may result in part from slope instability caused by wind erosion or faulting. Also, Mars is believed to have permafrost below the surface, and localized melting of that permafrost could cause slope failure and solifluction.

One of the more striking Martian surface features, observable even with Earth-based telescopes, is the presence of polar ice caps (figure 22.11). The thinness of the atmosphere and lack of circulating ocean helps to account for the maintenance of strong temperature gradients between the poles and equator, and extreme cold at the poles. In winter, polar temperatures may drop to $-125°$ C (about $-200°$ F). This is so cold that atmospheric carbon dioxide condenses as dry ice at the poles; CO_2, not water ice, is the major constituent of the Martian polar caps. Their extent varies seasonally. In the warmer summer, the

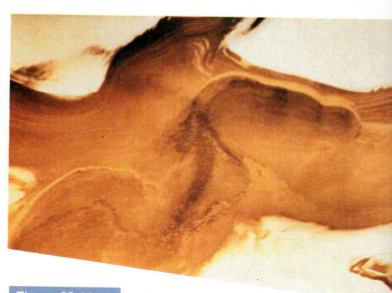

north polar cap vaporizes completely. A small residual cap sometimes persists at the south pole through the summer. This residual ice cap may consist of water ice, which is stable to temperatures about 40° C (70° F) warmer than CO_2 ice.

Martian Tectonism?

Much of the Martian surface, like that of Mercury, Venus, and Earth's Moon, is heavily cratered. However, there are broad, relatively featureless areas that cannot be explained entirely as a result of the smoothing effects of erosion. On the basis of relative crater density, it can be deduced that volcanic rocks of many ages are present. Mars may still be tectonically active, although not to the extent that the Earth is.

Past volcanic activity has produced some spectacular features. Prominent among them is *Olympus Mons,* a huge shield volcano that dwarfs Mauna Loa (figure 22.12). Olympus Mons is 600 kilometers (375 miles) across, and rises 25 kilometers above the adjacent plain. (To put its size in perspective, a volcano that size would just about cover the entire state of

Colorado—or, alternatively, New York, most of Pennsylvania, and all of New England except Maine!) Summit calderas have formed by collapse after eruptive phases. The lack of a depression around the massive edifice of Olympus Mons suggests that by the time the volcano formed, the Martian lithosphere was sufficiently thick and strong to support its weight, rather than sinking in isostatic compensation. The existence of a thick modern lithosphere is consistent with limited tectonic activity at present.

Life on Mars?

Early observations of the Martian surface, together with the fact that its climate, while colder than Earth's, is not prohibitively so over the whole planet, have long prompted speculation that life might exist on Mars. The channels were originally attributed to present surface water. Seasonal color changes were optimistically interpreted as resulting from seasonal plant growth. Now we know that there is no surface water; the color changes are believed to result from redistribution of light and dark dust by winds, and some seasonal frost. Nevertheless, it is not impossible that microorganisms at least might exist on Mars. Several of the Viking lander experiments were designed to test this possibility.

In one experiment, samples of soils were heated, little by little, to 500° C, and the compounds released were analyzed by very sensitive instruments. No organic compounds were found. This was surprising, since organic compounds, including amino acids (the building blocks of proteins), have been found in meteorite samples, and over time meteorites would certainly have fallen on Mars, even if there were no indigenous organic compounds.

If microscopic plants were present, and if they manufactured food by photosynthesis, they would very likely consume carbon monoxide (CO) or CO_2, just as terrestrial plants use CO_2. Another experiment indicated uptake of CO or CO_2 by a Martian soil sample, but the results were not reproducible and may have been due to the effects of some unusual iron compound in the soil rather than to life forms.

In still another experiment, Martian surface samples were "fed" a rich nutrient mixture, to see if anything would consume the food and produce detectable waste products. Some of the nutrients were broken down early in the experiment, but then reactions stopped and the further addition of nutrients had no effect. (With terrestrial microorganisms, growth would continue with the addition of more nutrients.) Again, unexpected chemical reactions with the Martian soil probably account for the experimental results.

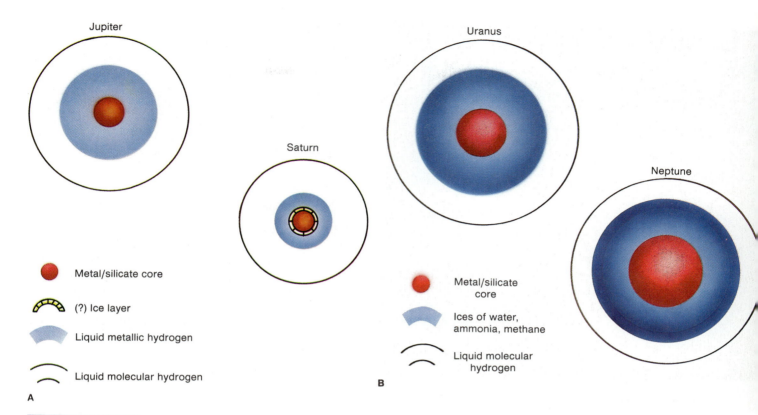

Jupiter

Saturn

Uranus

Neptune

■ Metal/silicate core

▨ (?) Ice layer

▨ Liquid metallic hydrogen

⌒ Liquid molecular hydrogen

A

■ Metal/silicate core

▨ Ices of water, ammonia, methane

⌒ Liquid molecular hydrogen

B

Figure 22.13

Cross sections of the Jovian planets. *(A)* Jupiter and Saturn have both liquid metallic hydrogen and liquid molecular hydrogen zones. *(B)* Uranus and Neptune lack the metallic hydrogen and have icy mantles.

In short, no clear evidence for life on Mars has been found. One remaining possibility would be to check the polar regions. There considerable water is trapped as ice, so the environment is somewhat different from that over most of the planet, and from the previous Viking lander sites.

The Outer Planets

The Jovian Planets

The next four planets outward from Mars present a marked compositional contrast to the terrestrial planets. They consist predominantly of fluids, rather than solid silicates and metal. Their compositions more closely approach that of the primordial solar nebula. This is partly a consequence of their great size and correspondingly strong gravity. They range from 15 to over 300 times the mass of the Earth, and therefore have sufficiently strong gravitational fields to retain even the lightest gases, hydrogen and helium. Another factor is the very low temperatures pre-

vailing at such vast distances from the Sun: not only are water and CO_2 ice condensed, but even methane, ammonia, and other substances that are gaseous at Earth temperatures solidify into ices on the Jovian planets.

These planets are believed to have concentric compositional zones, but the zones do not resemble the corresponding ones of the Earth (figure 22.13). All of the Jovian planets may have small cores of metal and silicate. In the larger planets, Jupiter and Saturn, the core is probably mantled by liquid metallic hydrogen, which is in turn surrounded by a thick layer of liquid molecular hydrogen. The smaller planets Uranus and Neptune are believed to have mantles of ices of water, methane, and ammonia, surrounded by liquid molecular hydrogen. In all cases, the liquid hydrogen surface layers grade outward, with decreasing density, into a hydrogen-rich atmosphere. The planetary surfaces are thus less well defined than are the surfaces of rocky planets. They also naturally lack the kinds of surface features found on the terrestrial planets (craters, volcanoes, channels, and so on); there is no firm surface to preserve such physical features.

In the case of Jupiter and Saturn, the surfaces are further obscured by thick clouds in the atmospheres. The structure of the atmosphere of Jupiter at least is believed to be known in some detail. Above a foggy

A

B

layer of water-ammonia droplets is a layer of water snowflakes, overlain by ammonia snowflakes. The atmosphere is convecting, with light zones of warmer, rising gas alternating with dark belts of colder, sinking gas. The varied colors of the cloud bands (figure 22.14) are attributed to small amounts of sulfur, phosphorous, or organic molecules. The existence and longevity of the Great Red Spot (figure 22.14B) have not been satisfactorily explained, given the apparent extent of atmospheric mixing and the fact that there is no fixed, solid surface below.

One notable characteristic of the Jovian planets is that, to the extent that we can make appropriate measurements, we find that they are radiating more heat than they receive from the Sun. This heat is believed to be left over from the planets' early history, from heating of their interiors during compression by gravity and from energy released as the metallic and rock components fell in to form the planets' cores. They seem cold when one considers the water and ammonia snowflakes, but the temperatures of the deep interiors are extremely high—ranging from about 6750° C (about 12,200° F) for Neptune to about 25,000° C (45,000° F) for Jupiter.

Diversity in Satellites: The Galilean Moons of Jupiter

Recall from table 22.1 that each of the Jovian planets has several satellites, or moons. Though sometimes dwarfed by the planet around which they orbit (figure 22.15), the largest of these moons are themselves the size of planets. Thus these giant planets and their encircling moons are really much like solar systems in miniature. Even around a single planet, the various moons may show considerable variation in density, chemistry, and extent of present activity. The four largest moons of Jupiter illustrate this well (table 22.2).

These four moons are also known as the *Galilean* moons, for they were first observed and reported by Galileo in the year 1610. They range from Pluto-sized to Mercury-sized. All have much lower densities than the Earth, suggesting that they lack significant metallic iron. However, Io and Europa at least have densities appropriate to rocky planets, while the densities of Ganymede and Callisto suggest a mix of rock and ice in comparable proportions. Little else was known about them before the Voyager flybys produced high-quality pictures.

© NSSDC/Goddard Space Flight Center

Figure 22.15

Jupiter dwarfs even its Galilean moons.

Table 22.2

Properties of the Galilean moons of Jupiter

Moon	Radius (km)	Radius/ (radius of Earth)	Density (g/cm³)	Mean distance from Jupiter (km)
Callisto	2410	0.378	1.79	1,809,000
Ganymede	2640	0.414	1.93	999,000
Europa	1560	0.244	3.03	600,000
Io	1820	0.285	3.53	350,000

Perhaps the least surprising was Callisto's appearance (figure 22.16). Pocked with craters, like many planets and moons that lack protective atmospheres, Callisto is a cold, dead world. Callisto's ice and rock crust overlies a water or water-ice mantle and a rocky core. Fresh craters punch holes in the crust to reveal fresh ice; older craters may be slowly reduced in size through plastic flow in the ice. The absence of high mountains supports the theory that Callisto has a weak, ice-rich crust.

Ganymede, similar in density to Callisto, was expected to resemble it closely. In overall structure, the two moons are probably similar. The surface of Ganymede, however, is much more varied in appearance (figure 22.17A). There are some dark, ancient-looking, cratered areas. There are also complex ridged or grooved terranes, with relief similar to that of the Appalachian Mountains (figure 22.17B). Indeed, these

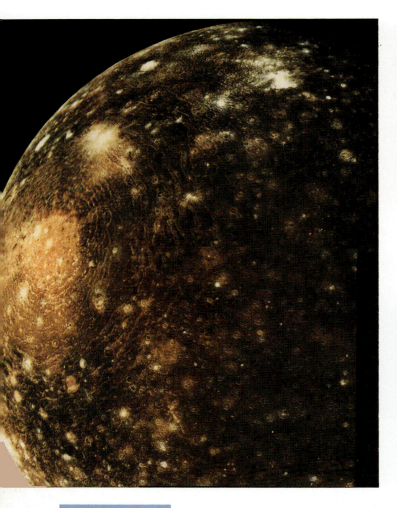

are interpreted as mountain ridges. They suggest that in the past, temperatures were sufficiently high to initiate flow in the mantle of Ganymede, deforming the crust. The periods of deformational activity must have spanned long periods of time: there are intersecting sets of ridges obviously formed at different times, and ridges cutting and cut by craters.

Europa (figure 22.18) is different yet. This bright planet has an icy crust, possibly floating on a thin, watery mantle, and a large rocky core (to account for its density). The surface is crisscrossed with complex patterns of dark streaks that resemble fracture patterns. Yet these "rifts" do not show central depressions like rift valleys; there is very little surface relief on the planet. In addition to the dark cracklike markings, Europa has arcuate bright ridges that make scalloped patterns on the surface (look near the horizon on figure 22.18). The similarity in shape of some of Europa's surface features to spreading ridges, transform faults, island arcs, and the like has inspired some

A

B

Figure 22.17

Ganymede. (A) Panoramic view showing surface variations. (B) Closeup of the grooved terrane that is interpreted as consisting of mountain ridges.

A

Figure 22.18

Europa's surface resembles a cracked eggshell.

© NSSDC/Goddard Space Flight Center

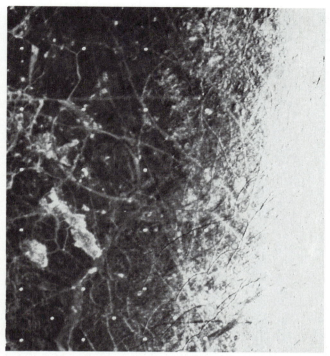

B

speculation that processes analogous to plate tectonics may exist on Europa, with a watery slush from its mantle rising to fill fractures formed in the brittle, icy crust. Such an interpretation is still tentative.

Perhaps the biggest surprise of the Galilean moons was Io. Its size and density are comparable to those of the Earth's Moon, and it must consist mostly of silicate material. There the similarity ends. Io is a brilliantly colored body of yellows, reds, and browns (figure 22.19). It also happens to be the most geologically active body found in the solar system. Voyager 2 images captured volcanic eruptions in progress (figure 22.20A). Volcanic features are abundant on the surface (figure 22.20B), while craters are nearly absent, confirming the vigorous reworking of Io's surface.

The dramatic colors suggest that the composition of Io's volcanoes differs considerably from terrestrial analogues. In fact the surface and erupted material appear to consist predominantly of sulfur,

Figure 22.19

The colorful moon Io owes its color mainly to sulfur.

© NSSDC/Goddard Space Flight Center

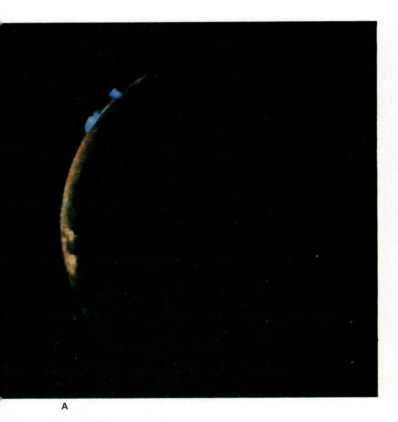

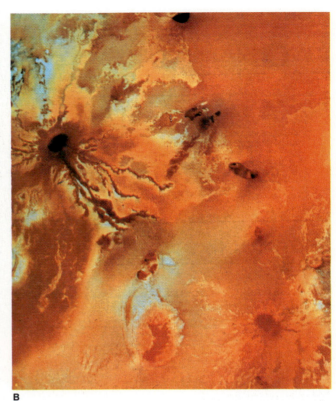

A

B

Figure 22.20

Volcanism on Io. *(A)* Eruption in profile captured by Voyager.
(B) Lavalike flows of sulfur surround a volcanic vent.

© NSSDC/Goddard Space Flight Center

which can vary greatly in color with differences in temperature and crystal structure. Apparently the rocky core is surrounded by a zone of molten sulfur, topped with a sulfur-rich crust. The heat required to maintain that molten sulfur layer is thought to result from the strong tidal stresses of giant Jupiter acting on Io, closest of the Galilean moons. Tidal stresses may also weaken the crust, helping to promote the eruption of the molten sulfur from below.

Rings

The beautiful rings of Saturn (figure 22.21) were also observed by Galileo. With the Voyager missions, it has been revealed that Jupiter (figure 22.22) and Uranus have rings too, though less spectacular ones. Neptune may likewise have faint rings, but observations to date have been made from such great distances that any rings have not been detected.

There are some differences in character among these three ring systems. The narrow bands of Uranus's rings are made up principally of small rocky chunks, and the rings are individually rather narrow.

The rings of Jupiter are also made of rocky particles, but for the most part these particles are very tiny, and the rings are much more tenuous. The broad, much brighter rings of Saturn probably contain some rocky fragments, but they also contain abundant chunks of water ice. The particles in the principal rings of Saturn range in size from about 10 meters downward. The structure of Saturn's ring system, particularly, is very complex. More than a hundred distinct rings have now been identified. Collisions among particles must be frequent, and tend to break up the fragments. Within the ring system are several "moonlets" that may serve a shepherding function, maintaining the sharply defined boundaries of the rings.

How do rings arise? Broadly, there are two classes of theories.

As the gaseous planets contracted gravitationally, during the early history of the solar system, they would have rotated faster, and may have flung off a gaseous, rotating disk. Condensation of fragments from this gas, followed by collisions between fragments, could gradually have produced a flattened disk of particles.

Figure 22.21

The rings of Saturn.

© NSSDC/Goddard Space Flight Center

Figure 22.22

Jupiter's rings are much thinner and less spectacular.

© NSSDC/Goddard Space Flight Center

Another possibility is related to the planetary phenomenon known as the **Roche limit.** For each planet, there is a distance inside which any sizeable moon will tend to be torn apart by tidal stresses. This is the Roche limit. Perhaps one or more moons, formed or captured inside the Roche limit, disintegrated under stress, and over time the fragments were distributed into uniform rings. Some support for this idea comes from the observation that all the ring systems of the Jovian planets lie *inside* their systems of significant satellites.

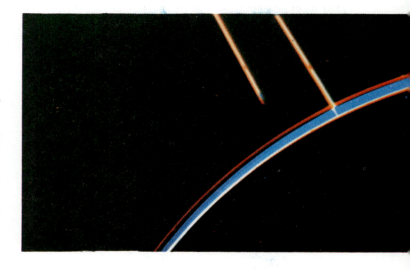

The Roche-limit phenomenon may also explain why small terrestrial planets, such as Mercury, do not have moons. For small planets, the *capture radius,* or proximity of approach required for a planet to hold a moon, is inside the Roche limit. Any satellite close enough to be held as a moon would be so close that it would be torn apart.

Pluto and the Lesser Planetary Objects

Pluto

The odd member of the outer satellite group is Pluto. Its remoteness means that it is extremely difficult to observe or measure, and therefore very little is known about it. It seems to be icy, rather than gaseous; it is by far the smallest of the planets. Its orbit is somewhat peculiar, for it is significantly inclined to the ecliptic plane in which most of the planets' orbits lie. That orbit is also so elliptical that at some points Pluto's orbit cuts inside that of Neptune, although on average Pluto is much farther from the Sun. Pluto's diminutive size and eccentric orbit have led to the suggestion that it is not a proper major planet at all, and it should instead be classed as another minor solar-system object, like the asteroids. It has even been suggested that it did not form in its present location, but may have been deflected from elsewhere in the solar system—perhaps under the influence of a large planet like Jupiter—to settle into its present orbit.

Asteroids

The **asteroid belt** lies between Mars and Jupiter, thus separating the terrestrial from the gaseous planets. It spans a broad area, with most asteroids orbiting between 300,000,000 and 500,000,000 kilometers from the Sun. The **asteroids** derive their name from the Greek *asteroides,* meaning "starlike," for through a telescope they appear as small points of light. In reality they are rocky objects, their light being only reflected sunlight.

Asteroids are distinguished from the planets simply on the basis of size. The largest of the asteroids, Ceres, has a radius of only 487 kilometers, less than one-third that of Earth's Moon, and a mass about 0.02% that of the Earth. A precise count of the number of asteroids is impossible—there are too many, and many are too small to distinguish telescopically. Current estimates put the number of asteroids with diameters over 1 kilometer at 400,000 to 500,000, with many more smaller asteroids. Through time, too, collisions among these many objects break up larger ones into smaller ones, so that the numbers and sizes are continually changing.

When the first asteroids were discovered, they were named for lesser gods or other mythological figures—Ceres, Vesta, Pallas, and so on. As telescopes improved, these names were quickly used up. Now, as a new asteroid's identification is confirmed (when its orbit is determined, and it is sighted on two successive orbits), it is designated simply by a number, indicating the order of its discovery. A name—nowadays, commonly the name of the astronomer, a friend or relative, or a public figure—may be appended. So, for example, asteroid 1580 Betulia was the 1580th to have its orbit confirmed, and has also been named Betulia.

The compositions of asteroids are imperfectly known. They have not been sampled in place. It is risky to send a spacecraft into the asteroid belt because of the probability of collision and damage to the craft. One can, as with the larger planets, look at the reflected light spectra, identifying some of the elements and minerals present on the basis of the wavelengths of light reflected, but asteroids are so small and distant that this is difficult to do with precision. However, asteroids have been found to span a range of compositions, some rich in metal (iron-nickel), some rich in silicates (especially ferromagnesians), some containing minor carbon, clays, or other minerals.

At one time, it was believed that the asteroids were once a single planet that somehow disintegrated into bits. But the total mass in the asteroid belt—about 0.04% of the Earth's mass—is far too small to equal a former planet. Also, the compositional diversity of asteroids is somewhat inconsistent with their formation as one body. It now appears that the asteroids are bits of early accreted material that never did assemble into a single planet. The perturbing effects of huge Jupiter's gravity may have been a factor. Instead, many separate, smaller bodies formed. Some differentiated; some apparently did not heat sufficiently to do so. The iron-nickel-rich asteroids are interpreted as the cores of differentiated asteroids, exposed by collision and fragmentation.

Earth's Moon

Largely as a result of its proximity and comparative accessibility—to observation, to measurement, and finally to direct sampling—we know more about Earth's Moon than about any other major solar-system object beyond the Earth. The increased information gained from lunar missions has, in some respects, raised as many questions as it has answered.

At the Lunar Surface

Early telescopic surveys of the Moon showed light-colored **highland** regions and lower, flatter, darker regions, the latter named **maria** (the plural of the Latin *mare,* "sea"; see figure 22.23). Closer investigation has revealed that these "seas" are utterly dry: there is no free surface water on the Moon. The resemblance of lunar terranes to terrestrial continents and oceans is a superficial one. The lunar highlands are composed not of granitic or granodioritic rock, but of a plagioclase-rich gabbro called **anorthosite.** (Recall that *anorthite* is a calcic plagioclase.) The highlands are thus much less silica-rich than the terrestrial continental crust. Within the highlands are some flat plains blanketed with fragmented, brecciated highland materials, shattered by meteorite impacts. The maria are basins, excavated by meteorite impacts. They have been flooded with basalts, not unlike terrestrial seafloor basalts, so here at least there is some limited resemblance to terrestrial ocean basins.

Not only is there no surface water on the Moon, there are essentially no hydrous minerals in the rocks either—no amphiboles, micas, or clays, for example. This restricts the range of lunar rock types relative to the Earth. The Moon has an atmosphere of sorts, but it is an extremely thin one. The composition of the atmosphere, mostly hydrogen and helium, suggests that it is gravitationally captured from the **solar wind** streaming outward from the Sun, as Mercury's atmosphere is. Any volatiles outgassed from the lunar interior early in its history were apparently lost to space.

The predominant surface process on the Moon has therefore been meteorite impact. There is no chemical weathering; there is no erosion by wind, water, or ice. No life has been detected. Thus the lunar regolith has simply formed by progressive physical breakup of the surface rocks through time, with minor additions of meteoritic material; redistribution of the regolith has been accomplished mainly as a result of the larger meteorite impacts. The extensive preservation of craters large and small testifies to a long period of geologic/tectonic quiescence at the lunar surface.

Figure 22.23

Dark and light areas on the Moon were once interpreted as land and seas.

Lick Observatory photo.

The Interior of the Moon

A generalized model of the lunar interior is shown in figure 22.24. The anorthositic crust extends to an average depth of about 65 kilometers, the outer 25 kilometers heavily fractured by impacts. Note that the basalt-filled maria are not regions of thinner crust but craters within the anorthosite partially filled with a thin layer of lava. Below the crust is a very thick mantle, about 1000 kilometers thick, which seismic evidence indicates is composed principally of ferromagnesian silicates. Density constraints would allow a small lunar core; how small depends on the proposed composition. If the core is iron-nickel, it must be smaller than if it is made up of less dense iron sulfide. In the latter case, a core of up to 700 kilometers in diameter is possible.

There is no volcanism on the Moon at present, but there are "moonquakes." These are few—only a few thousand a year, compared to the nearly one million quakes annually on Earth—and they are very

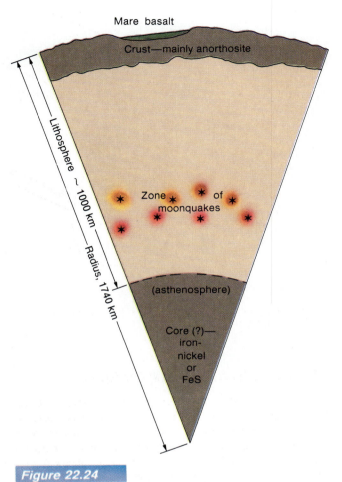

Mare basalt

Crust—mainly anorthosite

Lithosphere ∼ 1000 km

Radius, 1740 km

Zone * of moonquakes

(asthenosphere)

Core (?)— iron- nickel or FeS

Figure 22.24

Cross section of the lunar interior. The existence and size of the core are uncertain.

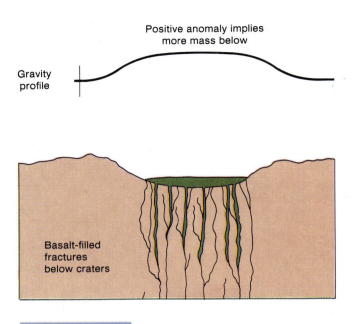

Positive anomaly implies more mass below

Gravity profile

Basalt-filled fractures below craters

Figure 22.25

Mascons cause positive gravity anomalies over negative relief.

weak; most moonquakes have Richter magnitudes less than 2. They are confined to a deep (600 to 800 kilometers) zone of the mantle, and are believed to be caused by tidal stresses of the Earth acting on the Moon. The weakness of the moonquakes limits their usefulness as planetary probes. However, it has been found that the seismic waves from those moonquakes are attenuated, or weakened, by the lowermost lunar mantle or core, and S-wave transmission is blocked. This suggests a lunar asthenosphere, but a much deeper one than on Earth. The fact that the lunar lithosphere is ten times as thick as that of Earth precludes any semblance of plate-tectonic activity.

Core or no, the Moon has no substantial magnetic field at present. It does have some rocks that retain a remanent magnetization that suggests that they formed in the presence of a significant magnetic field. It may be that the Moon *had* a magnetic field earlier in its history, at the time those rocks were formed,

which was generated as the Earth's field is. Since then, the Moon may have cooled to such an extent that the necessary fluid motions in the core are lacking.

Gravity surveys over basalt-filled maria have shown positive gravity anomalies, despite the negative relief of the basins. This suggests a subsurface concentration of high-density material. The anomalous features were named **mascons** (for "mass concentrations"). At one time, mascons were believed to reflect burial during impact of iron-rich meteorites that excavated the basins. Further studies of impact dynamics suggest that this is highly unlikely. Now the mascons are interpreted as resulting from the presence of many basalt-filled fractures below the basins, the relatively high-density basalt accounting for the gravity anomaly (figure 22.25).

Lunar History

Prior to the Apollo missions, lunar chronology was based largely on density of cratering: the more plentiful the craters, the older the terrane, and the correlation was thought to be basically linear. Once samples were returned and dated, one flaw in that scheme emerged: it appears that the intensity or rate of cratering was much higher during the first half-billion or billion years of lunar history, declining sharply after 4 billion years ago. Perhaps the intensive early cratering actually reflects the last stages of accretion, as remaining condensed material was swept out of the solar system under the influence of the gravitational pull of the larger accreting bodies.

When samples of lunar material were dated radiometrically, several patterns emerged. First, lunar rocks are uniformly old—all over 3 billion years in age, some as old as any solar-system material dated (4.6 billion years). This confirms the testimony of the abundant craters that the Moon is a long-dead world geologically. Second, the highlands are systematically older than the mare basalts. The highland samples are, for the most part, 4 to 4.6 billion years in age; the basalts, 3.1 to 3.9 billion years. Third, the basalts are significantly younger than the basins they fill. This was a surprise, because it was once thought that the basalts were impact melts, formed and extruded at the time of cratering as the energy of impact was dissipated as heat. In fact, individual basalts may be hundreds of millions of years younger than their corresponding basins. Actually, there is geologic evidence consistent with these chronologic data. For one thing, impact debris is not generally found atop the basalts, suggesting that any debris had settled before the flows emerged. Also, a complete or partial melt of the anorthosite from which the basins were carved would not be basaltic in composition.

The history of the first half-billion years of lunar evolution is obscure. We have samples from that period, but heavy meteorite bombardment of the highlands continued until less than 4 billion years ago, and the resultant surface is so highly fractured, brecciated, and jumbled that structures and field relationships between rock units are hidden. It seems clear that differentiation of the Moon, like that of Earth, must have occurred early. The anorthositic crust may have formed during crystallization of an extensively molten primitive moon, with relatively low-density plagioclase tending to float as it crystallized, and the denser ferromagnesians remaining concentrated in the interior.

Cooling of the Moon would be expected to be more rapid than that of the Earth, as it is a smaller body with a higher surface-to-volume ratio. By the time the mare basalts were extruded, the lithosphere must have been sufficiently thick and stiff that isostatic compensation for this denser material did not occur; the mascons attest to this. The exact source of the basalts is uncertain, but must have been a partly molten zone within the lunar mantle, shallow enough that the melts could make their way to the surface. Cooling would have progressed from the outside in, with progressive thickening of the lithosphere. So far, no evidence of lunar rocks younger than 3.1 billion years has been found. By that time, apparently, any remaining molten material lay so deep that it could no longer reach the surface.

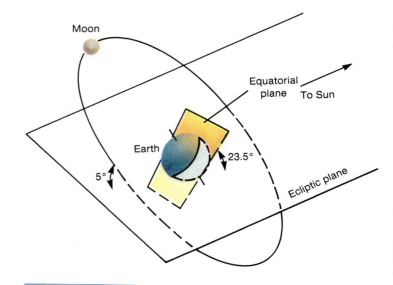

Figure 22.26

The Moon's orbit is inclined both to the ecliptic and to the earth's equatorial plane.

The Puzzle of Lunar Origin

A completely convincing theory for the origin of the Moon would have to explain both its composition and its position/orbit relative to the Earth. With respect to composition, a key point is that the Moon is much poorer in iron than the Earth, which accounts for its lower density (3.34 grams/cu cm versus 5.52 grams/cu cm for Earth). The Moon also seems to be poorer in volatile elements, and there are other compositional differences. With respect to orbit, the Moon is unusual in its orientation. Its orbit around the Earth lies neither in the ecliptic plane, nor in the equatorial plane of the Earth (figure 22.26). A variety of possible models of lunar origin have been proposed. Some of the principal ones are outlined below.

According to the *sister-planet hypothesis,* the Moon would have accreted contemporaneously with the Earth, in its present position. However, if the compositions of planetary objects are a reflection of their distances from the Sun at the time of accretion, it is not clear why the Moon and Earth should differ chemically. It has been argued that perhaps iron was selectively accreted into the larger Earth, but it is not obvious why that should occur. Also, there is no apparent reason that the Moon's orbit should be inclined to the ecliptic; all the terrestrial planets' orbits lie in that plane.

The *fission hypothesis* would explain the Moon as having been flung off from a rapidly rotating primitive Earth, perhaps while Earth was molten. If fission occurred after Earth's core formed, and the material split off was a chunk of mantle, that would partially account for the moon's relative lack of iron, although in fact the Moon seems to be even poorer in iron than Earth's mantle. Other problems with this theory include the following: (1) There is not enough momentum in the Earth–Moon system now to be consistent with the Moon's having been split off in that way, and (2) If fission had occurred, one would expect the Moon to have been flung off, and to be orbiting in, the Earth's equatorial plane, which it plainly is not.

Oddities in orbit and differences in composition might both be explained by the *capture hypothesis,* according to which the Moon formed elsewhere, then wandered near enough to the Earth to be captured by its gravitational field. But where did the Moon come from, and how did it then come to be passing by the Earth? There are also serious mechanical problems with the capture idea. The Moon is rather large relative to the Earth, and would have had to be moving very slowly relative to the Earth in order to be captured. Considering that the Earth is traveling around the Sun at about 100,000 kilometers per hour, it is difficult to picture the Moon approaching at just the right speed and angle to pass it moving at only a few kilometers per second relative to Earth, so as to be captured rather than to keep right on going. There is also some speculation that in order to be captured, the Moon would have to approach within the Earth's Roche limit, in which case one would expect it to have been broken up instead.

These are not all of the proposed models of lunar origins. Most recently, there has been growing interest in the possibility that the Moon was produced as a consequence of impact between a Mars-sized object and the early Earth; this model is still in the developmental stages. The models described do serve to illustrate the fact that no one hypothesis seems to account satisfactorily for all that we now know about the Moon's composition and orbital characteristics. The origin of the Moon is still a puzzle.

Figure 22.27

Example of an iron meteorite.

Meteorites

A **meteoroid** is a very small bit of interplanetary matter, too small to be seen until it enters the Earth's atmosphere. Meteoroids, then, are objects with diameters of about 10 kilometers or less. When meteoroids passing close to Earth are captured gravitationally and begin to fall toward the ground, they may be heated to incandescence by friction with the atmosphere. A **meteor,** also known (inaccurately) as a shooting or falling star, is a meteoroid to which this has happened, so that it is observed as a bright streak of light in the sky. If some of the object survives to reach the ground, the fallen object is called a **meteorite.**

Meteorite Compositions

The mineralogy of meteorites is relatively simple. The principal minerals found in meteorites include nickel-iron alloys, iron sulfide (troilite, FeS), olivine, pyroxenes, and plagioclase. Most meteorites fall into one of two broad compositional classes: the **irons,** which consist mainly of iron-nickel alloys, and the **stones,** which consist mainly of silicates (figure 22.27). There are also a relatively few *stony-iron* meteorites, containing roughly equal proportions of silicate and metal.

Box 22.1

Craters and Cratering

Craters are prominent features of many planetary surfaces. They are formed by the impact of chunks of interplanetary matter striking the surface (meteorites). The extent to which cratering occurs is a function of the number and velocity of incoming fragments. Velocity is moderated by the presence of an atmosphere; the most energetic impacts occur on planets or moons lacking such a shield. The extent to which impact craters are preserved depends on the extent of surface modification over time, through erosion, weathering, volcanism, and tectonism.

Upon first impact, shock waves dissipate the energy of the meteorite through the ground. The rock is rapidly and severely compressed, commonly fractured; some may be melted or vaporized. After passage of the shock wave the rock is rapidly decompressed, and material is flung violently out from the impact site to form a crater (figure 1). The materials thus displaced are collectively termed **ejecta.** A shock wave and subsequent decompression affect the meteorite, too, and it commonly shatters and becomes part of the debris.

The energy released during impact depends on the mass and velocity of the meteorite. A high-velocity meteorite can excavate a crater far larger than the meteorite itself. An iron meteorite only 20 meters in diameter striking the surface at a velocity of 10 kilometers per second would produce a crater nearly 600 kilometers across.

In cross section, a crater is a shallow feature in relation to its diameter, with an uplifted rim, *fallback breccia* at the bottom, and *ejecta blanket* on its flanks. If the crater has formed in layered rocks,

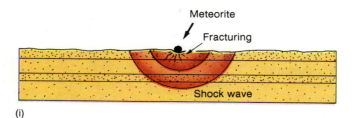

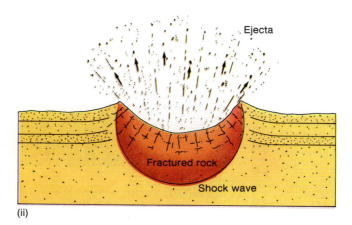

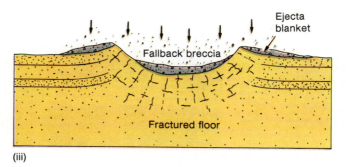

Figure 1
Formation of an impact crater (schematic).

the strata may be overturned outward around the rim of the crater. Some larger craters also have a central uplift (figure 2), the origin of which is not well understood.

The geological activity of the Earth's surface tends to erase older craters, and vegetation masks them further. The major impact craters that

have been recognized are most often in sparsely vegetated and tectonically stable areas. Often they have first been spotted in satellite photographs rather than from the ground (figure 3). Not all circular features are impact craters, of course; for example, some are calderas, some plutons or volcanic pipes. Once an impact origin

Box 22.1 *Continued*

is suspected for a particular circular depression, the possibility can be tested by looking for several features characteristic of impacts.

If the crater is young, meteorite fragments may be found nearby. The passage of the shock wave induces **shock metamorphism** in the rocks and minerals of an impact crater, which can include the following features: production of distinctive fracture patterns in the rocks; disruption of crystal structures (detectable microscopically or by X rays); production of ultra-high-pressure polymorphs, not normally found in the crust (conveniently, the very common mineral quartz has two such polymorphs). Impact melt may also have been produced. Impact melts can be distinguished from volcanic glasses in several ways: they may incorporate shocked minerals, high-pressure polymorphs, or bits of nickel-iron from the meteorite itself; they tend not to be very homogeneous; impact melts also have bulk compositions similar to those of the country rocks, which is not generally true of volcanics. Energetic eruption of a volcanic pipe may lift the strata around the rim of the crater, but will not fold them back upon themselves as an impact-related explosion may.

Fortunately for us, few large meteorites strike the Earth. There are only a few confirmed reports of meteorites striking people, and none of anyone being killed by a meteorite.

Figure 2
This Martian crater has a raised center as well as an uplifted rim.
© NSSDC/Goddard Space Flight Center

Figure 3
Satellite photographs reveal possible impact structures on earth—Clearwater Lakes, Quebec.
© NASA

More than 90% of meteorites seen to fall and collected thereafter (*falls*) are stony meteorites. However, among the other meteorites found, presumably some time after they have fallen (*finds*), the majority are iron meteorites. There are probably two reasons for this: first, iron meteorites look more unusual to the untrained observer, and second, irons are also more resistant to weathering after they fall, and therefore are better preserved.

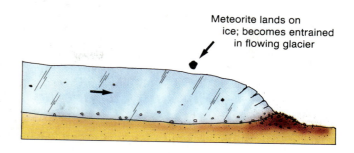

Meteorite lands on ice; becomes entrained in flowing glacier

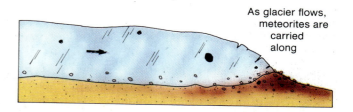

As glacier flows, meteorites are carried along

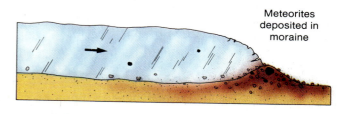

Meteorites deposited in moraine

Figure 22.28

Antarctic ice sheets collect and preserve meteorites falling there.

Further compositional subdivisions among the meteorites are complex, and will not be reviewed here. One class of stony meteorites does deserve special mention: the **carbonaceous chondrites.** These meteorites are characterized by nonvolatile element abundances relatively similar to those of the Sun (recall chapter 10), but they also contain the highest proportions of volatiles, especially water, of any meteorites. Densities of carbonaceous chondrites are as low as 2.2 grams/cu cm, with 20% water and high contents of carbonate and sulfate minerals. They lack metallic components, but do contain a variety of high- and low-temperature silicate and nonsilicate minerals that could not have formed together simultaneously. Nor could carbonaceous chondrites ever have been heated very much, as during planetary differentiation, without destruction of some of the lowest-temperature minerals. The carbonaceous chondrites are apparently the best samples we have of primitive solar-system material. They seem to be collections of material accreted from the solar nebula, assembled into larger bodies, then left unmelted, unmetamorphosed, and otherwise unchanged since that time.

Where Do Meteorites Come From?

The vast majority of meteorites that have been dated radiometrically give formation ages of 4.5 to 4.6 billion years. A few yield younger ages, which may represent the times of differentiation of their parent bodies, or episodes of shock metamorphism during collisions in space. In general, we can say that meteorites formed concurrently with the rest of the solar system.

The most likely candidate for a source region is the asteroid belt. Insofar as we know the compositions of the asteroids, their makeup is broadly similar to that of meteorites, although the exact proportions of materials of different compositions may not be identical. Relatively few meteorites' paths have been tracked prior to fall, but several that have could have originated in the asteroid belt. There are periodic collisions among asteroids that could knock bodies out

of the belt, allowing their motion to be further perturbed by Jupiter, and then deflected in such a way that they could be captured by the Earth.

Within the last decade, available samples of meteorites have been greatly augmented by returns from Antarctica. The Antarctic glaciers have been very effective at collecting, preserving, and concentrating meteorites (figure 22.28). Among the many new specimens are several that, on chemical grounds, have been tentatively identified as having come from the Moon or Mars. The presumption is that these are fragments of the bodies' surfaces jarred loose by the impact of large meteorites. The majority of meteorites could not have originated on the Moon or Mars, for several reasons. Both bodies are differentiated, and therefore could not be the sources of the very primitive meteorite types. Chemically, mineralogically, and texturally, there are substantial differences between Martian or lunar surficial materials and most meteorites (including all the irons, for example). And there are escape-velocity problems, especially with Mars: it would be the rare impact that would be strong enough to blast away chunks of the planetary surface violently enough that they would escape into space rather than just fall back to the ground.

Meteorites—Dinosaurs' Downfall?

The fossil record includes numerous examples of **mass extinctions,** geologically short events during which entire animal groups, diverse and apparently flourishing, faded out and disappeared from the subsequent record. The Mesozoic/Cenozoic boundary is one such time, and a prominent feature of that event is the extinction of the dinosaurs, which had been abundant through the Mesozoic.

One recent theory to account for the extinction of the dinosaurs involves a major meteorite impact at about 65 million years ago. A sufficiently large impact could put huge volumes of dust into the air and set off raging forest fires on land. The combined effects of the fires and the blocking of sunlight by dust and soot would devastate land-based plant life (at least until the next generation of seeds and spores could germinate), which would in turn deprive plant-eating dinosaurs of food, then meat-eaters who dined on the plant-eaters.

An important line of evidence in support of this suggestion—indeed, the observation that inspired it—was the discovery of a clay layer rich in the element *iridium* at about that point in the sedimentary rock record, in many places all over the world. Iridium is a platinum-group element that behaves geochemically like iron and nickel, and would be enriched in an iron meteorite. Most of the Earth's original supply of iridium is probably in its core. It is difficult to imagine a source for the iridium in this unusual clay layer, if it was not a meteorite. Further investigation has also revealed a carbon-rich layer in sediments of about the same age, interpreted as possible soot from the global fires.

The concept of extinction by meteorite impact is not universally accepted. Given the immense spans of time that can be represented by a few millimeters of sediment, there are questions about whether the iridium-rich layer and the extinctions actually coincide closely enough to support the cause-and-effect interpretation. There is, for example, evidence that some dinosaurs did survive the end of the Mesozoic. There are other unresolved questions that contribute to ongoing controversy regarding this explanation. Nevertheless, it is a plausible and intriguing way to account for a striking phenomenon in the fossil record.

Comets

The approach of Halley's Comet in 1985–86 has kindled new interest in **comets.** Of all of the objects discussed in this chapter, comets are the most remote in origin, and perhaps the most difficult to study, for they are small objects, encountered rarely, and never sampled directly. We do have some observational and spectral data from which to draw conclusions about the nature of comets.

The simplest way to describe a comet is as a "dirty snowball," a clump of ices and silicate dust. Most have radii of 1 to 2 kilometers, with the largest about 10 kilometers. This snowball is the **nucleus,** or center, of a comet as we observe it from Earth. As comets approach the sun, a portion of the ice in the nucleus begins to vaporize, and gas and dust disperse around it in a foggy **coma** that may be 10,000 to 100,000 kilometers across. Still closer to the Sun, the solar wind begins to streak out the matter of the coma, sweeping it away from the nucleus into a long **tail** that may be over 100,000,000 kilometers long (figure 22.29). It is this characteristic that gives comets their name, taken from the Greek *kometes,* "long-haired."

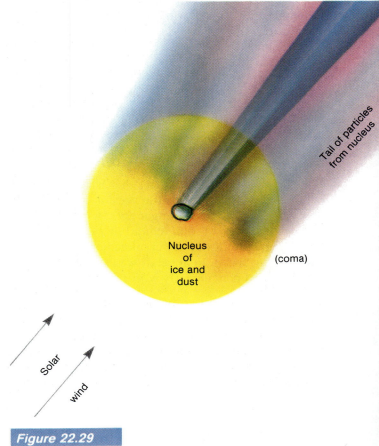

Nucleus of ice and dust

(coma)

Tail of particles from nucleus

Solar wind

Figure 22.29

The component parts of a comet as it approaches the sun.

Implicit in this discussion is the fact that comets lose mass every time they brush by the sun, so in time, even if they escape gravitational capture by the planets, those comets that orbit through the inner solar system will eventually be worn away altogether.

Comets come from the **Oort cloud,** a cloud of comets circling the Sun at an average distance 50,000 times the Earth's distance from the Sun. They may have formed there, during the early history of the solar system, or they may have formed not far from Neptune, then been perturbed out into the Oort cloud. (The presence of methane and ammonia ices in comet nuclei precludes formation closer to the Sun.) Most comets never approach the Earth closely enough to be seen. Occasionally, a comet will be perturbed in such a way that its orbit is deflected into the inner solar system, where we can observe it. About 625 comets have been detected from Earth. Those whose orbits also take them back out to the Oort cloud may have orbital periods of more than one million years. The so-called short-period comets, those that circle the Sun in less than 200 years (of which Halley's Comet is one), have probably been further perturbed out of long-period orbits by encounters with the Jovian planets' gravitational fields.

There is a relationship between comets and some meteors. When Earth's orbit carries it through a comet's tail, meteor showers often result. This is presumably a consequence of bits of rocky dust being heated in Earth's atmosphere. It is possible that larger rocky fragments of comets might survive to fall as meteorites. If so, those meteorites would be of the primitive, undifferentiated kind only, for comets could never have been hot since their formation.

Summary

There are nine identified planets in the solar system. The four inner planets (Mercury, Venus, Earth, and Mars) are the terrestrial planets. They are Earthlike in the sense that they consist principally of iron-rich metal and silicates in varying proportions. Internally, too, all are probably differentiated into a core, mantle, and crust analogous to the Earth's. Their surface properties differ markedly, as a function of the density and composition of any atmosphere, or the presence of surface water or nearsurface permafrost. The Jovian planets (Jupiter, Saturn, Uranus, and Neptune) are much larger, and include a high proportion of liquid or gaseous hydrogen in their makeup. At least three of the four have ring systems. All have several moons, which may show considerable diversity in composition. Io, one of Jupiter's moons, is the most active object, volcanically, in the solar system. Pluto, the outermost planet, is icy rather than fluid, and quite small, with an unusual orbit; it may not properly be a planet at all, but rather a large asteroid.

The asteroid belt, between Mars and Jupiter, consists of many small rock and/or metallic bodies. They probably never formed a planet, perhaps because of the disruptive influence of nearby Jupiter. Compositional and orbital constraints suggest that the asteroid belt is the source of most meteorites that fall to Earth. Primitive meteorites, like the carbonaceous chondrites, would come from undifferentiated asteroids; other stony and all iron meteorites would have come from differentiated bodies. Comets, balls of ice and dust that circle in toward the Sun from the Oort cloud, are a possible source only of the most primitive meteorites, and of meteor showers as Earth passes through the tail of a comet near the Sun.

Earth's Moon is an enigma. It is as old as the oldest known solar-system materials, rocky like the terrestrial planets. Major differences in composition between the Earth and Moon (including a much lower total iron content in the Moon), and the fact that its orbit is inclined both to the ecliptic and to the Earth's orbital plane, together make its origin difficult to explain. Internally, it is differentiated, with an anorthositic crust formed 4 to 4.6 billion years ago, a ferromagnesian silicate mantle, and a possible small iron-nickel or iron sulfide core. The maria are basins excavated from the crust by meteorites, and are shallowly filled with flood basalts ranging in age from about 3.1 to 3.9 billion years. With a 1000-kilometer-thick lithosphere, the Moon is tectonically very quiet, and has been so for billions of years: the youngest of the mare basalts are the youngest rocks known on the Moon. There is no surface water and virtually no lunar atmosphere. Meteorite impact is therefore the principal surface process shaping the lunar surface.

Terms to Remember

anorthosite	crater	maria	Oort cloud
asteroid	ecliptic	mascons	Roche limit
asteroid belt	ejecta	mass extinctions	shock metamorphism
carbonaceous chondrites	highland	meteor	solar wind
coma	irons	meteorite	stones
comet	Jovian planets	meteoroid	tail (comet)
		nucleus (comet)	terrestrial planets

Questions for Review

1. Name and briefly describe three methods used to investigate other planets. What kinds of information can be obtained from each?

2. What are the terrestrial planets? The Jovian planets?

3. How and why do the compositions of the planets vary with distance from the Sun?

4. Iron-rich Mercury has a weak magnetic field. Why is this somewhat difficult to explain? How might one account for it?

5. Describe the surface conditions on Venus. What is the origin of the high temperatures?

6. What is the nature of the Martian polar ice caps?

7. What evidence do we see for surface processes or tectonism on Mars?

8. The Jovian planets, especially Jupiter, may preserve a composition like that of the solar nebula. Explain.

9. What are the Galilean moons of Jupiter? Which of them is the most ''geologically'' active object in the solar system, and what is the nature of that activity?

10. Which planets have confirmed ring systems? Outline one theory for the origin of rings.

11. Pluto is an oddity among the planets. How so?

12. What are the asteroids, and what is the asteroid belt? How may the asteroids have arisen?

13. What are the two principal types of crustal regions on the Moon, and how do they compare to the Earth's continents and ocean basins?

14. Summarize the major features of the Moon's internal structure.

15. What are *mascons,* and how are they explained?

16. Which periods of lunar history are characterized by (a) intense meteorite bombardment; (b) formation of mare basalts? Is the Moon presently tectonically active? Why or why not?

17. What is the extent of surface processes presently acting on the Moon?

18. Name any two hypotheses for the origin of the Moon, and note what features each does and does not account for.

19. What are meteorites, and where are they believed to originate?

20. Describe the principal components of a comet as it approaches the Sun. Where do comets come from?

21. Name three ways in which an old impact crater on Earth might be recognized.

For Further Thought

1. Landers have been sent to Mars and to the Moon and have provided considerable information about these bodies' surfaces. Assess the feasibility of sending similar missions to Venus and to Jupiter, considering what we know of the surface conditions on those planets.

2. Trace the history of the Voyager missions. Identify the major advances in understanding, changes in interpretation, and new pieces of information about Jupiter, Saturn, and Uranus that have come out of the Voyager data.

Suggestions for Further Reading

Beatty, J. K., O'Leary, B., and Chaiken, A., eds. *The New Solar System*. 2d ed. Cambridge, Mass.: Sky Publishing Corp., 1982.

Cardogan, P. H. *The Moon: Our Sister Planet*. New York: Cambridge University Press, 1981.

Chapman, C. K. *Planets of Rock and Ice*. New York: Scribner, 1982.

Elliot, J., and Kerr, R. *Rings: Discoveries from Galileo to Voyager*. Cambridge, Mass.: M.I.T. Press, 1984.

Fielder, G., and Wilson, L. *Volcanoes of the Earth, Moon, and Mars*. New York: St. Martin's Press, 1975.

Glass, B. P. *Introduction to Planetary Geology*. New York: Cambridge University Press, 1982.

Greeley, R. *Planetary Landscapes*. London: George Allen and Unwin, 1985.

Kaufmann, W. J., III. *Universe*. New York: W. H. Freeman and Co., 1985.

Lewis, J. S., and Prinn, R. G. *Planets and Their Atmospheres: Origin and Evolution*. New York: Academic Press, 1984.

Morrison, D., and Samz, J. *Voyage to Jupiter*. Washington, D.C.: National Aeronautics and Space Administration, 1980.

Murray, B., Malin, M. C., and Greeley, R. *Earthlike Planets*. San Francisco: W. H. Freeman and Co., 1981.

Spitzer, C. R., ed. *Viking Orbiter Views of Mars*. Washington, D.C.: National Aeronautics and Space Administration, 1980.

Taylor, S. R. *Lunar Science: A Post-Apollo View*. New York: Pergamon Press, 1975.

The Planets. San Francisco: W.H. Freeman and Co. A collection of reprints from *Scientific American*, 1975–1983.

Wasson, J. T. *Meteorites*. New York: W. H. Freeman and Co., 1985.

Introduction to Topographic and Geologic Maps and Satellite Imagery

Maps and Scale

Many kinds of information can be presented in map form. Topographic and geologic maps are the two kinds most frequently used by geologists. This appendix presents a brief introduction to both types of maps.

A point to note at the outset is that in the United States, such maps commonly carry English rather than metric units. Many maps were drawn before there was any move toward adoption of metric units in this country. The task of redrawing topographic maps, in particular, in order to convert to metric units, is formidable. (A metric map series is being prepared by the U.S. Geological Survey, but it will be many years before its completion.)

A basic feature of any map is the map **scale,** a measure of the size of the area represented by the map. Map scales are reported as ratios: 1:250,000, 1:62,500, and so on; or equivalently, in words, "One to two hundred fifty thousand," and so on. A 1:250,000 scale map is one on which a distance of one inch on the map represents 250,000 inches (almost 4 miles) in reality; on a 1:62,500 scale map, one inch equals about one mile of actual distance. The larger the scale factor, the more real distance is represented by a given distance on the map. As a result, the fineness of detail that can be represented on a map is reduced as the scale factor is increased. The choice of scale factor often involves a compromise between minimizing the size of the map (for convenience of use) or the number of maps required to cover the area, and maximizing the amount of detail that can be shown.

Topographic Maps

Topographic maps primarily represent the form of the earth's surface. Selected other features, both natural and artificial, may be included for information. Once one becomes accustomed to reading them, topographic maps can make excellent navigational aids.

Contour Lines, Contour Intervals

The problem of representing three-dimensional features on a two-dimensional map is addressed through the use of **contour lines.** A contour line is a line connecting points of equal elevation, measured in feet (or meters) above or below sea level. The **contour interval** of a map is the difference in elevation between successive contour lines. For instance, if the contour interval is 10 feet, contour lines would be drawn at elevations of 1000′, 1010′, 1020′, 1030′, and so on (in whatever range of elevations is appropriate to that map). If the contour interval is 20 feet, contours will be drawn for elevations of 1000′, 1020′, 1040′, and so on. The contour interval chosen for a particular map will depend on the overall amount of relief in the map area. If the terrain is very flat, as in a Midwestern floodplain, a 10′ or even 5′ contour interval may be appropriate to depict what relief there is. In a rugged mountain terrain, with several thousand feet of vertical relief, using a 10′ contour interval would make for a very cluttered map, thick with contour lines; a 50′ or even 100′ contour interval may be used in such a case.

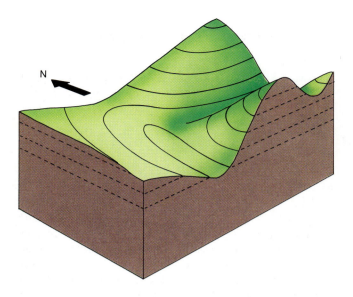

Figure A.1

Contours and relief. *(A)* Perspective view of a hilly area, with contours superimposed for reference. *(B)* The same contours as they would appear on a topographic map.

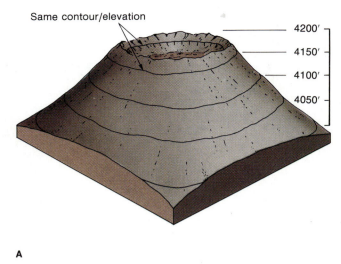

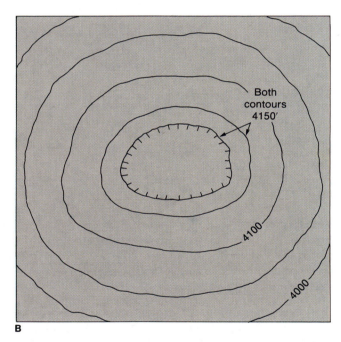

Figure A.2

Volcanic caldera with summit depression. *(A)* Actual relief. *(B)* As the volcano would appear on a topographic map.

The relationship between the spacing of contours and the steepness of a slope can be seen by referring to figure A.1. Comparing the actual relief, as shown in figure A.1A, to the resultant arrangement of contour lines on the corresponding topographic map, shown in figure A.1B, illustrates the fact that on a given map, the more closely spaced the contours, the steeper the corresponding terrain. In other words, if there are many closely spaced contours, this means that there is a great deal of vertical relief over a limited horizontal distance. Note also that contours run along the face of a slope; the upslope and downslope directions are perpendicular to the contours.

Contours do close, although they may not do so on any one given map. A series of concentric closed contours indicates a hill. If there is a local depression, so that one encounters the same contour twice, once going up in elevation and again within the depression, the repetition of the contour within the depression is marked by a hachured contour, as shown in figure A.2.

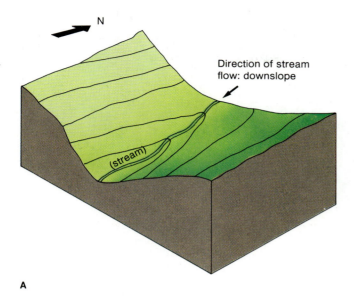

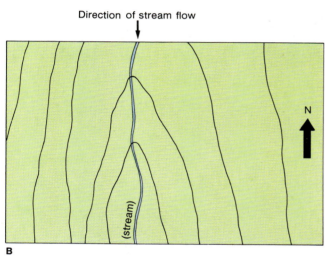

Figure A.3

Contours crossing stream valleys. *(A)* Relief with superimposed contours. *(B)* On topographic map, the contours point upstream.

Where contour lines cross a stream valley, they point upstream, toward higher elevations (figure A.3). Contour lines corresponding to different elevations should not ordinarily cross each other (that would imply that the same point has multiple elevations). The only exception to this would occur in the case of an overhanging cliff, where a range in elevations exists at one spot on the map.

Other Features on Standard Topographic Maps

The most extensive topographic maps of the United States have been compiled by the U.S. Geological Survey. They have adopted a uniform set of symbols for various kinds of features, which facilitates the reading of topographic maps. Some of these are illustrated in figure A.4.

Contour lines are drawn in brown. Major contours (typically, those contours corresponding to multiples of 100 feet, except on maps with large contour intervals) are drawn more boldly and labeled with the corresponding elevation. This is especially convenient in rugged terrain, where it would otherwise be easy to lose count of numerous closely spaced contours. Roads are drawn in black and/or red. Town limits are shown in light red, and the names of towns, cities, mountains, or any other labels are printed in black. Watery features are drawn in blue: lakes and perennial streams in solid blue, intermittent streams in dashed blue lines, swampy areas denoted by blue symbols resembling tufts of low grasses. The background color for wooded areas is green; for open fields, brushy areas, deserts, bare rock, or any area lacking overhead vegetative cover the background color is white. Revisions based on aerial photography ("photorevisions") are shown in purple.

Some Notes for Hikers

Topographic maps can be extremely helpful to anyone hiking off established roads in undeveloped areas. Where hiking trails are well defined, as for example in state or national parks, those trails are mapped in dashed black lines. In planning a hike, it can be extremely helpful to have some idea of the nature of the terrain. Trails running parallel to contour lines will be fairly level, trails running across contour lines involve more climbing up and down, with the spacing of the contours providing a measure of the steepness of the trail. (In comparing trails on several different maps, be alert to possible differences in contour interval.)

It is also useful to be forewarned of streams or swampy land to be crossed. The fact that a trail crosses a stream does not necessarily mean that a bridge exists, unless one is marked. The stream may have to be forded, and how safe or difficult that is often depends

Figure A.4

Representative section of standard topographic map: a portion of
Yellowstone National Park. Note the closely spaced contours
along the Grand Canyon of the Yellowstone and edges of
plateaus, the shapes of contours that cross streams, the closed
contours around hills (e.g., Junction Butte), and the swampy flat
area (along Slough Creek).

Source: United States Department of the Interior, Geological Survey.

strongly on the season. If in doubt, inquire locally before starting out. Also, trails and other features may change over time. Keep the date of the map in mind in planning a route, and if the map was made several decades ago, don't be surprised if the trails' routes or other artificial features have been changed somewhat. (At least with topographic map in hand, you can probably figure out where the new route is taking you, if not where the old trail is shown.)

Obtaining Topographic Maps

U.S. Geological Survey maps are sometimes identified as "15-minute" or "7 1/2-minute" *quadrangle* maps. Such a designation means that the map area is a rectangle corresponding to so many minutes of latitude and longitude. Because it covers a smaller area, a 7 1/2-minute quadrangle map shows more detail than a 15-minute quadrangle. Many parts of the country have been mapped at both scales; for others, only 15-minute quadrangles are available. Individual maps are named for a town, mountain, or other prominent geographic feature within the map area: for example, "Cooke City 15-minute quadrangle," "Cutoff Mountain 7 1/2-minute quadrangle." There are also special maps covering areas of particular interest, such as individual national parks.

Topographic maps may be ordered directly from the U.S.G.S. at nominal cost (variable with map size and scale). If you are uncertain of what map(s) you want, you can request free *index maps* of the corresponding state(s), along with price lists, before ordering. For example, if you were interested in obtaining quadrangle maps of selected parts of Yellowstone National Park, you would request a topographic map index of Wyoming, then choose the appropriate quadrangle(s) from that. The index map shows what part of the state is covered by which quadrangle map, the name by which each map is identified, and the date of the map or last revision thereof. A cautionary note: the same place name may be used both for a 15-minute quadrangle and for a 7 1/2-minute quadrangle within it. Be alert to this when ordering, and specify quadrangle size if necessary.

Maps are obtained from two principal locations. For maps of areas east of the Mississippi River, contact

Branch of Distribution
U.S. Geological Survey
1200 S. Eads Street
Arlington, Virginia 22202

For areas west of the Mississippi, maps can be obtained from

Branch of Distribution
U.S. Geological Survey
Box 25286 Federal Center
Denver, Colorado 80225

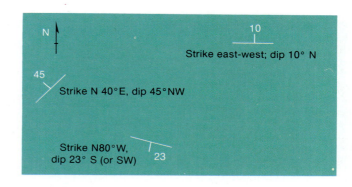

Figure A.5

Strike and dip markings used in mapping. (Recall from chapter 11 that strike is conventionally measured in degrees east or west of north.)

Geologic Maps

Topographic maps show the form of the land surface; geologic maps are one way of representing the underlying geology. The most common kind of geologic map is a map of **bedrock geology.** Such a map shows the geology as it would appear with soil stripped away. From a well-prepared geologic map, aspects of the subsurface structure can often be deduced.

Geologic Maps—Some Basic Concepts

Fundamental to making a geologic map is identifying a suitable set of **map units.** These may be individual sedimentary rock formations or members, distinguishable lava flows, single plutonic or metamorphic rock units. The main requirement for a map unit is that it be identifiable by the presence or absence of some characteristic(s), and thus distinct from other map units chosen. The mapper then marks which of the map units is found at each place where the rocks are exposed.

Additional information, such as strike and dip of beds or the location of contacts between map units, may also be recorded. Where obvious, contacts are drawn as solid lines; where only inferred, as dashed lines. (If one finds granite at point *A* and limestone at *B* a short distance away, one can infer a contact between the granite and limestone somewhere between *A* and *B,* even if the exact spot is covered by soil.) Strike is indicated by a line segment oriented in the appropriate compass direction. A shorter line is drawn perpendicular to the strike in the direction of dip, with a small number added beside the dip line to indicate the number of degrees of dip from the horizontal (see figure A.5).

A

B

Figure A.6

Rock exposure, good and bad. *(A)* In this gently rolling grassland, bedrock geology is hard to determine. *(B)* In rugged terrain or above the tree line, soil is sparse and plants are few; rocks are far better exposed.

How easy it is to produce an accurate and complete map of the geology depends on several factors (aside from the competence of the mapper!). Exposure is one (figure A.6). Thick soil, water or swamps, and vegetation all can obscure what lies below. In some areas, the geology is well exposed only along stream valleys; in others, the rocks are completely exposed and mapping is greatly simplified. In glaciated areas, too, one need not only deal with the cover of glacial till but also be cautious about mistakenly identifying a large, partially buried glacial erratic as a small exposure of bedrock.

Another factor is the ease with which different map units can be distinguished in the field. If the only rock units in the area are a sandstone, a granite, and a basalt, the mapper's task is relatively straightforward. If, on the other hand, the local geology consists of two dozen sedimentary rock formations, including several each of sandstones, shales, and limestones, with different units of the same rock type distinguished on the basis of subtle differences in mineralogy or the presence of certain microfossils, the mapping will be correspondingly more difficult, and the final map may not be drawn until many samples have been collected and analyzed in the laboratory.

A completed geologic map is generally much more colorful than a topographic map, as different units are represented in different colors for clarity (figure A.7). Units of similar age may be shown in different shades of the same color. The map will be accompanied by a key showing all the map units, arranged in chronological order (as far as their ages are known), with the youngest at the top. Ordinarily, a brief description of each map unit will be given; alternatively, standard patterns may be used to indicate the general rock type (figure A.8). Each unit will also be assigned a symbol. The first part of the symbol consists of one or two letters corresponding to the unit's age (generally the era or period). This is usually followed by one to three lowercase letters corresponding to the rock type or unit's name (if any). For example, "pЄqm" might be used for an unnamed Precambrian quartz monzonite, "Dl" for the Devonian Littleton Formation, "Qa" for Quaternary alluvium along a stream channel. These symbols are useful for distinguishing units mapped in similar colors, as well as providing a general indication of the age of each unit directly on the map.

Devonian

D   Dolomite and shale

Silurian

Sd   Dolomite

Ordovician

Om   Maquoketa Formation —shale and dolomite

Os   Sinnipee Group— dolomite, limestone, shale

Osp   St. Peter Formation— predominantly sandstone

Opc   Prairie du Chien Group – dolomite with sandstone and shale

Cambrian

€   Predominantly sandstone

Proterozoic

ss   Sandstone

V   Volcanics

g   Wolf River granite

q   Quartzite

gr   Granite

Vo   Volcanics

S   Siltstone and other sediments

Archean

gn   Gneisses and metavolcanics

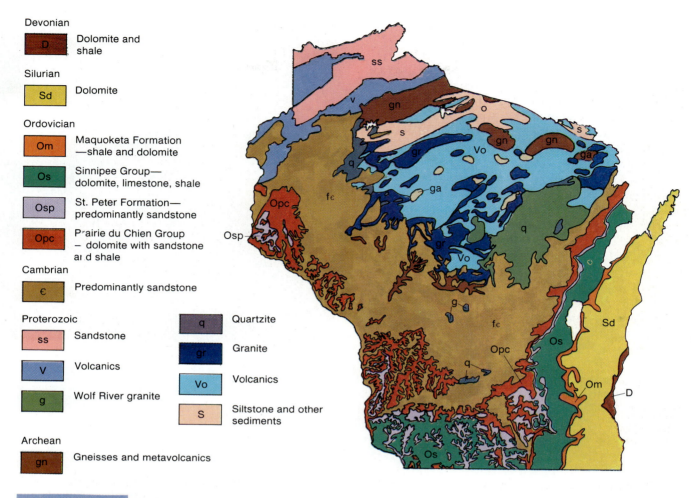

Figure A.7

Sample of geologic map with key: simplified bedrock geology of Wisconsin.

After Wisconsin Geological and Natural History Survey, 1981.

Figure A.8

Selected standard map symbols for various rock types.

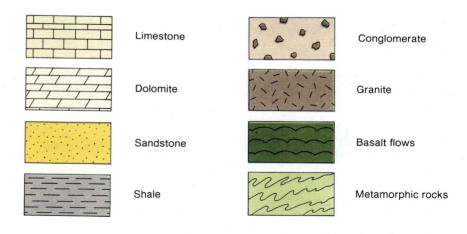

Limestone

Dolomite

Sandstone

Shale

Conglomerate

Granite

Basalt flows

Metamorphic rocks

Interpretation from Geologic Maps—Some Examples

Interpretation of the resultant geologic map likewise varies in difficulty, depending on the fundamental complexity of the geology, and the extent of exposure (completeness of map). Sometimes, too, how much of the geology can be seen will depend not only on cover or lack of it, but on topography. That is, in rugged terrain with considerable vertical relief, one has more of a three-dimensional look at the geology than in flat terrain. For example, consider the vicinity of the Grand Canyon. The Paleozoic sedimentary sequence in the region is quite flat-lying. So, for the most part, is the land surface. Outside the canyon, only the Permian Kaibab limestone is exposed, forming a nearly level plateau and giving no evidence of what lies below. The cutting of the canyon has exposed many more rock units, down to the Precambrian. This is apparent both in a satellite image (figure A.9A) and on a geologic map (figure A.9B).

A

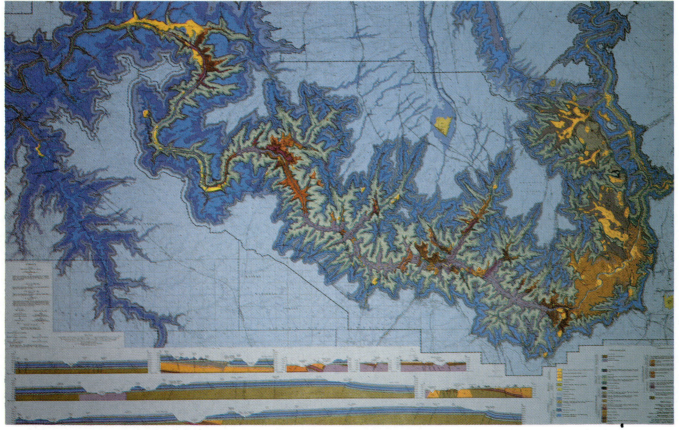

B

Figure A.9

The presence of the Grand Canyon allows much better knowledge of the area's subsurface geology. *(A)* Computer-enhanced satellite photograph shows different rock units in different colors. Note monotony outside canyon. *(B)* The same result can be seen on a geologic map.

(A) © NASA *(B)* Used with permission of Grand Canyon Natural History Association.

Figure A.10

A dome as it would appear on a geologic map.

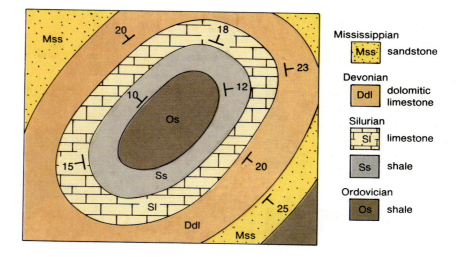

Mississippian
Mss — sandstone

Devonian
Ddl — dolomitic limestone

Silurian
Sl — limestone
Ss — shale

Ordovician
Os — shale

Figure A.11

Synclines on a geologic map. *(A)* Syncline with horizontal axis. *(B)* Plunging syncline.

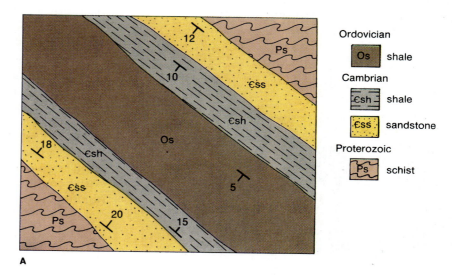

Ordovician
Os — shale

Cambrian
Є sh — shale
Є ss — sandstone

Proterozoic
Ps — schist

A

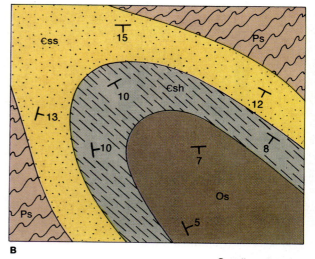

B

Syncline plunges to southeast

The Grand Canyon has a regular, layer-cake geology that is straightforward to interpret. Where the geology is more complex, one must rely on patterns of repetition of units, and strikes and dips of beds, to interpret the structure. Consider, for example, figure A.10. The map pattern shows circular exposures of rock units. All units dip away from the center of the circle, and the key shows that the oldest rocks are at the center. This, then, is a dome.

Figure A.11A shows a sequence of beds symmetrically repeated to either side of the map. In this case, the dips are toward the center of this linear feature, and the key reveals that the beds at the center are the youngest. This is a synclinal structure. If a fold plunges, the nose of the fold will intersect the surface, resulting in a curving map pattern such as is shown in figure A.11B.

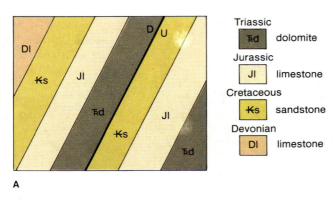

Triassic
Ŧd dolomite

Jurassic
Jl limestone

Cretaceous
Ks sandstone

Devonian
Dl limestone

A

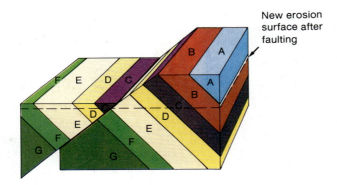

Normal faulting
of tilted sediments

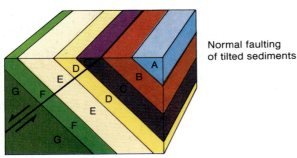

New erosion
surface after
faulting

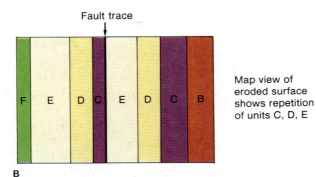

Fault trace

Map view of
eroded surface
shows repetition
of units C, D, E

B

Faulting can produce repetition of map units. *(A)* Example in
map pattern. *(B)* How normal faulting might have produced that
map pattern.

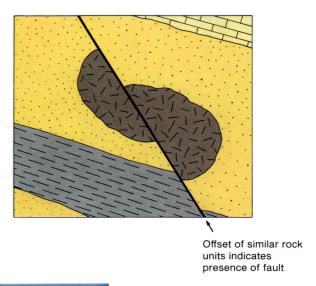

Offset of similar rock
units indicates
presence of fault

Pluton and surrounding rocks offset by faulting.

When sets of rocks are repeated *in the same
order,* rather than in reversed order, this may indicate
the presence of a fault. Consider the map pattern
shown in figure A.12A. The same sequence of units is
repeated; moreover, all are dipping in the same direction. It is difficult to create such a pattern by
folding. A normal fault, however, can account for the
result (figure A.12B). Offset of features that are otherwise continuous is another sign of possible faulting
(figure A.13).

When located, the fault trace will be drawn on
the map. If the fault is a high-angle fault, and the sense
of motion can be determined, the uplifted and downdropped blocks may be identified by markings of "U"
and "D" respectively on either side of the fault (as in
figure A.12A). With a thrust fault, a sawtooth pattern
may be drawn along the fault trace, with the pattern
pointing toward the overthrust block. (The same representation is commonly used for the overthrust, or
overriding, plate of a subduction zone.) Naturally it
is easier to recognize a fault if the fault zone is exposed, in which case one may be able to see fault
breccia or other obvious signs of the faulting. Even if
the fault is concealed, however, the regional geology
may give its presence away.

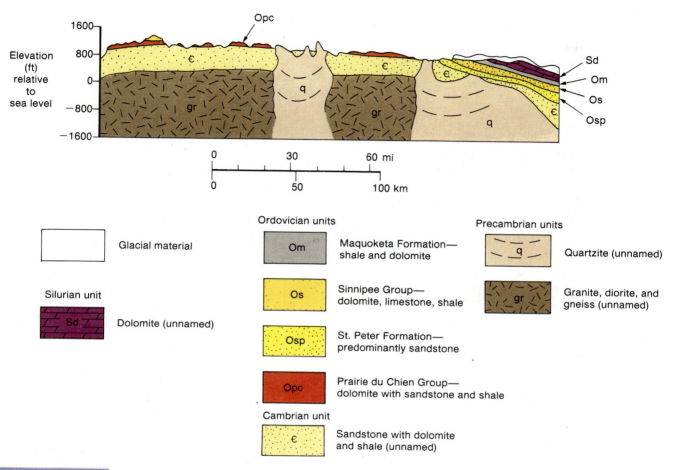

Elevation (ft) relative to sea level

Opc

Sd
Om
Os
Osp

q
gr
gr
q

0   30   60 mi
0   50   100 km

Glacial material

Ordovician units

| Om | Maquoketa Formation—shale and dolomite |

Silurian unit

| Sd | Dolomite (unnamed) |

| Os | Sinnipee Group—dolomite, limestone, shale |

| Osp | St. Peter Formation—predominantly sandstone |

| Opc | Prairie du Chien Group—dolomite with sandstone and shale |

Cambrian unit

| ∈ | Sandstone with dolomite and shale (unnamed) |

Precambrian units

| q | Quartzite (unnamed) |

| gr | Granite, diorite, and gneiss (unnamed) |

Figure A.14

An example of a geologic cross section, indicating subsurface structures. Simplified section across southern Wisconsin. Compare with figure A.7.

After Wisconsin Geological and Natural History Survey (1981).

Cross Sections

Interpreting structure from geologic map patterns and strikes and dips requires some practice and the ability to visualize in three dimensions. Often the maker of a geologic map will provide some assistance to the map reader by supplying one or more geologic **cross sections** (figure A.14). A cross section is a three-dimensional interpretation of the geology seen at the surface. The line along which the cross section is drawn is indicated on the map. The cross section will use the same map units and symbols as the map proper, and will attempt to show the geometric relationships inferred to exist among them—folds, faults, intrusive relationships, and so on.

One draws a cross section by starting with a topographic profile along the chosen line, and marking on it the geology as seen from the surface (figure A.15A). One then devises a structural interpretation that is consistent with all the known data (figure

A.15B). Depending on the complexity of the geology and the completeness of exposure on which the original geologic map has been based, it may or may not be possible to develop a unique structural interpretation for the observed map pattern. If it is not, several plausible alternatives might be presented.

Obtaining Geologic Maps

A variety of geologic maps are available from the U.S. Geological Survey, through the distribution offices given in the section on topographic maps. As with the latter, index maps are available to assist you in selecting the geologic map(s) of interest.

Geologic mapping is a more time-consuming business than topographic mapping, and the U.S.G.S. has not had the personnel or funding to produce geologic maps of every part of the country. However,

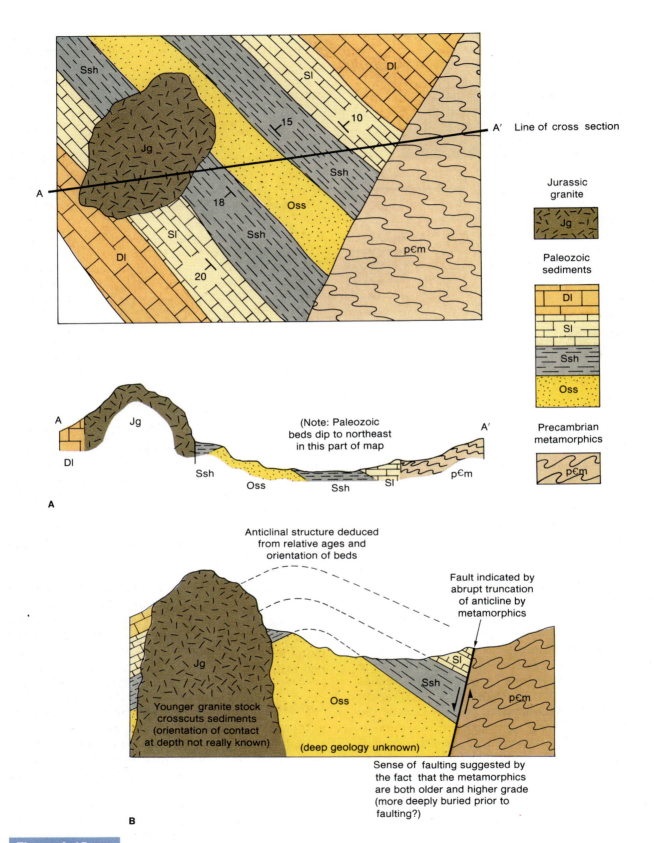

A   Line of cross section

Jurassic granite

Paleozoic sediments

Precambrian metamorphics

(Note: Paleozoic beds dip to northeast in this part of map)

Anticlinal structure deduced from relative ages and orientation of beds

Fault indicated by abrupt truncation of anticline by metamorphics

Younger granite stock crosscuts sediments (orientation of contact at depth not really known)

(deep geology unknown)

Sense of faulting suggested by the fact that the metamorphics are both older and higher grade (more deeply buried prior to faulting?)

Figure A.15

Constructing a cross section. *(A)* Geology as seen at surface is sketched onto a topographic profile. *(B)* The pattern is interpreted, and a set of structures consistent with the pattern seen at the surface is sketched in.

many states have very active geological surveys that have undertaken extensive mapping programs of their own within their home states. The information office of the state of interest should be able to supply you with information about the state geological survey.

Once one has developed an interest in geology, one may begin asking, "What's that?" when observing rocks while traveling, especially in unfamiliar parts of the country. For a quick answer, it may be helpful to consult a *geological highway map*. The American Association of Petroleum Geologists has produced a set of such maps that collectively cover all of the contiguous United States. Given that each map covers several states, the level of detail possible on each map is necessarily limited, but the maps do provide an overview of a region's geology. Information, prices, and maps can be obtained from

American Association of Petroleum Geologists
P.O. Box 979
Tulsa, Oklahoma 74101

Remote Sensing and Satellite Imagery

Remote sensing methods encompass all of those means of examining planetary features that do not involve direct contact. These methods instead rely on detection, recording, and analysis of wave-transmitted energy—visible light, infrared radiation, and others. Radar mapping of surface topography, using airplanes or spacecraft, is one example. Another involves analyzing the light reflected from the surface of a body. In the case of many planets, remotely sensed data may be the only kind readily available (chapter 22). In the case of the earth, remote sensing, especially using satellites, is a quick and efficient way to scan broad areas, to examine regions having such rugged topography or hostile climate that they cannot easily be explored on foot or with surface-based vehicles, and to view areas to which ground access is limited for political reasons. Probably the best known and most comprehensive earth satellite imaging system is the one initiated in 1972, known until recently as Landsat.

Landsat

The Landsat satellites orbit the earth in such a way that images can be made of each part of the earth. Each orbit is slightly offset from the previous one, with the areas viewed on one orbit overlapping the scenes of the previous orbit. Each satellite makes 14 orbits each day; complete coverage of the earth takes 18 days.

Therefore one ought to be able to obtain images of any given area every 18 days, although in practice, shifting distributions of clouds obscure the surface some part of the time at any point. Five Landsat satellites have been launched; a sixth is planned.

The sensors in the Landsat satellites do not detect all wavelengths of energy reflected from the surface. They do not take photographs in the conventional sense. They are particularly sensitive to selected green and red wavelengths in the visible light spectrum, and to a portion of the infrared (invisible heat radiation, with wavelengths somewhat longer than those of red light). These wavelengths were chosen particularly because plants reflect light most strongly in the green and the infrared. Different plants, rocks, and soils reflect different proportions of radiation of different wavelengths. Even the same feature may produce a somewhat different image under different conditions: wet soil differs from dry; sediment-laden water looks different from clear; a given variety of plant may reflect a different spectrum of radiation depending on what trace elements it has concentrated from the underlying soil or how vigorously it is growing. Landsat images can be powerful mapping tools.

Landsat Images and Applications

A common format for Landsat imagery is photographic prints at 1:1,000,000 scale. At that scale, a 23-centimeter (9-inch) print covers 34,225 square kilometers (13,225 square miles). The quality of the resolution is indicated by the fact that the smallest features that can be distinguished in the image are about 80 meters (250 feet) in size. Multiple images can be joined into mosaics covering whole countries or continents.

Images are typically presented either in black and white or as *false-color composites*. The latter are produced by projecting the data for individual spectral regions through colored filters, and superimposing the results. The false-color images are so named because the resulting pictures, though superficially resembling color photographs, do not present all features in the colors they would appear to the human eye. The most striking difference is in vegetation, which appears in shades of red, not green. Rock and soil usually show as white, blue, yellow, or brown, depending on composition; water will be blue to bluish-black, snow and ice white. Examples of false-color Landsat images are used throughout this text. Several more are presented below. Landsat image data can also be further processed by computer, to produce images in more "realistic" (expected) colors, or selectively to enhance particular features by emphasizing certain wavelengths of radiation.

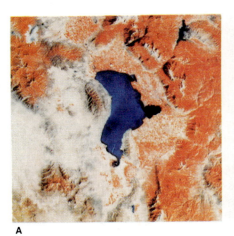
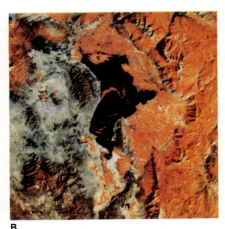
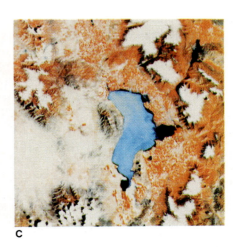

A   B   C

Figure A.16

Seasonal variations in Utah Lake. *(A)* In early August, some incoming sediment clouds the lake. *(B)* By mid-September, dry conditions have greatly reduced the extent of sediment input; water appears dark and clear. *(C)* The high runoff of spring, and the corresponding increased sediment input, is reflected in the very turbid water in this image, taken in late May.

© NASA

Figure A.17

Shifting patterns of sea ice breaking up off Prince Edward Island affect local ferry service.

© NASA

Dozens of applications of Landsat imagery exist in the natural sciences. Uses in basic geologic mapping, identification of geologic structures, and resource exploration have already been noted. The fact that Landsat scans the same area repeatedly over time allows monitoring of seasonal changes (figures A.16, A.17) and of the progress and extent of occasional events such as flooding (figure A.18). Landsat imagery can be used to monitor the development of and changes in surface features such as stream channels and currents (figure A.19). They are helpful in identifying patterns of land use (figure A.20), and in monitoring the progress of crops and the extent of damage to vegetation from fires, insects, or disease. In all cases, Landsat imagery is especially useful when some *ground truth* can be obtained. This is information gathered by direct surface examination (best done at the time of imaging, when vegetation is involved), which can provide critical confirmation of interpretations based on remotely sensed data.

An overview of global diversity as seen through Landsat imagery, and a survey of many applications of remotely sensed data, is given by the volume *Mission to Earth: Landsat Views the World* (by N. M. Short, P. D. Lowman, Jr., S. C. Freden, and W. A. Firch, Jr.), published by NASA in 1976, and available through the U.S. Government Printing Office.

Figure A.18

Extent of flooding can be assessed quickly with satellite images. Here, the 1973 Mississippi River flooding near St. Louis (right; contrast with normal flow, left). Landsat images can contribute to preparation of more accurate flood-hazard maps.

© NASA

Figure A.19

Swirling currents in James Bay, Canada, are marked by patterns of suspended sediment.

© NASA

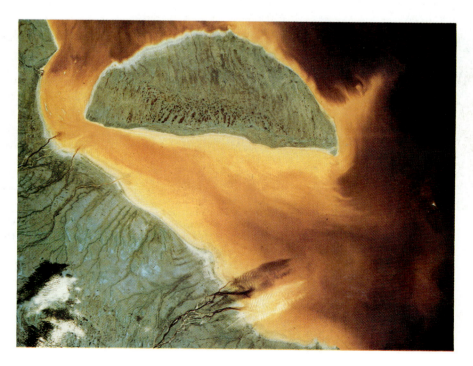

Figure A.20

The extent of municipal and agricultural development in the Imperial Valley of California is clearly shown in this image. Cultivated fields make up the bulk of the bright checkerboard pattern. The U.S./Mexican border (lower right) is sharply defined as a result of differences in land use and agricultural practices in the two countries.

© NASA

Mineral and Rock Identification

Mineral Identification

The following identification chart lists many of the more common minerals that you might encounter. Representative chemical formulas are provided for reference. Some appear complex because of opportunities for solid solution; some have actually been simplified by limiting the range of compositions represented, although additional elemental substitutions are possible.

A few general identification guidelines and comments:

1. Minerals showing metallic luster are usually sulfides (or native metals, but these are much rarer). Native metals have been omitted from the table. Those few that are likely to be encountered, such as native copper or silver, may be identified by their resemblance to household examples of the same metals.
2. Of the nonmetals, the silicates are generally systematically harder than the nonsilicates. Hardnesses of silicates are typically over 5, with exceptions principally among the sheet silicates; many of the nonsilicates, such as sulfates and carbonates, are much softer.
3. Distinctive luster, cleavage, or other identifying properties are listed under "Other Characteristics." In a few cases, this column notes restrictions on the occurrence of certain minerals as a possible clue to identification: for example, "found only in metamorphic rocks," or "often found in pegmatites."

Rock Identification

One approach to rock identification is to decide which of the three major rock types (igneous, sedimentary, metamorphic) the sample is, then look at the detailed descriptions in the corresponding chapters. How does one identify the basic rock type? Here are some general guidelines:

1. Glassy or vesicular rocks are volcanic.
2. Coarse-grained rocks with tightly interlocking crystals are likely to be plutonic, especially if they lack foliation.
3. Coarse-grained sedimentary rocks will differ from plutonic rocks in that the grains in the sedimentary rocks will tend to be more rounded and to interlock less closely. A breccia does have angular fragments, but the fragments in a breccia are typically rock fragments, not individual mineral crystals.
4. Rocks that are not very cohesive, that crumble apart easily into individual grains, are generally clastic sedimentary rocks. One exception would be a poorly consolidated volcanic ash, but this should be recognizable by the nature of the grains, many of which will be glassy shards. (Note, however, that extensive weathering can make even a granite crumble.)

5. More cohesive fine-grained sedimentary rocks may be distinguished from fine-grained volcanics by the fact that the sedimentary rocks are generally softer, and more likely to show a tendency to break along bedding planes. Phenocrysts, of course, indicate a (porphyritic) volcanic rock.

6. Foliated metamorphic rocks are distinguished by their foliation (schistosity, compositional banding). Also, rocks containing abundant mica, garnet, or amphibole are commonly metamorphic rocks.

7. Nonfoliated metamorphic rocks, like quartzite and marble, resemble their sedimentary parents but are harder, denser, and more compact. They may also have a shiny or glittery appearance on broken surfaces, due to recrystallization during metamorphism.

The foregoing guidelines are not infallible, but they should lead to the correct preliminary classification of most rocks commonly encountered.

Mineral	Formula	Color	Hardness	Other characteristics
amphibole	$(Na,Ca)_2(Mg,Fe,Al)_5Si_8O_{22}(OH)_2$	green, blue, brown, black	5–6	often forms needlelike crystals; two good cleavages forming 120° angle
apatite	$Ca_5(PO_4)_3(F,Cl,OH)$	usually yellowish	5	crystals hexagonal in cross section
azurite	$Cu_3(CO_3)_2(OH)_2$	vivid blue	3 1/2–4	often associated with malachite
barite	$BaSO_4$	colorless	3–3 1/2	high specific gravity, 4.5 (denser than most silicates)
beryl	$Be_3Al_2Si_6O_{18}$	aqua to green	7 1/2–8	usually found in pegmatites
biotite	$K(Mg,Fe)_3AlSi_3O_{10}(OH)_2$	black	5 1/2	excellent cleavage into thin sheets
bornite	Cu_5FeS_4	iridescent blue, purple	3	metallic luster
calcite	$CaCO_3$	variable; colorless if pure	3	effervesces in weak acid
chalcopyrite	$CuFeS_2$	brassy yellow	3 1/2–4	
chlorite	$(Mg,Fe)_3(Si,Al)_4O_{10}(OH)_2$	light green	2–2 1/2	cleaves into small flakes
cinnabar	HgS	red	2 1/2	earthy luster; may show silvery flecks
covellite	CuS	blue	1 1/2–2	metallic luster
dolomite	$CaMg(CO_3)_2$	white or pink	3 1/2–4	powdered mineral effervesces in acid
epidote	$Ca_2FeAl_2Si_3O_{12}(OH)$	green	6–7	
fluorite	CaF_2	variable; often green or purple	4	cleaves into octahedral fragments; may fluoresce in ultraviolet light
galena	PbS	silver-grey	2 1/2	metallic luster; cleaves into cubes
garnet	$(Ca,Mg,Fe)_3(Fe,Al)_2Si_3O_{12}$	variable; often dark red	7	glassy luster
graphite	C	dark grey	1–2	streaks like pencil lead
gypsum	$CaSO_4 \cdot 2H_2O$	colorless	2	
halite	$NaCl$	colorless	2 1/2	salty taste; cleaves into cubes
hematite	Fe_2O_3	red or dark grey	5 1/2–6 1/2	red-brown streak regardless of color
kaolinite	$Al_2Si_2O_5(OH)_4$	white	2	earthy luster
kyanite	Al_2SiO_5	blue	5–7	found in high-pressure metamorphic rock; often forms bladelike crystals

Mineral	Formula	Color	Hardness	Other characteristics
magnetite	Fe_3O_4	black	6	strongly magnetic
malachite	$Cu_2CO_3(OH)_2$	green	3 1/2–4	often forms in concentric rings of light and dark green
molybdenite	MoS_2	dark grey	1–1 1/2	cleaves into flakes; more metallic luster than graphite
muscovite	$KAl_3Si_3O_{10}(OH)_2$	colorless	2–2 1/2	excellent cleavage into thin sheets
olivine	$(Fe,Mg)_2SiO_4$	yellow-green	6 1/2–7	glassy luster
phlogopite	$KMg_3AlSi_3O_{10}(OH)_2$	brown	2 1/2–3	mica closely resembling biotite
plagioclase	$(Na,Ca)(Al,Si)_2Si_2O_8$	white to grey	6	may show fine striations on cleavage surfaces
potassium feldspar	$KAlSi_3O_8$	white; often stained pink	6	
pyrite	FeS_2	yellow	6–6 1/2	metallic luster; black streak
pyroxene	$(Na,Ca,Mg,Fe,Al)_2Si_2O_6$	usually green or black	5–7	two good cleavages forming a 90° angle
quartz	SiO_2	variable; commonly colorless or white	7	glassy luster; conchoidal fracture
serpentine	$Mg_3Si_2O_5(OH)_4$	green to yellow	3–5	waxy or silky luster; may be fibrous
sillimanite	Al_2SiO_5	white	6–7	occurs only in metamorphic rocks; often forms needlelike crystals
sphalerite	ZnS	yellow-brown	3 1/2–4	glassy luster
staurolite	$Fe_2Al_9Si_4O_{20}(OH)_2$	brown	7–7 1/2	found in metamorphic rocks; elongated crystals may have crosslike form
sulfur	S	yellow	1 1/2–2 1/2	
sylvite	KCl	colorless	2	cleaves into cubes; salty taste, but more bitter than halite
talc	$Mg_3Si_4O_{10}(OH)_2$	white to green	1	greasy or slippery to the touch
tourmaline	$(Na,Ca)(Li,Mg,Al)(Al,Fe,Mn)_6(BO_3)_3Si_6O_{18}(OH)_4$	black, red, green	7–7 1/2	elongated crystals, triangular in cross section; conchoidal fracture

Unit Conversions

Common Prefixes

deci-	:	one-tenth	one deciliter (dl)	=	0.1 liter
centi-	:	one-hundredth	one centimeter (cm)	=	0.01 meter
milli-	:	one-thousandth	one milliliter (ml)	=	0.001 liter
kilo-	:	one thousand	one kilogram (kg)	=	1000 grams

Units of Length

1 centimeter	=	0.394 inches
1 meter	=	39.37 inches = 1.09 yards
1 kilometer	=	0.621 miles
1 inch	=	2.54 centimeters
1 yard	=	0.914 meters
1 mile	=	1.76 yards = 1.61 kilometers

Units of Area

1 square centimeter	=	0.155 square inches
1 square meter	=	1.20 square yards = 1550 square inches
1 square kilometer	=	0.386 square miles
1 square inch	=	6.45 square centimeters
1 square yard	=	1.296 square inches = 0.836 square meters
1 square mile	=	2.59 square kilometers
1 acre	=	4840 square yards = 4047 square meters

Units of Volume

1 cubic centimeter	=	0.061 cubic inches
1 cubic meter	=	1.31 cubic yards
1 cubic kilometer	=	0.240 cubic miles
1 cubic inch	=	16.4 cubic centimeters
1 cubic yard	=	0.765 cubic meters
1 cubic mile	=	4.17 cubic kilometers

Units of Liquid Volume

1 milliliter	=	0.0338 fluid ounces
1 liter	=	1.06 quarts
1 fluid ounce	=	29.6 milliliters
1 quart	=	0.946 liters
1 gallon	=	4 quarts = 3.78 liters
1 acre-foot	=	326,000 gallons = 1220 cubic meters

Units of Weight or Mass

1 gram	=	0.0353 ounces
1 kilogram	=	2.20 pounds
I metric ton	=	1000 kilograms = 2200 pounds
1 ounce (avoirdupois)	=	28.4 grams
1 pound	=	454 grams = 0.454 kilograms
1 ton	=	2000 pounds = 909 kilograms
1 (troy) ounce	=	1.10 ounce (avoirdupois) = 31.2 grams

Energy Equivalents and Conversions

1 calorie = amount of heat required to raise the temperature of 1 milliliter of water by 1° C

1 BTU (British Thermal Unit) = amount of heat required to raise the temperature of 1 pound of water by 1° F

1 BTU = 252 calories

Average Energy Contents of Various Fossil Fuels

Fuel	Calories	BTU
1 barrel crude oil	1,460,000,000	5,800,000
1 ton coal	5,650,000,000	22,400,000
1 cubic foot natural gas	257,000	1020

Glossary

aa
Rough, blocky lava flow.

ablation
Loss of material from a glacier, by melting, evaporation, or calving.

abrasion
Grinding erosion by rocks entrained in glacial ice, or by windblown sand.

absolute age
Old name for *radiometric age.*

abyssal hills
Low hills, several hundred meters high, found in the abyssal plains.

abyssal plains
The flat areas of the deep ocean basins.

accretionary prism
Wedge of sediment affixed to the edge of a continent at a subduction zone.

acid rain
Rainfall more acidic than typical precipitation, especially as due to sulfur pollution in the atmosphere (forming sulfuric acid).

active margin
A continental margin at which there is significant volcanic and earthquake activity; commonly a convergent plate margin.

aftershocks
Smaller earthquakes that follow a major earthquake.

A horizon
The topmost soil horizon, also known as *zone of leaching;* includes topsoil.

alluvial fan
A wedge-shaped deposit of sediment deposited where a tributary flows into a more slowly flowing stream, or where a mountain stream flows into a desert.

alluvium
Stream-deposited sediment.

alpine glacier
A small glacier found in a mountainous region; commonly a valley glacier.

amphiboles
Hydrous, ferromagnesian double-chain silicates.

amphibolite
A metamorphic rock rich in amphiboles.

amphibolite facies
A medium-grade regional-metamorphic facies, commonly characterized by the production of abundant amphiboles.

angle of repose
The steepest angle at which a slope of unconsolidated material is stable.

angular unconformity
An unconformity at which the bedding of rocks above and below is oriented differently.

anion
Negatively charged ion.

anorthosite
A plutonic rock rich in plagioclase feldspar.

anticline
An arching fold, in which the limbs dip away from the axis.

aquiclude
Rock that is effectively impermeable on a human time scale.

aquifer
Rock sufficiently porous and permeable to be useful as a source of water.

aquitard
Rock of low permeability, through which water flows very slowly.

arête
Ridge left as parallel valley glaciers erode to either side of it.

artesian
A system in which groundwater in a confined aquifer is under extra hydrostatic pressure, so that it can rise above the aquifer containing it.

ash
A fine pyroclastic material.

assimilation
Process by which magma incorporates and melts bits of country rock.

asteroid
A rocky solar-system object, much smaller than a planet; most asteroids are found in the asteroid belt.

asteroid belt
The region between Mars and Jupiter in which most asteroids are found.

asthenosphere
Weak, plastic, partly molten layer of the upper mantle directly below the lithosphere.

atom
The smallest particle into which a chemical element can be subdivided.

atomic mass number
The sum of the number of protons and the number of neutrons in a particular atomic nucleus.

atomic number
The number of protons characteristic of a particular element.

aureole
The contact-metamorphic zone around a pluton.

axial plane
The plane dividing the two limbs of a fold.

axial trace
The intersection of the axial plane of a fold with the land surface.

axis
Line of intersection of the axial plane with the surface of a fold.

backwash
The return flow of swash to the sea.

bankfull
The condition in which stream stage just equals stream-bank elevation.

barchan
A crescent-shaped transverse dune, with arms pointing downwind.

barrier island
A long, low, narrow island parallel to a coast.

baseflow
Stream flow supported by groundwater in adjacent rock or soil.

base level
Ordinarily, the level of the water surface at a stream's mouth; the lowest level to which a stream can cut down.

basin
A syncline in which rocks dip inward on all sides; also, any depression in which sediments are deposited.

batholith
A massive, discordant pluton, often produced by multiple intrusions.

beach
A gently sloping shore covered with sediment, washed by waves and tides.

beach face
That portion of the beach exposed to direct wave and swash action.

bedding
Depositional layering, as in sedimentary rocks.

bed load
The amount of material moved by a stream along its bed, by rolling or by saltation.

bedrock geology
The geology as it would appear with the overlying soil and vegetative cover stripped away.

Benioff zone
Zone of earthquake foci dipping into the mantle away from a trench, resulting from subduction of lithosphere.

berm
A flat or gently sloping zone behind a beach face.

B horizon
The soil layer at intermediate depth, also known as *zone of accumulation*.

biomass
The total mass of all living organisms; also, fuel derived from modern organisms.

block
A pyroclastic fragment coarser than cinders.

blueschist
A high-pressure metamorphic facies, not encountered under conditions of normal geothermal gradient and burial pressures.

body waves
Seismic waves that travel through the earth's interior (P-waves and S-waves).

bomb
A pyroclastic fragment formed from an erupted blob of magma, which may take on a streamlined shape during flight.

Bowen's Reaction Series
The predicted sequence of crystallization of principal silicates from a magma.

braided stream
A stream with multiple channels that divide and rejoin.

breeder reactor
A nuclear fission reactor designed to produce additional fissionable fuel while it operates to generate energy.

brittle
Tending to rupture rather than to deform under stress.

caldera
A large bowl-shaped summit depression in a volcano; may be formed by explosion or collapse.

calving
The formation of icebergs as chunks of ice break off a glacier that terminates in water.

capacity
The maximum total load a given stream can move.

capillary action
The movement of water toward drier soil through fine pores in the soil and along the grain surfaces.

carbonaceous chondrites
Stony meteorites rich in volatiles, including carbonates, sulfates, and hydrous minerals.

carbonate
A nonsilicate mineral containing carbon and oxygen in (CO_3) group.

catastrophism
A now-discredited theory that explained earth's history as a static one, punctuated by global catastrophes that were the only agents of change.

cation
A positively charged ion.

cementation
The process by which sediments are stuck together through the deposition of mineral material between grains.

Cenozoic
The most recent era of geologic time, from 66 million years ago to the present.

chain reaction
Sequence of fission events in which each fission event triggers further fission of atomic nuclei.

chain silicates
Silicates in which silica tetrahedra are linked in one dimension by the sharing of oxygen atoms.

channelization
The modification of a stream channel; for example, the straightening of meanders or dredging of the channel to deepen it.

chemical maturity
The extent to which sediment has been depleted in soluble or easily weathered minerals.

chemical sediment
A sediment precipitated directly from solution.

chemical weathering
The solution or chemical breakdown of minerals by reaction with water, air, or dissolved substances.

chilled margin
The fine-grained rock at the margin of a pluton, showing the effects of rapid cooling.

C horizon
The deepest soil layer, consisting mainly of coarse chunks of bedrock.

cinder cone
A volcanic structure built of pyroclastic materials.

cinders
Pyroclastic material of intermediate size, with fragments ranging up to several centimeters across.

cirque
A bowl-shaped depression formed at the head of an alpine glacier.

clastic
Rock or sediment made of fragments of pre-existing rocks and minerals.

cleavage (mineral)
The tendency of a mineral to break preferentially along planes in certain directions in the crystal structure.

cleavage (rock)
The tendency of rock to break along parallel planes, corresponding to planes along which platy minerals are aligned.

closed system
A system that neither gains nor loses matter; an isolated system with respect to mass transfer.

coastline
The zone at which land and water meet; also, the geometry of this zone.

columnar jointing
The development of polygonal columns in a lava flow during cooling.

coma
The foggy zone around a comet nucleus, formed as some of the ices vaporize upon heating.

comet
A small clump of ice and silicate dust that originates in the Oort cloud; its orbit may carry it through the inner solar system.

compaction
The compression and consolidation of sediment under compressive stress.

competence
The largest size of particle a stream can move as bed load.

composite volcano
A volcanic cone formed of interlayered lava flows and pyroclastics.

compound
A chemical combination of two or more elements, in specific proportions, with a distinct set of physical properties.

compressive stress
Stress tending to compress or squeeze an object.

concentration factor
Enrichment of an ore in a metal of interest, relative to that metal's concentration in average crustal rock.

concordant
Having contacts parallel to the structure in adjacent rocks.

cone of depression
A conical depression of the water table or potentiometric surface caused by pumped extraction of groundwater.

confined aquifer
An aquifer overlain by an aquiclude or aquitard.

confining pressure
The directionally uniform pressure to which rocks at depth are subjected.

contact metamorphism
The metamorphism characteristic of wallrocks surrounding a pluton, subjected to locally increased temperature only.

continental drift
The concept that continents have shifted in position over the earth.

continental glacier
A thick, extensive ice cap or ice sheet covering a significant portion of a continent.

continental rise
Gently sloping region between the foot of the continental slope and the abyssal plains.

continental shelf
The nearly level, shallowly submerged zone immediately offshore from a continent; water depths on the shelf are typically less than 100 meters.

continental slope
The continental-marginal zone extending from the outer edge of the continental shelf down to the more gently sloping ocean depths (continental rise or abyssal plains).

contour interval
The difference in elevation represented by two successive contours on a topographic map.

contour line
A line joining points of equal elevation on a map.

convection cell
A circulating mass of material (in air, water, or asthenosphere) in which warm material rises, moves laterally, cools, sinks, and is re-heated, cycling back to rise again.

convergence (circulation)
A zone in which opposing water currents meet.

convergent boundary
A plate boundary at which two plates are moving together: subduction zone or zone of continental collision.

coral atoll
A ring-shaped coral reef structure, formed around an island now submerged or eroded away.

core
The innermost zone of the earth, rich in iron; the outer core is molten, the inner core solid.

correlation
Determination that two or more distinct rock units are of the same age or are related in origin or history.

cosmic abundance curve
A graph depicting the relative abundances of the chemical elements as a function of atomic number.

country rock
Rock into which a pluton is intruded.

covalent bond
A bond formed by the sharing of electrons between atoms.

crater (impact)
A structure excavated by the force of a meteorite striking a planetary surface.

craton
The stable continental interior.

creep (fault)
A slow, gradual, more-or-less continuous slippage along a fault zone.

creep (mass movement)
Very slow mass movement, not noticeable during direct observation.

crest (flood)
The maximum stage reached during a flood event.

crest (wave)
The highest point on a wave.

crevasses
Deep vertical cracks in brittle glacier ice.

cross-bedding
A sequence of inclined sedimentary beds deposited by flowing wind or water.

cross section
An interpretation of geology and structure in the third (vertical) dimension based on rock exposures and attitudes at the surface; drawn in a particular vertical plane.

crust
The outermost compositional shell of earth, 10 to 40 kilometers thick, consisting predominantly of relatively low-density silicates.

crystal form
External shape of crystals; distinguished from internal crystal structure.

crystalline
Solid having a regular, repeating geometric arrangement of atoms.

Curie temperature
Temperature above which a magnetic material loses its magnetization; different for each such material.

daughter nucleus
A nucleus produced by radioactive decay.

debris avalanche
A flow involving a wide range of types and sizes of objects.

deep layer
The deepest water in the oceans, with temperatures often near freezing; circulates very slowly.

deflation
The removal of sediment by wind.

delta
A sediment wedge deposited at a stream's mouth.

dendritic drainage
An irregular, branching drainage pattern.

desert
A region having so little vegetation that it is incapable of supporting a significant population.

desertification
The rapid conversion of marginally habitable arid land into true desert, typically accelerated by human activities.

desert pavement
A surface of coarse rocks protecting finer sediment below; formed by the selective removal of fine surficial material.

dewatering
The release of water from pores and/or breakdown of hydrous minerals under conditions of increasing pressure or temperature.

diagenesis
Set of processes by which lithification is accomplished; occurs at lower temperatures than metamorphism.

dike
A tabular, discordant pluton.

dilatancy
The model according to which cracks open in stressed rock, allowing fluids to seep in, before an earthquake occurs and the rocks snap back elastically to their unstressed condition.

diorite
A plutonic rock of intermediate composition, consisting of ferromagnesian minerals and feldspar, with little quartz, and lacking olivine.

dip
The angle made by a line or plane with the horizontal.

dip-slip fault
A fault along which movement is vertical (parallel to dip).

directed stress
Stress that is not uniformly intense in all directions.

discharge
The volume of water flowing past a point in a given period of time; equal to flow velocity multiplied by the cross-sectional area of the channel.

disconformity
An unconformity at which the bedding of rocks above and below lie parallel.

discordant
Having contacts that cut across or are set at an angle to the structure of the adjacent rocks.

dissolved load
The quantity of material carried in solution by a stream.

divergent boundary
A plate boundary at which the plates are moving apart: an ocean spreading ridge or continental rift zone.

divide
A topographic high separating two drainage basins.

dolomite
A carbonate mineral, $CaMg(CO_3)_2$, or the chemical sedimentary rock made predominantly of that mineral.

dome
An anticline dipping radially in all directions.

dormant
Said of a volcano presently inactive, but believed capable of future eruption.

downcutting
The downward erosion by a stream toward its base level.

downstream floods
Prolonged flooding affecting broad areas or a whole drainage basin.

drainage basin
The area from which a stream system draws its water.

drowned valley
A stream or glacial valley partially flooded by seawater; occurs on a submergent coastline.

drumlins
Elongated mounds of till oriented parallel to ice flow.

dune
A low mound or ridge of sediment deposited by wind.

dynamic equilibrium
The condition in which two opposing processes are in balance; for a stream, the condition in which erosion and deposition in the channel are equal.

dynamothermal metamorphism
Regional metamorphism.

earthquake
Ground displacement associated with the sudden release, in the form of seismic waves, of built-up stress in the lithosphere.

ecliptic
The plane in which the sun and the orbits of all the planets in the solar system (except Pluto) lie.

ejecta
Fragments of clastic material flung out of an impact crater during its formation.

elastic
Deforming in such a way that deformation is proportional to applied stress; the material will return to its original dimensions when the stress is removed.

elastic limit
The stress beyond which material no longer behaves elastically.

elastic rebound
The phenomenon whereby stressed rocks behave elastically before and after an earthquake, returning afterward to an undeformed, unstressed condition.

electron
A negatively charged subatomic particle, found outside the atomic nucleus.

element
The simplest kind of chemical substance; elements cannot be decomposed further by chemical or physical means.

end moraine
A ridge of till accumulated at the end of a glacier.

enhanced recovery
Any of several methods used to increase the proportion of oil and/or gas extracted from a petroleum reservoir.

eolian
Formed by or related to wind action.

ephemeral stream
A stream that flows only occasionally in direct response to precipitation.

epicenter
The point on the earth's surface directly above an earthquake's focus.

era
Major subdivision of the geologic time scale.

erratic
An isolated large boulder not derived from local bedrock; a depositional feature of glaciers.

esker
A winding ridge of till deposited by a stream flowing in and under a melting glacier.

estuary
A coastal body of brackish water, open to the sea.

eustatic
Describes a simultaneous worldwide rise or fall of sea level.

evaporite
A mineral deposit formed by the evaporation of seawater in a restricted basin; also, the minerals of such deposits.

Exclusive Economic Zone
The territory extending 200 miles outward from a nation's shoreline, within which it has the exclusive right to exploit marine resources.

exfoliation
The breakup of exposed plutonic rocks in concentric sheets; due to release of stress by unloading.

extinct
Said of a volcano expected never to erupt again.

facies (metamorphic)
The set of pressure-temperature conditions that leads to a particular, distinctive metamorphic mineralogy or rock type.

facies (sedimentary)
The set of conditions that leads to the formation of a particular type of sediment or sedimentary rock; also, the rock or sediment so formed.

fall
A free-falling mass movement in which the moving mass does not always remain in contact with the land surface.

fault
A planar break in rock along which there is movement of one side relative to the other.

feldspars
One group of framework silicates, containing aluminum and calcium, sodium, or potassium; collectively, the most abundant minerals in the crust.

felsic
Rock rich in feldspar and silica (quartz).

ferromagnesian
Silicate containing significant iron and/or magnesium.

firn
Dense, coarsely crystalline snow partially converted to ice.

first-motion studies
Seismic studies that identify the sense of displacement along a fault by noting whether the first motion at various surrounding stations was compressional or dilatational.

fission
The process by which large atomic nuclei are split into smaller ones.

fissure eruption
An eruption of lava from a crack rather than from a pipelike vent.

flash flooding
A rapid rise of stream stage, common with ephemeral streams or after intense local precipitation events.

floodplain
The nearly flat area around a stream channel, into which the stream overflows during floods.

flood stage
The condition in which stream stage is above channel bank elevation, so that the stream overflows its banks.

flow
A mass movement in unconsolidated material in which the material moves in a chaotic or disorganized fashion, rather than as a coherent unit.

fluid injection
A proposed means of increasing pore fluid pressure and decreasing shear strength along locked faults, in order to release built-up stress.

focus
The point of first break along a fault during an earthquake.

foliation
A texture, usually metamorphic, involving parallel alignment of linear or planar minerals, or compositional banding.

fossil
The remains or evidence of ancient life.

fossil fuels
Any of the carbon-rich fuels produced through heat, pressure, and time from the remains of organisms.

fractional crystallization
The crystallization of magma with early crystals removed or isolated from later reaction with the remaining melt.

fracture
Irregular breakage; contrasted with *cleavage*.

framework silicate
A silicate in which silica tetrahedra are linked in three dimensions by shared oxygen atoms.

frost wedging
The breakup of rock by the expansion of water freezing in cracks.

fumarole
A steam vent, caused when subsurface water is heated by shallow magma or hot rock.

fusion
The process by which small atomic nuclei are combined to form larger ones.

gabbro
A mafic plutonic rock rich in ferromagnesians and plagioclase feldspar.

gasification
The conversion of coal to a gaseous fuel such as methane.

geopressurized zones
Deeply buried regions in which natural gas is dissolved in pore waters under pressure.

geosyncline
A large syncline, of regional scale.

geothermal
Related to the heat of the earth's interior; geothermal power involves extraction of that heat through circulating subsurface water.

geothermal gradient
The increase in temperature with depth in the earth.

geyser
A feature characterized by intermittent ejection of hot water and steam, heated by shallow magma or hot rocks.

glacier
A mass of ice, on land, that moves under its own weight.

glass
A solid lacking a regular crystal structure, in which atoms are randomly arranged.

gneiss
A metamorphic rock showing banded texture, the banding usually defined by differences in mineralogy between bands.

graben
A downdropped block bounded by steeply dipping faults.

graded bedding
Vertical progression of grain sizes within a sediment layer, from coarse to fine or vice versa.

graded stream
A stream in dynamic equilibrium.

gradient
The steepness or slope of a stream channel along its length.

granite
A plutonic rock rich in quartz and potassium feldspar.

granulite facies
The highest-grade regional-metamorphic facies.

gravity anomaly
A local or regional variation in the force of gravity as measured near the earth's surface.

greenhouse effect
Atmospheric heating resulting from the trapping of heat by CO_2 and other gases in the atmosphere.

greenschist
A low-grade regional-metamorphic facies, named for the common presence of the green minerals chlorite and epidote in rocks of this facies.

groundmass
The finer-grained matrix of a porphyritic rock.

ground moraine
A sheet of moraine left by a melting glacier.

ground truth
Check of the interpretation of remotely sensed data by direct contact (observation on the ground).

groundwater
Water in the saturated zone, below the water table.

group
A set of related sedimentary rock formations, usually having a common history.

gullying
The formation by water of large erosional channels on a sloping soil surface.

guyot
A flat-topped seamount.

gyre
A nearly closed, oval or circular water circulation pattern.

half-life
The length of time required for half of an initial quantity of a radioisotope to decay.

halide
A nonsilicate containing a halogen element (Cl, F, Br, I).

hanging valley
The valley of a tributary glacier; smaller than the main glacier valley, and having a higher floor.

hardness
The ability to resist scratching; measured on the Mohs scale of relative hardness.

hard water
Water containing high concentrations of dissolved calcium, magnesium, and iron.

headward erosion
The cutting back of a stream channel at its source.

heap leaching
The extraction of metals from tailings by the use of percolating solutions.

heat flow
The outward flow of heat from the earth's interior to the surface.

height (wave)
The difference in elevation between a wave's crest and trough.

highland (lunar)
The light-colored, higher-elevation region of lunar crust, composed largely of anorthosite.

horn
A peak formed by headwall erosion of several alpine glaciers diverging from the same topographic high.

hornblende-hornfels facies
A moderate-temperature contact-metamorphic facies characterized by abundant amphiboles.

hornfels
A contact-metamorphic rock formed under conditions of low to intermediate temperature.

horst
An uplifted block bounded by high-angle faults.

hot dry rock
A potential geothermal resource; characterized by above-average heat flow, but lacking abundant subsurface water.

hot spot
An isolated area of active volcanism not associated with a plate boundary.

hot springs
Springs heated by shallow magma bodies or young, hot rocks.

hydrograph
A graph of stream stage or discharge as a function of time.

hydrologic cycle
The cycle of precipitation, evaporation, infiltration, and migration of the water in the hydrosphere.

hydrosphere
All water at and near the earth's surface not chemically bound in rocks.

hydrostatic pressure
Fluid pressure.

hydrothermal
Literally, "hot water"; said of processes or ore deposits related to circulating subsurface water warmed by shallow magma or hot rock; the hydrothermal fluids commonly contain dissolved minerals and gases.

hydrothermal vents
Areas along spreading-ridge systems where waters heated by reaction with new lithosphere emerge into the colder ocean.

hydrous
Containing water or hydroxyl (OH^-) ions.

hypothesis
A conceptual model or explanation for a set of data, measurements, or observations.

ice age
Period of very extensive continental glaciation; when capitalized ("Ice Age"), it refers to the last such episode, 2 million to 10,000 years ago.

ice wedging
Frost wedging.

igneous
Formed from or related to magma.

incised meanders
Meanders cut deeply into rock, with little or no floodplain at channel level.

index minerals
Minerals stable over a restricted range of pressure and/or temperature conditions, useful in evaluating metamorphic grade attained.

inert
Not tending to bond or form compounds with other elements.

infiltration
The process by which water percolates into the ground.

intensity
The size of an earthquake as measured by its effects on structures; one earthquake may have several intensities, decreasing with increasing distance from the epicenter.

internal drainage
Stream drainage into an enclosed, landlocked basin.

ion
An electrically charged atom.

ionic bond
A bond formed by attraction between cations and anions.

irons
Meteorites composed predominantly of iron-nickel alloy.

island arc
A line or arc of volcanic islands formed over, and parallel to, a subduction zone overlain by oceanic lithosphere.

isograd
A line on a map connecting points of equal metamorphic grade, as determined by index minerals.

isostasy
The tendency of crust and lithosphere to float at an elevation consistent with the density and thickness of the crustal rocks relative to underlying mantle.

isostatic equilibrium
The condition in which the mass of rock above a given level in the earth is everywhere the same.

joint
A planar break in rock without relative movement of rocks on either side of the break.

joint set
A set of parallel joints.

Jovian planets
The large, gaseous planets Jupiter, Saturn, Uranus, and Neptune.

karst topography
Topography characterized by abundant sinkholes and other solution features.

kettle
A hole in glacial outwash, formerly occupied by a block of stranded ice.

kimberlite
A pipelike igneous rock body, originating up to 200 kilometers deep in the mantle, in which diamonds may be found.

knickpoint
An abrupt change in streambed elevation—for example, at a waterfall.

laccolith
A concordant pluton with a flat bottom and domed country rocks above.

lahar
A volcanic mudflow deposit.

lamination
Very fine or thin bedding.

landslide
Any rapid mass movement; contrasted with *creep*.

lateral moraine
Moraine deposited at the sides of a valley glacier.

lateritic soil
Extensively leached soil, characteristic of tropical climates.

lava
Magma that flows out at the earth's surface.

law
A basic concept or mathematical relationship that is invariably found to be true.

Law of Faunal Succession
The concept that organisms, and thus fossil forms, change through time, each specific form corresponding to a unique period of earth history.

leaching
The removal of soluble chemicals by infiltrating or percolating water.

levees
Ridges along the bank of a stream; may be natural or artificial.

limbs
The two sides of a fold, on either side of its axial plane.

limestone
A carbonate-rich (especially calcite-rich) chemical sedimentary rock.

liquefaction (coal)
The conversion of coal to a liquid hydrocarbon fuel such as gasoline.

liquefaction (seismic)
A quicksand-like condition with loss of soil strength that occurs when water-saturated soil is shaken by seismic waves.

lithification
The conversion of sediment into rock.

lithosphere
The rigid outermost layer of the earth, 50 to 100 kilometers thick, encompassing the crust and uppermost mantle.

littoral drift
Sand movement along the length of a beach, which occurs in the presence of longshore currents.

load
The total amount of material moved by a stream.

locked fault
A fault on which friction is preventing stress release through creep; no displacement is occurring.

loess
Silt-sized sediment deposited by wind.

longitudinal dunes
Dunes elongated parallel to the direction of wind flow.

longitudinal profile
A diagram of the elevation of a stream bed along its length.

longshore current
The net current parallel to a coastline, caused when waves approach the shore at an oblique angle.

lopolith
A concordant pluton with sagging floor, one that is concave upward.

low-velocity layer
The zone within the upper mantle characterized by lower seismic velocities than layers immediately above and below it, a result of plastic behavior and/or partial melting of rocks in this zone.

luster
The surface sheen exhibited by a mineral.

mafic
A rock, magma, or mineral rich in iron and magnesium.

magma
A silicate melt, usually containing dissolved volatiles, sometimes also containing crystals.

magma mixing
The process by which two compositionally dissimilar magmas are combined into one.

magnetic anomaly
A local or regional variation in magnetic orientation and/or strength of magnetization of rocks.

magnitude
The size of an earthquake as measured by vertical ground displacement near the epicenter.

manganese nodules
Lumps of manganese and iron oxides and hydroxides, with other metals, found on the sea floor.

mantle
The zone of the earth's interior between crust and core; rich in ferromagnesian silicates.

map unit
A distinct, identifiable rock unit used in preparing a geologic map.

marble
Metamorphosed limestone.

maria
Lunar "seas": darker, low-relief crustal regions, often floored with basalt (singular: *mare*).

mascons
"Mass concentrations": lunar craters having positive gravity anomalies.

mass extinctions
Events in the fossil record characterized by the rapid disappearance of many types of organisms within a geologically short time.

mass spectrometer
An instrument used in radiometric dating to measure relative quantities of different isotopes.

mass wasting
The downslope movement of material under the influence of gravity.

matrix
The fine-grained sediment filling spaces between coarse clasts in poorly sorted clastic sediment or rock.

meanders
Lateral bends in a stream channel.

mechanical weathering
The physical breakup of rocks, without change in composition.

medial moraine
Moraine formed by the joining of lateral moraines as tributary glaciers flow into a valley glacier.

Mesozoic
The middle era of the Phanerozoic; the time from about 245 to 66 million years ago.

metamorphic grade
Measure of the intensity of metamorphism to which a metamorphic rock was subjected.

metamorphism
Literally, "change in form" of rocks, brought about particularly through the application of heat and pressure.

metasomatism
The introduction of ions in solution into a rock, and the resulting alteration of that rock.

meteor
A meteoroid that has become visible as a result of heating during its fall through the earth's atmosphere.

meteorite
A meteor that has survived its fall through earth's atmosphere to reach the surface.

meteoroid
A rocky solar-system object too small to be seen until it strikes the earth's atmosphere.

micas
A group of sheet silicates characterized by the tendency to cleave well between sheets of silica tetrahedra.

migmatite
"Mixed rock," partly melted during metamorphism, having a mix of igneous and metamorphic characteristics.

milling
Erosion by water-borne sediment.

mineral
A naturally occurring, inorganic solid element or compound, with a definite composition or compositional range, and a regular internal crystal structure.

mineralogical maturity
The extent to which a sediment has been depleted in easily weathered minerals and enriched in resistant ones.

mineraloid
A material satisfying the definition of a mineral except that it lacks a regular internal crystal structure.

Moho
The Mohorovičić discontinuity.

Mohorovičić discontinuity
The boundary between the crust and mantle.

moraine
A landform composed of till.

mouth
Where a stream reaches its base level, or terminates.

mud cracks
Cracks in fine sediment caused by shrinkage during dehydration.

mudstone
A very fine-grained clastic sedimentary rock; siltstone or claystone.

native element
A mineral consisting of a single chemical element.

neap tides
The least extreme tides, which occur when sun and moon are at right angles relative to the earth.

neck (volcanic)
A pipelike, discordant pluton.

neutron
An electrically neutral subatomic particle, generally found in an atomic nucleus.

nonrenewable
Said of a resource that is not being produced at a rate comparable to that at which it is being consumed.

normal fault
A dip-slip fault in which the hanging wall moves down relative to the footwall.

nose
The most sharply curved part of a fold.

nucleus (atomic)
The central unit of an atom, containing protons and neutrons.

nucleus (comet)
The solid central mass of dust and ice in a comet.

nuée ardente
A hot, glowing cloud of volcanic ash and gas, so dense that it flows down the volcano's slopes.

obduction
The process by which a segment of lithosphere is emplaced atop another.

oblique-slip fault
A fault involving both strike-slip and dip-slip movements.

oil shale
Sedimentary rock containing a waxy solid hydrocarbon, kerogen.

olivine
A ferromagnesian silicate with a structure consisting of individual silica tetrahedra.

oolites
Spheroidal carbonate grains formed in concentric layers.

Oort cloud
A swarm of comets circling the sun at an average distance of 7½ trillion kilometers (about 5 trillion miles).

ooze
A fine-grained, water-rich siliceous or calcareous pelagic sediment of biogenic origin.

ophiolite
A complex assemblage of marine sediments, mafic and ultramafic rocks, found on a continent; generally believed to be a piece of obducted oceanic lithosphere.

order
Hierarchical rank of a stream based on the existence and complexity of tributaries: a first-order stream has no tributaries, a second-order stream has only first-order streams as tributaries, and so on.

ore
A rock in which a valuable or useful metal occurs in sufficient concentration to be economic to mine.

orogenesis
The set of plutonic, metamorphic, and tectonic processes involved in mountain building.

orogeny
The period of time over which orogenesis occurs in a particular mountain range.

outwash
Glacial sediment moved and deposited by meltwater.

overbank deposits
Alluvium deposited outside a stream channel during flooding.

overturned fold
The extreme of a recumbent fold, in which the axial plane is tilted past the horizontal.

oxbows
Cut-off meanders.

oxide
A nonsilicate containing oxygen with one or more metals.

pahoehoe
A lava flow with a smooth, ropy appearance.

paired metamorphic belts
Parallel belts, one of high-pressure, low-temperature metamorphic rocks and the other of moderate-pressure, moderate- to high-temperature metamorphic rocks, formed at convergent plate boundaries.

paleomagnetism
"Fossil magnetism" preserved in rocks, reflecting the orientation of the earth's magnetic field at the time magnetization was acquired.

Paleozoic
The oldest era in the Phanerozoic; the time from approximately 570 to 245 million years ago.

Pangaea
A single supercontinent that existed approximately 200 million years ago.

parabolic dunes
Crescent-shaped dunes with arms pointing upwind.

parent (radioactive)
An unstable, decaying nucleus.

parent (rock)
The original rock from which a metamorphic rock was formed.

partial melting
The melting of only a portion of a rock.

passive margin
A geologically quiet continental margin, lacking significant volcanic or seismic activity.

pedalfer
A moderately leached soil rich in iron and aluminum oxides and hydroxides.

pediment
A gently sloping bedrock surface at the foot of mountains bordering a desert.

pedocal
A soil retaining many of its soluble minerals, especially calcite.

pegmatite
A very coarse-grained igneous rock.

pelagic sediments
Fine-grained sediments of the open ocean.

perched water table
A water table locally elevated above the regional water table, as by the presence of an impermeable lens of rock.

periodic table
The regular arrangement of chemical elements that reflects patterns of chemical behavior related to the electronic structure of atoms.

period (time)
A subdivision of an era of the geologic time scale.

period (wave)
The length of time between the passage of two successive wave crests or troughs by a fixed point.

permafrost
The permanently frozen soil found in many cold regions.

permeability
A measure of the ease with which fluids move through rocks or sediments.

Phanerozoic
A major division (eon) of the geologic time scale, 570 million years ago to the present, spanning the time over which complex life forms have been abundant.

phase change
A change in mineralogy, crystal structure, or physical state, with no gain or loss of chemical elements.

phenocryst
A coarse crystal in a porphyritic rock.

photovoltaic cell
A device for the direct conversion of sunlight to electricity; also known as *solar cell*.

phreatic eruption
A volcanic steam explosion caused when magma heats subsurface water.

phreatic zone
The zone of saturation, in which pores in rock or soil are filled with water.

phyllite
A fine-grained metamorphic rock formed by progressive metamorphism of slate, in which cleavage planes shine with light reflected from small mica flakes.

physical geology
That branch of geology concerned particularly with the materials of and physical processes that shape the earth.

pillow lava
A lava flow with a bulbous surface, made up of ''pillows'' or lobes with a glassy, quenched rind; formed from lava extruded under water.

placer
A deposit of dense or resistant minerals concentrated by stream action.

plastic
Behavior in which deformation is not proportional to applied stress, and is permanent; the material stays deformed when stress is removed.

plateau (oceanic)
A broad topographic high, shallowly submerged or slightly exposed, within the ocean basins; often underlain by continental crust.

plate tectonics
The theory according to which the lithosphere is broken up into a series of rigid plates that can move over the earth's surface.

playa
A ''dry lake,'' floored by fine sediment, formed in a desert having internal drainage.

plucking
Glacial erosion caused as ice freezes onto rock, then moves on, tearing away rock fragments.

plume
A rising column of magma in the asthenosphere.

plunging fold
A fold with a dipping (nonhorizontal) axis.

pluton
A body of plutonic rock.

plutonic
Igneous rock crystallized at depth.

pluvial lakes
Lakes formed through abundant rainfall during ice ages.

point bar
A sedimentary feature built in a stream channel on the inside of a meander, or anywhere the water slows.

polar-wander curve
A curve mapping past magnetic pole positions relative to a given region or continent.

polymorphs
Minerals having the same chemical composition but different crystal structures.

porosity
The proportion of void space (cracks, pores) in a rock.

porphyry
An igneous rock with coarse crystals in a fine-grained groundmass.

potentiometric surface
A theoretical surface indicating the elevation corresponding to hydrostatic pressure in a confined aquifer; analogous to the water table in an unconfined aquifer.

Precambrian
The eon spanning the time from the formation of the earth to the start of the Phanerozoic.

precursor phenomena
Detectable changes that occur prior to volcanic eruptions or earthquakes, which might be used in prediction efforts.

Principle of Original Horizontality
The concept that sedimentary rocks are generally deposited in horizontal layers, so that deviations from the horizontal reflect post-depositional disturbance.

Principle of Superposition
The concept that in an undisturbed sedimentary section, the rocks on the bottom are the oldest, and the overlying rocks progressively younger toward the top of the sequence.

proton
A positively charged subatomic particle, generally found in an atomic nucleus.

P-waves
Compressional seismic body waves.

pyroclastics
Fragments of rock and lava emitted during an explosive volcanic eruption.

pyroxene-hornfels facies
A high-temperature contact-metamorphic facies, characterized by the production of abundant pyroxene.

pyroxenes
Single-chain silicates, mostly ferromagnesian.

quartz
The simplest framework silicate, with formula SiO_2.

quartzite
Metamorphosed quartz-rich sandstone.

quick clay
An unstable, failure-prone clay sediment, derived from glacial rock flour deposited in a marine setting, weakened by later flushing with fresh water.

radial drainage
Streams radiating outward from a topographic high, such as a mountain.

radioactivity
The spontaneous decay or breakdown of unstable atomic nuclei.

radiometric age
A numerical date determined by the use of radioisotopes.

rain shadow
The dry zone landward of coastal mountain ranges, caused by loss of moisture from air passing over mountains.

recessional moraine
End moraine deposited by a retreating glacier during stationary periods.

recharge
Set of processes by which groundwater is replenished.

rectangular drainage
A fracture-controlled drainage pattern in which streams make right-angle bends.

recumbent fold
A fold in which the axial plane is close to horizontal.

recurrence interval
The average length of time between floods of given severity on a given stream.

refraction
The deflection or change in direction of body waves as they move across a boundary between two materials of different properties.

regional metamorphism
Metamorphism on a large, or regional, scale; may be associated with mountain building; involves increases of both pressure and temperature.

regolith
Surficial sediment deposit in place, analogous to soil but not capable of supporting plant life; generally lacks organic matter.

regression
A long-term seaward retreat of the shoreline.

relative dating
Determining the sequence of rocks or events indicated by a particular rock section.

remote sensing
Investigation or examination using light or other radiation rather than by direct contact; examples are the use of aerial photography, satellite imagery, or radar.

reserves
The quantity of a mineral or fuel that has been located and can be exploited economically with existing technology.

reservoir rock
Rock in which oil and gas deposits are found.

resources
Reserves, plus that quantity of a useful mineral or fuel known or believed to exist but not exploitable economically with existing technology.

retention pond
A basin used to hold water back from a stream temporarily after rain or a melting event, to reduce the risk of flooding.

reverse fault
A dip-slip fault in which the hanging wall moves up relative to the footwall.

rift valley
A depression formed by grabens, along the crest of a seafloor spreading ridge or on a continent.

rill erosion
Soil erosion on sloping land caused by water forming very small channels.

ripple marks
The rippled surface formed on sediment by wind or water.

Roche limit
The distance from a planet within which any satellite will tend to be torn apart by tidal stresses.

rock
A solid, cohesive aggregate of one or more minerals or mineral materials.

rock cycle
The concept that all rocks are continually subject to change, and that any rock can be transformed through appropriate geologic processes into another type of rock.

rock flour
Silt-sized sediment produced by glacial abrasion.

rupture
Breakage or failure under stress.

saltation
The process in which sediment is transported in a series of short jumps along the ground.

salt dome
A stocklike feature formed from salt rising from buried evaporite beds.

saltwater intrusion
The replacement of fresh pore water by saline water as the fresh water is depleted.

sandstone
Clastic sedimentary rock made up of sand-sized particles.

sanidinite
The highest-temperature contact-metamorphic facies.

scale (map)
The ratio of a unit of length on the map to the corresponding actual distance represented.

scarp
A steep cliff resulting from vertical displacement along a fault; also, a similar feature formed by mass movement.

schist
A medium- to coarse-grained metamorphic rock displaying schistosity.

schistosity
The growth of coarse, platy minerals, especially micas, in parallel planes, due to directed stress.

seafloor spreading
The process through which plates diverge and new lithosphere is created at oceanic ridges.

seamounts
Volcanic hills rising 1 kilometer or more above the sea floor.

sediment
An unconsolidated accumulation of rock and mineral grains and organic matter.

seismic gap
A seismically quiet section of an active fault zone, where the fault is presumed to be locked.

seismic shadow
An effect of the liquid outer core, which blocks S-waves and partially deflects P-waves originating on one side of the earth from reaching the opposite side of the earth.

seismic waves
The form in which energy is released during earthquakes; divided into *body waves* and *surface waves*.

seismograph
An instrument for detecting and measuring ground motion.

sensitive clay
A weak, failure-prone clay sediment similar in behavior to quick clay, but derived from other materials (e.g., weathered volcanic ash).

shale
A clastic sedimentary rock made of clay-sized particles, having a tendency to break along parallel planes.

shearing strength
The ability of a solid material to resist shearing stress.

shearing stress
Stress tending to cause different parts of an object to slide past each other across a plane.

sheet silicate
A silicate in which tetrahedra are linked in two dimensions by shared oxygen atoms.

sheet wash
Water flow over a sloping land surface, not confined to a channel.

shield
A large, stable continental region consisting of exposed Precambrian igneous and metamorphic rocks.

shield volcano
A volcano with a low, flat, broad shape, formed by the buildup of many thin lava flows.

shock metamorphism
Metamorphism characteristic of impact events, in which very high pressures are imposed abruptly and briefly; the extent of accompanying heating is variable.

shoreline
The line along which the land and water surfaces meet.

silicate
A mineral containing silicon and oxygen, with or without other elements.

sill
A tabular, concordant pluton.

sinkhole
A circular depression formed by ground collapse into a solution cavity.

slate
A low-grade, fine-grained metamorphic rock exhibiting cleavage along parallel planes, due to alignment of clays, micas, and other sheet silicates.

slaty cleavage
The rock cleavage characteristic of slate.

slide
The movement of a coherent mass of rock or soil along a well-defined plane or surface.

slip face
The downwind side of a dune; assumes a slope equal to the dune sediment's angle of repose.

slump
A short-distance slide.

soil
The surface accumulation of weathered rock and organic matter, overlying the bedrock from which it formed; generally also defined as capable of supporting plant growth.

soil moisture
Water in the vadose zone.

solar wind
The flow of particles and energy that streams outward continually from the sun.

solid solution
The phenomenon of substitution of one element for another in a mineral, within some compositional limits; also, a mineral in which this occurs.

solifluction
A flow in wet soil above the permafrost layer, in alpine terrain.

sorting
Separation of minerals in a sediment by grain size; also, a measure of the extent to which this has occurred.

source
The point at which a stream originates.

specific gravity
The density of a mineral divided by the density of water.

spheroidal weathering
The chemical weathering of a rock having a spheroidal shape, in a series of concentric layers.

spoil banks
Piles of waste rock and soil displaced during strip-mining.

spring
A site where the water table intersects the ground surface, so that water flows out at the surface.

spring tides
The most extreme tides, which occur when sun, moon, and earth are aligned.

stage
The elevation of the water surface of a stream at a given point along the channel.

stock
A pluton similar to, but smaller than, a batholith.

stones
Meteorites composed predominantly of silicate minerals.

strain
Deformation resulting from stress.

stratovolcano
A *composite volcano*.

streak
The color of a mineral when powdered.

stream
Any body of flowing water confined within a channel.

stream piracy
The process by which a stream undergoing headward erosion cuts through a divide and begins to drain part of an adjacent drainage basin.

stress
The force applied to an object.

striations (glacial)
The parallel grooves cut in rock by rock fragments frozen into glacial ice.

strike
The compass orientation of a line or plane as measured in the horizontal plane.

strike-slip fault
A fault along which movement is horizontal only (parallel to strike).

strip-mining
Mining by stripping off overlying rock, soil, and vegetation to expose the desired mineral or fuel; used for shallow, tabular bodies, especially coal beds.

subduction zone
A convergent plate boundary at which a slab of oceanic lithosphere is being pushed beneath another (continental or oceanic) plate, and carried down into the mantle.

submarine canyon
A V-shaped canyon cut in the continental slope.

subsurface water
Any water below the ground surface.

sulfate
A nonsilicate mineral containing sulfur and oxygen in sulfate (SO_4) groups.

sulfide
A nonsilicate mineral containing sulfur but lacking oxygen.

surface waves
Seismic waves that travel along the earth's surface.

surge
A localized increase in the water level of an ocean or large lake, commonly associated with storms.

suspect terrane
A region or set of rocks apparently unrelated to the adjacent regions in geology or history.

suspended load
The quantity of material moved in suspension by a stream.

suture
The zone along which two continental landmasses become joined.

swash
The rush of water up the beach face following the breaking of waves.

S-waves
Seismic body waves characterized by shearing motion or displacement.

syncline
A trough-shaped fold in which the limbs dip toward the axis.

tail (comet)
The stream of matter trailing from the comet nucleus away from the sun, streaked out by the solar wind.

tailings
Crushed rock waste from ore-mineral processing.

talus
Coarsely broken rock debris from rock falls or slides.

tarn
A lake occupying a cirque.

tar sand
Sand or sandstone containing a deposit of viscous, asphaltic petroleum.

tectonic
Relating to large-scale movement and deformation of the earth's crust; *tectonics* is the study of tectonic phenomena.

tensile stress
Stress tending to pull an object apart.

terminal moraine
The end moraine marking the farthest advance of a glacier.

terraces (stream)
Steplike plateaus surrounding the present floodplain of a stream.

terrane
A region or group of rocks with similar geology, age, or deformational style.

terrestrial planets
The earthlike, rocky planets Mercury, Venus, Earth, and Mars.

terrigenous sediment
Clastic sediment derived from the continents.

theory
A hypothesis that has been tested sufficiently against further observations or experiments that it has gained general acceptance.

thermocline
A layer of rapidly decreasing temperature in the oceans, extending from the surface mixed layer to 500-1000 meters depth.

thin section
A thin slice of rock, mounted on a glass slide; typically so thin that light passes readily through most minerals in it.

thrust fault
A reverse fault with a shallowly dipping fault plane.

tides
The slow rise and fall of water level at a point, as the earth rotates through bulges of water caused by the earth's rotation and the gravitational pull of the moon.

till
Glacial sediment deposited directly by melting ice.

tillite
A rock formed from till.

total dissolved solids (TDS)
The total concentration of all dissolved substances in water.

trace (fault)
The line of intersection of a fault plane with the earth's surface.

trace fossils
Fossils that preserve evidence of organisms, rather than the organisms themselves; for example, dinosaur tracks.

transform fault
A strike-slip fault between offset segments of an oceanic spreading ridge.

transgression
Landward encroachment of the sea.

transition zone
The zone of the upper mantle below the low-velocity layer, characterized by several abrupt (though small) increases in seismic-wave velocities; extends to a depth of about 700 kilometers.

transverse dunes
Dunes elongated perpendicular to the direction of wind flow.

traps
Sites of localization or concentration of migrating oil and/or natural gas.

trellis drainage
A rectilinear drainage pattern in which tributaries join the main stream at right angles.

trench
An elongated, steep-walled valley on the sea floor, characteristic of and parallel to a subduction zone.

tributary stream
A stream that flows into a larger stream.

trough
The lowest point of a wave.

tsunami
A seismic sea wave set off by a major earthquake in or near an ocean basin.

tuff
Rock formed of consolidated volcanic ash.

turbidity
Cloudiness of water, caused by suspended sediment.

turbidity current
A density current of sediment-laden water that flows along the ocean bottom.

ultramafic
Igneous rock extremely rich in ferromagnesians and poor in silica.

unconfined aquifer
An aquifer overlain by permeable rocks or soil.

unconformity
A surface within a sedimentary sequence at which there has been a period of nondeposition or erosion.

uniformitarianism
The theory that the same physical laws have always operated, and that by observing present processes we can understand the past history and development of the earth.

upconing
A saltwater intrusion in a conical pattern below (and in reflection of) a cone of depression in an overlying freshwater lens.

upstream flood
A localized flood affecting a small area upstream in a drainage system.

upwelling
The rising of deep, cold waters to shallower depths in response to reduced surface pressures.

vadose zone
An unsaturated zone below the ground surface, in which pores are filled partly with water, partly with air.

varve
A sediment couplet representing one year's deposition in a glacial lake.

ventifact
A rock sculptured by wind abrasion.

vesicles
Gas bubbles in igneous rock.

volcanic
Pertaining to volcanoes; also, describes an igneous rock crystallized at or near the earth's surface, as from lava.

volcanic breccia
Rock formed of a mix of ash, cinders, and coarser angular volcanic blocks.

volcanic dome
A compact, steep-sided volcanic structure formed of very viscous lava.

water table
The top of the saturated (phreatic) zone.

wave-cut terrace
A flat erosional surface in rock, cut at water level by wave action, exposed by coastline emergence.

wavelength
The horizontal distance between two successive wave crests or two troughs.

wave refraction
Deflection (change in direction) of a wave; for water waves, caused by variable water depth near shore.

weathering
The set of physical, chemical, and biological processes by which rock is broken down in place.

xenolith
A rock caught up in a magma as an inclusion.

zeolite facies
The lowest-temperature contact-metamorphic facies.